AF683381

MICROCELLULAR INJECTION MOLDING

WILEY SERIES ON POLYMER ENGINEERING AND TECHNOLOGY

Richard F. Grossman and Domasius Nwabunma, Series Editors

Polyolefin Blends / Edited by Domasius Nwabunma and Thein Kyu

Polyolefin Composites / Edited by Domasius Nwabunma and Thein Kyu

Handbook of Vinyl Formulating, Second Edition \ Edited by Richard F. Grossman

Total Quality Process Control for Injection Molding, Second Edition \ M. Joseph Gordon, Jr.

Microcellular Injection Molding \ Jingyi Xu

MICROCELLULAR INJECTION MOLDING

JINGYI XU
Engel Machinery
York, Pennsylvania

WILEY
A JOHN WILEY & SONS, INC., PUBLICATION

Copyright © 2010 by John Wiley & Sons, Inc. All rights reserved

Published by John Wiley & Sons, Inc., Hoboken, New Jersey
Published simultaneously in Canada

No part of this publication may be reproduced, stored in a retrieval system, or transmitted in any form or by any means, electronic, mechanical, photocopying, recording, scanning, or otherwise, except as permitted under Section 107 or 108 of the 1976 United States Copyright Act, without either the prior written permission of the Publisher, or authorization through payment of the appropriate per-copy fee to the Copyright Clearance Center, Inc., 222 Rosewood Drive, Danvers, MA 01923, (978) 750-8400, fax (978) 750-4470, or on the web at www.copyright.com. Requests to the Publisher for permission should be addressed to the Permissions Department, John Wiley & Sons, Inc., 111 River Street, Hoboken, NJ 07030, (201) 748-6011, fax (201) 748-6008, or online at http://www.wiley.com/go/permission.

Limit of Liability/Disclaimer of Warranty: While the publisher and author have used their best efforts in preparing this book, they make no representations or warranties with respect to the accuracy or completeness of the contents of this book and specifically disclaim any implied warranties of merchantability or fitness for a particular purpose. No warranty may be created or extended by sales representatives or written sales materials. The advice and strategies contained herein may not be suitable for your situation. You should consult with a professional where appropriate. Neither the publisher nor author shall be liable for any loss of profit or any other commercial damages, including but not limited to special, incidental, consequential, or other damages.

For general information on our other products and services or for technical support, please contact our Customer Care Department within the United States at (800) 762-2974, outside the United States at (317) 572-3993 or fax (317) 572-4002.

Wiley also publishes its books in a variety of electronic formats. Some content that appears in print may not be available in electronic formats. For more information about Wiley products, visit our web site at www.wiley.com.

Library of Congress Cataloging-in-Publication Data:

Xu, Jingyi.
Microcellular injection molding / Jingyi Xu.
p. cm. – (Wiley series on plastics engineering and technology)
Includes index.
ISBN 978-0-470-46612-4 (cloth)
1. Molding (Chemical technology) 2. Foamed materials. 3. Microfluidics. I. Title.
TP156.M65X83 2010
668.4′12–dc22

2010004436

Printed in Singapore

10 9 8 7 6 5 4 3 2 1

CONTENTS

FOREWORD

It is a great privilege to write the Foreword for a book that provides so much detailed and insightful information about microcellular injection molding technology. Injection molding is one of the most versatile and important polymer-processing methods for mass production of complex plastic parts. In addition to thermoplastics and thermosets, injection molding has been extended to such materials as fibers, ceramics, and powdered metals, with polymers as binders. Among polymer-processing methods, injection molding accounts for one-third by weight of all polymeric materials processed. New variations and emerging innovations of conventional injection molding have been continuously developed to further extend the applicability, capability, flexibility, productivity, and profitability of this versatile, mass production process. These special and emerging injection molding processes introduce additional design freedom, new application areas, unique geometrical features, enhanced part strength, and sustainable economic benefits, as well as improved material properties and part quality that cannot be accomplished by conventional injection molding processes. Among these special and emerging techniques, microcellular injection molding holds great promise for improving process economics while enhancing the characteristics and functionality of molded parts.

The author, Mr. Jingyi Xu, is a world-renowned expert in microcellular injection molding due to his early and continuous involvement in the developments of this novel process that originated from the work of Dr. Suh and his colleagues at MIT in the early 1980s. The potential benefits of microcellular injection molding go beyond the apparent and attractive savings of material, cycle time, and energy, as it also greatly boosts dimensional stability and facilitates part consolidation and innovation. A wide range of commodity and engineering resins ranging from amorphous, semicrystalline, and thermosetting materials to thermoplastic elastomers and bioplastics have been successfully used with the microcellular injection molding process.

While there are many publications and patents related to new developments and applications of microcellular injection molding, there remains a huge gap between process fundamentals and real-world applications. This book bridges this gap by providing comprehensive coverage in a format that is easy to understand and apply. It also touches on a wide range of subjects, including historical developments, fundamental principles, applicable materials, design guidelines, process comparisons, equipment design, modeling, material properties, and commercial applications. I admire the painstaking commitment and sheer effort of Mr. Xu in making this book a reality. His contribution to the illumination and dissemination of process know-how and guiding principles will be applicable for years to come. This book will undoubtedly assist readers of various backgrounds and levels of expertise to better understand, implement, and benefit from this novel and promising technology.

Polymer Engineering Center
University of Wisconsin—Madison

LIH-SHENG (TOM) TURNG

PREFACE

Microcellular polymers can replace solid polymers with 5% or more material reductions without compromising a significant amount of material properties. In addition, this technology yields even more benefits of microcellular foam such as dimensional stability, short cycle time, elimination of sink marks and warpage, and stress-free parts. Therefore, microcellular injection molding provides a revolutionized way to save materials, and protect the environmental. Consequently, microcellular injection molding has become the fastest developing technology in all microcellular processes. However, most published papers and books on foaming are not often concerned with the practical aspects of applying this technology. This book is intended to bridge these gaps and provides everybody engaged in the design, research, and professional training with a reference book that covers the design and production of microcellular processing in a comprehensive manner.

This book takes the available respected literature into account as well as the real results from extended research and development works in the world. It is my intent to include sufficient detailed discussions for the student pursuing, or just beginning to pursue, a career in the broad microcellular processing arena.

The greatest appreciation should go to everyone who worked hard for this technology. I thank Professor Lih Sheng Turng for reviewing part of my book and for providing the Foreword for this book. Special thanks also go to Professor Chul B. Park for providing valuable information; Levi Kishbaugh for providing courtesy information from Trexel Inc.; and Peter Kennedy, Sejin Han, and Xiaoshi Jin for simulation information, My co-worker Ben Keur read this manuscript and offered excellent opinions, and many others provided their expertise.

I wish to express my appreciation to editor Jonathan T. Rose at John Wiley & Sons, Inc., particularly for his initial idea to write this book and for his splendid efforts during this difficult time to publish this book.

Additionally, I thank my daughter Jiayun Xu as a first general reader with many excellent suggestions for readability of this book. My wife Jufen Guan also gave me unconditional support, as always.

Engel Machinery
York, Pennsylvania

JINGYI XU

1

INTRODUCTION

1.1 HISTORY OF MICROCELLULAR PLASTICS

Historically, microcellular plastics are not new: They existed more or less in the thin transition layer of structural foams. It can be found partially in sections with thin thickness, as well in the high shearing zone of structural foam parts. However, as an idea to develop microcellular plastics, Dr. Nam Suh and his students at the Massachusetts Institute of Technology invented microcellular processing in the early 1980s. This technology proposes two goals: One is to reduce the material, and another is to promote the material toughness by tiny spherical cells that act as crack arrestors by blunting the crack tip [1]. Furthermore, the rigidity of the material in resisting the buckling of the cell walls has been improved through the formation of spherical closed cells. Concentrated research and development efforts of microcellular foams began in the late 1980s, with a focus on the batch process and the topics mentioned above.

The microcellular batch processing technology was invented at the Massachusetts Institute of Technology (MIT) from 1980 to 1984 [1], and the first U.S. patent on microcellular technology was issued in 1984 [2]. Jonathan Colton showed a heterogeneous nucleation mechanism from the effects of additives in the polymers at certain levels of solubility [3]. Jonathan Colton also investigated the methodology of foaming for semicrystalline polymers such as polypropylene (PP) [4]. The gas can be dissolved into the amorphous

Microcellular Injection Molding, Edited by Jingyi Xu
Copyright © 2010 John Wiley & Sons, Inc.

structure because raising the temperature beyond its melting point eliminates the crystalline phase of PP. This heterogeneous nucleation is now dominating today's industry processing. On the other hand, the crystalline material, such as PP, has been used for microcellular foam by Jonathan's method in the industry practice now. Chul Park and Dan Baldwin studied the continuous extrusion of microcellular foam. Chul Park investigated both (a) the dissolution of gas at the acceptable production rate and (b) the application of a rapid pressure drop nozzle as the nucleation device [5]. Dan Baldwin studied the microcellular structure in both crystalline and amorphous materials [6]. Sung Cha investigated the application of supercritical fluid, such as CO_2, to dissolve the gas faster and to create more cells [7, 8]. With supercritical fluid, the cell density was increased from 10^9 cells/cm^3 to 10^{15} cells/cm^3. Vipin Kumar also used thermoforming supersaturated plastic sheets to study the issues of shaping three-dimensional parts [9]. Sung Cha also found that the large volume of gas in polymers decreases significantly with the glass transition temperature of plastics. Therefore, simultaneous room temperature foaming is possible. All of these pioneer contributions are fundamental to microcellular foam technologies. Through many people's creative research, this technology has completed the laboratory stage and transitioned to industry application.

The commercial application of microcellular technology began in 1995 by Axiomatics Corp., which was later renamed Trexel Inc. Trexel continued to develop microcellular technology through extrusion first. Then, the first injection molding machine with plunger for injection and extruding screw for plasticizing and gas dosing was developed in Trexel Inc. with the help from Engel Canada in mid-1997. After successful microcellular injection molding trials were carried out in this plunger-plus-extruder injection molding machine, the first reciprocating screw injection microcellular molding machine was built by Trexel and Engel together in 1998 [10]. This machine marks the milestone of the commercialization of microcellular injection molding and is now the most popular microcellular injection molding machine in the world. Trexel also modified a Uniloy Milacron machine to the first microcellular blow-molding machine in 2000.

One important term, supercritical fluid, is abbreviated as SCF. SCF is the name of the state condition of a gas when the gas is above both its critical pressure and critical temperature; this is discussed in more detail in Chapter 2. It is critical to use SCF to describe a gas if the gas is at a supercritical state. Otherwise, use the general term, gas, if the gas is at any condition from normal atmospheric to supercritical state. Unless otherwise specified, the term of SCF and gas will be used with the conditions above in the entire book.

The injection molding aspect of microcellular foam processing has developed the fastest. The main developed technologies of microcellular injection molding are listed in Table 1.1. The most popular trade name for this technology is MuCell® and is licensed by Trexel Inc. since 2000 (MuCell® is a Registered Trademark of Trexel Inc., Woburn, Massachusetts). Several other injection molding companies and research groups in the world were

TABLE 1.1 Main Developed Microcellular Injection Molding Technologies

Type of Technology	Trade Name	Comment
Microcellular plasticizing unit with special reciprocating screw and barrel to carry out the SCF dosing and injection.	MuCell®	Most popular technology was developed by Trexel, Inc., and has been widely applied worldwide.
Microcellular equipment with special nozzle sleeve for SCF dosing; regular reciprocating screw for injection.	Optifoam®	It was developed by IKV and has been commercialized by Sulzer Chemtech. There are some applications worldwide.
Microcellular dynamic mixer for SCF dosing plus plunger for injection, later modified with reciprocating screw for injection.	Ergocell®	It was developed by Sumitomo-Demag; it has not been common usage on the market yet.
Microcellular equipment with special gas dosing unit in hopper of the regular reciprocating screw for injection.	ProFoam®	It has been invented and tested fully by IKV, and it is still is in the development stage.
Microcellular extruder for SCF dosing plus plunger for injection.	None	It was developed by Trexel and Engel in 1997, and it is not available on market yet.

developing this technology prior to Trexel's announcement of MuCell®. However, they did not finish the commercialization of their technologies for real applications. The MuCell® technology uses a reciprocating screw as the SCF dosing element, and the SCF is injected into the reciprocating screw through the barrel. It makes full use of the shearing and mixing functions of the screw to quickly finish the SCF dosing and to maintain the minimum dosing pressure in the barrel and screw for the possible continuing process of microcellular injection molding. In addition, two other trade names of this technology were found later on: (a) Optifoam® licensed by Sulzer Chemtech [11] and (b) Ergocell® licensed by Demag (now Sumitomo-Demag in 2008) [12]. Optifoam® is a microcellular technology that uses a nozzle as the SCF dosing element. It is a revolutionary change to the traditional SCF dosing method, which adds gas into the barrel. This unique, innovative idea has a special nozzle sleeve made of sintered metal with many ports to let gas go through as tiny droplets. On the other hand, the melt flow through the nozzle is divided into a thin film between the nozzle channel and the sintered metal sleeve. As a result, the gas can diffuse into the melt in a short amount of time. The gas-rich melt is then further mixed in a static blender channel that is located in the downstream of the nozzle dosing sleeve. The advantage of this technology is that the regular injection screw and barrel do not need to be changed. The regular injection molding machine in existence can be easily changed to use the Optifoam® process. However, only some of these applications have been successful [11]. At K2001, Demag Ergotech introduced its Ergocell® cellular foam system [12]. Ergocell® technology has reached an agreement with Trexel to have their customers pay a reduced price to the MuCell® license when using Ergocell® technology legally. The Ergocell® system is essentially an assembly of an accumulator, a mixer, a gas supply, and a special injection system that is mechanically integrated between the end of the barrel and the mold to put gas into the polymer and create the foam upon injection into the mold. A special assembly needs to be created for each screw diameter. Additional hydraulic pumps and motor capacity must be added to operate the mixer and accumulator injection system. The system only uses carbon dioxide as the blowing agent.

The latest developing foam technology from IKV is the ProFoam® process [13]. It is a new and cheap means of physically foaming injection molding technology. The gas, either carbon dioxide or nitrogen, as the blowing agent is directly added into the hopper and diffuses into the polymer during the normal plasticizing process. The plasticizing unit of the molding machine is sealed off in the feeding section of screw for gas adding at pressure, but feeding of pellets of material occurs at normal conditions without pressure. With this ProFoam® process the part can reduce up to 30% weight via the foaming.

Trexel continues to develop and support the microcellular injection molding process worldwide. There are already over 300 MuCell® injection microcellular molding machines in the world. Through the efforts of many more organizations, more and more advances are being made for the microcellular injection molding process. These organizations include not only original equipment manufacturers (OEMs) licensed from Trexel but also numerous unlicensed organizations, such as universities, and university/industry consortia. All of them are contributing to further advances in microcellular technology.

1.2 ADVANTAGES AND APPLICATIONS OF MICROCELLULAR PLASTICS

The microscopic cell size and large number of cells in microcellular material can reduce material consumption as well as improve the molding thermodynamics, which results in a quicker cycle time. Additionally, the process is a low-pressure molding process and produces stress-free and less warped injection molding products. The major differences between conventional foam and microcellular foam are cell density and cell size. The typical conventional polystyrene foam will have an average cell size of about 250 microns, and a typical cell density in the range of 10^4–10^5 cells/cm^3. Microcellular plastic is ideally defined with a uniform cell size of about 10 μm and with a cell density as high as 10^9 cells/cm^3 [1]. It is possible to make this kind of microstructure

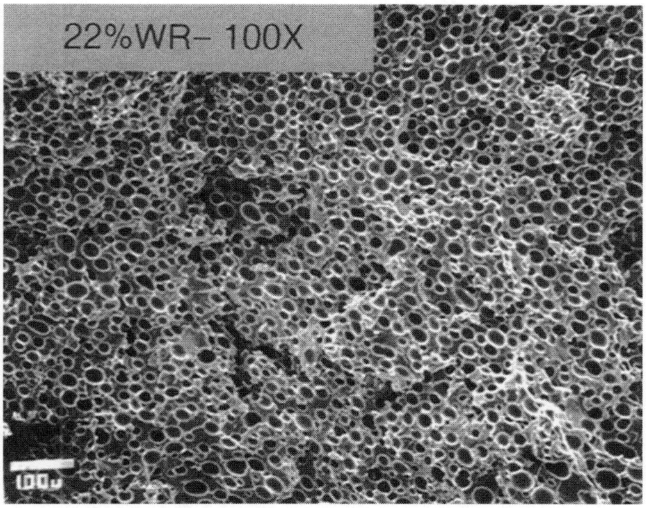

Figure 1.1 Morphology of polystyrene microcellular foam (white bar indicates 100 μm). Average cell size: 25 μm. Cell density: 8.1×10^7 cells/cm^3.

cell density with microcellular injection molding if material and processing are controlled very well. The scanning electron microscope (SEM) morphology of glass-fiber-filled PBT is an excellent example of microcellular injection molding that almost matches the ideal definition of microcellular plastics made by batch process. It is made by using 30% glass fiber and reinforced polybutylene terephthalate (PBT) with a 15% weight reduction (see Chapter 3, Figure 3.12). The cell density is about 8×10^8 cells/cm^3, with an average of 15 μm of uniform cell distribution. However, this microstructure is not always the result of microcellular injection molding. The SEM picture in Figure 1.1 is a more typical microcellular unfilled polystyrene foam made by injection molding that has an average of 25 microns, and has a cell density of about 8.1×10^7 cells/cm^3. The microstructures of industrial parts from microcellular injection molding are characterized by an average cell size on the order of 100 μm, although the real cell size can be varied from 3 μm to 100 μm. However, the cell structure of the microcellular part with microcellular injection molding might not necessarily be defined as the cell density of 10^9 cells/cm^3. The microstructure of ABS has a cell density of about 10^6 cells/cm^3, and it definitely shows a microcellular structure with an average cell size of about 45 μm. The comparisons of average cell sizes between microcellular foam and conventional foam are summarized in Table 1.2. The data in Table 1.2 show that the minimum cell size of conventional foam is about the same size as the maximum cell size of microcellular foam; the maximum cell size of conventional foam is about twice as large as the maximum cell size of microcellular foam. Usually the cell density of the conventional foam is about 10^2 to 10^6 cells/cm^3. However, the cell density of the microcellular foam is 10^6 cells/cm^3 or higher.

TABLE 1.2 Comparisons Among Conventional Foam, Microcellular Foam, and Regular Solid

Item of Comparison	Conventional Foam	Microcellular Foam	Regular Solid
Average cell density (cells/cm^3)	10^2–10^6	10^6–10^9 or higher	NA[a]
Average cell size (μm)	250 or larger	3–100	NA
Sink mark	No	No	Yes
Cell structure	Open or closed	Closed, or a few partially open	NA
Cell size (or density) distribution across the part of thickness	Nonuniform, the distribution pattern across the thickness: small (side near skin)–big (center)–small (side near skin)	Uniform	NA
Maximum flow path length-to-thickness ratio	50–100:1	Up to 350:1	Up to 300:1
Hold	Short time	No	Yes
Wall thickness	Thick wall, 4–9 mm and up to 50 mm	Thin wall, 0.5–3 mm, possibly up to 6 mm with short flow ratio and fast injection speed	0.5–6 mm, up to 9 mm
Residual stress after molding	Nil for ≤4 mm thickness, 0.5–3 MPa in the thick part of 10–20 mm	Nil	Yes
Surface finish	Poor	Between conventional foam and solid	Class A
Cycle time reduction	Long cycle time because of thick part, 1–8 minutes, depending on the wall thickness	Up to 50% reduction versus solid part for ≤4 mm. Thickness. Over 4 mm is similar to traditional foam on the left.	NA

TABLE 1.2 ***Continued***

Item of Comparison	Conventional Foam	Microcellular Foam	Regular Solid
Weight reduction	Some weight reduction is possible. NA if the stiffness is needed to match the solid thin part.	0–15% weight reduction	NA
Maximum injection pressure reduction	20–50%	20–60%	NA
Clamp tonnage reduction	Up to 50%	Up to 60%	NA
Energy saving	About 15%	Up to 30%	NA
Dimension stability, such as warpage, shrinkage, etc.	Good	Excellent	Poor
Toughness	Increase compared to the same solid material	Increase significantly compared to the same solid material	The same initial solid material served as baseline here
Stiffness	Stiff with extra thickness	Flexible	Between microcellular and conventional foam
Other mechanical properties	Decrease with the weight reduction %	Decrease with the weight reduction %	100% for comparison with foam properties
Mold wearing	Less	Less	Normal
Mold cost	Cheap	Cheap	Expensive
Material	PE, PP, PC, POM, ABS, PS, etc.	Any material	Any material
Color	Natural white color may save white color cost	Natural white color may save white color cost, or light color	No savings for white color cost
Postprocess	Needed	May need	None
Weld line strength	Poor	Good	Excellent

TABLE 1.2 ***Continued***

Item of Comparison	Conventional Foam	Microcellular Foam	Regular Solid
Insulation of heat, sound	Good	Excellent	Normal
Application	Insulator, structure part with stiffness, impact absorber, wood replacement, non-sink-mark appearance part, dimensional stable part.	Same as both conventional foam and regular solid parts. In addition, for precise molding parts with sink mark and no warpage; fiber disorientation requirement, tonnage saving, cycle time saving, and material saving, difficult mold filling parts, and soft touch surface with strength.	Widely used except for insulator

[a]NA, not available.

The cell size in the foam mainly determines the property differences between conventional foam and microcellular foam. Table 1.1 shows the comparisons among injection molding parts made by conventional foam, microcellular foam, and regular solid. It is clear that microcellular foam has more advantages than conventional foam. Microcellular foam overcomes the major disadvantages of conventional foam, such as a long cycle time and a thick wall. The most important advantages of microcellular foam can be summarized as follows:

- The main advantage of structural foam molding (one of the conventional foams) is to increase stiffness without increasing the weight of the component. Microcellular foam can be made for this target as well, by redesigning thin wall structures and by creating a nice cell structure to save material (weight reduction by a thin wall) and cost (shorter cycle time).
- The microcellular process can be used for thin-wall solid parts that are difficult to make full mold filling from flow restrictions, which results in either clamp tonnage shortage or injection pressure limit.
- Microcellular technology allows mold filling without foaming because the gas-rich melt reduces viscosity significantly.

- The microcellular process almost eliminates all dimension stability problems, such as sink mark, flatness defects, warp, and residual stress after molding due to the elimination of pack and hold phases during molding.
- The microcellular process dramatically reduces cycle time if the part is designed properly.
- Microcellular processing equipment can be designed to save more energy since the peak of injection pressure is not necessary and also saves up to 50% of clamp tonnage.

The disadvantages of microcellular foam are the same as conventional foam, such as poor surface finish, strictly balanced runner system for multicavity mold, nontransparent application only, and complicated processing technology.

1.3 PATENTS AND PUBLICATIONS COVERING MICROCELLULAR INJECTION MOLDING TECHNOLOGY

There have been many patents issued for microcellular injection molding since 1998. The major patents, directly or indirectly related to microcellular injection molding technology, are listed here:

Pierick, D. E., et al., International Patent Application WO 98 31 521 A2 (1998)

Park, C. B., et al., U.S. Patent No. 5,866,053 (1999)

Pierick, D. E., et al., International Patent Application WO 00 26 005 A1 (2000)

Xu, J., International Patent Application WO 00 59 702 A1 (2000)

Michaeli, W., et al., German Patent DE 19 853 021 A1, (2000)

Anderson, J. R., et al., International Patent Application WO 01 89 794 A1 (2001)

Xu, J., U.S. Patent No. 6,322,347 (2001)

Burnham, T. B., et al., U.S. Patent No. 6,284,810 (2001)

Anderson, J. R., et al., U.S. Patent No. 6,376,059 (2002)

Gruber, H., et al., U.S. Patent Application No. 0,056,935 A1 (2002)

Pierick, D. E., et al., International Patent Application WO 02 090 085 A1 (2002)

Kim, R. Y., et al., International Patent Application WO 02 081 556 A1 (2002)

Vadala, J. P., et al., International Patent Application WO 02 026 484 A1 (2002)

Kishbaugh, L. A., et al., International Patent Application WO 02 026 485 A1 (2002)

Kishbaugh, L. A., et al., International Patent Application WO 02 072 927 A1 (2002)

Xu, J., U.S. Patent No. 6,579,910 B2 (2003)

Anderson, J. R., et al., U.S. Patent No. 6,593,384 (2003)

Dwivedi, R. K., U.S. Patent No. 6,759,004 (2004)

Cardona, J. C., et al, U.S. Patent No. 6,926,507 (2005)

Anderson, G., et al., U.S. Patent No. 7,172,333 (2007)

Xu, J., U.S. Patent No. 7,267,534 (2007)

Xu, J., et al., U.S. Patent No. 7,318,713 (2008)

Kishbaugh, L.A., et al., U.S. Patent No. 7,364,788 B2 (2008)

Xu, J., et al., U.S. Patent No. 7,615,170 B2 (2009)

There are many publications regarding the technology behind microcellular injection molding. They cover both the fundamentals and real practices in industry. However, it is well known a huge gap exists in fundamentals and realities. Hopefully, this comprehensive coverage in the book will help bridge this gap and will enable readers to apply the concepts in a straightforward manner.

1.4 OUTLINES OF THE BOOK

This book presents the microcellular history and a specific short history of microcellular injection molding in Chapter 1. Then, in Chapters 2 and 3, the fundamental knowledge of microcellular injection molding is covered. With the understanding of the principles of microcellular processing, a review of materials and details of design for microcellular injection molding are well discussed in Chapters 4 and 5. Moreover, injection molding makes the foaming process more complex. Therefore, both theory and experiments are needed for good analyses of microcellular process. Chapter 6 uses the fundamental guidelines in previous chapters to analyze the specific processing procedures one by one with a combination of theory and empirical data. Some comparisons among different gas-entrained processes, such as gas assistant, microcellular extrusion, microcellular blow molding, and structural foam molding are discussed in Chapter 6. It is also important to know the differences between regular injection molding and microcellular injection molding, which is discussed briefly in Chapter 6. To realize the processing requirements in Chapter 6, the equipment designing rules are introduced in Chapter 7. It will generate further insight on both the future development and the efficient operation. After understanding normal microcellular injection molding, more specialized microcellular injection molding processes are discussed in Chapter 8. All commercialized special processes and most developing special processes are covered in this chapter. In addition, the modeling of microcellular injection molding is also presented in Chapter 9. Some PVT data and rheology data of the gas-laden polymer melt are given in Chapter 9. The necessary postprocesses and basic test procedures are briefly introduced in Chapter 10. Finally, application in the market is covered in Chapter 11, and cost analyses are presented in Chapter 12.

REFERENCES

1. Suh, N. P. *Innovation in Polymer Processing*, edited by James F. Stevenson, Hanser/Gardner Publications, Cincinnati, 1996, Chapter 3, pp. 93–149.
2. Martine-Vvedensky, J. E., Suh, N. P., and Waldman, F. A. U.S. Patent No. 4,473,665 (1984).
3. Colton, J. S. Ph.D. Thesis, Massachusetts Institute of Technology, Cambridge, MA, 1985.
4. Colton, J. S., and Suh, N. P. U.S. Patent No. 4,922,082 (1990).
5. Park, C. B. Ph.D. Thesis, Massachusetts Institute of Technology, Cambridge, MA, 1993.
6. Baldwin, D. F. Ph.D. Thesis, Massachusetts Institute of Technology, Cambridge, MA, 1994.
7. Cha, S. W. Ph.D. Thesis, Massachusetts Institute of Technology, Cambridge, MA, 1994.
8. Cha, S. W., Suh, N. P., Baldwin, D. F., and Park, C. B. U.S. Patent No. 5,158,986 (1992).
9. Kumar, V. Ph.D. Thesis, Massachusetts Institute of Technology, Cambridge, MA, 1988.
10. Xu, J., and Pierick, D. *J. Injection Molding Technol.* **5**, 152–159 (2001).
11. Pfannschmidt, O., and Michaeli, W. *SPE ANTEC, Tech. Papers*, 2100–2103 (1999).
12. Witzler, S., *Injection Molding Mag.* **December**, 80 (2001).
13. Defosse, M. *Modern Plastics Worldwide* **December**, 14–15 (2009).

2

BASICS OF MICROCELLULAR INJECTION MOLDING

The fundamental theory for microcellular injection molding has been developed for a decade and is still one of major research topics in the plastics industry. The basics of microcellular plastics introduced in this chapter will serve as the general guidelines to both fundamental research and technology development for the microcellular injection molding.

2.1 BASIC PROCEDURES OF MICROCELLULAR INJECTION MOLDING

Typical analyses of gases as the supercritical fluids (SCF) focus on the solubility and dissolution capability in different plastics. To make a nice mixture of gas–polymer solution is a real challenge in the industrial plasticizing unit. After this solution is ready, the nucleation will be the next key technology for success of microcellular injection molding. Finally, how to control the cell growth and distribution throughout the molding part becomes the value of the microcellular part. If the injection time is too long, the bubble collapse and coalescence in the flow front may occur. However, the injection time usually is very short for microcellular injection molding. Therefore, the bubble collapse and coalescence will not be considered as the normal defects in injection molding. To summarize the key issues for successful microcellular injection molding, there are four basic steps of microcellular injection molding: SCF

Microcellular Injection Molding, Edited by Jingyi Xu
Copyright © 2010 John Wiley & Sons, Inc.

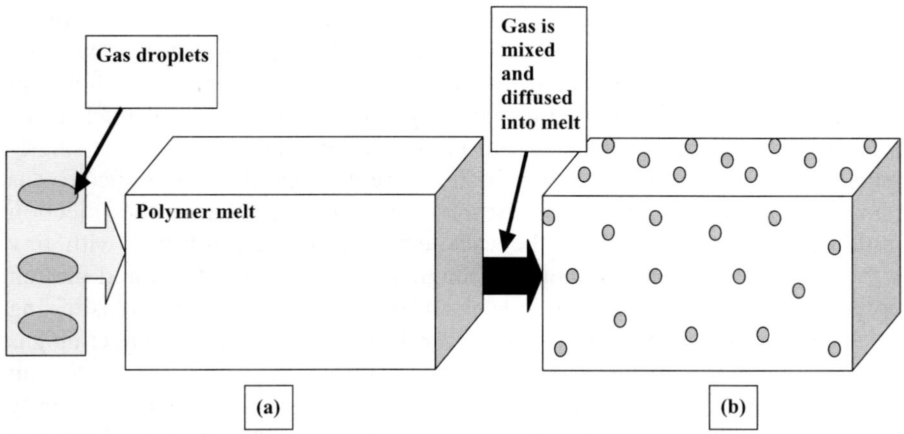

Figure 2.1 Schematic of the gas–polymer solution. (**a**) Gas injected into polymer melt. (**b**) Mixed gas-polymer solution.

mixing and dissolution in the melt of polymer; nucleation of cells; cell growth; and shaping in the mold [1–3].

The concept of the continuous process was successfully tried in an extruding process [4]. It basically needs to create the gas–polymer solution first in the barrel. The gas at the supercritical fluid state is metered and injected into the barrel and then dissolved into molten polymer, as shown in Figure 2.1a. The gas is pressurized up to 20.7 MPa before flowing into the barrel. As the gas flows into molten polymer, it forms big gas droplet in the molten polymer since the flow of the gas is briefly interrupted every time the screw flight wipes over the barrel. The size of the gas droplet in the molten polymer is determined by five major factors: gas pressure and molten polymer pressure; gas flow rate; viscosity of molten polymer; wiping frequency of the flights (screw rotation speed); and diameter of orifice in the gas injector. Then, the large gas droplet is elongated in the barrel through the shear deformation induced by the screw rotation. The elongated gas droplet will be broken up forming many small gas droplets above a critical value of the Weber number W_e, which is a ratio of shear force to the surface force (refer to Chapter 7). These gas droplets may be stabilized in the screw channels to form bubbles in the molten polymer matrix. These bubbles, in turn, undergo elongation with additional shear deformation that increases the area-to-volume ratio of each gas bubble. Then, the gas in the bubble diffuses quickly into the molten polymer due to the increased polymer–gas interfacial area and decreased striation thickness of polymer between the gas bubbles.

However, the shearing rates are varied from the different layers in the channel of screw so that the bubble sizes are different from top (inside diameter of barrel) to bottom (root diameter in the screw) in the flow channel of the screw. The screw mixing section must be designed with the mixing elements to alter the positions of bubbles from top to bottom and vice versa, which will be discussed in Chapter 7. Eventually, the gas droplets must be small and uniformly distributed in the molten polymer matrix, as shown in Figure 2.1b. It may be defined as gas–polymer mixture ready for nucleation. Ideally, the final gas–polymer solution should become the so-called single-phase solution [3, 5]. In other words, there are no separate phases such as gas phase and polymer melt phase. However, the real practice in the injection molding machine can create the excellent gas–polymer solution with tiny bubbles in the molten polymer. The single-phase solution of gas and molten polymer may never truly form in such short recovery time in the plasticizing screw [1], and even shorter mixing time with new technology of injecting gas through the nozzle during injection [6, 7]. Therefore, the single-phase solution may be defined in this book as the gas–polymer solution with a uniformity distribution of many tiny bubbles, which has been proven to be a good mixture of gas–polymer solution ready for the next step of microcellular processing in the most current technologies of microcellular injection molding processes [1–7].

Then, the gas–polymer solution needs to be induced by a rapid thermodynamic instability of this mixture of gas and molten polymer for cell nucleation. The thermodynamic instability is generated by either a pressure drop or a temperature change with high rate. Practically, a quick change in the pressure is much easier than a fast change in the temperature in a very short period. Therefore, a very high pressure drop rate occurs in either the nozzle orifice or the valve gate, where the narrow orifice causes a high pressure drop rate up to 1 GPa/sec or higher.

Once enough nuclei are created, the nucleated gas–melt mixture is still kept warm for cell growth in the center layer of the part when the skin begins to cool down. In addition, the short shot of injection leaves enough space for cell growth. On the other hand, enough gas is available to provide the necessary gas supply around the nuclei, which is growing further to form a stable cell.

Finally, the part in the mold not only conforms to the shape of the mold but also builds up the skin-cell structure. The cell growth in the part results in a perfect microcellular part. The cells retain their shape and size during the cooling, and also the residual gas pressure inside the cells pushes the part, thereby expanding the cells to overcome the shrinkage of polymer. Then, the expanding of cells inside the part helps the part to contact the cold wall of mold to copy the mold shape exactly and to form the solid skin quickly.

2.2 SUPERCRITICAL FLUIDS (SCF)

The gas must be dissolved into molten polymer in the limited time so that the best condition of the gas state to satisfy the rate-limiting process is the gas at the supercritical state. Figure 2.2 shows the gas-phase diagram. The shadowed area represents the region of supercritical state of the gas where the gas is at

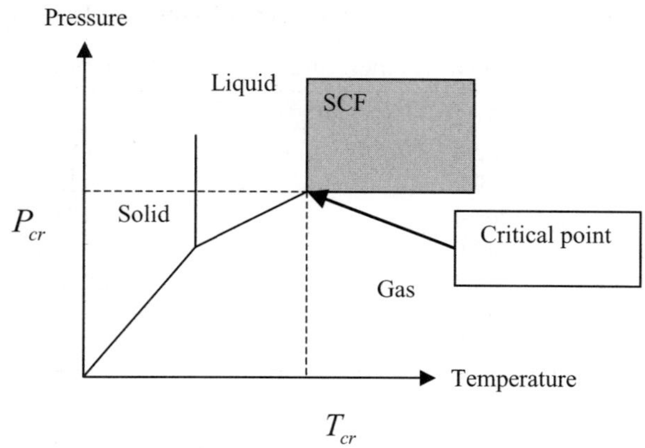

Figure 2.2 Diagram of material phases. (Courtesy of Trexel Inc.)

a liquid-like state, the so-called supercritical fluid (SCF). This SCF region is located beyond both critical pressure P_{cr} and critical temperature T_{cr}. Therefore, the processing setup parameters for microcellular processing must have high pressure and temperature to produce the gases used for microcellular processing at the supercritical state. There are also two critical points in this diagram that must be set up accordingly during the process. One is the critical pressure and another is critical temperature. The microcellular process usually works with the processing conditions above both critical pressure and temperature to quickly diffuse the gas into molten polymer.

The supercritical fluid is neither gas nor liquid in a certain temperature and pressure regimen higher than the critical pressure and critical temperature. In this state the gas will have both gas-like and liquid-like properties. The gas-like property of SCF is the low viscosity similar to the air viscosity even if the viscosity of SCF may increase several times higher than the original one at gas phase. On the other hand, SCF has a liquid-like property—that is, a heavy liquid density compared to gas density. For example, the density of carbon dioxide (CO_2) is 0.001 g/cm^3 at the gas phase, but 0.7 g/cm^3 at SCF state that is close the density of liquid CO_2 0.8 g/cm^3. Both properties are important to precisely meter the SCF weight percentage to match the molten polymer output rate. In addition, both properties are necessary for possible mixing of SCF into molten polymers.

The list of critical point data for various fluids that can be used for blowing agents is presented in Table 2.1 [4, 8]. Although many of them can be used as supercritical fluids in microcellular processing, their shortcomings may restrict their applications for microcellular process. Nitrogen (N_2) has a low solubility but can make very small cells. This is the main reason why N_2 gas is widely used as one of the common physical blowing agents. Carbon dioxide (CO_2) at a critical state can enhance the solubility and diffusion rate. The great percentage of CO_2 can be added into the polymer and can be quickly degassed

TABLE 2.1 Critical Points of Pure Components of Potential Blowing Agents [4, 8]

Solvent	Critical Temperature (°C)	Critical Pressure (MPa)
Carbon dioxide	31.1	7.22
Argon	−122.4	4.83
Helium-3	−269.9	0.12
Helium-4	268.0	0.23
Nitrogen	−147.0	3.4
Water	374.2	21.76
Hydrogen	−240.2	1.28
Propane	96.7	4.19

after the molding. Therefore, CO_2 is also a common physical blowing agent for microcellular processing. Water is corrosive and has very low solubility in the molten polymer. Argon (Ar) is expensive and also has relatively low solubility. Lots of other organic liquids are used for blowing agents, but they are abandoned because of environmental consequences also because they are hazardous. Both CO_2 and N_2 gases are the most common blowing agents used for microcellular processing. They are environmentally benign blowing agents and are inexpensive to obtain from the air. In addition, they are not ozone-depleting and are a viable alternative to other volatile blowing agents.

There are huge viscosity differences between SCF and all molten polymers. Typically, molten polymers normally exhibit high viscosity in the range of 10–1,000,000 poise (g/cm ·sec). However, the gaseous blowing agent will normally exhibit viscosity in the very low range of 0.00005–0.05 poise (g/cm·sec) [9]. Hence, the relatively very low resistance for the blowing agent flows easily in the gas injector and helps to clean up the high-viscosity material in the flow lines of the gas injector. In addition, a low-viscosity liquid such as SCF makes the mixing between SCF and molten polymer easier. However, every time the injector of low-viscosity SCF is opened for injecting gas into high-viscosity material, it is kind of a surge, and an initial large gas pocket may form in the gas dosing position. It must be controlled by the small pressure difference between gas pressure and melt pressure at dosing position in the barrel.

2.3 GAS SOLUBILITY AND DIFFUSION CAPABILITY IN POLYMER MELT

There are many gases available to be used as blowing agents. As the common physical blowing agent sources, some gases have been used for many professionals in industry for a long time, such as nitrogen, CO_2 and even air. Some exotic gases such as argon (Ar), helium (He), and hydrogen (H_2) have been

TABLE 2.2 Properties of Popular Blowing Agents

Physical Properties	Carbon Dioxide, CO_2	Nitrogen, N_2
Molecular weight	44	28
Gas density at 21 °C, g/cm^3	0.00183	0.00116
Thermal conductivity (W/m · K)/°C	0.0166/30	0.0261/27
Boiling point (°C)	−78.5	—
Vapor specific heat (cal/g K)/°C	0.204/25	0.243/25
Heat of vaporization (cal/g)	137	—
Global warming potential (GWP)	0.00025	—
Triple point, temperature (°C)/pressure (MPa)	−57/0.52	—
Density at 25 °C (g/cm^3)	1.811	1.146

Source: Throne [12], with permission of Sherwood Publishers.

tested in the laboratory by some researchers, but none of them are used as commercialized applications. Water (H_2O) is also a possible blowing agent source that was used for both practices and researches. However, for all foaming industries, including microcellular injection molding, only CO_2 and nitrogen gases are by far the most widely used physical blowing agents. Therefore, nitrogen and CO_2 will be discussed in more detail in this book. In addition, argon and helium gases will be introduced for their solubility and diffusion capability since recent studies include them as potential new sources for microcellular processing. Some basic physical data for CO_2 and N_2 gases are listed in Table 2.2 since only these two gases are used now in the microcellular foam industry. The data in Table 2.2 are useful for calculating the gas flow rate to match the processing requirements.

2.3.1 Gas Solubility in Polymer Melt

All gases can dissolve in all liquids to some extent. The measure of a gas dissolving potential in a liquid is defined as solubility. Solubility is measured in standard gas volumetric uptake per unit weight of liquid, such as (cm^3 (STP)/g polymer).

2.3.1.1 General Relationship Between Solubility and Processing Conditions in the Gas–Polymer System. Within the plasticizing stage of injection molding equipment or within an extruder, the pressure and temperature are high and the solubility of the gas blowing agent is high; at this point, the polymer is saturated with the gas blowing agent. In the gate of a mold, the pressure drops quickly, and the gas blowing agent becomes supersaturated within the polymer; at this point, the gas blowing agent will begin to precipitate out in the form of gas, thereby foaming the polymer. If the drop in blowing agent solubility is

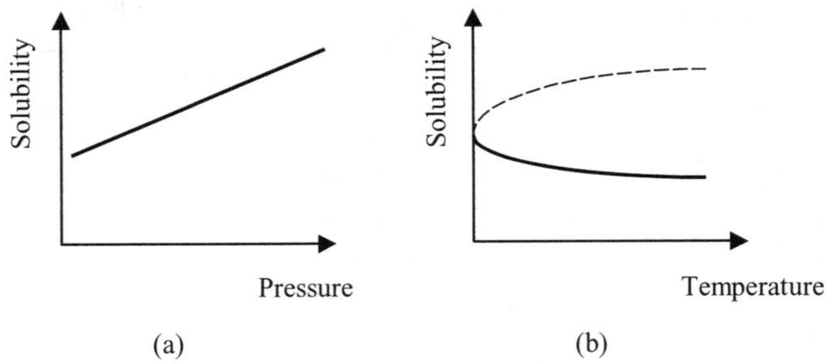

Figure 2.3 Solubility of gas in a molten polymer. (**a**) Solubility versus pressure [4]. (**b**) Solubility versus temperature: solid line for most materials, dashed line for some "reverse solubility" materials.

sufficiently large and sufficiently fast, then conditions will exist for homogeneous nucleation of cells, and a large number of evenly distributed microscopic cells will form and grow uniformly.

The conditions required for homogeneous nucleation are best illustrated by plotting the solubility of a blowing agent as a function of pressure in a typical polymer system (see Figure 2.3). As the general trend the gases solubility in the molten polymer will increase with the increasing pressure and will decrease with the increasing temperature, as shown in Figure 2.3 [4, 8]. In most gas–polymer systems, the solubility increases almost linearly with the melt pressure [10]. Usually, the pressure and temperature have opposite tendencies to promote the gas solubility in the polymer. The processing parameters can be set up for high solubility based on clear trends shown in Figure 2.3. The results in Figure 2.3 indicate that a higher solubility of a blowing agent in a polymer is obtained by either increasing the processing pressure or decreasing the processing temperature. Sometimes, to maximize the solubility of gas in molten polymer, both pressure increasing and temperature decreasing can be used simultaneously. For the general rule of processing, both high pressure and low temperature will increase the gas solubility in the molten polymer.

The experimental data from Sato and others verify the trends of solubility in Figure 2.3a and the solid line in Figure 2.3b. For example, the same trend of solubility changes with pressure and temperature as shown in Figure 2.3 is verified by the CO_2–PP system in Table 2.3, CO_2–HDPE system in Table 2.4, and N_2–PS system in Table 2.5, respectively [10]. It is obvious that the solubility of CO_2 in PP and HDPE and the solubility of N_2 in PS will increase with the increasing pressure and will decrease with the increasing temperature.

However, there are some exceptions for the temperature effect on the gas solubility in some polymers (see dashed line in Figure 2.3b, represented as "reverse solubility"). The solubility of gases, like other solubility, can increase or decrease with temperature, as determined by two contributions:

TABLE 2.3 Solubility of Carbon Dioxide in Unfilled Polypropylene Material[a]

Pressure (MPa)	Solubility × 10^2 (g-gas/g-polymer)
At 433.2 K	
7.400	5.03
11.803	9.48
14.558	12.53
16.355	14.51
17.529	15.87
At 453.2 K	
5.419	3.23
7.160	4.41
8.797	5.94
10.953	7.87
11.558	8.46
12.352	9.17
14.342	11.19
14.889	11.58
16.170	13.06
17.376	14.18
At 473.2 K	
6.204	3.02
10.113	6.01
12.637	8.30
14.296	9.68
15.397	10.86

[a]PP with $\bar{M}_w = 4.51 \times 10^5$, $\bar{M}_w/\bar{M}_n = 7.05$, $T_m = 431\,K$.

Source: Sato et al. [10], with permission of Elsevier Publishers.

- Energy is absorbed to open a pocket in the solvent. Solvent molecules attract each other. Pulling them apart to make a cavity will require energy, and heat is absorbed in this step for most solvents.
- Energy is released when a gas molecule is popped into the pocket. Intermolecular attractions between the gas molecule and the surrounding solvent molecules lower its energy, and heat is released. The stronger the attractions, the greater the amount of heat released.

There is usually net absorption of heat when gases are dissolved in organic solvent, such as molten polymer, because the pocket-making contribution is bigger. Le Chatelier's principle predicts that when heat is absorbed from the dissolution process, it will be favored at higher temperature. Solubility is expected to increase when temperature rises, like the dashed line of the profile in Figure 2.3b [10].

TABLE 2.4 Solubility of Carbon Dioxide in Unfilled High-Density Polyethylene Material[a]

Pressure (MPa)	Solubility × 10^2 (g-gas/g-polymer)
At 433.2 K	
6.936	4.07
11.066	7.47
15.286	11.42
16.347	12.30
17.453	13.20
At 453.2 K	
7.055	3.50
11.326	6.71
14.013	8.90
15.762	10.05
16.896	10.94
18.123	11.98
At 473.2 K	
6.608	3.19
10.731	5.51
13.344	7.46
15.034	8.73
17.019	10.33

[a]HDPE with $\bar{M}_w = 1.11 \times 10^5$, $\bar{M}_w/\bar{M}_n = 13.6$, $T_m = 402$ K.

Source: Sato et al. [10], with permission of Elsevier Publishers.

TABLE 2.5 Solubility of Nitrogen in Unfilled Polystyrene Material[a]

Pressure (MPa)	Solubility × 10^3 (g-gas/g-polymer)
At 313.2 K	
4.934	2.20
8.378	4.40
10.771	5.95
12.450	6.31
16.136	8.25
At 333.2 K	
5.037	1.82
6.149	2.15
7.882	3.06
9.868	3.56
11.674	4.30
13.457	4.86
16.542	6.34

TABLE 2.5 *Continued*

Pressure (MPa)	Solubility × 10^3 (g-gas/g-polymer)
At 353.2 K	
2.989	1.06
3.645	1.31
5.053	1.88
6.476	2.33
7.945	2.96
9.042	3.09
11.660	3.97
14.071	4.99
17.521	6.31

[a]PS with $\bar{M}_w = 1.87 \times 10^5$, $\bar{M}_w/\bar{M}_n = 2.67$, $T_m = 373.6\,K$.

Source: Sato et al. [10], with permission of Elsevier Publishers.

TABLE 2.6 Solubility of Nitrogen in Unfilled Polypropylene Material

Pressure (MPa)	Solubility × 10^3 (g-gas/g-polymer)
At 453.2 K	
4.233	4.39
7.115	7.58
9.031	9.74
10.898	11.87
12.699	13.70
14.938	17.04
17.999	19.28
At 473.2 K	
4.013	4.52
6.726	8.55
8.545	10.59
9.878	12.21
11.998	15.14
14.819	19.99
17.838	22.49

Source: Sato et al. [10], with permission of Elsevier Publishers.

For example, the solubility changes with pressure for both N_2–PP and N_2–HDPE with the same trend in Figure 2.3a, and the detailed data are listed in Table 2.6, and 2.7. However, the solubility changes with temperature show the exception that matches the dashed line profile in Figure 2.3b. It is named "reverse solubility" since it is not a common phenomena. Both N_2–P and N_2–HDPE systems show the reverse solubility in Tables 2.6 and 2.7. It is more obvious for PP with N_2 gas at high pressure. However, the HDPE will have a

TABLE 2.7 Solubility of Nitrogen in Unfilled High-Density Polyethylene Material

Pressure (MPa)	Solubility × 10^3 (g-gas/g-polymer)
At 433.2 K	
2.541	2.58
4.278	4.32
5.463	5.52
7.935	7.79
10.089	9.54
11.555	11.57
14.629	14.65
At 453.2 K	
3.743	3.72
6.330	6.56
8.192	8.49
9.401	9.66
11.809	12.61
At 473.2 K	
2.818	2.88
4.704	5.19
5.958	6.70
6.792	7.60
8.386	9.85
10.695	11.84
12.682	14.94
15.214	17.24

Source: Sato et al. [10], with permission of Elsevier Publishers.

small increase in solubility with the increasing temperature at the same pressure. As a brief comparison, CO_2 in PP and HDPE do not have this "reverse solubility" trend for the temperature effects. Therefore, to know the solubility changes, both plastics and gases need to be checked, and the right choices for the processing conditions need to made accordingly.

Although the conclusions above are from the experiments, the general trends in Figure 2.3 are still good standards for the industry. In most cases, there will be more mechanical heat generated from higher processing pressure, so the processing temperature will be automatically increased when the processing pressure is increased. Based on the trend of solubility varied with pressure and temperature in Figure 2.3, if the processing temperature and pressure increase simultaneously, then the final solubility will be unknown. However, the experiences in real practices show that the blowing agent solubility in a polymer increases more quickly with the increasing pressure whereas the blowing agent solubility in a polymer decreases with the increasing temperature from mechanical heat. Therefore, the solubility in a polymer will be increased with the increasing pressure in most cases.

2.3.1.2 Gas Concentration Calculation. To estimate the gas concentration in certain molten polymers, Henry's law provides an equation. At thermodynamic equilibrium, the external pressure and the gas concentration are related to each other [4, 11]:

$$C = H(P_m, T_{\text{poly}})P_m \tag{2.1}$$

where C is gas concentration (cm^3 (STP)/g polymer), H is Henry's law constant (cm^3 (STP)/g atm), P_m is molten polymer pressure (also means gas pressure), and T_{poly} is molten polymer temperature (K).

At low melt pressure and low gas concentrations, H is constant. At high pressure, H depends on both pressure and temperature. It is well known that the temperature dependence follows the Arrhenius-type rate equation [4]. For most gases in polymers, the amount of gas dissolved in a polymer is linearly dependent on the imposed gas pressure at a given temperature [12]. Then, Henry's law constant is temperature-dependent according to

$$H = H_0 e^{-E_s/R_g T_{\text{poly}}} \tag{2.2}$$

where H_0 is the preexponential constant for Henry's law constant R_g is the gas constant.

Some experimental values for Henry's law constant for several polymer–gas combinations can be found in reference 12.

Some of the heats of solution and energies of activation for solution are listed in Table 2.8 [12]. An empirical relationship for the Henry's law constant has been developed [12]:

$$\ln H = -2.338 + 2.706(T_{\text{cr}}/T_{\text{poly}})^2 \tag{2.3}$$

where T_{cr} is the critical temperature of gas.

There are more empirical formulae for different gas–polymer systems listed in Chapter 9. They are used for simplified simulation model for gas solubility, concentration, nucleation, and cell growth.

This correlation has been proposed for the solubility of gases in any amorphous polymer [12]. The solubility of semicrystalline plastics is a function of the extent of crystallinity, X_c:

$$C = (1 - X_c)X_a \tag{2.4}$$

where X_c is the extent of crystallinity in semicrystalline material and X_a is the solubility of the gas in an amorphous portion of the semicrystalline material.

The measure for the rate at which molecules move through solids or liquids is diffusivity. Diffusion coefficients are related to the specific molecule moving through a specific liquid at a specific temperature. They are measured as unit area per unit time and are listed in Table 2.8 [12].

TABLE 2.8 Diffusivities and Diffusional Energy of Activation (kcal/mol)

Polymer	Diffusional Energy of Activation of CO_2 (kcal/mol)	Diffusional Energy of Activation of N_2 (kcal/mol)	Diffusion Coefficient × 10^{-5}(cm^2/sec), CO_2	Diffusion Coefficient × 10^{-5}(cm^2/sec), N_2
Low-density polyethylene	9.2	9.9	5.69	6.04
High-density polyethylene	8.5	9.0	4.74	5.10
Polystyrene	8.7	10.1	—	0.001
Polypropylene	7.85[a]	8.2[a]	4.25	3.51
Polybutadiene	7.3	7.4[a]	2.42	2.04[a]

[a]Calculated value.

Source: Throne [12], with permission of Sherwood Publishers.

2.3.1.3 Nitrogen Gas (N_2). Nitrogen (N_2) gas is an inexpensive, nonflammable, nontoxic permanent gas. It can be easily made from the air and is chemically inert, which results in an environmentally safe blowing agent to replace some ozone depletion chemical blowing agents. The gas state of nitrogen is available at 13.8 MPa (2000 psi) to 20.7 MPa (3000 psi) as compressed gas in the steel cylinder. Therefore, the pressure of N_2 gas in the vendor's tank already remains higher than the critical pressure. The liquid state of N_2 is stored in the dewars as a cryogenic liquid at about −196 °C. For a heavily N_2 flow rate application the cryogenic N_2 is preferred. However, the N_2 vapor needs to be boiled off from liquid state during the real usage, and the temperature will be near room temperature prior to metering and injection into the machine since the gas temperature needs to be back to critical state as well before injecting into the barrel. In other words, the N_2 blowing agent is used only in the gas state in the delivery equipment except in the storage equipment. Overall, N_2 is preferred in many, if not most, technical applications because it results in a more consistent and uniform microcellular part.

Polyolefin resins typically require significantly higher N_2 levels to achieve good cell structure than do most other materials. These materials are also more likely to have significant cell structure variation from the gate to the end of fill. This situation will be aggravated by increased maximum wall thickness, greater than about 3.0 mm. It should be expected that the final nitrogen levels when running unfilled HDPE or unfilled PP will be 1% or higher. In fact, N_2 gas levels as high as 2% already ran with these materials successfully with high processing pressure.

Some experimental values for Henry's law constant for several polymer–nitrogen gas combinations are as follows [12]:

Materials	H (cm^3 (STP)/g atm)
Polyethylene	0.111
Polypropylene	0.133
Polyisobutylene	0.057
Polystyrene	0.049
Polymethyl methacrylate	0.045

As the data show above, the dosage of N_2 gas for polyethylene, or polypropylene, is more than double that of the N_2 gas dosage for polystyrene, which is the result from the laboratory at ideal conditions. The weight gain percentage of N_2 gas in different plastics at real processing conditions, 200 °C and 27.6 MPa, are estimated as follows [4, 13–15]:

Materials	Nitrogen Maximum Weight Gain (%)
Polyethylene	3
Polypropylene	4
Polystyrene	2
Polymethyl methacrylate	1

The real gas dosage used in industry is very close to the ideal data; this verifies that the ideal data represent good guidelines for the real process. Also, for the heats of solution and the energies of activation, the data for N_2 in some polymers are as follows [12]:

Materials	E_s (kcal/mol)	E_d (kcal/mol)
Polyethylene (125–188 °C)	0.95	2.0
Polystyrene (120–188 °C)	not available	10.1

2.3.1.4 Carbon Dioxide Gas (CO_2). Carbon dioxide (CO_2) can be useful for a number of special cases when gas diffusion, or viscosity, is the primary challenge. It is similar to N_2 for the usage of an ideal foaming agent. It is also an inexpensive, chemically inert, environmentally acceptable, and intriguing physical blowing agent. However, CO_2 gas has some inherent handling problems, such as a relatively low critical point of 31 °C and 7.29 MPa (1027 psi). Therefore, CO_2 will be a vapor above the critical point. It may be used in the delivery system with either gas state or liquid state.

Some experimental values for Henry's law constant for several polymer–carbon dioxide gas combinations [12] are as follows:

Materials	H (cm^3 (STP)/g atm)
Polyethylene	0.275
Polypropylene	0.228
Polyisobutylene	0.210
Polystyrene	0.220
Polymethyl methacrylate	0.260

Also, for the heats of solution and energy of activation, the data for carbon dioxide in different polymers are as follows [12]:

Materials	E_s (kcal/mol)	E_d (kcal/mol)
Polyethylene (188–224 °C):	−0.80	4.4
Polypropylene (188–224 °C):	−1.7	3.0

The weight gain percentages of carbon dioxide gas in different plastics at real processing conditions, 200 °C and 27.6 MPa, are estimated as follows [4, 13–15]:

Materials	Carbon Dioxide Maximum Weight Gain (%)
Polyethylene	14
Polypropylene	11
Polystyrene	11
Polymethyl methacrylate	13

As the general trend the gas dosage of CO_2 in polyethylene, or polypropylene, is just slightly higher than the gas dosage of CO_2 in polystyrene. On the other hand, total weight percentage of CO_2 in the same plastic material will be much higher than the N_2 gas dosage, which is about 3–4 times higher with the exception of PMMA. However, the experimental data in the laboratory of Trexel Inc. also show that an acrylic has the CO_2 solubility of (a) 4.25 weight percent at 177 °C (350 °F) and 12.4 MPa and (b) 5.15 weight percent at 177 °C (350 °F) and 18.2 MPa.

Li et al. [16] at the University of Toronto reported the solubility of CO_2 in PP at different temperature and pressure. They also matched the trends of the solubility of CO_2 in polymer: high at high pressure and low temperature.

Figure 2.4 [17] shows the CO_2 gas absorption weight percentage at different saturation percentages and different melt temperatures. The solubility of CO_2 in polystyrene (PS) has been plotted as a function of pressure and temperature. As shown, the solubility of CO_2 increases with increasing pressure, but it decreases with increasing temperature. It verifies the trends shown in Figure 2.3 regarding the solubility of CO_2 in PS material. In addition, one more test result in Figure 2.4 is the solubility of CO_2 in PS under the shearing. It is

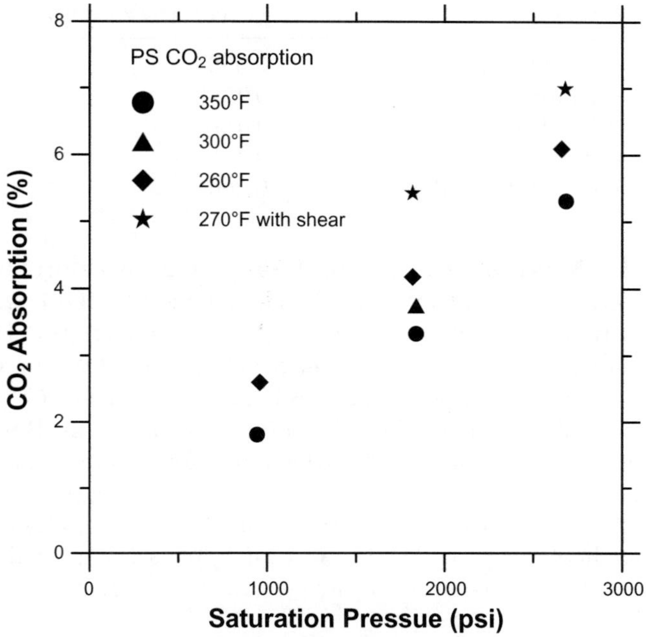

Figure 2.4 Gas solubility in PS melt (1 MPa=145 psi) at different pressure, temperature, and with shear [17], (Courtesy of Trexel Inc.)

obvious that the shearing helps to increase the solubility of CO_2 in PS material.

2.3.1.5 Argon Gas (Ar). Some experimental values for Henry's law constant for several polymer–argon gas combinations [12] are as follows:

Materials	H (cm^3 (STP)/g atm)
Polyethylene	0.133
Polypropylene	0.176
Polyisobutylene	0.102
Polystyrene	0.093
Polymethyl methacrylate	0.105

Based on the test performed by Wong et al. [18], the solubility of Ar gas in PP copolymer is the highest compared to N_2 and He inert gases tested in their paper.

2.3.1.6 Helium Gas (He). Wong and others also tested He gas in the PP copolymer melt. It has the lowest solubility compared to N_2 and Ar gases. Some experimental values for Henry's law constant for several polymer–helium gas combinations [12] are as follows:

Materials	H (cm^3 (STP)/g atm)
Polyethylene	0.038
Polypropylene	0.086
Polyisobutylene	0.043
Polystyrene	0.029
Polymethyl methacrylate	0.066

2.3.1.7 Filled Materials. Chen et al. [19] reported gas absorption percentages with different filled and unfilled polymer systems: (a) HDPE with/without talc and (b) rigid PVC with/without calcium carbonate. The HDPE was Equistar LP5403. The filler was Talc LG445, with a normal size of about 5 μm. The Rigid PVC uses Geon pipe-grade resin with a *K* value of 67, and filler is calcium carbonate with a nominal diameter of about 3 μm. Both talc and calcium carbonate were coated with surfactant before compounding in a twin screw. CO_2 gas is the only one used in this test. A foaming process simulator has been built to study the gas absorption, and it can be pressurized up to 34.5 MPa and heated up to 232 °C. A rotor applies shear to the polymer melt in the pressurized chamber to investigate the shear effects on the gas absorption [20, 21].

The micropore theory is adapted by many researchers. For a porous surface, it not only offers surface energy but also is the residence for the gas molecules in the cavities that may cause cavitations locally. Figure 2.5 is the schematic of micropore model. The hypothesis of this model is that gas accumulation occurs at micropore [19]. The size of micropore is proportional to the size of filler. This explains why the filled material increases the solubility.

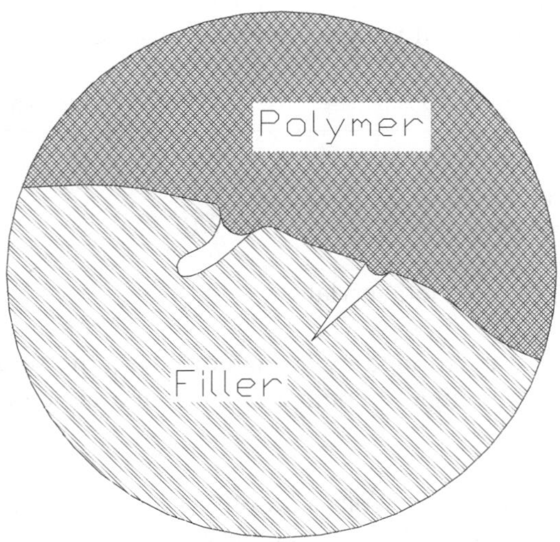

Figure 2.5 Micropore model for filled material [19]. (Reproduced with copyright permission of Society of Plastics Engineers.)

TABLE 2.9 Solubility of CO_2 (Weight Percentage) in HDPE with/without Talc

Filler Level, wt %	Temperature, 177 °C	Temperature, 149 °C
0	4.35	4.7
5	4.85	5.2
10	5.24	5.42
28	5.55	5.75

Source: Chen et al. [19], with permission of Society of Plastics Engineers.

TABLE 2.10 Solubility of CO_2 (Weight Percentage) in Rigid PVC Calcium Carbonate

Filler Level, wt %	Temperature, 177 °C	Temperature, 149 °C	Temperature, 121 °C
0	2.77	2.95	3.20
2	3.02	3.20	3.35
10	3.08	3.40	3.47

Source: Chen et al. [19], with permission of Society of Plastics Engineers.

Table 2.9 shows the effect of filler level on the CO_2 gas absorption in high-density polyethylene (HDPE) at different melt temperatures: 149 °C and 177 °C at fixed pressure 18.6 MPa [19]. The high filler level will help to increase the gas solubility in HDPE. In addition, the high temperature causes the gas solubility decrease in the HDPE melt. This trend verifies the general trend of effects of temperature on the solubility in Figure 2.3.

The data in Table 2.10 are the results of the effect of filler level on the gas absorption in rigid PVC (RPVC) at different melt temperatures: 121 °C, 149 °C, and 177 °C at fixed pressure 18.6 MPa [19]. It is clear that the filled materials absorb more gas than the unfilled materials (see the data of filler lever = 0 in Table 2.9), and the gas absorption increases with increasing filler level. However, the dependence of gas absorption on the filler level is not linear. The trend for filled RPVC to absorb the CO_2 matches the trend in Figure 2.3 with high pressure and low temperature for high solubility.

The CO_2 gas absorption at different gas pressures is shown in Figure 2.6 for filled and unfilled HDPE samples [19]. As expected, the gas absorption is basically a linear function of the gas saturation pressure. It also shows the significant difference of gas absorption at the same pressure between filled and unfilled materials. It also verifies that filled material gains more gas than unfilled material does at different pressures. The results in Figure 2.6 explain why more cells are created with filled material. The fillers do not absorb gas; the polymer–filler interface is the only place that absorbs extra gas. The conclusions regarding to the sources of gas accumulation are listed below [19]:

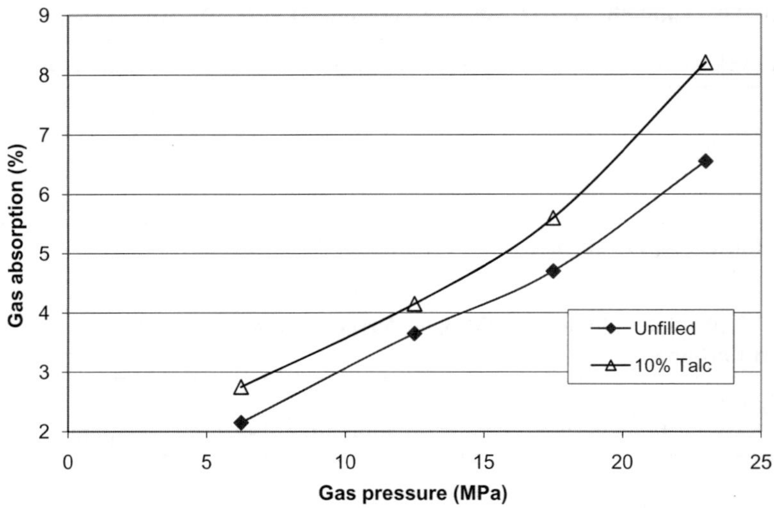

Figure 2.6 CO_2 gas absorption as a function of pressure [19]. HDPE with/without filler. (Reproduced with copyright permission of Society of Plastics Engineers.)

TABLE 2.11 CO_2 Gas Absorption (Weight Percentage) at Different Filler Level and Shear, HDPE with/without Filler

Filler Level, wt %	With Shear	Without Shear
0	6.35	4.65
5	6.55	5.2
10	7.56	5.5
20	7.85	5.65

Source: Wong et al. [18], with permission of Society of Plastics Engineers.

- Preexisting microgaps between polymer and fillers after compounding.
- Convex areas on filler surface where higher interface energy is required for the polymer to fill in. There is a tendency for the polymer to be replaced by the gas after being melted.
- If polymer–filler bonding is not strong enough, there is a tendency for the interface to be separated by the gas because the total surface energy of polymer and filler is smaller than the interfacial energy of the polymer–filler combination.

Chen et al. [19] also discussed the results of gas absorption with and without shearing. Table 2.11 shows that gas absorption is different between the tests conditions with and without shearing. In addition, much higher gas absorption was observed in all the tests with shearing. It explains why there are some good results of gas dosing in high shear rate (high rotation speed of screw).

With all materials, the addition of fillers can improve the efficiency of the nitrogen added to the polymer. The most common filler with polypropylene is talc. As talc levels approach 20% or more, the N_2 gas level will be in the range of 0.5% to 0.75%. Compared to talc and other fillers, glass fiber is a more efficient filler to reduce the gas level with good microcellular structure. The N_2 gas level can be decreased to about 0.5% with glass-fiber-filled material.

Park and his group studied talc-enhanced PS foaming with CO_2 gas as a blowing agent. At low weight percent of CO_2 (2.1%) the onset time of cell nucleation decreases and the cell density increases with the higher talc content [22]. However, at higher CO_2 content up to 4.0 weight percent, cell density is almost invariant with increase of talc content [22]. This conclusion is consistent with the research results on the extrusion foaming of the PS–CO_2 system [23]. On the other hand, both the onset time of cell nucleation and the cell density are virtually unaffected by the mean size of talc particles [22]. Furthermore, Park found that the increasing CO_2 content weakened the effect of the bubble expansion on the promotion of the cell nucleation since high CO_2 gas may reduce the viscosity and the elasticity of the polymer–gas solution. It then suppresses the induction of the negative pressure around the talc particles, and it results in no promotion of cell generation around the expanding bubbles [22].

The most important observations for the talc-filled material foaming is that the generation of new cells propagated outward in the radial direction as the nucleated bubbles grew, and new cells grow even more with the increase of talc content at 2.1 weight percent of CO_2 [22]. Park proposed a series of hypotheses for this observation [22]:

- With addition of talc particle, the free energy barrier to initiate cell nucleation is reduced.
- The rugged surface of talc particles may serve as the sites to trap CO_2 as preexisting nuclei at the PS–talc interface. As the pressure drops, the critical radius of cell nucleation also decreases continually until it is smaller than this preexisted nuclei. Then, those preexisting nuclei will be activated and will start to grow.
- Some cells grow and push the surrounding polymer gas solution outside the growing cells. As a result, local stretching of the polymer–gas solution may generate a negative pressure in some sections at the surface of the talc to promote the nucleation of new cells around the growing bubbles.

2.3.1.8 Comparison among Different Inert Gases. There are a few published papers that shows the results of different inert gases as blowing agents in different materials. Chen et al. [24] did some tests with CO_2, N_2, and Ar gases in filled HDPE and PVC materials. Table 2.12 shows the gas absorption percentages of three different gases in HDPE with different filler levels. At

TABLE 2.12 Solubility of CO_2, Ar, and N_2 (Weight Percentage) in HDPE with Different Filler Percentages, 270 °C

Filler Level, wt %	CO_2	N_2	Ar
0	4.72	1.2	1.92
5	5.20	1.32	2.20
10	5.42	1.45	2.30
20	5.78	1.55	2.40

Source: Chen et al. [20], with permission of Society of Plastics Engineers.

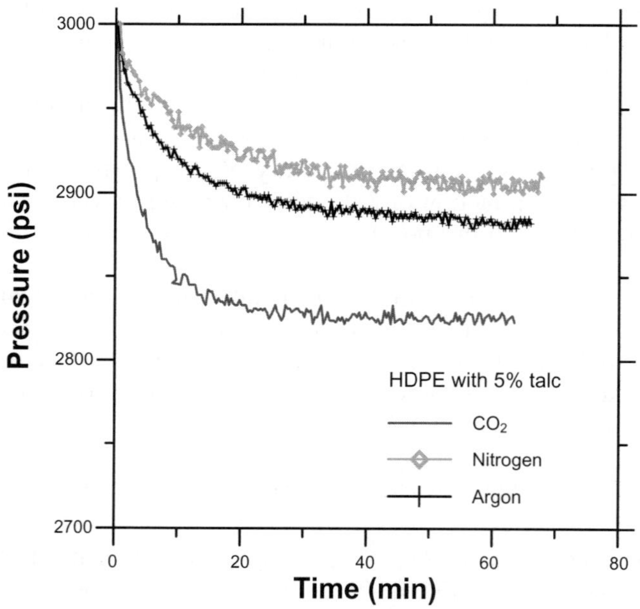

Figure 2.7 Saturation time for 5% talc-filled HDPE with different gases. (Courtesy of Trexel Inc.)

the same temperature and pressure the gas absorption percentage for all gases in HDPE increases with the filler level increasing. However, only CO_2 shows an obvious increase in the gas absorption with the filler level increasing. A similar result is from the RPVC test, which presents the following trend: The gas absorptions of all three gases increasing with the filler level become higher. However, the rate of RPVC solubility changing with different filler levels and different temperature is not as obvious as that of HDPE. There are other test results in Figure 2.7 showing that the shortest saturation time (which is the time for the horizontal line of the saturation pressure in the figure) among all three different inert gases tested for solubility is for CO_2—that is, only about

TABLE 2.13 Comparison of Laboratory Result and Industry Gas Percentage in the Polymer

Solvent	Weight of Nitrogen Used in Continue Process	Weight % of Nitrogen Used in Batch Process
HDPE-unfilled, LDPE-unfilled, LLDPE-unfilled	1–2	3
PP-unfilled	1–2	4
PP talc filled, 5% or more	0.5–0.75	NA[a]
PP glass-fiber-filled	0.5	N/A
PS and PC	0.4–0.6	2
PMMA	0.6–0.8	1
Glass-filled amorphous	0.3–0.5	NA
PA-unfilled	0.5–0.7	
Glass fiber PA, PBT, PET	0.2–0.4	NA
Polysulfone, PEEK, etc.-unfilled	0.5–0.7	
Polysulfone, PEEK, etc.-glass-filled	0.3–0.4	NA

[a]NA, not available.

35 minutes. The saturation time is about 60 minutes required for both N_2 and Ar gases.

The conclusion is that the solubility in polymer is generally much lower with N_2 and Ar compared to CO_2; viscosity reduction is also lower with N_2 and Ar; cell density is similar at high saturation pressure, but much lower with N_2 or Ar at low pressures; the cell size is smaller and density is higher with N_2 and Ar; N_2 and Ar are good candidates for high-density foams. However, CO_2 can reduce the viscosity significantly because it has greater blowing agent concentration in the molten polymer than do other inert gases. It results in a greater reduction in density as well.

The acrylic material has the largest difference between the solubility for CO_2 gas and N_2 gas. At 177 °C of temperature, and 12.4 MPa of pressure the solubility of CO_2 gas in acrylic is 4.25 weight percent. However, at 177 °C of temperature and 13.2 MPa of pressure, the solubility of N_2 gas in acrylic is only 0.34 weight percent.

On the other hand, the weight percentage of gas to be added into the molten polymer for practical process is much less than the solubility measured in the batch process. The data in Table 2.13 show the gas dosing percentage differences among different materials with typical batch process and continue process. It is because the batch process may take many minutes (see Figure 2.7), or even hours, to saturate the material whereas the continue process must finish the gas dosing in the molten polymer in less than 1 min. However, the continue process accelerates the gas diffusion process by the high shearing, which is not easy to do in the batch process. Chen et al. [20, 21] proved that the shearing can speed up the gas diffusion process. The processing pressure and temperature also promote the gas solubility, which is discussed in Chapter 6 for processing and in Chapter 7 for equipment designing.

The data in Table 2.13 provide typical operating levels for N_2 in various materials. While it is always best to use the minimum amount of N_2 necessary to achieve the desired results, cell structure is critical to property retention. Unfortunately, weight reduction and cycle time reduction are much easier to evaluate than cell structure, and these are normally achieved at as much as 50% lower SCF levels than cell structure.

2.3.1.9 Amorphous Versus Crystalline Materials. There is substantial difference in gas solubility in amorphous and crystalline materials. Crystalline material usually has less solubility than the amorphous material does. Even in the same semicrystalline material, the solubility in the crystalline region is less than the solubility in the amorphous region. Baldwin reported that the CO_2 uptake decreases as the volume fraction of the crystallized PET increases [4, 25]. The data he obtained in experiments shows the crystallinity in PET as a function of the gas concentration. The crystallization begins during the primary gas saturation process after a threshold concentration is reached.

Amorphous resins can be split into (a) resins such as polystyrene, polycarbonate, acrylic, and SAN, which do not contain an impact modifier, and (b) ABS, HIPS, and impact-modified PC, which contain an impact modifier. For those materials that do not contain an impact modifier, nitrogen levels will be about 0.4%. These materials typically achieve excellent cell structure at relatively low levels of supercritical fluid. Cell structure will be essentially uniform from gate to end of fill and will be microcellular. Adding an impact modifier has the effect of increasing cell size at equivalent SCF level. In order to achieve a cell structure that is microcellular or close to microcellular, nitrogen levels need to be closer to 0.7%.

As with polyolefin, fillers significantly improve the cell structure in amorphous resins regardless of whether the materials are impact-modified. This is particularly true because most amorphous resins are filled with glass fibers that are highly effective as a nucleating agent and in controlling cell structure. The addition of as little as 10% glass fibers will allow the nitrogen level to be decreased to 0.3% to 0.5% while still maintaining a microcellular structure.

Semicrystalline engineering resins show similar behavior as polyolefin resins. Unfilled versions show larger cell structure variation from gate to end of flow and require higher nitrogen levels to achieve good cell structure, 0.5% to 0.7%. The addition of 20% or more of glass fiber will allow the supercritical fluid level to be dropped to 0.25% to 0.3%. Other filler types such as mineral will also act as a nucleating agent and allow for low SCF levels but will inhibit weight reduction. An example of this would be that glass-fiber-reinforced PA that achieves 20% weight reduction may only get 15% weight reduction with an equivalent amount of mineral filler.

In addition to the specific information above, there are some general trends that can be applied across all materials. Fillers act as nucleating agents and have the effect of improving cell structure and increasing the efficiency of the given level of supercritical fluid. Glass fiber is the most beneficial filler in terms of controlling cell structure and achieving weight reduction. Talc and mineral are less effective in terms of both cycle time reduction and weight reduction.

Amorphous resins will almost always require lower nitrogen levels than semicrystalline resins, although the presence of impact modifiers will require a higher SCF level. This will apply regardless of whether it is an impact-modified amorphous resin or a semicrystalline resin such as a TPO or toughened PA.

The guidelines above apply to general injection molding applications, with wall thickness of 1.5–4.0 mm. Cell structure control becomes easier in thinner parts due to higher cavity pressure. The trends discussed above still hold, but cell structure variation from gate to end of fill will increase.

The barrel temperature for microcellular injection molding is usually the same as the regular injection molding. However, the high back-pressure requirement may bring more heat in the screw. Based on the theory in this chapter, the high melt temperature will reduce the solubility of gas in the molten plastic. In terms of the difference in the specific volume change between gas and molten plastic with the temperature change, it may be one of the reasons to explain this phenomena. Gas goes within polymer with two possibilities based on the theory from MIT. First, gas may occupy the free volume sites in the polymer. However, the free volume should increase with the increasing temperature. The second possibility is that the gas can go into interstitial sites of polymer. Then, it will form the secondary bond with the polymer molecules. This kind of possibility will be decreased with the temperature rise due to higher vibration of the polymer molecules as well as the gas molecules. Therefore, the result of high melt temperature is to reduce the gas dosing percentage to avoid the extra gas out of solution.

2.3.2 Gas Diffusivity in Polymer Melt

To make a single-phase solution for microcellular process diffusivity is another key factor that needs to be understood. For the continue process, it is more important than the solubility since gas dosing must be finished in a time period as short as possible. The solubility gives the range of possible gas dosing percentage shown in Table 2.13. If solubility is the static measure of maximum uptake, the diffusivity is the time-dependent mobility or mass transfer of molecules through a system. Therefore, diffusivity will determine if the gas–molten polymer system can do the job in time economically. Diffusivity is the rate at which molecules move through molten polymer [12]. Diffusion coefficients are always associated with a specific molecule moving through the specific molten polymer at specific temperature. Diffusion coefficient D_a is

TABLE 2.14 Estimated Diffusion Coefficients of Gases in Polymer at Elevated Temperature [4, 10–14] with Units of cm^2/sec

Polymer	D_a of CO_2 at 188 °C	D_a of CO_2 at 200 °C	D_a of N_2 at 188 °C	D_a of N_2 at 200 °C	D_a of Ar at 188 °C	D_a of He at 188 °C
PS	—	1.3×10^{-5}		1.5×10^{-5}	—	—
PP	4.2×10^{-5}	—	3.51×10^{-5}	—	7.4×10^{-5}	10.51×10^{-5}
PET	—	2.6×10^{-6}	—	8.8×10^{-7}	—	—
HDPE	5.69×10^{-5}	2.4×10^{-5}	6.0×10^{-5}	2.5×10^{-5}	9.19×10^{-5}	17.09×10^{-5}
LDPE	—	1.1×10^{-5}	—	1.5×10^{-4}	—	—
PTFE	—	7.0×10^{-6}	—	8.3×10^{-6}	—	—
PVC	—	3.8×10^{-5}	—	4.3×10^{-5}	—	—
	—					

measured as unit area per unit. Table 2.14 shows the diffusivity coefficient D_a from different references [4, 11–15].

It is well known that the gas diffusion time is very slow at room temperature. The estimated diffusivity of CO_2 in most thermoplastics is in the range of 5×10^{-8} cm^2/sec. The diffusivity of N_2 is nearly the same as the diffusivity of CO_2 at room temperature. However, at elevated temperature up to 200 °C the diffusivities of both gases are three to four orders of magnitude greater than at room temperature. For both gases, the estimated diffusivity coefficient in polymer is in the range of 10^{-4} cm^2/sec to 10^{-6} cm^2/sec at 188 °C to 200 °C without any shearing, and at atmospheric pressure, as shown in Table 2.14. It may be used for a qualitative estimation or comparison to select the processing conditions. However, it is not ready to be the useful data used for a quantitative calculation of practical injection molding process since the shearing will greatly influence the data in Table 2.14.

The general trend is the diffusion rate of gases in the molten plastics increasing with the temperature rise, as opposed to the effect of changing direction on the solubility in the plastics. However, the data in the reference show a decrease of gas diffusivity for HDPE with the temperature increasing. For CO_2 gas it is about 5.7×10^{-5} cm^2/sec at 188 °C, while it is about 2.4×10^{-5} cm^2/sec at 200 °C. The N_2 gas shows the similar trend of gas diffusivity changing of carbon dioxide—that is, about 6.0×10^{-5} cm^2/sec at 188 °C and about 2.5×10^{-5} cm^2/sec at 200 °C. The explanation is the complex of crystalline material that may have a significant effect on the gas diffusion process, causing crystalline change and temperature change.

The small molecules of gases may move through polymers by migrating from free domain to free volume domain [12]. For many popular polymers, such as PS, PE, and PVC, there are no apparent changes in the temperature dependency of the diffusion coefficient. Therefore, the cell size may change in the popular polymers discussed above, but the number of cells remains without dramatic change as the polymer passes through transitional temperatures. On the other hand, for only a few polymers, such as polyethylene terephthalate (PET) and polyvinyl acetate (PVA), the diffusion coefficient increases with the temperature increasing to over glass transition temperature [12].

The theory of gas diffusion in the molten polymer has been discussed in many books [4, 12]. It is assumed that the polymer is suddenly exposed to highly pressurized gas at a certain elevated temperature. The polymer will approach the thermodynamic equilibrium by gas diffusion in the polymer. The driving force is the free energy of the polymer–gas system. The diffusion phenomenon can be expressed with a partial differential equation as [4]

$$\nabla(\alpha \nabla C) = \partial C / \partial t_d \tag{2.5}$$

where C is gas concentration, α gas diffusivity, and t_d is gas diffusion time.

The equation of diffusion can be written as follows:

$$\frac{DC}{Dt_d} = \frac{D}{r^2}\frac{\partial}{\partial r}\left(r^2 \frac{\partial C}{\partial r}\right) = \frac{2D}{r}\frac{\partial C}{\partial r} + D\frac{\partial^2 C}{\partial r^2} \tag{2.6}$$

With initial and boundary conditions, it can be expressed as

$$\left\{C(r,0) = C_0, C(R,t) = C_\theta, \left(\frac{\partial C}{\partial r}\right)_{r=R_0} = 0\right\} \tag{2.7}$$

The solution results in Equation (7.9), which shows that the diffusion coefficient is Arrhenius-dependent on temperature. The important relationship is that the diffusivity increases with temperature in an Arrhenius relationship:

$$\alpha \sim \exp[-\Delta G/(kT)] \tag{2.8}$$

where ΔG is activation energy, k is Boltzmann's constant, and T is absolute temperature.

The implication for the relationship in Equation (2-8) is that the diffusing time t_d is proportional to the square of thickness of plastic layer where gas must be diffused into it, as shown in Equation (7.9). The thickness of the plastic layer between gas droplets is the diffusion distance [4]:

$$l \approx \sqrt{t_d \alpha} \tag{2.9}$$

where t_d = SCF diffusion time (seconds), l = thickness of plastic for gas diffusion (millimeters), and α = gas (SCF) diffusivity [see Equation (7.10)].

Some values of ΔG as the activation energy are listed in Table 2.8. Throne [12] suggested using the constant relationships of ratios of activation energies for diffusion coefficient and permeation. The ratio can be considered constant for a given gas, such as nitrogen in this case, through a series of polymers belonging to the homologous series. These ratios can be given as

$$\frac{E_p(g)}{E_p(\mathrm{N}_2)} = \text{constant [12]} \tag{2.10}$$

where E_p is the activation energy through polymer p for either diffusion or permission. In Equation (2.10), (g) represents the unknown gas and (N_2) means that N_2 gas is the one known gas to be used here. Similarly, if one polymer is known for diffusion or permeation of a certain gas, the constant ratio can be written as

$$\frac{E_g(p)}{E_g(p')} = \text{constant [12]} \tag{2.11}$$

where E_g is the activation energy of gas g for either diffusion or permeation, p is the known polymer, and p' is the unknown polymer.

For any case in Equation (2.10) or (2.11), the unknown gas diffusion or permeation can be estimated with one of either known gas or polymer, as long as the polymer belongs to a homologous series. For example, if the diffusional energies of activation for HDPE are 8.5 kcal/mol for CO_2 and 9.0 kcal/mol for N_2 respectively, then the ratio from Equation (2.10) is 0.94. From Table 2.8, the diffusional energy of activation for N_2 in PS is 10.1 kcal/mol. With average ratio of 0.94 from Equation (2.10), the diffusional energy of activation for CO_2 in PS is expected to be $10.1 \times 0.94 = 9.45$ kcal/mol. A value of measurement is 8.7 kcal/mol in Table 2.10, and the estimated error is about 9%. However, if it is just for estimation, this error is acceptable from an engineering point of view. It is also recommended for the simple modeling program in Chapter 9 if some activation energy is not available.

Diffusion is so important for microcellular injection molding since it determines whether the first step is successful or not. Throne [12] summarized the generalities of the gas diffusion, and some of them related to injection molding are listed as the following:

- Increasing the pendant group size decreases the diffusion rate. However, the diffusion energy of activation of gases increases with the increasing pendant group size.
- Increasing the number of methyl groups on linear olefins decreases the diffusion rate but increases the energy of activation. This was found for an ethylene–propylene copolymer with benzene as the penetrant.
- The high concentration of polar groups decreases the diffusion rate but increases the energy of activation for diffusion rate.
- It is well known that increasing the level of crystallinity of the crystalline material decreases the gas diffusion rate in polymer. The explanation

of physics in reference 12 is that the formation of lamellae restricts the diffusional paths and makes them substantially more tortuous. The decrease in diffusion rate can be by as much as a factor of four or five.

The single-phase solution is really defined by the gas being fully diffused into the molten polymer. The real result may be that the gas molecule occupies most of the free space in the molecule of the polymer. However, there may be lots of microbubbles that are not in the free space of the molten polymer but which are mixed with molten polymer to form a gas–polymer mixture. As long as this mixture is uniform and the gas does not form a big gas pocket, this gas–polymer mixture will be sufficient enough to make a microcellular structure during molding. Only this gas–polymer mixture, not true single-phase solution, can explain the successful gas dosing in a molten polymer with only 6 seconds (20 seconds as the theoretical minimum time for gas diffusion to form single-phase solution based on current theory) total residence time or less for gas mixing and diffusing [26, 27]. A reasonable assumption is that the microcellular part is made by gas injected into the nozzle during the short time period of injection stroke with an extra static mixer downstream [6, 7]. This is a challenge for all researchers to investigate this dynamic processing. This is an example that the industry technology is way ahead of research in universities. It may be necessary to have a new or modified theory with new approaches for both gas diffusion models and the experimental equipment that must be closer to the real processing conditions, instead of traditional equipment in current batch processing.

2.3.3 Physical Properties of the Polymer–Gas Solution

The polymer–gas mixture obviously becomes a different material. The theory is that as the gas occupies the interstitial areas between molecules, the distance between these polymeric molecules will increase. Then, the polymeric molecules moves relative to each other and to deform the polymer–gas solution [4, 28]. Consequently, the glass transition temperature and viscosity decrease with the increase in gas concentration. This change is substantial at high gas concentrations.

An important point to emphasize is that the gas–molten polymer mixture is unlike the platicizers used to lower the viscosity and increase the ductility of polymers. The dissolved gas does not permanently alter the properties of the plastics. Once the gas diffuses out of the solution, the plastic returns to its original state immediately, which is one way to explain why cooling is much faster for microcellular processing than for regular processing. On the other hand, the low viscosity of the gas-laden polymer melt benefits the faster and farther mold filling for the thin-wall mold filling. A detailed discussion of viscosity changes from the gas-laden polymer melt can be found in Chapter 6, 7, and 9.

2.4 NUCLEATION OF CELLS

After the single phase solution or uniform gas–molten polymer system is made in the plasticizing unit, which is defined as the first stage of microcellular process, the next critical step in the microcellular injection molding is the nucleation through the injection unit. If the second stage of nucleation does not create enough nuclei, the microcellular process may not be successful. Therefore, the nucleation must create a larger number of cells for the target of cell density of about 10^9 cells/cm^3 [4]. However, the real microcellular part will have the variation of cell density from 10^6 cells/cm^3 to 10^9 cells/cm^3, or higher, with the cell sizes being in the range of 100 microns or smaller. This is acceptable microcellular structure in most of the microcellular part made by the injection molding process. The literature related to nucleation is quite extensive, and not all are directly useful for the injection molding process because the injection process provides an extremely high pressure drop rate (up to 1 GPa/sec) and a very high shear rate (up to 40,000 1/sec) so that the experiments cannot simulate those conditions in most of the laboratory equipments and in most normal extruder dies. On the other hand, the principle of classical nucleation theory is still good to help for the nucleation device design of the microcellular injection molding process. In fact, most nucleation processes for the injection molding materials are heterogeneous nucleation since more or less additives are always in the molding materials. From the processing point of view, a heterogeneous nucleation can result in better microcellular structure than a homogeneous nucleation can do. In other words, if the result of nucleation is satisfied for homogeneous nucleation, then heterogeneous nucleation will have better results of nucleation. In addition, the injection molding process provides a much better driving force to create a better nucleation result than any result based on batch process and extruding process. Therefore, the nucleation in injection molding seems like it is not an issue at all. The selected literature reviews and analyses will give the fundamental knowledge of nucleation of cells. Most of the modern microcellular technologies for injection molding process are developed based on the theory discussed in the following sections.

2.4.1 Nucleation Theory

As one of the physical phenomena, nucleation is a new phase formation. It can originate from self-structural adjustment or from foreign "seeds" as a way to release an outside change-induced load [29, 30, 31, 32]. The microcells are nucleated either homogeneously or heterogeneously. When the driving force for nucleation is very high, the difference in the activation level is so much smaller than the driving force that both homogeneous and heterogeneous nucleation can occur simultaneously. The condition for this case is the that degree of supersaturation of the gas in the molten polymer is large [4]. Youn and Suh [30] presented a model for the nucleation of thermosetting resins.

However, the most popular thermoplastic nucleation model has been summarized well by Colton and Suh [31] with all three possible nucleation models.

2.4.1.1 Homogeneous Nucleation Theory. The classical nucleation theory is widely used as the basis for the development of a model of other nucleation mechanisms. It is modified for homogeneous nucleation model including free volume effects [31, 32]. Homogeneous nucleation occurs only if the material is entirely homogeneity because the nucleation occurs in the matrix where the activation energy required for cell nucleation is uniform [4]. Then, the change in Gibbs free energy for homogeneous nucleation can be written as

$$\Delta G_{\text{hom}} = -V_b \Delta P + A_{\text{bp}} \gamma_{\text{bp}} \tag{2.12}$$

where ΔG_{hom} is the change in Gibbs free energy for homogeneous nucleation, V_b is the volume of the bubble nucleus, ΔP is the pressure of the gas in the bubble, A_{bp} is the surface area of the bubble, and γ_{bp} is the surface energy of polymer–bubble interface.

The spherical shape of bubble can minimize the excess energy. It is actually the right shape of bubble for ideal nucleation without shearing. Therefore, the bubble shape is spherical in this chapter unless specified. Then, if the γ_{bp} is isotropic, then Equation (2.12) becomes

$$\Delta G_{\text{hom}} = -(4/3)\pi r^3 \Delta P + 4\pi r^2 \gamma_{\text{bp}} \tag{2.13}$$

where r is the radius of the bubble in nucleation.

This relationship results in an important diagram (Figure 2.8) to illustrate the free energy change with radius r. The second term in Equation (2.13) is the interfacial energy that increases with r^2, shown in Figure 2.8 as the increasing line upward. The first term in Equation (2.13) is the volume free energy that is proportional to $r^3 \Delta P$ and which decreases downwards in Figure 2.8. Then, the free energy $\Delta G(r)$ is the function of r and is the combination value between interfacial energy and volume free energy. The creation of a small bubble results in an increase of free energy until the radius of the bubble nucleated is larger than a critical size. The critical size of cell growth is associated with the maximum free energy $\Delta G^*(r^*)$. The r^* and $\Delta G^*(r^*)$ can be found with $dG/dr = 0$. In addition, at r^* the bubble nucleus is in unstable equilibrium with its environment since $d^2G/dr^2 < 0$. If $r < r^*$, the system can lower its free energy by the dissolution of the gas in the polymer; and if $r > r^*$, the growth of the bubble leads to a reduction in the free energy [30]. On the other hand, if the cell size is larger than critical cell size with r^*, the cell usually becomes stable and grows. However, if the cell size is below the critical cell size, the cell embryo will collapse.

The differentiation of Equation (2.13) can reveal the answer of critical radius r^*,

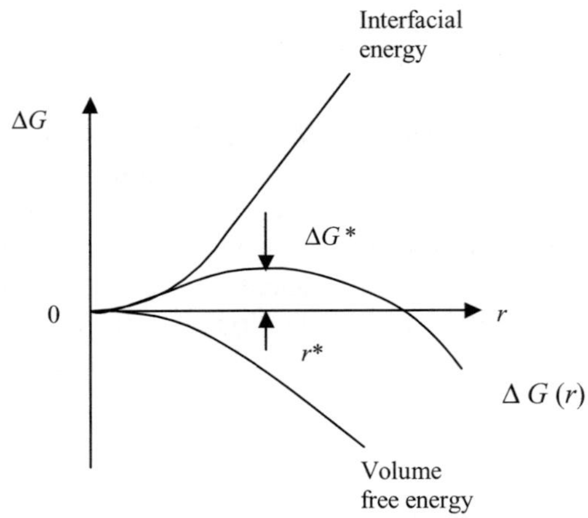

Figure 2.8 Free energy change associated with the homogeneous nucleation of a sphere of radius r [29]. (Reproduced with copyright permission of Society of Plastics Engineers.)

$$r^* = 2\gamma_{\mathrm{bp}}/\Delta P \tag{2.14}$$

It is also important to note that the r^* is independent of the nucleation rate. Then, the Gibbs free energy for the homogeneous nucleation of a critical nucleus is given by

$$\Delta G^*_{\mathrm{hom}} = \frac{16\pi}{3\Delta P^2}\gamma^3_{\mathrm{bp}} \tag{2.15}$$

In Equation (2.15) for the batch process, ΔP (to be taken as a first-order approximation) equals the pressure used to saturate the polymer with gas, with the assumption of idea gas. However, for the injection molding process, ΔP is the instant injection pressure, or an average injection pressure in the whole injection stroke. It is important to know that as either the surface energy of the interface decreases or the pressure (saturation pressure in batch process, or injection pressure in injection molding) increases, the Gibbs free energy is decreased, which has been shown to increase the nucleation rate and the number of bubbles produced.

The surface energy of the solution can be determined by every component in the polymer. It can be modeled in many ways. One of them is to use a rule of mixture. It is valid for small concentrations of additives, as follows:

$$\gamma_s = \gamma_a\omega_a + \gamma_p\omega_p \tag{2.16}$$

where γ_s is the surface energy of polymer solution, γ_a is surface energy of additive, γ_p is the surface energy of polymer, ω_a is the weight percentage of additive, and ω_p is the weight percentage of the polymer.

The γ_s may be substituted into previous equations for γ_{bp} if the dissolved additives are added into polymer solution. Colton also makes modification of classical nucleation theory with the free volume change influence to the change of Gibbs free energy [31]. The free volume of the polymer can be changed by many methods, such as thermal expansion of polymer, or by the presence of a dissolved gas or a dissolved component such as additives. Since the injection pressure usually is high up to over 138 MPa (20,000 psi) even with low viscosity gas-laden material, the volume change may be neglected compared to the pressure for injection molding process.

Once the activation energy barrier is determined, the nucleation rate of gas bubbles can be calculated. The number of gas clusters that have reached a critical size to form a stable nuclei, C^*, can be given by assuming a Boltzmann distribution:

$$C^* = C_0 \exp(-\Delta G^*_{\mathrm{hom}}/kT) \tag{2.17}$$

where C^* is the concentration of gas clusters that have reached a critical size and C_0 is concentration of gas molecules in solution.

For nucleation to occur, the energy barrier has to be overcome. Generally, the energy barrier depends on two competing factors: (a) the energy available in the gas diffused into the embryo of the cell and (b) the surface energy that must be supplied to form the surface of the cell [4]. Typically, the nucleation formula is given as [4]

$$dN/dt = N_0 f \exp(-\Delta G/kT) \tag{2.18}$$

where N is the number of cells, N_0 is the number of available sites for nucleation, and f is the frequency of atomic or molecule lattice vibration.

Assume that the addition of one more gas molecule to the critical nuclei will convert it to stable nuclei and that this occurs with a frequency f_0; the homogeneous nucleation rate is written as [31]

$$N_{\mathrm{hom}} = f_0 C_0 \exp(-\Delta G^*_{\mathrm{hom}}/kT) \tag{2.19}$$

where f_0 is frequency factor for homogeneous nucleation and N_{hom} is homogeneous nucleation rate.

There are very few data available to have homogeneous nucleation in microcellular injection molding. Even homogeneous nucleation is not truly homogeneous in terms of the energy level involved for nucleation [4]. Some of the reference data for homogeneous nucleation calculation are listed in Table 2.15 [31].

2.4.1.2 Heterogeneous Nucleation Theory. On the other hand, heterogeneous nucleation occurs at the position where it takes less energy for nucleation. The low energy requirement for nucleation is located at an interface

TABLE 2.15 Data for Nucleation Calculation

Polymer	Surface Energy, Homogeneous (dynes/cm)	Frequency Factor for Homogeneous Nucleation (1/sec)	Surface Energy, Heterogeneous (dynes/cm)	Frequency Factor for Heterogeneous Nucleation (1/sec)
LDPE	9.2	9.9	9.2	9.9
HDPE	8.5	9.0	8.5	9.0
PS	34	10^{-5}	25	10^{-5}
PP	7.85[a]	8.2[a]	7.85[a]	8.2[a]

[a]Calculated value.

Source: Colton and Suh [31], with permission of Society of Plastics Engineers.

between different materials where the interfacial energy is high. In the practical process of injection molding, heterogeneous nucleation actually dominates most of the microcellular injection molding process because the plastic material will have lots of additives, and the plastic itself is not a pure homogenous material. In addition, the cells do not nucleate at the same time if injection speed is slow or if pressure drop rate is low.

The classic nucleation theory describes the heterogeneous nucleation of a third phase at the interface of two other phases [31]. Figure 2.9 illustrates the nucleation of a gas bubble at the interface of a polymer and a solid particle. A balance of the interfacial surface tensions yields [31]

$$\gamma_{\text{ap}} = \gamma_{\text{bp}} + \gamma_{\text{ab}} \cos(\theta_w) \tag{2.20}$$

where γ_{ap} is interfacial tensions of the solid particle–polymer, γ_{bp} is the interfacial tensions of the bubble–polymer, γ_{ab} is the interfacial tension of the solid particle–bubble, and θ_w is the wetting angle. Then, compared to Equation (2.12), the change in Gibbs free energy for heterogeneous nucleation can be written as [31]

$$\Delta G_{\text{het}} = -V_b \Delta P + A_{\text{bp}}\gamma_{\text{bp}} + A_{\text{ab}}\gamma_{\text{ab}} - A_{\text{ap}}\gamma_{\text{ap}} \tag{2.21}$$

where ΔG_{het} is the change in Gibbs free energy for heterogeneous nucleation, A_{ap} is the surface area of the additive particle–polymer interface, and A_{ab} is the surface area of the additive particle–bubble interface. With similar assumptions in Equation (2.13) and some algebraic manipulation, Equation (2.21) is rearranged as

$$\Delta G_{\text{het}} = [-(4/3)\pi r^3 \Delta P + 4\pi r^2 \gamma_{\text{bp}}] S(\theta_w) \tag{2.22}$$

where

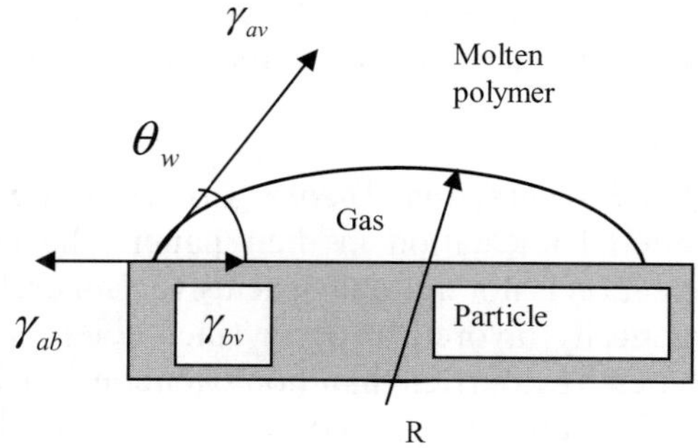

Figure 2.9 Force balance on a gas nucleus at a solid–liquid interface [28]. (Reproduced with copyright permission of Society of Plastics Engineers.)

$$S(\theta_w) = (1/4)[2 + \cos(\theta_w)][1 - \cos(\theta_w)]^2 \tag{2.23}$$

$S(\theta_w)$ is the function that depends on the wetting angle between the polymer, the gas, and the second-phase particle as shown in Figure 2.9. It is equal to 1 in the case of homogeneous nucleation and is less than 1 for heterogeneous nucleation [4].

The differentiation of Equation (2.22) yields expressions for the radius of a critical nucleus that is the same as a homogeneous radius of a critical nucleus. However, compared to Equation (2.15), the critical ΔG_{het} becomes

$$\Delta G^*_{\text{het}} = \frac{16\pi}{3\Delta P^2} \gamma^3_{\text{bp}} S(\theta_w) \tag{2.24}$$

Colton and Suh [31] also pointed out that the typical wetting angle is about 20°, and then $S(\theta_w)$ is only on the order of 10^{-3}. This means that the energy barrier for heterogeneous nucleation can be greatly reduced by this presence of an interface. It explains why heterogeneous nucleation is much easier than homogeneous nucleation. In other words, filled material will be preferred from the nucleation point of view.

The rate of heterogeneous nucleation is similar to the homogeneous nucleation and is given [31] as

$$N_{\text{het}} = f_1 C_1 \exp(-\Delta G^*_{\text{het}}/kT) \tag{2.25}$$

where f_1 is the frequency factor for heterogeneous nucleation, N_{het} is the heterogeneous nucleation rate, and C_1 is the concentration of heterogeneous nucleation sites.

In reference 31, C_1 is 10^{10} sites/cm^3 (based on a particle size of 0.1 μm), γ_a is about 24 dynes/cm for zinc stearate particles, and γ_p is 25.5 dynes/cm for the PS with heterogeneous nucleation [31]. The increase in gas saturation pressure in the batch process or an increase in injection pressure in injection molding, along with an increase in the number of nucleation sites, will increase the nucleation and, hence, will increase the number of bubbles in the final microcellular part.

2.4.1.3 Mixed-Model Nucleation Theory. Colton and Suh [31] also discussed the mixed-model nucleation in their paper. The homogeneous and heterogeneous nucleation is not mutually exclusive. Generally, heterogeneous nucleation is energetically favored to occur since heterogeneous nucleation has lower activation energy barrier than homogeneous nucleation. However, some homogeneous nucleation may still occur in regions of the material where there are some heterogeneous nucleation sites.

It is easy to understand that the heterogeneous nucleation can also affect homogeneous nucleation since the gas available around solid particle will be reduced by heterogeneous nucleation. The gas near the solid particle will tend to diffuse into the bubble nucleated by heterogeneous nucleation because it is thermodynamically favored. Furthermore, the existed bubble will attract more gas in the matrix since gas diffusion into larger bubbles is also thermodynamically favored. Therefore, the gas concentration C_0 in the vicinity of the solid particle is significantly reduced and it can be modeled as a first approximation by

$$C_0' = C_0 - N_{\text{het}} n_b t_b \tag{2.26}$$

where C_0 is the concentration of gas molecules in solution of mixed model, n_b is the number of gas molecules in a bubble nucleus, and t_b is the time since the first heterogeneous nucleation has occurred.

Thus, the homogeneous nucleation rate in the presence of heterogeneous nucleation is given by substituting Equation (2.26) into Equation (2.19) and is as follows [31]:

$$N'_{\text{hom}} = f_0 C_0' \exp(-\Delta G^*_{\text{hom}}/kT) \tag{2.27}$$

The rate of total nucleation for mixed homogeneous nucleation and heterogeneous nucleation is given by combining Equation (2.25) and Equation (2.27), yielding

$$N = N'_{\text{hom}} + N_{\text{het}} \tag{2.28}$$

It was suggested by Colton to add soluble additives at levels slightly below, but close to, their solubility limit and to saturate the polymer with gas at high pressure in order to make microcellular foam.

2.4.2 Experimental Nucleation Results

Many researchers have investigated the nucleation results with different processing conditions for different materials. Although both pressure and temperature changes can result in nucleation changes the pressure change is the easiest way in injection molding process. Some typical processing parameters are investigated for the effects on the nucleation rate and are discussed in the following sections.

2.4.2.1 Effects of Pressure on Cell Nucleation Density. It is well known that the pressure has a significant effect on the cell nucleation density regardless of the type of material and gases. Figure 2.10 shows the effect of saturation pressure on the cell nucleation density for HDPE with three different gases [24]. Overall, CO_2 has the highest cell nucleation density in the whole pressure range. However, the cell nucleation density of N_2 and Ar gases are increased with the increasing pressure much faster than that of CO_2. It means that change in saturation pressure for N_2 and Ar may be very effective for increasing cell nucleation density. Similar experimental results shown in Figure 2.11 show the pressure effect on the cell nucleation density for rigid PVC. It has similar trends for all three gases to affect the cell nucleation density. In addition, the N_2 gas is so effective that the cell nucleation density of the N_2 sample is greater than the cell nucleation density of CO_2. All the results in Figures 2.10 and 2.11 verify the theory above that the high pressure is the key factor to increase the cell nucleation density. The conclusions can be summarized below:

- The cell nucleation density is similar at high saturation pressure for all three different gases. However, the N_2 gas and Ar gas have lower cell nucleation density at low pressure ranges.

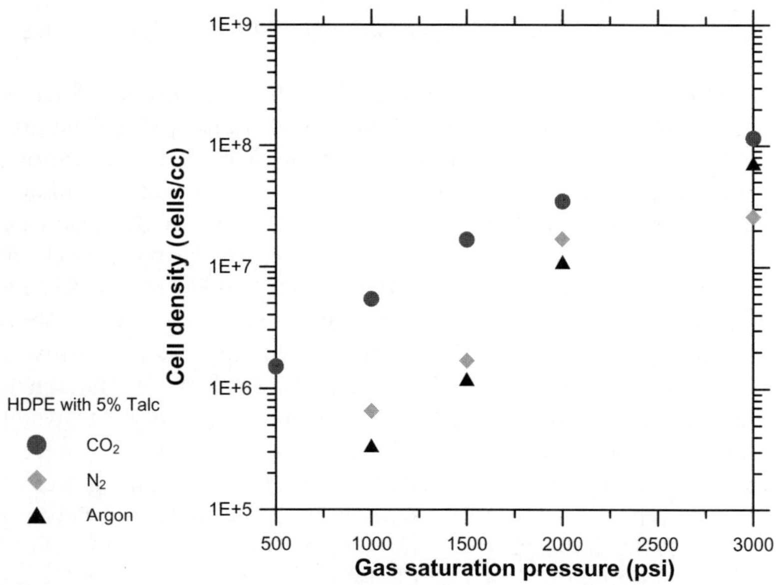

Figure 2.10 Effects of gas saturation pressure on cell nucleation density for HDPE. (Courtesy of Trexel Inc.)

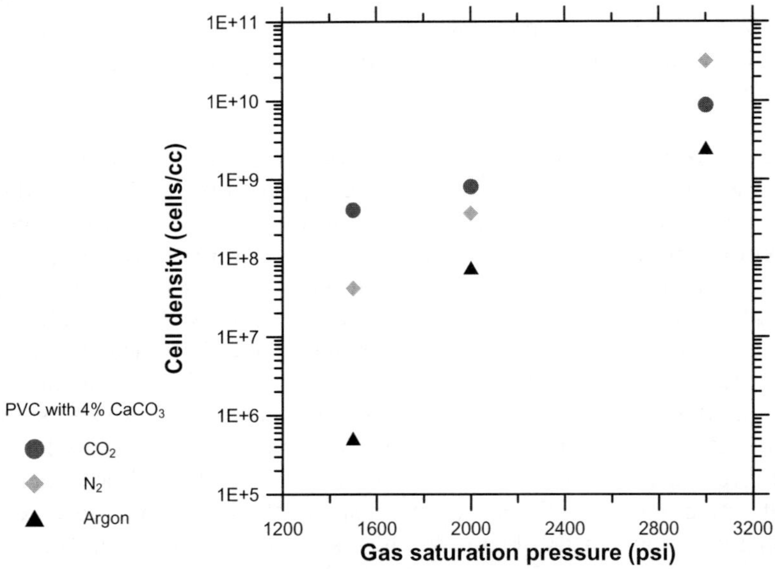

Figure 2.11 Effects of gas saturation pressure on cell nucleation density for rigid PVC. (Courtesy of Trexel Inc.)

- The morphology shows that the cell size is smaller with N_2 and Ar gases (see Chapter 3, Figures 3.20 and 3.21).
- Ar gas is another potential good blowing agent; it makes smaller cell size than CO_2 in rigid PVC, but bigger cell size than CO_2 in HDPE.
- N_2 gas makes the smallest cell size among all three different gases.

2.4.2.2 Effect of Shear Stress on the Cell Nucleation Density. Shear stress may have a significant effect on the cell nucleation density when the pressure drop rate is low. It is even more critical when the SCF, or gas, saturation pressure is low [20, 21]. It is well known that the cell size is usually small near the interface between skin and foamed core in structural foam. It is also true that there are very fine cells near the skin in the microcellular part, specifically in the thick part (3 mm or thicker wall thickness). It is clear that the highest shearing rate occurs in the interface layer that has the finest cells. There are some published papers that examine in good detail the concept of stress nucleation in the extrusion process and in other processes [33–35]. The conclusion is that the cell nucleation increases with the increase of shear rate, which can be calculated from the throughput and die-end opening.

Chen et al. [21] designed a special simulator to investigate the effects of shear stress and pressure drop rate separately. Three materials were investigated: HDPE (Aquistar LP5403), HDPE (same as unfilled one) with 5% of talc (5 microns nominal size of Talc LG445), and PS (Fina 585). CO_2 gas was used as the blowing agent in the tests. Figure 2.12 shows the effects of shear

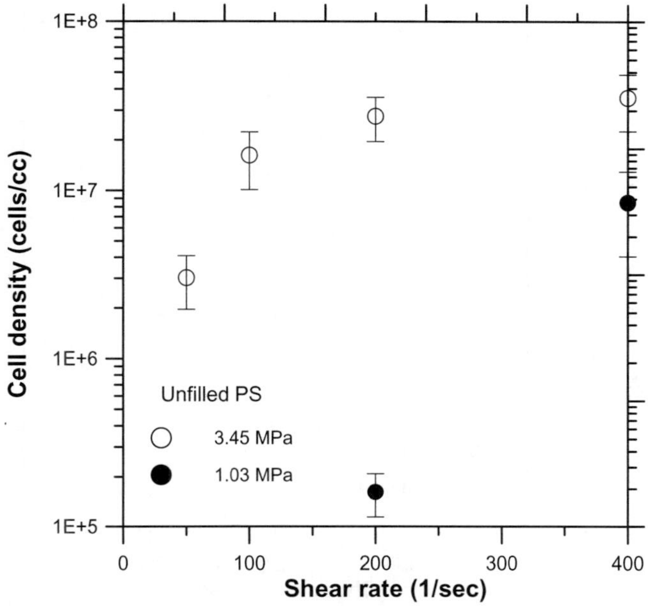

Figure 2.12 Effect of shear stress on the cell nucleation density of PS [20]. (Reproduced with copyright permission of Society of Plastics Engineers.)

stress on the cell nucleation density of PS at two saturation pressures: 3.45 MPa and 1.03 MPa, respectively. It is definitely true that the cell nucleation density increases with the shear stress increasing at low saturation pressure 1.03 MPa. The cell nucleation density is increasing with the shear stress at 3.45 MPa until the shear stress reaches a higher level of about 200 1/sec or above. It seems like the cell nucleation density increase is not significantly increasing at 3.45 MPa pressure and high shear rate. Similar results are illustrated in Figure 2.13. The filled PS shows that the shear stress reaches the maximum (almost horizontal profile line) at the 100 1/sec shear rate at 3.45 MPa. Because the filler helps the nucleation, the shear stress effect on the nucleation is not as strong as the effect for unfilled PS. On the other hand, at low saturation pressure 1.03 MPa, the trend of cell nucleation density increasing is maintained even at high shear rate. It again verifies that the shear stress helps nucleation at low pressure more than at high pressure.

The mechanism of stress nucleation may be the mechanical energy transformation into surface energy, along with orientation of free volumes [21]. The shear-stress-induced nuclei may last long enough for flow rearrangement. It is interesting that the shear effect decays with the relaxation time while the relaxation of polymer occurs after shearing deformation. It is clear that the shear stress effect decaying is slower for filled material than for unfilled material, as shown in Figure 2.14. Therefore, the filled material can keep the shear stress effect on the nucleation longer than unfilled material.

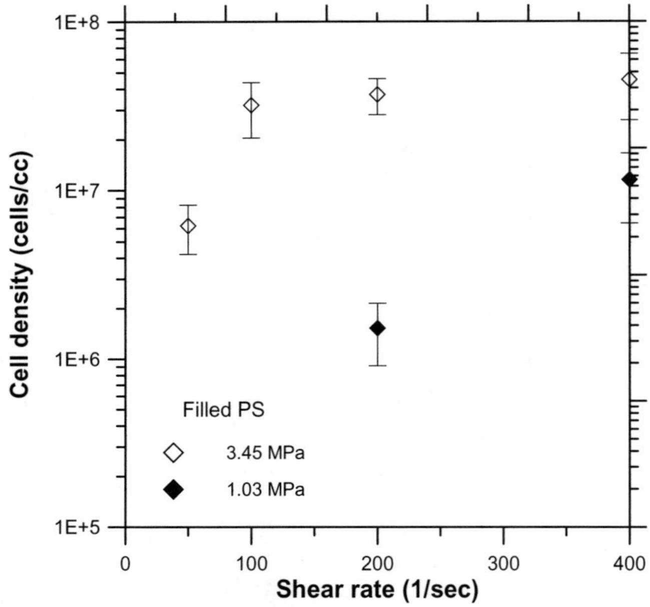

Figure 2.13 Effect of shear stress on the cell nucleation density of filled PS [20]. (Reproduced with copyright permission of Society of Plastics Engineers.)

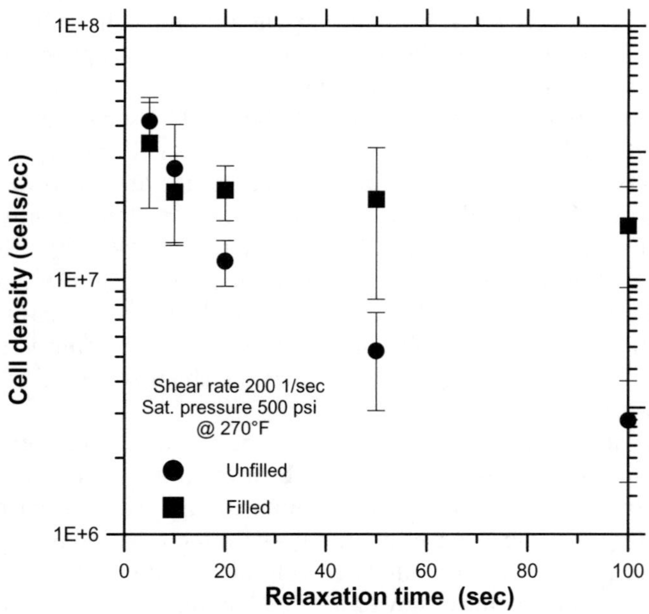

Figure 2.14 Shear effect decay with relaxation time [20]. (Reproduced with copyright permission of Society of Plastics Engineers.)

2.4.2.3 Effect of Pressure Drop and Drop Rate on the Cell Nucleation Density. Both pressure drop and drop rate influence the cell nucleation density. In fact, the method of controlling pressure drop and drop rate is the major way to promote cell nucleation and is widely used for microcellular injection molding technology [1, 17, 20]. This is the advantage of injection molding since the injection volume rate can be set up so high that the pressure drop rate easily reaches more than 1 GPa/sec, which is commonly used as minimum pressure drop rate for general-purpose polystyrene in industry. However. The filled material can significantly reduce the minimum pressure drop rate requirement because the heterogeneous nucleation requires a lower pressure drop rate than does the homogeneous nucleation. The minimum pressure drop rate may be reduced to 10^8 Pa/sec for some filled materials.

Park, Baldwin, and Suh studied the effect of pressure drop rate on the cell nucleation in a continuous process of microcellular foam [4, 36]. Different pressure drop rate nozzles are tested with 10% N_2 gas in the HIPS, and the pressure is 34.5 MPa. The slowest pressure drop rate is 0.076 GPa/sec, and the highest pressure drop rate is 3.5 GPa/s. The result shows the high cell density at high pressure drop rate condition since the number of cells nucleated increases exponentially.

Chen and others used a special simulator to verify the pressure drop effect on the cell nucleation without any influence from shearing [20]. For unfilled HDPE with CO_2 gas, the result shows that the highest *dp/dt* rate is always related to the highest cell density, as shown in Figure 2.15. It is obvious that at the low saturation pressure of 10.35 MPa, the cell density is significantly lower than the cell density at the high saturation pressure for all three pressure drop rates. This result of cell density related to saturation pressure points out that the gas saturation is an important factor for nucleation that must be high

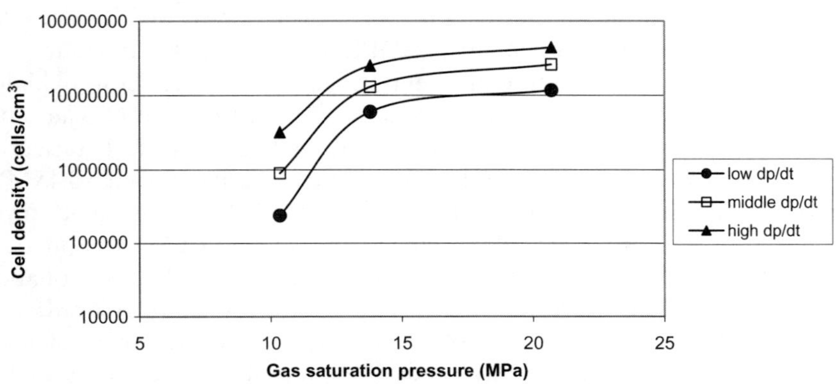

Figure 2.15 Cell density as a function of saturation pressure and pressure drop rate for unfilled HDPE. Low *dp/dt*: 0.0015 GPa/sec. Middle *dp/dt*: 0.02 GPa/sec. High *dp/dt*: 0.15 GPa/sec.

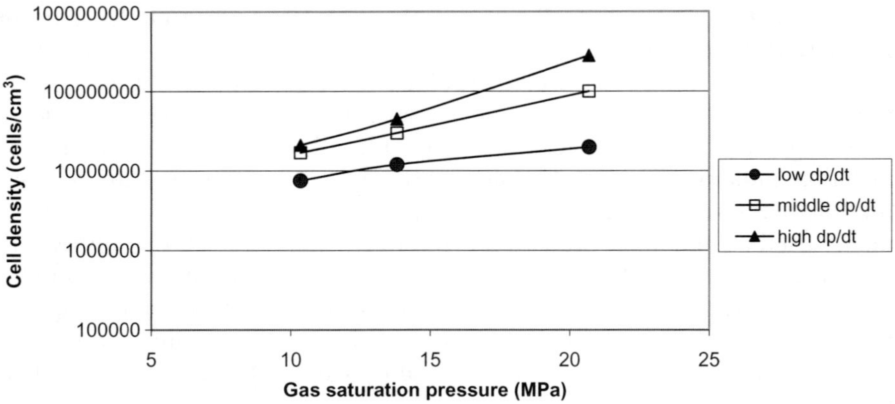

Figure 2.16 Cell density as a function of saturation pressure and pressure drop rate for filled HDPE with 5% talc. Low *dp/dt*: 0.0015 GPa/sec. Middle *dp/dt*: 0.02 GPa/sec. High *dp/dt*: 0.15 GPa/sec.

enough to effectively reach the certain level of cell nucleation. Cell nucleation at high saturation pressure above 13.79 MPa seems significantly higher than cell nucleation at lower saturation pressure. However, for saturation pressure increases from 13.79 MPa to 20.69 MPa the cell nucleation density does not increase dramatically. If the microcellular cells need to be defined as 100 μm or smaller, the density for industry is about 10^7 cells/cm^3, which is achieved only at a saturation pressure of 20.69 MPa at the lowest pressure drop rate. The same density of about 10^7 cells/cm^3 can be produced at a saturation pressure of 13.79 MPa with the medium or high pressure drop rate.

As a comparison, the filled HDPE with 5% talc has been tested for the similar relationships among cell density, gas saturation pressure, and pressure drop rate. The filled HDPE shows significantly saturation pressure reducing requirement for the similar nucleation result of cell density at the same pressure drop rate as the test for unfilled HDPE above, as shown in Figure 2.16. The nucleation results for three different pressure drop rates show an almost linear relationship between the cell density and gas saturation pressure. It is because the filler acts as a nucleation agent, and then the cell nucleation density remains relatively high until saturation pressure goes down to 3.45 MPa. A close comparison between filled and unfilled HDPE at the same lowest pressure drop rate is that the density of about 10^7 cells/cm^3 can be made at 12.43 MPa for filled HDPE in Figure 2.16, while the same cell density of about 10^7 cells/cm^3 needs to be achieved at a saturation pressure of 20.69 MPa for unfilled HDPE in Figure 2.15. In other words, to make the same quality of cell density, the filled HDPE only needs 60% saturation pressure of unfilled HDPE. This becomes an advantage of filled material to save energy and be a more stable process as a result of the lower saturation pressure requirement.

It is noted that the effect of pressure drop rate becomes more significant when the saturation pressure is low. In other words, the pressure drop rate becomes critical when the driving force is insufficient for cell nucleation.

Suh [4] explained the cell nucleation will have the competition for gas. During the cell nucleation stage, there is some instant cell growth as well. It then becomes the competition for gas between cell nucleation and cell growth. If the cell does not nucleate instantaneously, some cells nucleate before others. Thus, the gas in the solution will diffuse into the nucleated cells to lower the free energy of the system. Consequently, the low gas concentration regions are generated adjacent to the stable nuclei. There are two possibilities when the solution pressure drops further: either (a) both nucleation of additional cells and expansion of the existing cells by gas diffusion or (b) only expansion of the existing cells without new nucleation. When the solution pressure drops quickly, the system will create more uniform small cells distribution because there is the gas-depleted region where nucleation cannot occur. Therefore, the cells in this gas-depleted region at quick pressure drop will be even smaller.

There are important dimensionless groups to be checked when the nucleation process is to be designed properly. They are helpful to understand the competition between cell nucleation and growth. The first one is the dimensionless group [4, 25]:

$$\frac{\text{Chracteristic} - \text{nucleation} - \text{time}}{\text{Characterisitic} - \text{diffusion} - \text{time}} << 1$$

or mathematically to be written as [4, 25]

$$\frac{D_a}{\frac{dN}{dt_d} d_{\text{cell}}^5} << 1 \tag{2.29}$$

The nucleation rate dN/dt_a has the unit of number of cells/sec. The coefficient D_a represents the gas diffusivity, and d_{cell} is the average cell diameter that can be obtained from the SEM picture. The second dimensionless number is given by considering the length associated with the nucleation and growth of cells over finish processing times. The competition between microcell nucleation and cell growth is negligible when [4, 25]

$$\frac{\text{Charateristic} - \text{gas} - \text{diffusion} - \text{distance}}{\text{Charateristic} - \text{spacing} - \text{between} - \text{stable} - \text{nuclei}} << 1$$

or, mathematically [4, 25],

$$2\rho_c^{1/3}(D_a t_d)^{1/2} << 1 \tag{2.30}$$

where ρ_c is cell density. The diffusion time t_d corresponds to the characteristic nucleation time. These two dimensionless numbers and the relationships must be satisfied in all continuous processes [4, 25] to make the part with microcellular structure. The characteristic nucleation time for microcellular processing is about a factor of 10^{-4} smaller than for conventional foam processes [4].

Strauss and others found the same trend that in a nanocomposite of GPPS the high temperature and pressure rate changes will help significantly for the cell density, shape, and size [35]. The montmorillonite-layered silicate (MLS) was added into GPPS material and foamed in a batch supercritical CO_2 chamber at various temperatures and pressures. They found that the presence of nanoclay in the GPPS sample will have a highly accelerated absorption rate with supercritical CO_2. The other research in the literature usually shows the saturation of supercritical CO_2 taking about 3–24 hr, without nanoclay. However, with the nanoclay in GPPS, they finish the supercritical CO_2 gas saturation in 3% clay of a nanocomposite GPPS in only 5 min. [37]. There may be a secondary mechanism of mass transfer of CO_2 in the nanocomposite foams that is significantly faster than the linear diffusion model predicts.

The practical details of cell nucleation density promoted by pressure drop rate are discussed in Chapter 6 at different injection rates with different morphologies that result from some optimized processing conditions (see Figure 6.19).

The presence of cell nucleation agents creates discontinuities in the foaming mixture. Knowledge of the pressure variation and distribution around a nucleation site is a key to understand the underlying mechanism that promotes cell nucleation. The area that involves negative pressure distribution can induce cell propagation, and the cell nucleation rate can be high [38].

2.5 CELL GROWTH

Cells are expanded by diffusion of gas into already existing bubbles. Processing conditions provide the pressure and temperature necessary to control cell growth. This work concerns the initial stage of bubble growth during the polymer foaming process. Specifically, the difference between filled and unfilled polymer has been investigated. It was found that a certain amount of gas accumulates in the polymer–filler interface (1). The accumulated gas has a significant impact on initial cell growth.

2.5.1 Cell Growth Model

Ramesh summarized theoretical and experimental analyses of bubble growth models since 1917 [39]. A single bubble growth model was the focus in most of the literature between 1917 and 1984. It focuses on a single bubble surrounded by an infinite sea of fluid and is supplied with an infinite amount of gas available for cell growth. Realistically, the cell grows with numerous bubbles expanding in close proximity to one another with actually limited gas available. Hence, the industries practical requirement led to the development of a new model called "Cell Model" between 1984 and 1998. Amon and Denson [40] introduced the first cell model in 1984. It considered a group of gas bubbles growing that are separated by a thin film of polymer and dissolved gas during the injection molding process. Since then, the two groups have been developed. One is the cell model for a closed system that is good for the injection molding process. Another one is the modified cell model with gas loss effects that is suitable for the extrusion process. All references are well summarized in reference 39.

2.5.1.1 A Dimensionless Cell Growth Model [40–43]. Arefmanesh and others developed a cell growth model for injection molding. It is used for a large number of bubbles growing in a closed proximity without gas loss that has adequate assumption for injection molding [41, 42] since the injection mold is closed all the time during molding. These studies presented a qualitative agreement between experimental data and theoretical estimation. It accommodates the viscoelastic nature of the polymer. This model employs the system of simultaneous equations for the mass and momentum transfer. The polymer is treated as a simple Maxwell element with a single relaxation time constant. The dimensionless numbers are defined in the model of cell growth in a viscoelastic fluid that has been proposed and used for many papers [40–43]. These equations have been used in microcellular injection molding for an estimate of the bubble growth as well, and they are represented as follows:

$$P_g^* - \frac{2D_1^*}{R^*} - P_f^* + \frac{2}{3}\int_0^1 \frac{\tau_{rr}^* - \tau_{\theta\theta}^*}{y^* + (R^{*3}/D_4^*)} dy^* = 0 \tag{2.31}$$

where P_g^* is the reduced bubble pressure, P_f^* is the reduced applied pressure, R^* is the dimensionless instantaneous bubble radius, y^* is the reduced Lagrangian coordinates (a combination of space and time coordinates), τ_{rr}^* is the dimensionless normal stress in radial direction, and $\tau_{\theta\theta}^*$ is the dimensionless normal stress in the circumferential direction. D_1^* and D_4^* are dimensionless parameters that incorporate system properties and operating conditions, and they arise in the process of nondimensionalization of the variables. They are defined as

$$D_1^* = \frac{\sigma_b}{R_0 P_a} \tag{2.32}$$

$$D_4^* = (S_0/R_0)^3 - 1 \tag{2.33}$$

where S_0 is the radius of the shell within which one cell grows (see Figure 9.7), P_a is the atmospheric pressure, R_0 is the initial bubble radius, and σ_b is the bubble surface tension.

$$\frac{d\tau_{rr}^*}{d\tilde{t}} + \left(\frac{1}{D_2^*} + \frac{4R^{*2}\dot{R}^*}{D_4^* y^* + R^{*3}}\right)\tau_{rr}^* = -\frac{4}{D_3^*}\left(\frac{R^{*2}\dot{R}^*}{D_4^* y^* + R^{*3}}\right) \tag{2.34}$$

where $\tilde{t}$ is educed time and $\dot{R}^*$ is the time derivative of dimensionless instantaneous bubble radius R^*.

D_2^* and D_3^* are dimensionless parameters and are defined as

$$D_2^* = \frac{\lambda D_a}{R_0^2} \tag{2.35}$$

$$D_3^* = \frac{\lambda P_a}{\eta_T} \tag{2.36}$$

where λ is the relaxation time of the polymer and η_T is the non-Newtonian gas-laden polymer viscosity.

$$\frac{d\tau_{\theta\theta}^*}{d\tilde{t}} + \left(\frac{1}{D_2^*} - \frac{2R^{*2}\dot{R}^*}{D_4^* y^* + R^{*3}}\right)\tau_{\theta\theta}^* = \frac{2}{D_3^*}\left(\frac{R^{*2}\dot{R}^*}{D_4^* y^* + R^{*3}}\right) \tag{2.37}$$

$$\frac{d}{d\tilde{t}}(P_g^* R^{*3}) = 9\left(\frac{D_6^*}{D_4^*}\right) R^{*4} \left.\frac{\partial^2 \phi^*}{\partial y^{*2}}\right|_{y^*=0} \tag{2.38}$$

$$D_6^* = \frac{\rho_{\text{matrix}} P_g}{\rho_g P_a} \tag{2.39}$$

where ρ_{matrix} is the gas density in the polymer–gas matrix and ρ_g is the gas density inside of the growing bubble.

$$\frac{\partial \phi^*}{\partial y^*} = 9\left(\frac{1}{D_5^*}\right)^{2/3} \left(y^* + \frac{1}{D_4^*} R^{*3}\right)^{4/3} \frac{\partial^2 \phi^*}{\partial y^{*2}} \tag{2.40}$$

$$D_5^* = 6\rho_{\text{matrix}}^2 H^2 (R_g T)^2 \tag{2.41}$$

where H is Henry's constant, R_g is the gas universal constant, T is absolute temperature, ϕ^* is the reduced form of the gas concentration potential function.

The initial condition is expressed as

$$\phi^*(y^*, \tilde{t} = 0) = 0 \tag{2.42}$$

and the boundary conditions are represented as

$$\frac{\partial \phi^*}{\partial y^*}(y^*=0,\tilde{t}) = D_7^*(P_g^* - P_{g0}^*) \tag{2.43}$$

$$\frac{\partial^2 \phi^*}{\partial y^{*2}}(y^*=1,\tilde{t}) = 0 \tag{2.44}$$

$$D_7^* = HP_a \tag{2.45}$$

Equation (2.31) is the momentum balance and is coupled with Equations (2.34), and (2.37), which are the component forms of the constitutive equation for a Maxwell-type viscoelastic fluid. Equation (2.38) represents the mass balance across the bubble interface. The Equation (2.40) displays the mass balance on the gas diffusing from the shell of polymer–gas matrix to the surface of the growing bubble [40–43]. The boundary condition of cell growth in injection molding can be simplified by the assumption that there is no gas loss because of the closed mold and quick injection.

The system properties, such as viscosity, diffusivity, and matrix density, must be carefully evaluated before modeling of cell growth. The viscosity for gas-laden material can be referred to in Chapter 9. In addition, the initial cell radius for the cell growth in heterogeneous nucleation will be simplified to use the homogeneous nucleation model. Usually, the CO_2 in PS material will have a 0.1- to 1-micron diameter as the initial critical radius [39]. Although the cell growth modeling has been studied since 1917, more data and theory modification are still required to describe real processing.

2.5.1.2 A Simple Model to Quick Estimate the Final Cell Size. The specific simple model for injection processing has been developed. It will provide two results: One is to describe the initial cell growth, and another is to predict the final cell size after the injection process has finished. Although the model above provides universal equations, it is still not good enough for quick injection process.

It is found that with the gas accumulated in the polymer–filler interface, the cell growth is faster than an unfilled system at the initial stage of cell growth. The estimation of the cell growing to a certain size that can cause visible surface defects may be used to find the minimum injection time. Within the minimum time the cell size growing in the surface may not reach the visible size to make obvious rough surface. The surface defect issues with filled and unfilled polymer system, in injection molding process, can be explained using the initial cell growth model in Chapter 9.

The cell growth will be ended with three different conditions: The material with expanded cell growth completely fills the mold, the material develops (cools) enough rigidity to withstand the pressure of the gas in the cells, or there is no longer sufficient gas to continue to supply for expanding the cells. These three cell growth ending conditions can be used either for a theoretical model or for a simplified model of cell growth calculations.

2.5.2 Cell Size Distribution

The cell size distribution in a microcellular part is not truly uniform and will be in a certain range of different cell sizes. In order to compare the cell size distribution breadth in a microcellular part, a polydispersity index (PDI) is defined as follows [41]:

$$\text{PDI} = \frac{\bar{d}_n}{\bar{d}_w} \tag{2.46}$$

where PDI is the polydispersity index of the cell sizes, $\bar{d}_n$ is the number average cell diameter and $\bar{d}_w$ is the weight average cell diameter.

The number average cell diameter is defined as [41]

$$\bar{d}_n = \frac{\sum d_i n_i}{\sum n_i} \tag{2.47}$$

$$\bar{d}_w = \frac{\sum d_i^2 n_i}{\sum d_i n_i} \tag{2.48}$$

where n_i is the number of cells with equivalent diameter and d_i is the cell diameter in the defined same size group.

2.5.3 Effect of Pressure on Cell Size

Some data from the batch process for the pressure effect on the cell size is quite useful as the reference of continuous processing guidelines. Goel and Beckman [43] reported the pressure-induced phase separation method to make foam of PMMA material saturated by CO_2 gas. The cell size is in a very narrow cell size distribution with average cell size of 0.5–10 μm. The cell size decreases sharply with increasing saturation pressure at 13.8–20.7 MPa, and it levels out at low values (approximately 0.5 micron) at a pressure of 27.6 MPa and above [39]. However, temperature increasing only leads to a gradual increase in the average cell size [41]. It is why most of the industry application of microcellular processing will focus on the pressure-induced foaming, not temperature change.

The same effect has been verified in microcellular injection molding. The SEM pictures of PS with N_2 gas displays (a) small cell size with high back pressure of 82.76 MPa in the barrel and (b) big cell size with low back pressure 6.9 MPa in the barrel, respectively. It is tested with air shot only without any mold effect. Therefore it results from two possible factors: (a) pressure drop and (b) back pressure to precompress the gas-laden material (single phase solution) before air shot. The high back pressure in the barrel will create a larger pressure drop once the nozzle opens. It is another effect on the cell size because it creates more nucleation as discussed above. Therefore, sometime the batch process is necessary to show the individual effect from each parameter during processing. However, the real processing will have some mixed results so that it usually shows a combined result caused by several factors.

2.5.4 Effect of Extensional Viscosity on Cell Growth

During the final stages of cell growth, the melt strength is important to maintain the cells. An increase in the extensional viscosity has been shown to be a good quality of microcellular foam [43, 44]. The extensional viscosity to become higher during mold filling is also called strain hardening or extension thickening. It is defined as the transient elongation viscosity; it rises above the linear viscoelastic curve, at a constant strain rate [45]. Different materials have a different strain hardening effect during the mold-filling stage. For example, it was reported that the blend of linear and branched PP foam has better strain hardening than the linear resin itself [43]. Chaudhary and Jayaraman [46] used two different PP–clay nanocomposites in the same linear polypropylene. One of the clay is 7.2 weight percent of a treated organoclay, and another one of clay is 8 weight percent of treated organoclay. The clay with 7.2 weight percent of treated organoclay in linear PP successfully made a microcellular foam with all closed cells and more uniform cell size distribution. This nanocomposite has two distinct characteristics: The melt displays strain hardening in uniaxial extensional flow, and its crystallization temperature is much higher. However, the other nanocomposite with 8 weight percent of clay did not make as good a microcellular structure as the one with 7.2 weight percent of clay. However, all nanocomposites made a better cell structure than that in pure PP material [46].

2.6 SHAPING IN THE MOLD

Once the polymer melt is injected into the mold, the part shape will be controlled by the mold. In other words, injection mold provides the means for shaping the part finally. Although mold modifications are not required, in some cases the design modifications are recommended for optimal results.

Mold design controls part shape. All the effects for final quality of microcellular injection molding parts are part of shaping the part in the mold. They are summarized as the following topics:

- Surface finish control
- Warpage control
- Weight reduction
- Venting control

- Skin–core structure
- Morphology of the microcellular foam
- Welding line

Overall, a microcellular injection molding part can be shaped much better than the regular injection molding part since a microcellular part expands more than it shrinks. However, the venting and welding line may need to be paid a special attention since a nonsufficient venting and weak welding line may cause quality issues for the microcellular part.

REFERENCES

1. Xu, J., and Pierick, P. *J. Injection Molding Technol.* **5**, 152–159 (2001).
2. Pierick, D., and Jacobsen, K. *Plastics Eng.* **57**(5), 46–51 (2001).
3. Okamoto, T. K. *Microcellular Processing*, Hanser Publications, Cincinnati, 2003, pp. 38–60.
4. Suh, N. P. *Innovation in Polymer Processing*, edited by James F. Stevenson, Hanser/Gardner Publications, Cincinnati, Chapter 3, 1996.
5. Suh, N. P., Baldwin, D. F., Cha, S. W., Park, C. B., Ota, T., Yang, J., and Shimbo, M. Synthesis and analysis of gas/polymer solutions for ultra-microcellular plastics production, in *Proceedings of the 1993 NSF Design and Manufacturing Systems Conference*, Charlotte, NC, January, 1993.
6. Michaeli, W., et al. German Patent No. DE 19 853 021 A1 (2000).
7. Habibi-Naini, S., and Schlummer, C. *SPE ANTEC, Tech. Papers*, 470–474 (2002).
8. Cha, S. W. Ph.D. Thesis, Massachusetts Institute of Technology, Cambridge, MA, 1994.
9. Johnson, D. E. U. S. Patent No. 4,124,336 (1978).
10. Sato, Y., Fujiwara, K., Takikawa, T., Takishima, S. S., and Masuoka, H. *Fluid Phase Equilibria* **162**, 261–276 (1999).
11. Van Krevenlen, D. W. *Properties of Polymers*, Elsevier, New York, 1976.
12. Throne, J. L. *Thermoplastic Foams*, Sherwood Publishers, Hertford, UK, 1996, pp. 126–180.
13. Durrill, P. L., and Griskey, R. G. *AIChE J.*, **1147** (1966).
14. Durrill, P. L., and Griskey, R. G. *AIChE J.*, **106** (1969).
15. Park, C. B. Ph.D. Thesis, Massachusetts Institute of Technology, Cambridge, MA, 1993.
16. Li, G., Li, H., Wang, J., and Park, C. B. *SPE ANTEC, Tech. Papers*, 2332–2336 (2005).
17. Trexel Web Site, http://www.sabic-ip.com/.
18. Wong, A., Leung, S. N., Hason, M. M., and Park, C. B. *SPE ANTEC, Tech. Papers*, 2551–2555 (2008).
19. Chen, L., Sheth, H., and Kim, R. *SPE ANTEC, Tech. Papers*, 1950–1954 (2000).
20. Chen, L., Sheth, H., and Wang, X. *FOAMS 2000*, 127–131 (2000).

21. Chen, L., et al. *Polym. Eng. Sci.* **42**, 1151–1158 (2002).
22. Leung, S. N., Wong, A., and Park, C. B. *SPE ANTEC, Tech. Papers*, 1397–1402 (2009).
23. Lee, S. T. *Foam Extrusion Principles and Practice*, Technomic Publishing Company, Lancaster, PA, 2000, pp. 81–123.
24. Chen, L., Straff, R., and Wang, X. *Effect of Gas Type on Microcellular Process Atmospheric Gases as a Blowing Agent, ASME* (2000).
25. Baldwin, D. F. Ph.D. Thesis, Massachusetts Institute of Technology, Cambridge, MA, 1994.
26. Xu, J., et al. U. S. Patent No. 7,318,713 (2008).
27. Xu, J. U. S. Patent No. 7,267,534 (2007).
28. Wissinger, R. G., Paulaitis, M. E., *Ind. Eng. Chem. Res.* **30**, 842 (1991).
29. Xu, X., Park, C. B., Xu, D., and Pop-Iliev, R. *Polym. Eng. Sci.* **43**, 1378 (2003).
30. Youn, J. R., and Suh, N. P. *Polym. Compos.* **6**, 175 (1985).
31. Colton, J. S., and Suh, N. P. *Polym. Eng. Sci.* **27**, 485–492 (1987).
32. Porter, D. A., and Easterloing, K. E. *Phase Transformations in Metals and Alloys*, Van Nostrand Reinhold Workingham, UK, 1981.
33. Favelukis, M., Tadmore, Z., and Semiat, R. *AIChE J.* **45**, 691–695 (1999).
34. Lee, S. T. *Polym. Eng. Sci.* **33**, 418 (1993).
35. Lee, S. T., and Kim, Y. *SPE ANTEC, Tech. Papers*, 3527 (1998).
36. Park, C. B., Baldwin, D. F., and Suh, N. P. *Polym. Eng. Sci.* **35**, 432 (1995).
37. Strauss, W., Ranade, A., D'Souza, N. A., Reidy R. F., and Paceley, M. *SPE ANTEC Tech. Papers*, 1812–1816 (2003).
38. Wang L. C., Leung, S. N., Park, C. B., and Bussmann, M. *SPE ANTEC Tech. Papers*, 1403–1409 (2009).
39. Ramesh, N. S. *Foam Extrusion Principles and Practice*, edited by S. T. Lee, Technomic Publishing Company, Lancaster, PA, 2000, Chapter 5, pp. 125–144.
40. Amon, M., and Denson, C. D. *Polym. Eng. Sci.* **24**, 1026 (1984).
41. Arefmanesh, A. Ph.D. Thesis, Department of Mechanical Engineering, University of Delaware, 1991.
42. Arefmanesh, A., and Advani, S. G. *Rheol. Acta* **30**, 274 (1991).
43. Goel, K. S., and Beckman, E. J. *Polym. Eng. Sci.* **34**, 1148–1156 (1994).
44. Stange, J., and Munstedt, H. *J. Rheol.* **50**, 907 (2006).
45. Okamoto, M., Nam, P. H., Maiti, P., Kotaka, T., Hasegawa, T., and Okamato, H. *Nano Lett.* **1**, 503 (2001).
46. Chaudhary, A. K., and Jayaraman, K. *SPE ANTEC Tech. Papers*, 1910 (2008).

3

MORPHOLOGY OF MICROCELLULAR MATERIALS

The injection molding process is sufficiently versatile to allow wide variations in morphology. These variations include the different morphological cell structures, such as open cell, close cell, cell size, cell distribution, and solid skin thickness. However, some rules are the same for different morphologies, and usually an average cell size is used for the theoretical calculation. The average cell volume is inversely proportional to the cell density for a given overall density of the microcellular foam. The morphology study can mostly explain well the reasons for the variations of mechanical properties for different materials under different processing conditions.

The reliable method to check microstructure of the microcellular part is to examine a broken cross section of the part. The scanning electron microscope (SEM) is the most popular way to identify the morphological changes of different microcellular parts. Examination of cross sections should be a carefully broken section, not a cut section. These broken sections are usually cryogenically broken. Liquid nitrogen is used to deep-freeze the sample first and, then, to break it along the predetermined direction for a flat fracture section view. The sample needs to be carefully scored with a sharp knife on the surface along the decided direction of the breaking section. In this way the direction it breaks can be controlled very well. However, the scored side must be marked without affecting the side skin thickness, which will be checked because it may misrepresent the skin thickness by the scored surface even though the marks are shallow.

Microcellular Injection Molding, Edited by Jingyi Xu
Copyright © 2010 John Wiley & Sons, Inc.

The cross-sectional views need to be carefully reviewed in all different mold flow directions. The cross-sectional views may have different results depending on whether the views are made from the parallel or perpendicular directions of mold flow. These might result in different conclusions about the cell structure in microcellular part. If the correct views are made in two directions of mold flow, then solutions for the processing changes may be proposed to improve the morphology of a microcellular part.

For a quick visual inspection of cell structure, a sharp knife is used to score one side of the sample. Then, the scored sample is broken to check the section by a quick check of the center area of the cell structures. Usually, an 8× eye loop may be enough to see 100-micron cells.

Most of the published morphologies of microcellular are the ideal homogeneous morphology from batch processes. Suh [1] proposed some ideal microstructures. One is the closed-packed structure with interstitial space occupied by the original plastics. Mathematically, the minimum density of this microcellular plastic is 0.26 of the original solid density [1]. However, this cell structure is only possible for the batch process and is never made in the real production of a microcellular part. It is because the batch process is a slow gas diffusion process, and it may take hours or longer to finish the gas saturation in the sample [1]. As a production method, the injection molding for a microcellular part takes only seconds to finish the gas mixing and the diffusion into the polymer melt [2]. In addition, the injection molding process has shearing deformation and temperature differences in both thickness direction and mold filling direction. A complicated morphology from injection molding is much different from the ideal homogeneous morphology shown in a batch process. A composite structure of a foamed core, a transition layer (usually negligible), and a solid skin layer on the outside is the typical morphology of the part made from microcellular injection molding. It is important to understand the real morphology of microcellular injection molding because it will significantly influence the properties of the part made by microcellular injection molding.

Different materials will have different morphologies of microcellular injection molding. It is well known that there are differences between crystalline material and amorphous material. If the material consists of two different materials that are finely mixed at a molecular level, the density of the nucleation site may be much larger [1]. It is also interesting to know that filled materials will have different morphologies from the unfilled same material. This is because the heterogeneous nucleation in a filled material creates a much better nuclei structure than an unfilled material. In addition, the filled material requires less weight percentage of gases to make same microcellular structure in the part compared to the unfilled same material. Therefore, the size, the number of cells, and the uniformity of cell distribution are different between filled and unfilled same materials. There are also different cell architectures from different blowing agents, such as nitrogen (N_2), argon (Ar), and carbon dioxide (CO_2).

3.1 MORPHOLOGY DIFFERENCES BETWEEN BATCH PROCESS AND INJECTION MOLDING PROCESS

Figure 3.1 shows an ideal morphology of a microcellular sample of high-impact polystyrene (HIPS) made by a batch process that matches the ideal model of close-packed microstructure proposed by Suh [1]. Some of the cells are expanded beyond the close-packed structure, and they stretched the material in the interstitial sites with a uniform wall thickness among cells. The so-called hexagonal honeycomb structure only exists for a very low density foam made by a batch process without material flowing. In other words, for creating this kind of structure the nucleation sites must be uniformly distributed throughout the structure, and the cells must have dilatational expansion without any shear or distortional deformation. Only a batch process can match these requirements. On the other hand, this kind of structure may have too thin wall thickness between cells to be used for wide application. It is because the buckling load is a function of the wall thickness of the cells and the diameters of the cells. The realistic target of the ideal microcellular morphology is the uniform small cell size with wall thickness similar to the cell size that satisfies both weight reduction and resistance to cell collapse due to buckling of walls between cells.

Figure 3.2 shows a typical SEM photograph of a microcellular HIPS part produced by the injection molding process. Although the same material is used for the samples shown in Figure 3.1 and Figure 3.2, two different processes have many differences:

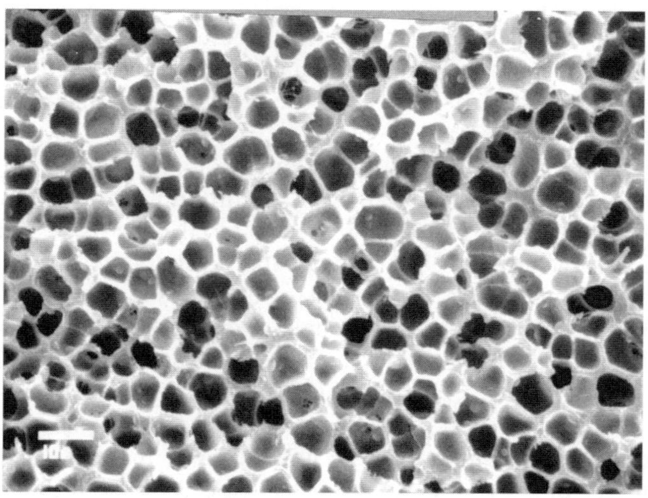

Figure 3.1 Typical SEM of general-purpose HIPS microcellular of batch process (white bar indicates 10 μm), saturated by N_2 gas. Average cell size: 7 μm (Reproduced with courtesy of Trexel Inc.)

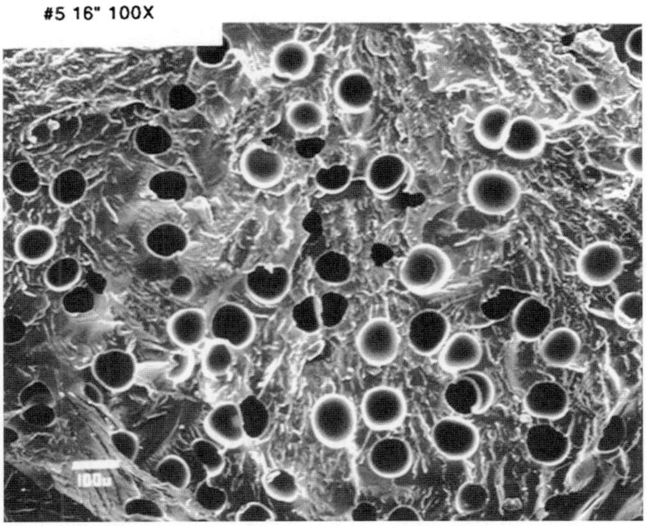

Figure 3.2 Typical SEM of center core for a HIPS microcellular injection molding process (white bar indicates 100 μm). Weight percentage of N_2 gas: 0.8. Average cell size: 90 μm.

- The cell density (number of cells per unit volume) from the injection molding process in Figure 3.2 is much less than the cell density from the batch process in Figure 3.1.
- The shapes of cells are different because of different processes. The part from the batch process in Figure 3.1 has irregular shape and uniform thin wall thickness among the cells. However, the part shown in Figure 3.2 made by the injection molding process has the cells with spherical shape and much thicker and nonuniform wall thickness among cells. In addition, there are several cells as a pie of close-packed cells, or as a stream of cells in Figure 3.2. It may indicate that the gas-rich area is not uniform in the injection molding process.
- The average cell size of injection molding part in Figure 3.2 is about 12 times larger than the cell size of the batch process part in Figure 3.1.
- The micrograph of the sample from the batch process shows some partial open cell structure, whereas the micrograph of the sample from injection molding process displays entirely close cell structures.
- The microstructures of the overall cell structure between the batch process and the injection process are also distinctly different. The cell structure of the sample from the batch process is uniform across the whole thickness as shown in Figure 3.1. However, the sample from injection molding has obvious skin–core architecture (see Figure 3.3) [3]. There are possibly three different layers existing in the microcellular part made by injection molding, center core layer, skin layer, and transition layer between skin

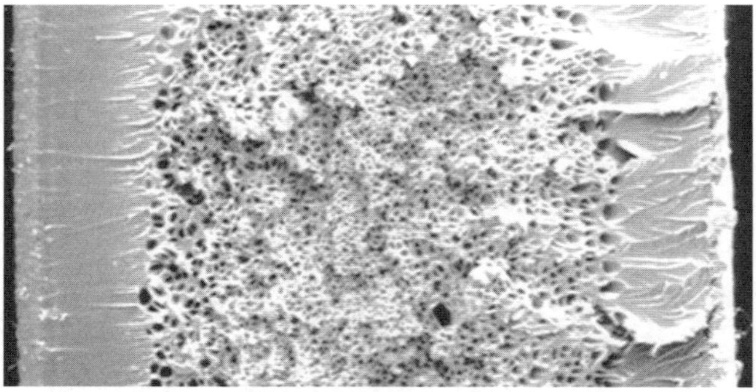

Figure 3.3 Typical SEM with skin–core structure of a microcellular part, a cross-section view for PC [3], N_2 gas. Average cell size: 45 μm. Sample thickness: 3.7 mm. Skin thickness: 0.65 mm. Mold temperature: 160 °F. Melt temperature: 580 °F. Pressure drop rate *dp/dt*: 1.7×10^{11} Pa/sec. Weight reduction: 13%. (Reproduced with copyright permission of Society of Plastics Engineers.)

and center core. Most of good parts made by microcellular injection molding have clearly two layers: solid skin and foam core. The transition layer between the skin and the center core is not obvious for most of the microcellular part from injection molding, so that it is usually negligible in the morphological analysis for the microcellular part.

- The injection molding parts may have some big cells surrounded by nice small cells in the center core (see Figure 3.3). It is because the gas mixing and diffusion may not be uniform in the melt. On the other hand, the quality of nucleation and the number of cells in the part are determined finally by molding conditions as well. The batch process does not have this kind of void in the center since the gas diffuses uniformly from both sides of surfaces to the center.

3.2 MORPHOLOGIES OF DIFFERENT MATERIALS

There are many different materials suitable for microcellular injection molding. Their morphologies may be different from each other. Some of the typical morphologies for different materials made by special injection molding machines are presented in this chapter. On the other hand, the morphologies from physical blowing agents such as CO_2 and N_2 represent most of the morphologies available from current industry technologies including the morphologies of microcellular parts made with chemical blowing agent. Therefore, only physical blowing agents CO_2 and N_2 are discussed for morphological differences. All microcellular materials can be divided into four groups of materials for the comparisons of different morphologies: unfilled amorphous material, unfilled semicrystalline material, material blend, and filled material.

3.2.1 Amorphous Materials

Typical amorphous materials for microcellular of injection molding are general-purpose polystyrene (GPPS), polycarbonate (PC), acrylonitrile/butadiene/styrene (ABS), and high-impact polystyrene (HIPS). Usually, amorphous material will have a wide processing window for microcellular injection molding, along with excellent cell architecture. However, ABS and HIPS will have wide variations since the different rubber phases in the base material may influence the results significantly. Therefore, the cell architectures for different grades of ABS and HIPS may be different as well. Usually if the rubber phase in ABS and HIPS is small, it helps for nucleation because of more heterogeneous nucleation from the small rubber phase.

GPPS is the most popular material for this initial equipment development of microcellular injection molding. It is transparent before foaming, and it becomes white after foaming. Refer to Figure 1.1 (Chapter 1), which is (a) more typical microcellular unfilled polystyrene foam made by injection molding that has an average of 25 microns and has a cell density of about 8.1×10^7 cells/cm^3.

GPPS is also the easy material to work with to create an excellent microcellular structure. Different cell structures with different gases in the GPPS microcellular foam with 16 weight percent of CO_2 gas as blowing agent create larger cells and thin wall thickness among the cells. There is no picture shown here because the cell structure of this sample is similar to the cell structure of HIPS of the batch process shown in Figure 3.1, but the cell size is as large as 50 µm. As the typical disadvantage with this kind of cell structure, the quality of part surface is not as good as the quality of part surface in Figure 3.2, and the strength loss of the part in the cell structure with ideal close packing closed-cell morphology [1] will be significant. Generally, this quality of cell structure will not be acceptable for the microcellular part to be used to replace the part made by regular injection molding process. However, it may the perfect solution for the packing industry as the energy absorption material or insulator for either heat or sound without strength requirement.

It is obvious that the morphological cell structure of HIPS in Figure 3.2 is not as good as the morphological cell structure of GPPS in Figure 1.1. The size and shape of rubber phase in HIPS will significantly influence the final morphological cell structure. The filler effects on the morphology will be discussed more in Chapter 4.

PC is also an easy amorphous material for microcellular injection molding. A cross-sectional view for microcellular PC (white bar is 1 mm) [3] shown in Figure 3.3 is typical of SEM with a clear skin–core structure of the microcellular part. This PC sample is made by N_2 gas with average cell size 45 µm. The sample thickness is 3.7 mm, and skin thickness is 0.65 mm. The processing conditions are mold temperature 160 °F, melt temperature 580 °F, pressure drop rate *dp/dt* 1.7×10^{11} Pa/sec, and weight reduction 13%. It shows acceptable cell structure with the small cells in the center area while big cells appear near the skin. Usually, for this thickness of foamed part the cell size distribution is small cells in the interface near the skin and big cells in the center of the part. However, the cell size distribution shown in Figure 3.3 is opposite to the trend of cell size distribution across the thickness of the part with the structure foam, or regular foam in the injection molded part. It is an important reason why the microcellular part is different from traditional foam structure and has more advantages than disadvantages. The other typical amorphous sample showing similar microstructure is PC/ABS that has clear boundary between foamed core and solid skin.

Overall, ABS is an excellent material for the microcellular part since it is the kind of blend among three different components. The blend of different materials creates a heterogeneous nucleation so that high density of cells becomes possible. However, similar to the HIPS is that the size and distribution of rubber phase in the ABS will be the factor to determine the final morphology of the ABS microcellular part. Figure 6.19b displays optimum cell structure occurring with 0.102 m/sec of injection speed that creates a 15 μm uniform cell size. The processing conditions for this ABS microcellular part are 127 rpm of screw (30-mm diameter, 26:1 L/D) speed, 0.5 weight percent of N_2 gas, 13.8 MPa of back pressure of screw recovery, and 480 °F of melt temperature. With 0.102-m/sec injection speed the calculated pressure drop rate through the nozzle orifice is about 3.4×10^9 Pa/sec. The cell structure of the morphology can be improved further for the cell structure by optimizing the processing conditions [4]. An experimental result verifies that optimum cell structures are possible at different injection speeds. The details of optimizing process conditions will be discussed in Chapter 6.

Sometimes, the flame retardant in the ABS will play different roles for nucleation as well. The particle size of this additive will determine if it helps for better nucleation or creates voids during microcellular processing.

Generally, the amorphous material has thick skin. It can be estimated that about 15–20% of the whole thickness for the microcellular part is made by amorphous material. The skin thickness of amorphous material will be 1.5–2 times as thick as that of crystalline material. Thick skin will help to maintain most of the flexural strength of microcellular parts. On the other hand, thick skin will reduce the total weight reduction percentage of the microcellular part.

3.2.2 Crystalline Material

Crystalline and semicrystalline materials have important microcellular injection molding applications and have become more popular since they are widely used for different industries. The typical materials are polypropylene (PP), polyethylene terephthalate (PET), and polyamide (PA). The crystallization during cooling may expel the gas near the crystalloid so that the cell structure of crystalline material may not be as uniform as the cell structure of amorphous material [5]. On the other hand, crystalline materials will have a

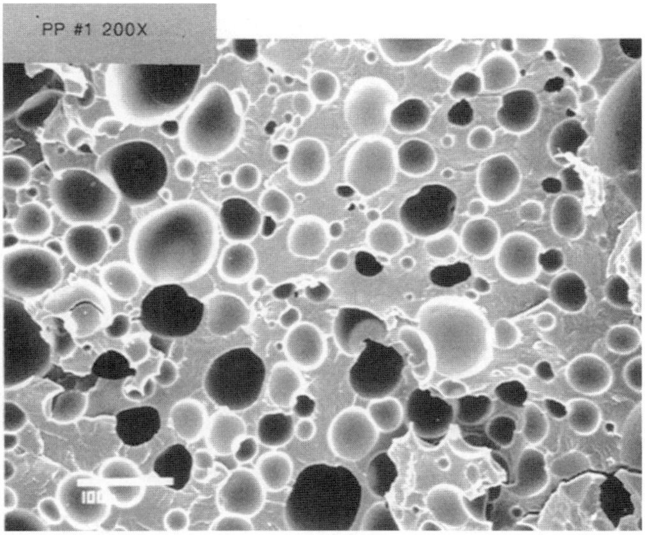

Figure 3.4 Morphology of unfilled PP (white bar indicates 100 μm), CO_2 gas.

nice microstructure with fine cells in the core area but will have no clear boundary between skin and core.

PP is the semicrystalline material. For unfilled PP, it is difficult to make a good part of microcellular injection molding. Figure 3.4 is the morphology of unfilled PP microcellular made by injection molding. The cell sizes vary from 5 microns to 80 microns. The distribution of cells showing in Figure 3.4 is not uniform, either. However, this is a typical morphology for unfilled PP with microcellular structure. One of the reasons for this cell structure is the different generations of nucleation of cells. Some of the big cells may be from first generation of nucleation that will grow in time and subtract gas from its surroundings. Then, a second generation of nuclei also creates more cells but cannot grow as much as the first generation of cells because of less available gas in the polymer matrix left. Another reason for this cell structure is that the heterogeneous nucleation exists even in unfilled materials since no pure polymer exists in the real processing materials. The heterogeneous nucleation itself will create a different size of cells. Finally, the gas distribution in the real process is not uniform after all since the industry equipment has a time limit and a mixing capability limit for the perfect gas dosing process in the screw. Cells may grow faster in the gas-rich region but may grow slower in the region with lower gas concentration.

On the other hand, the filled PP will have much better cell structure than the cell structure of unfilled PP. The cell sizes for filled PP are more uniform, and the cell distribution is more even with regard to cell sizes and the distance among cells. The heterogeneous nucleation from filled materials in PP helps to create many small nuclei simultaneously at the beginning of first nucleation.

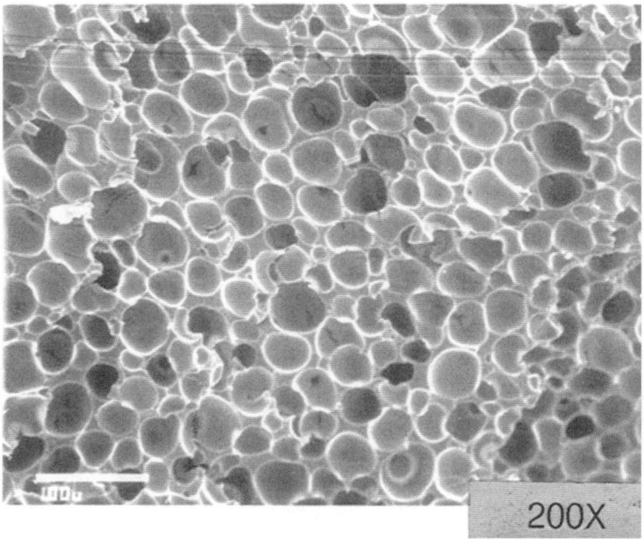

Figure 3.5 Morphology of unfilled PET (white bar indicates 100 μm), CO_2 gas.

Once the first-generation nuclei are good enough to take almost every region of the gas for their cell growth, the second-generation nuclei may not have much chance to be grown. It may explain why filled PP can increase the cell density uniformly and reduce the cell size effectively.

Unfilled PET material is widely used for blowing injection molding because of perfect transparency. However, it may be used for foamed bottles as well, specifically with recycled PET material that enables white or silvery colored bottles to be made without additives which may limit the package recycling for food industry. The morphology in Figure 3.5 shows the sample of unfilled PET with microcellular structure. This PET microcellular part has excellent cell structure with uniform average cell size of about 40 microns, and it also has a uniform wall thickness among the cells.

PA material is easier material to be processed with microcellular injection molding. The morphology in Figure 3.6 shows the sample of unfilled PA 10 with a microcellular structure. This PA 10 microcellular part has good cell structure with uniform average cell size of about 40 microns. However, the cell distribution in the matrix is not truly uniform because the wall thickness among the cells is not uniform. The thickness of the thickest wall is about the same size of the cell. In addition, the thickness of the thinnest wall is so thin that the adjacent cells almost touch each other to be an open cell. However, the cell structure in Figure 3.6 is still good enough to be acceptable as an industry part. On the other hand, the glass-fiber-reinforced PA will have much better microstructure than the micrograph shown in Figure 3.6 [6].

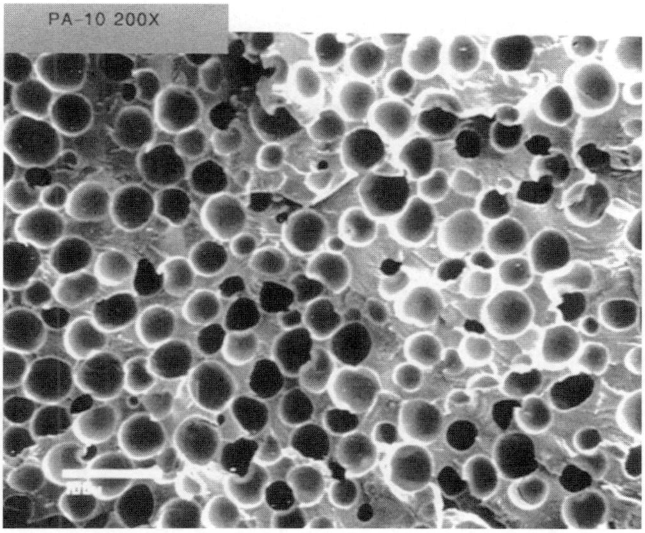

Figure 3.6 Morphology of unfilled PA 10 (white bar indicates 100 μm), N_2 gas.

3.2.3 Blend and Compound of Materials

The blend and compound materials are multiphase materials. The cell may preferentially nucleate at the interface of the phases when the degree of supersaturation of the gas in the molten polymer is relatively low [1]. However, when the degree of gas supersaturation is high enough, both homogeneous and heterogeneous nucleation will occur simultaneously in this multiphase material, although the heterogeneous nucleation will dominate the process.

The typical blend of material on the market is PC/ABS. It is actually good material to be used for a microcellular part because it helps with heterogeneous nucleation. Figure 3.7a shows an excellent cell structure with average cell size of 10 microns, distributed uniformly in the molding part of PC/ABS. With the same blend of material of PC/ABS in Figure 3.7a and the same other processing conditions, an air shot sample is made by injecting the gas-rich material into a air, instead of into a mold. The morphology of air shot in Figure 3.7b shows very fine 3-micron cells distributed uniformly in the whole sample. It proves that the ideal microcellular foam can be made much better than the microcellular part with current molding technology if a constant pressure drop rate is kept during the whole injection period, and cell growth is controlled well. However, the cells distribution in the whole part in Figure 3.7 is still not uniform even if the cell size seems uniform. This is very common morphology in even a very good microcellular injection molding part. If the cell size is uniform but the cells distribution in the matrix of polymer is not uniform, then the wall thickness among the cells are usually not uniform. It may be related

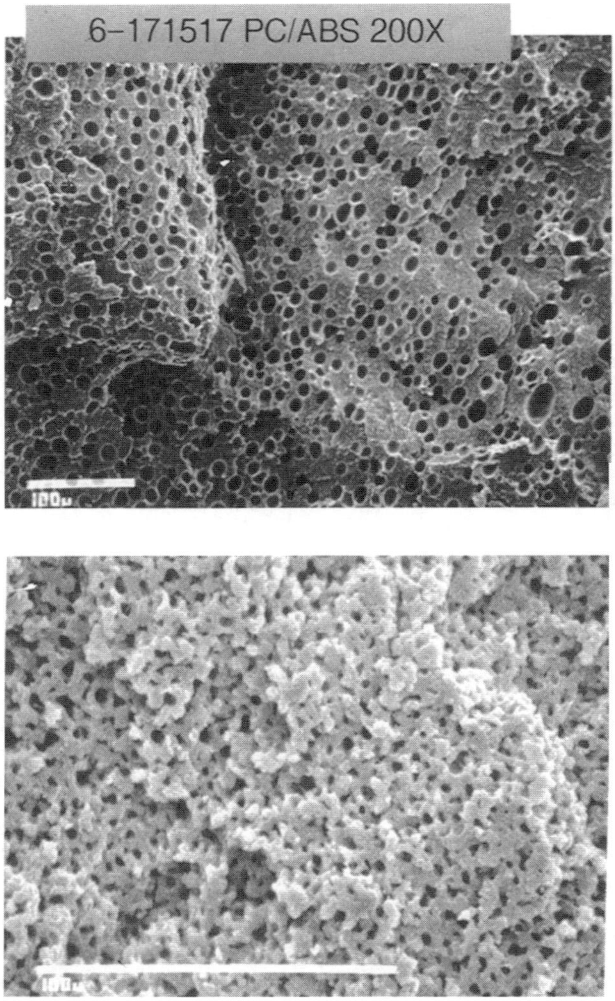

(a)

(b)

Figure 3.7 Morphology of PC/ABS of injection molding. (**a**) Molding part and (**b**) air-shot sample (white bar indicates 100 μm), with N_2 gas.

to both screw mixing performance and nucleation uniformity during mold filling.

The use of compound material is a growing application. The Kraton G-7722 is a typical TPE compound used for medical parts. A microcellular injection molding has been tried for the issue of high cost, poor autoclavability, and striction. The morphology of Kraton G-7722 compound is shown in Figure 3.8. The cell size is not uniform, and the cell shape is not spherical. However, the overall cell structure in Figure 3.8 is still in the range of microcellular definition. This sample is made with about 20% weight reduction for the cost saving. The processing conditions are: melt temperature 420 °F, mold tempera-

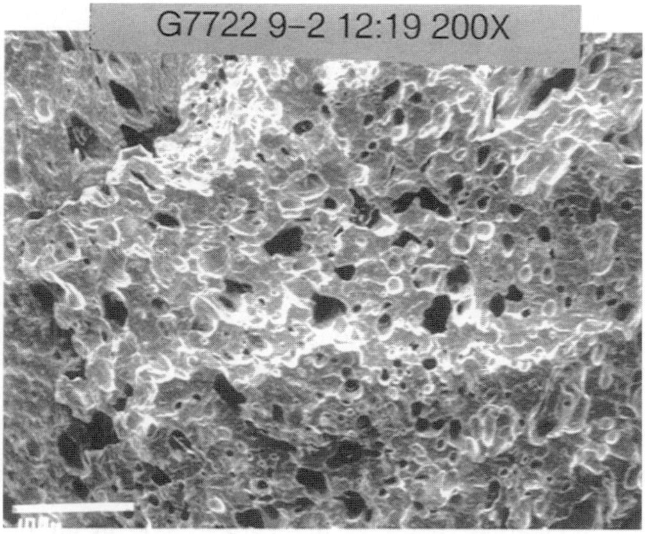

Figure 3.8 Morphology of TPE, KRATON G-7722 (white bar indicates 100 μm), N_2 gas. Weight reduction: 20%.

ture 100 °F, N_2 gas weight percentage ~0.8, and injection speed 0.076 m/sec (3 in./sec) with 40-mm diameter of plunger.

Another popular compound material is PVC. There are many different kinds of compounds of PVC that make the morphologies of PVC much different from each other. Figure 3.21 is one of the examples of PVC microcellular batch process samples with three different gases. All compounds are usually the mixture of incompatible polymers. The cells in the compound may form along the interfacial boundary. Therefore, the uniform mixture of compound may result in uniform cell distribution as well.

3.2.4 Reinforced and Filled Materials

The reinforced material usually means a glass-fiber-filled material. The typical materials filled with glass are PA, PBT, and so on. Also, the carbon fibers, natural fibers, metal fibers are special reinforcement materials in plastics. The filled material is material filled with the fillers, such as talc, mineral, and so on. Overall, all filled materials are good for the microcellular injection molding. Only reinforced material morphology is discussed with SEM in this chapter because the glass fiber is visible among the cells. The morphology of filled material is similar to that of reinforced material. Generally, the cell structure of filled material is simply better than the cell structure of unfilled material.

The filled and reinforced materials are definitely benefited by the heterogeneous nucleation and cell growth. The final microstructure may be affected by the size, distribution, orientation, and interfacial energy of the second or more phases. However, there are still some homogeneous regions in the heterogeneous dominated materials. The nucleation in this homogeneous region may not be as competitive as the nucleation in a heterogeneous region because the gas diffusing easily into the low-energy region. In fact, even homogeneous materials may behave as a heterogeneous material in the presence of interfaces due to the molecular orientation, crystallinity, and highly strained region.

The morphology for the case with 33 weight percent of glass-fiber-reinforced PA 6 material can be seen in Figure 6.16b. It has microcellular structure across the thickness direction. However, the cell size is varied from 5 microns up to 100 microns. Glass-fiber-reinforced PA 6 and PA 6/6 materials are usually good materials for the microcellular injection molding.

The SEM picture in Figure 3.9 is the morphology of POM with 25% glass fiber. It shows excellent cell structure as well. Although there are some big cells, most cells are uniformly distributed around the fibers. The difference between POM in Figure 3.9 and PA 6 in Figure 6.16b is the uniformity of cell sizes. The cell sizes of the POM sample are also varied, but the range is only half of the variation range of cell sizes in Figure 6.16b. Another obvious benefit from both the POM and PA 6 samples is the fiber disorientation. Both POM sample and PA 6 sample show excellent glass fiber disorientation.

There is complicated morphology for the carbon-fiber-reinforced PC compound that has been filled with Teflon material. The sample is cut with the

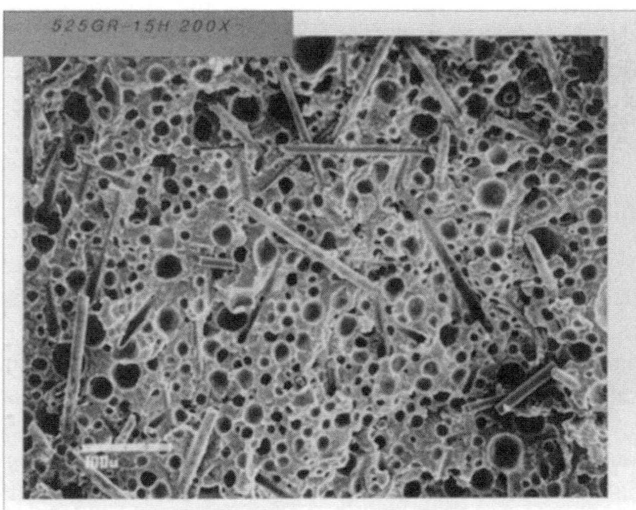

Figure 3.9 Morphology of POM with 25% glass fiber (white bar indicates 100 μm) [6], N_2 gas, $dp/dt = 2.6 \times 10^{10}$ Pa/sec, 15% weight reduction. (Reproduced with copyright permission of Society of Plastics Engineers.)

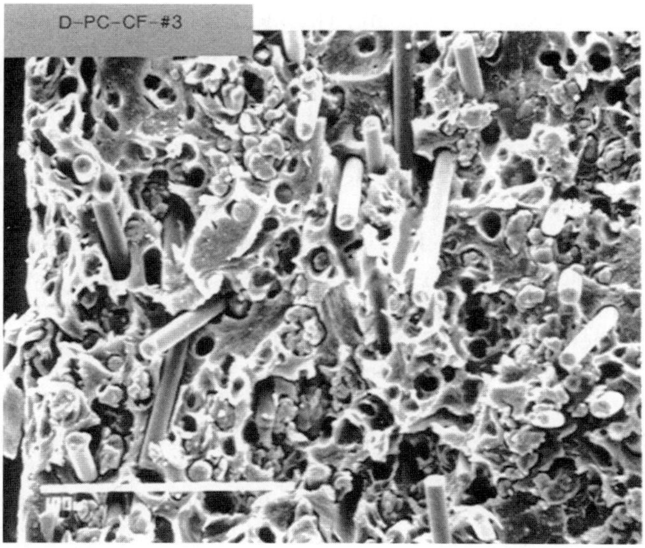

Figure 3.10 Morphology of PC with 20% carbon fiber and 1% Teflon (white bar indicates 100 μm), N_2 gas.

SEM shown in Figure 3.10. It still has excellent cell structure with the average cell size of about 15 μm. A white ring around the cells is the Teflon. The interesting phenomena from observation for this sample processing are that the Teflon is expelled during cell expansion and most of the Teflon content in the polymer moves to the surface. It is good for the surface of the part where Teflon is needed to get the lubrication and low friction for printing part function. However, it causes another problem when cleaning up the mold with too much Teflon coating on the mold surface every week. Teflon is very low friction material that may contribute the some extra lubrication between polymer and fibers. Some of the carbon fibers have a whiter color than other carbon fibers, which may indicate that the Teflon may coat some of carbon fibers. In addition, Teflon may also affect some of the cells nucleation and growing because some cells show obvious white color on its edge in this morphology picture. The skin is Teflon-rich in this sample based on the observation. Therefore, the skin structure looks complicated. It does not have obvious skin thickness in this sample compared to pure PC that has obvious skin thickness and boundary between skin and foamed core. This will be a new research topic for the future morphologic study with Teflon–carbon fiber mixed in the PC for microcellular foam.

Another ideal reinforced material is PBT. It is easy to make an excellent cell structure microcellular injection molding part no matter how much the percentage of fiber glass in it. If the processing conditions are kept the same except for the weight reduction, the SEM picture of PBT with 30% GF and 15% weight reduction shows an average cell size of about 15 microns and a cell density of $N_f = 8 \times 10^8$ cells/cm^3. The morphology of a sample with 30% glass-fiber-reinforced PBT has about 4% of weight reduction; also N_2 gas is the blowing agent, and the average cell size is about 45 μm because the space left for cells growing is less in 4% weight reduction. On the other hand, less weight reduction creates less cell density as well. The cell density of 4% weight reduction sample only has a cell density of about 4.2×10^6 cells/cm^3. The glass fibers in the PBT microcellular part are also disorientated significantly. The fiber disorientation in the PBT sample explains why sometimes the microcellular part made by glass-fiber-reinforced PBT offers better physical properties than does the solid glass-fiber-reinforced PBT part. Therefore, the reason for higher mechanical properties from microcellular PBT with glass fibers is actually from the glass fiber disorientation, not the foaming material region itself.

All glass-fiber-and carbon-fiber-reinforced microcellular parts made by injection molding show that the fiber always stays in the wall between cells. It means that the cells are only create around fiber, not in the same spot of fiber. In this way, fiber is still strongly held by solid material not in the cell or void of the part. This may explain why most glass-fiber- and carbon-fiber-reinforced microcellular parts maintain better original mechanical properties than do unfilled same materials.

Nanoclay-filled material is current growing quickly in the plastic industry. It was found that the nanoclay helps to facilitate the gas dissolution into the polymer melt. Gas in the melt also helps to further disperse the nanoclay platelets in the polymer–gas solution, according to the XRD data with PA 6 nanocomposites [7]. The comparison of morphological cell structure shows that the cell size of PA 6 nanocomposite is only 15 μm while the cell size of neat PA 6 is about 70 μm [8, 9]. This is because of the heterogeneous nucleation with nanoclay in the melt. It is the same reason for talc and mineral-filled material with similar results. The nanoclay in microcellular foam is so small that the magnification ratio of SEM micrographs for microstructure of cells is too small to see any nanoclay. Therefore, most of SEM micrographs show only the cell structure [8, 9]. However, the nanoclay distribution in the polymer can be illustrated by transmission electron microscopy (TEM) [8–10].

The clay only occupies a very small amount, in the range of 0.5–2 weight percent [10]. Both CO_2 gas and N_2 gas are successfully used for making good microcellular nanocomposites. However, N_2 gas is the more common blowing agent for biopolymer microcellular material because it is easily controlled for gas dosing in the barrel. Also, when comparing the weight percentage of each gas needed to make similar cell structure in the same material, CO_2 gas is required with much larger weight percentage than N_2 gas.

On the other hand, the crystallinity of plastics will play an important role in the nanoclay effect for cell morphology [10]. The effect of crystallinity on the cell morphology and mechanical properties in HDPE/clay nanocomposite foam with CO_2 cannot be neglected since it will be significant. Jo and Naguib [10] found that the nanoclay loading in HDPE/clay nanocomposite foam is effective only at a low crystallinity state. Cell size and distribution with nanoclay become more uniform as the crystallinity decreases. In addition, the cell size is decreased as clay content or crystallinity is increased in HDPE/clay nanocomposites. The toughness of HDPE/clay nanocomposites is also decreased as crystallinity increases. However, as clay content increases significantly, the influence of crystallinity on the variation of cell size is diminished [10].

3.2.5 Biopolymer

Biopolymers are made from PLA, PHA, and starch-based resins. They are attracting more and more growing market interests as green materials with no ties to petrochemical-based thermoplastics.

The biopolymer is difficult to process since it has a very narrow processing window. As one example of the basic methods to improve the rheological properties of PLA, two chain extenders are used to treat PLA in the melt stage. This results in higher viscoelastic modulus and improves the cell structure of PLA. The morphologic results of chain-extended PLA show the microcellular structure with 6.7×10^8 cells/cm^3 compared to neat PLA foam with 7.7×10^5 cells/cm^3 at the same processing conditions and, consequently, give rise to smaller cell sizes in higher cell density chain-extended PLA, too [11].

The inorganic particles in a polymeric matrix of PLA will affect not only the thermal properties of PLA but also the rheological properties of PLA. It is more significant to affect the PLA properties when the filler size has one nanometric dimension [12]. The ratio of nanometric particles is very high; as a consequence, the interaction between macromolecules and inorganic particles is enhanced to a great extent [11]. Any filler in polymer melt will act as heterogeneous bubble nucleation sites, increasing largely the number of nuclei.

The effect of a modified montmorillonite, Cloisite 30B, causes a great increase in the rheological properties. With 5 wt% of nanoclay filler in PLA, the cell size becomes truly microcellular at about 30 μm, while neat PLA shows the larger cell size of about 250 μm [11]. However, this behavior may ascribe both to increased cell nucleation, due to the presence of the heterogeneous bubble nucleation sites, and to the intrinsically high viscosity and elasticity. It may be caused by organoclay exfoliation and chain extension/branching. Finally, the fast crystallization rate for PLA nanocomposites is also helpful for stabilizing the cell structure during the foaming process [11].

3.3 CHARACTERIZATION OF CELL ARCHITECTURE

To clearly show the different morphologies of each layer of the part in Figure 3.3, the cell structure of microcellular injection molding is characterized by three different zones: skin, core, and interface between skin and core. Then, there are possibly three different layers in the injection molded microcellular part: center core layer, skin layer, and transition layer between skin and center core. Some typical SEM pictures of microcellular parts are selected for the characterization of cell structures of microcellular injection molding. These different cell structures in the parts from microcellular injection molding provide the complicated microstructures that are critical factors to determine the quality of the microcellular injection molding part. Unfortunately, in most cases the weight reduction is the simplest way to decide possible cell structure prediction because it is much easier to evaluate the density reduction than to check the real cell structure for the uniformity of density reduction distribution. Therefore, the simplest way to check only weight reduction without morphology views may result in a wrong conclusion about the process and part quality issues.

3.3.1 Skin–Core Structure Across the Section of the Microcellular Injection Molding Part

One of the unique characteristics of the part made by microcellular injection molding is the skin–core structure across the section of the part. Most of the good parts made by microcellular injection molding have distinct skin and core structure (see Figure 3.3). This cell distribution for microcellular foam has been repeated with a PC/ABS sample made by injection molding. The overall view of the cell distribution is clearly skin–core structure without obvious interface. The biggest cell is not in the center core, but in the interface of skin and core. There will be more samples of microcellular injection molding showing this cell structure if processing control runs well. Therefore, the model of microcellular cell distribution is proposed in reference 3, and it proves that it is right not only for PC but also for other amorphous materials. There are two zones illustrated in this model: One is a foamed core with uniform cells, and another is the solid skin.

However, if the processing condition is not right, there will be some big cell, or even voids, in the microcellular part. Then, these big cells, or voids, are usually in the center core (see the >100 μm big cell among the small cells in the center zone of the PC sample shown in Figure 3.3). On the other hand, the microcellular structure should be defined by an overall structure and not by a few big cells.

There is an unclear significant characteristic of microcellular foam with clear skin–core structure. The trend of cell size distribution across the section is opposite to that of traditional structural foam. The cell size in the center zone is fine cells, and the coarse cells occur near the skin or in the skin close to the interface between skin and core (see Figure 3.3). The physics behind this occurrence may be from the shear stress because the strongest shear stress is near the interface of skin and core. Additional evidence for this unique skin–core model of microcellular foam is the presence of some elongated cells in the interface of skin and core. This is an interesting research topic of stress foaming in the future, and it will be discussed in more detail in Chapter 8.

For a complicated or thick microcellular part (more than 4 mm thick) made by injection molding, there are possibly some structural foam structures with a bell shape of density distribution profile across the section [13]. When the average cell size is still about 100 microns, it may be still qualify as microcellular foam. The cell distribution across the section for the structural foam part is different from the microcellular part. The biggest cells are in the center zone, and then the cells get smaller when they are a further distance from the center until they disappear completely into the outer skin. If the thickness of microcellular foam is thicker than 4 mm, this kind of structural foam cell distribution may occur. Also, the critical flow ratio to decrease big cells in the center zone is 200 or less with total injection time 3 seconds or less.

3.3.2 Cell Structure of the Core Zone

Based on the cell distribution of microcellular parts in good microcellular foam, the shape of the cell in the center core is spherical. However, the cell distribution is determined by processing conditions. Dr. Nam Suh proposed an ideal cell distribution model for density calculation of the microcellular part. This ideal cell distribution model may very well define the microstructure for the batch process. The distance among cells may be closer, and the shape of cell may vary from spherical to polyhedron. The ideal cell model also matches a few cases of cell structure in the foamed core for microcellular injection molding with high weight percentage of CO_2 gas. Sometimes very few cases of microcellular parts with N_2 gas will have this kind of close-packed cell structure, such as the microstructures shown in reference 1. However, most of the microcellular parts made with N_2 gas as the blowing agent will not have the ideal close-packed cell structure. Therefore, there are more complicated cell architectures to be defined in detail in the future.

All non-close-packed microstructures shown above are still an ideal model of microcellular parts because of the uniform cell size and uniform cells distribution. Strictly speaking, none of the microcellular morphology matches the ideal model either by uniform cell size or by uniform distribution. The cell structure in Figure 3.4 may be acceptable for PP because of the difficulty of PP processing. The better results in Figures 3.6, and 3.7 are the final target of microcellular technology in injection molding. Most of the microcellular foams with such microcellular structure in Figures 3.6 and 3.7 are good enough to be used for real application. Even the cell shape is not spherical (Figure 3.8), but cell size is small enough to get the major benefits of microcellular parts: weight reduction, dimension stability, and cycle time reduction. Keeping the small size of cell is the most important target than the larger number of cells. In addition, weight reduction must be controlled with minimum strength loss. With N_2 gas as blowing agent, the reasonable weight reduction percentage is about 5–15 weight percent.

Nonuniform cell size distribution in the foam core of microcellular is common for some semicrystalline unfilled PP shown in Figure 3.4. The features of morphology shown in Figure 3.4 are listed as the following:

- Cell sizes are not uniform.
- Cell distribution is not uniform. Some of them are distributed as a stream of cells along the mold flow direction. Others spread out as a pie of cells collected together from a local rich gas spot or from local rich nuclei.
- Some cells are so close that they overlap each other with a very thin or deformed wall among cells and then, even local open cells with several broken cells joined together.
- The wall thickness among cells shown in Figure 3.4 is not uniform because the cell distribution is not uniform. It may create some weak points where failure occurs first.

Actually, this is the dominant cell architecture for all microcellular parts made by current commercialized injection molding equipment. This is because of the current limitations on the following: the materials, the mixing quality of the gas in the melt before injection, and different molding conditions. Usually the amorphous material and filled materials will have the cell architecture similar to that in Figures 3.5–3.7, Whereas the semicrystalline material more likely has the cell architecture shown in Figure 3.4. A detailed analysis of the cells distribution related to processing conditions and equipment design are discussed in Chapter 6 and 7.

The cell density is calculated in a modified method from Strauss et al. [14]:

$$N_j = (n_j M^2 / A)^{3/2} (r_c)^{1/2} \qquad (1/\text{cm}^3) \tag{3.1}$$

where n_j is the number of cells seen in a micrograph, M is the magnification factor, A is the area of the micrograph (usually as cm^2), r_c is the ratio of cell width over height (cell density), and N_j is the number of cells per cm^3.

For the model shown in Figure 3.6, r_c equals one since the shape of cell is spherical.

There is another way to calculate the cell density if the density of foam is known by the sample measuring, and assumption of perfect spherical cell shape, then, N_j will be given:

$$N_j = \frac{6\left(\dfrac{\rho_{\text{poly}}}{\rho_{\text{foam}}} - 1\right)}{\pi D_c} \tag{3.2}$$

where ρ_{poly} is the density of unfoamed polymer ρ_{foam} is the density of foam, D_c is the average diameter (cm) of cell seen in a micrograph picture. Then, the average morphological cell wall thickness t_j between cells is

$$t_j = \frac{1 - D_c \times (N_j)^{1/3}}{(N_j)^{1/3}} \tag{3.3}$$

It is well known that the injection molding process has strong shearing in the mold filling stage. The morphology for the section view perpendicular to the flow (mold filling) direction may be different from the morphology for the section view parallel to the flow direction. Therefore, the procedures to prepare the sample for microcellular injection molding are different from the preparation procedures to make the sample for the microcellular batch process. In order to have an accurate cell structure, it is suggested to have two samples prepared at the same position but with 90 degrees difference in orientation. The first one is a cut to check the section parallel to the flow direction. The second sample requires a cut at 90 degrees from the first one to check the section perpendicular to the flow direction. In Figure 3.11, the morphology shows only half of the thickness of the section view of sample (right side picture for skin layer, left side of picture for center layer). The sample in Figure 3.11 is made by ABS in the rectangular mold: 150 mm wide, 2 mm thick, and 150 mm long with fan gate. Figure 3.11a is the section view that is prepared in the direction perpendicular to the flow. Only reviewing this picture may cause misunderstanding that the bubble size is small near the skin without knowing that the bubble near skin is elongated in the flow direction and looks like small cells. However, elongated cells near skin are clearly shown in Figure 3.11b with the section view parallel to the flow direction. The results in Figure 3.11 clearly show the sheared cell along the flow direction in any layer away from the center. The real overall size of a sheared cell near the skin looks similar to the overall size of the nonsheared cell in the center. The strongest shearing is in the interface between mold and melt with an elongated cell, and the least shearing or nonshearing zone is in the center layer with an approximately spherical shape of cell.

It is obvious that the cell distribution in Figure 3.11b represents the velocity profile in the thickness direction. Therefore, it matches the schematic of the theoretical velocity profile in thickness direction during mold filling in Figure 3.11. This velocity profile in the mold filling channel results in a related cell distribution profile in Figure 3.11. The cell in the center layer is almost spherical because it has zero shearing. This cell will not only remain spherical but will also keep growing near the free flow front where there is no, or less, pressure compared to gate area in the part during mold filling. Then, the cells near the interface between skin and core elongate along the shearing direction, becoming ellipsoidal. The analyses here are useful to set up processing parameters correctly during injection molding. The bubble stretching in the surface during mold filling will be discussed more in Chapter 6.

Although the morphology in Figure 3.11 is not uniform, it is possible to simplify it to the ideal equal cell size of microstructure if the volume of ellipsoid is equal to the volume of sphere. This simplified model is good enough

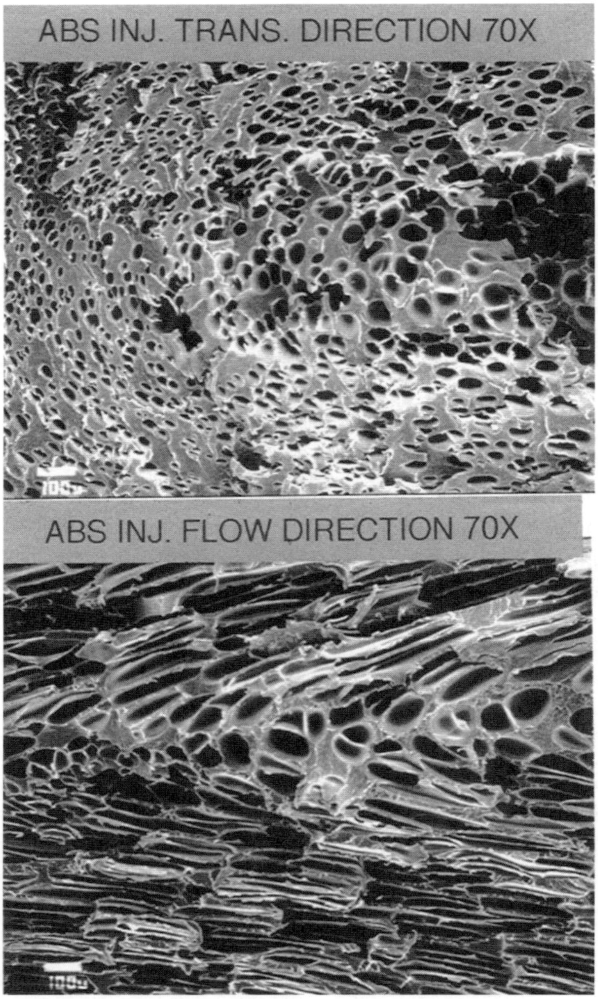

(a)

(b)

Figure 3.11 Morphology of cell distribution in thickness direction, 2 mm thick, ABS, N_2 gas, cold mold surface, and fast injection speed [16]. (**a**) Section view perpendicular to the flow direction. (**b**) Section view parallel to the flow direction. (Reproduced with copyright permission of Society of Plastics Engineers.)

for the density calculation. However, it may not be good for the strength calculation. Part design for preparing the cell shapes and sizes will be discussed in more detail in Chapter 5.

Another interesting observation of cell structure is the shape of elongated cells in the interface zone. The cell structure is not only elongated in the direction parallel to flow direction like Figure 3.11b but also has the shape of oval in the direction perpendicular to the flow direction showing in Figure 3.11a.

This indicates some deformation along the direction perpendicular to the shearing direction. It makes the flow analysis more complicated with this deformation.

3.3.3 Cell Structure in the Interface Zone

As discussed before, the ideal part of microcellular injection molding should have no interface zone at all. It has the uniform cells distributed across the center core until cells reaches skin. However, the interface zone exists if processing conditions are not controlled to remove it. If this interface exists according to the results of Figure 3.11, the shape of the cells in the interface zone is elliptical because the cell elongates severely by strong shearing. Suh proposed a model that fits the local view for the microstructure in the interface zone of the same sample in Figure 3.11b. Then, the cell density is calculated with Equation (3.1), where r_c is less than one.

There is an important clue about the shape in Figure 3.11. It indicates that the cells are not only sheared in the flow direction, but also are compressed by the force perpendicular to the flow direction. The detailed force balance will be discussed in Chapter 6.

The microstructure in Figure 3.11 shows that some of the cells are shaped like an ellipsoid, or like a teardrop. This cell's shape distribution across the section is similar to the morphology of structural foam of GPPS [15, 16]. However, microcellular foam has much more deformation of cells since the shearing deformation is higher than the shearing deformation in structural foam. It is also because of the thin thickness and high injection volume rate for microcellular foam. Some of the severely deformed cells near the skin are either almost flat or nearly broken. The cell wall ruptures so that it results in either creating an open cell or further making some almost flat cells. The extremely deformed cells finally become the ruptured layers near the skin, as shown in Figure 3.11b. These ruptured layers eventually will cause more rough surfaces; this is the disadvantage of microcellular parts made by injection molding. There are several ways to improve the rough surfaces; this will be discussed in Chapter 7.

If the mold temperature is cold and the injection speed is fast, there is a possibility of misunderstanding the SEM picture if the sample is cut without paying attention to flow direction. As shown in Figure 3.11a, the cell structure is cut perpendicular to the flow direction so that it shows very fine cell sizes near the skin. This mistake can be detected by the variation between (a) density calculation based on cell size and (b) the real measure density of the part. Then, the quick way to find the deviation between theory and the real part is to check again the sample section view with different directions of cutting the samples. The cell shapes shown in Figure 3.11b prove that cells are actually stretched so much that the shape of the cell becomes ellipsoid with the big ratio of long axis to short axis. The elongated long axis of ellipsoid is parallel to the flow direction near the skin. The hot mold surface and slow injection speed may help to reduce the deformation shown in Figure 3.11b; then, sample cutting in one direction can reflect the real shape of cells that should be spherical.

3.3.4 Cell Structure of the Skin Zone

The morphology of the skin in the smooth surface mold has been investigated by Xu [16]. This is a fragment from a broken cell on the surface (see Figure 6.25) because of strong shearing from the velocity profile shown in Figures 6.23 and 6.24. There may be some sliding action on the smooth surface mold. However, most cells on the skin will be broken because of strong shearing. It results in the rough surface of the microcellular part, along with silver streaks on the surface. The details of the processing improvements are discussed in Chapter 6.

3.4 INFLUENCES OF CELL ARCHITECTURE ON MICROCELLULAR QUALITY

Although the cell structure is an important factor to determine the final properties of a part in microcellular injection molding, the density reduction or weight reduction is still the dominant characteristic that influences mostly the properties of a microcellular foam. This relationship between weight reduction and property change is similar to, but not the same as, the conclusion for the structural foam from Throne [17]. Since the microcellular injection molding has unique skin–core architecture with related uniform cell distribution and fine cells compared to structural foam, some well-known knowledge of structural foam may not be true for the microcellular part anymore. Several new guidelines of cell structures (morphology) versus properties of the microcellular foam part have been given based on the studies of morphologies of skin–core architectures of microcellular foam.

3.4.1 Cell Size and Density

Cell size and cell density are the key factors for the final microcellular properties. A well-planned study for the cell structure versus mechanical properties was carried out to verify the relationships between cell structure (with SEM pictures) and mechanical properties (with detailed test data for each SEM picture of sample). The material used to make a microcellular sample is PBT material of Valox® 420 SEO with 30 wt% of glass fiber. The same processing conditions are applied for making three samples except for the different shot sizes. Three different shot sizes are 4%, 10%, and 15% weight reductions, respectively. Since the mold conditions are the same, the skin thickness is the same for all samples. Then, the weight reduction percentage here is the only

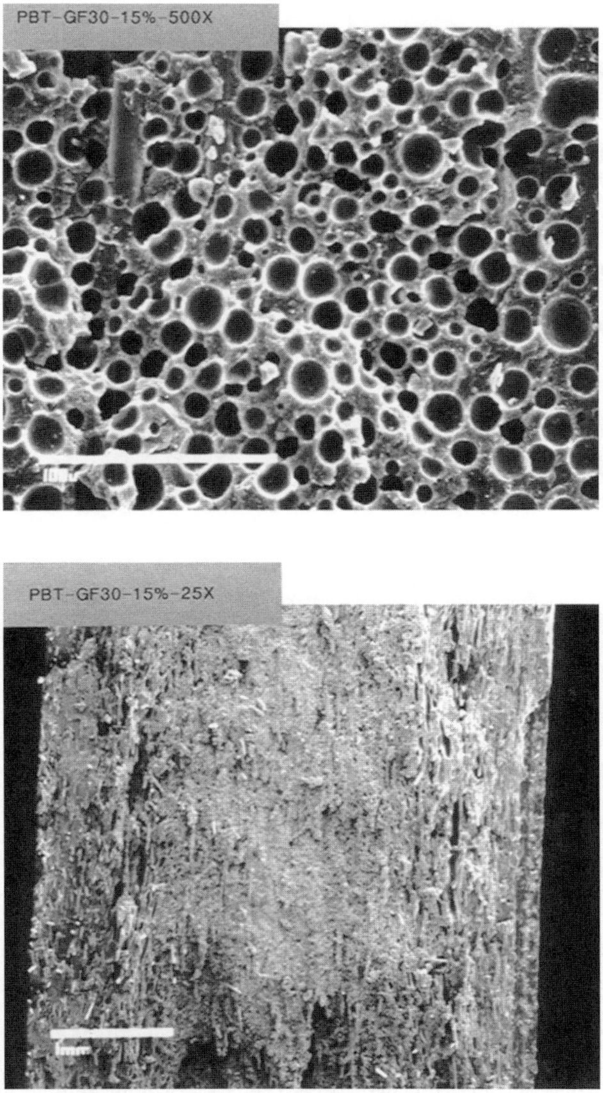

(a)

(b)

Figure 3.12 Morphology of PBT with 30% GF, 15% weight reduction. (**a**) Local center view with average cell size about 15 μm, $N_f = 8 \times 10^8$ cells/cm^3. (**b**) Across section view with some elongated voids in the interface between skin and core.

change, and it represents the change in core cell density. The samples in Figures 3.12 and 3.13 have almost the same average cell size (about 15 μm), but different weight reductions: 10% with cell density of about 2×10^8 cells/cm^3 and 15% with approximate cell density of 8×10^8 cells/cm^3, respectively. The 4% weight reduction sample shown in Figure 3.14 has an average cell size of about 45 μm and a cell density of about 4.2×10^6 cells/cm^3. All three SEM

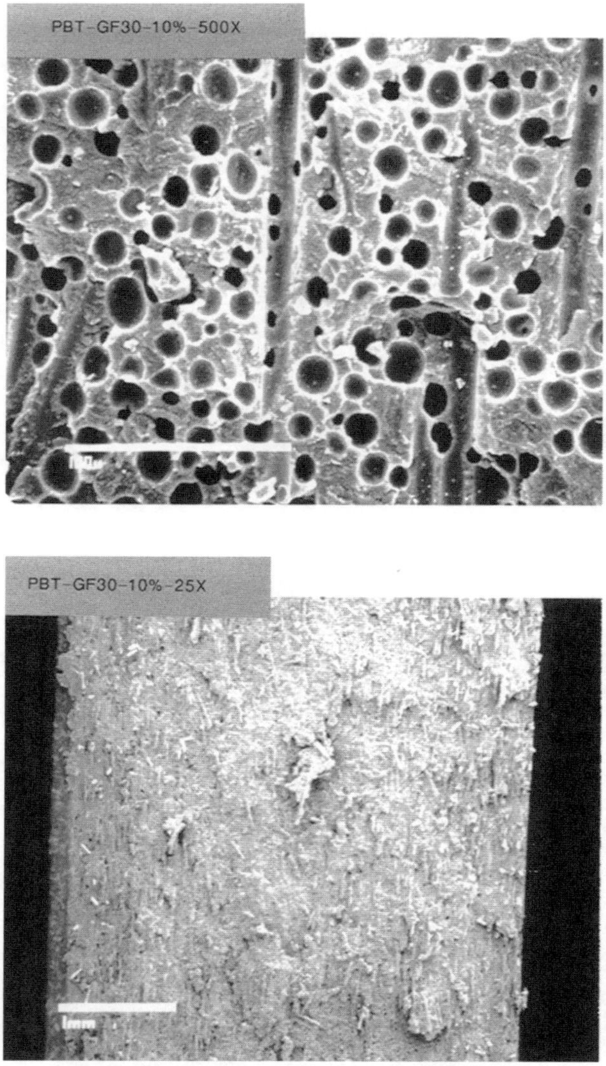

(a)

(b)

Figure 3.13 Morphology of PBT with 30% GF, 10% weight reduction. (**a**) Local center view with average cell size about 15 µm, 2×10^8 cells/cm^3. (**b**) Cross section view with no voids and big cells on the near skin.

pictures in Figures 3.12, 3.13, and 3.14 are shown with both overall view and local center view. They are trustable sample views because there are no voids and other defects in the foamed part. Therefore, the analysis below is reliable.

The test results in Figure 3.15 show the mechanical properties changing with the different weight reductions. In Figure 3.15, the property ratio of foam to solid is 1 for the sample with 4% weight reduction. It means no change of

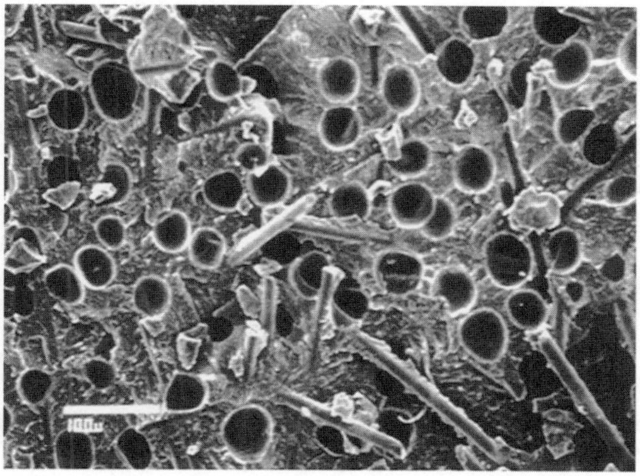

(a)

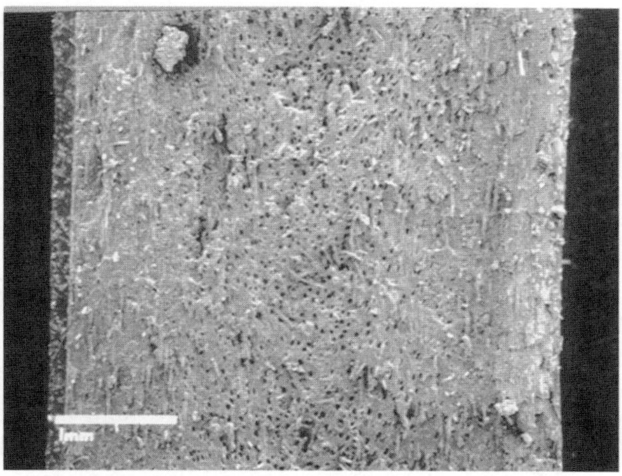

(b)

Figure 3.14 Morphology of PBT with 30% GF, 4% weight reduction. (**a**) Local center view with average cell size about 45 μm, $N_f = 4.6 \times 10^6$ cells/cm^3. (**b**) Cross section view with obvious big cells in the center core.

the elongation at break for the solid and foam with 4% weight reduction. It is interesting that the result in Figure 3.15 shows the elongation at break increasing to 1.06 with 10% and 15% weight reduction of the samples. The general trend is that the tensile strength and flexural strength decrease with the increase in weight reduction. However, with a nice cell structure the elongation at break will not decrease with the microcellular foam, but may even increase with the increase in weight reduction from 4% to 10%. A 4% weight reduction of 45-micron cells shown in Figure 3.14 has the same elongation at break as a solid. Both a 15% weight reduction with 15-micron cells in Figure 3.12 and a 10% weight reduction with 15-micron cells in Figure 3.13 have 6% more elongation at break than the elongation at break of the solid sample. It

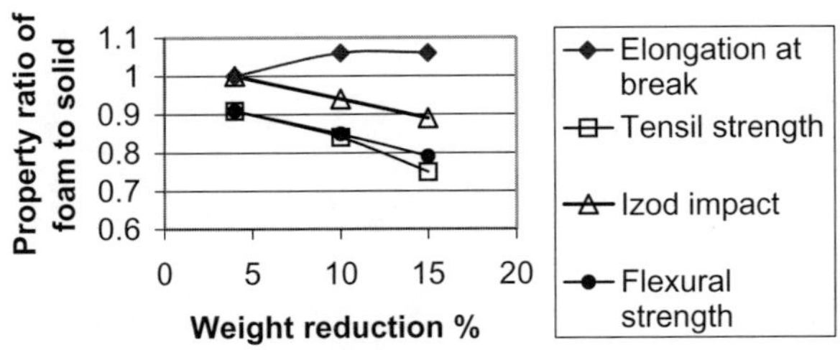

Figure 3.15 Property ratio of foam to solid versus weight reduction % for PBT with 30% glass fiber.

TABLE 3.1 Ratio of Strength Drop % to the Weight Reduction % R_{sw} in Three Different Cell Structures of PBT w/30% GF

Weight Reduction %	R_{sw} of Tensile Strength	R_{sw} of Flexural Strength	Cell Size (micron)	Cell Density (number of cells/cm^3)
4	2.25	2.25	45	4.2×10^6
10	1.60	1.50	15	2×10^8
15	1.67	1.40	15	8×10^8

seems that the elongation at break will be affected by cell size because of the same elongate value at break for both samples with the same average cell size but different weight reduction of 10% sample of 2×10^8 cells/cm^3, and 15% sample of 8×10^8cells/cm^3, respectively.

The general trend of mechanical properties changing with the weight reduction of microcellular foam is that the tensile strength and flexural strength decrease with the increasing weight reduction. However, it is not the same as the structural foam anymore. The rate of strength loss changing with weight reduction of microcellular foam is much less than rate of structural foam. More detailed analyses of these relationships are discussed in Chapter 5.

From the analysis of the relationship between cell structure and mechanical properties above, the conclusion is that the physical property test must have the results of the inspections for the cell structure. It is because the microstructure of the part can make big differences in the same material with same percentage weight reduction. On the other hand, the skin structure is an effective factor specifically for testing tensile strength of microcellular parts.

A special ratio R_{sw} is defined as the ratio of the strength drop percentage to the weight reduction percentage. R_{sw} of mechanical properties is strongly related to cell structure that can be proved by the data in Table 3.1. An advantage of the microcellular part is to have this ratio R_{sw} as low as possible by having a fine cell structure and high cell density. The lowest ratio R_{sw} is where a microcellular part will have the minimum strength drop for maximum weight reduction; this will be discussed further in Chapter 5.

The data in Table 3.1 show that 10% and 15% weight reduction samples have almost the same value of ratio R_{sw} because of the same size range of cells. However, even if there is a little difference between two samples, the trend shows a reasonable cell structure difference to explain the difference. For tensile strength, the 15% weight reduction sample has a ratio of 1.67 compared to a 1.60 ratio of the 10% weight reduction sample. This means that the 15% weight reduction will have a little more tensile strength drop than the 10% weight reduction sample, although both samples have almost the same cell sizes. Therefore, tensile strength is more sensitive for the weight reduction, although the cell size is important factor as well. On the other hand, the flexural strength shows a different trend from tensile strength: The cell size is more important than the weight reduction. For flexural strength the ratio of 15% weight reduction sample is 1.4 and the ratio of 10% weight reduction sample is 1.5. Therefore, the flexural strength drop for the morphology of nice cells and more cells in a 15% weight reduction sample is less than the morphology of a 10% weight reduction sample. Finally both the 10% weight reduction sample and the 15% weight reduction sample have a lower R_{sw} of all mechanical properties than the R_{sw} of all mechanical properties for 4% weight reduction sample. It is because the 4% weight reduction sample has bigger cells and a lower cell density compared to the results for both 10% and 15% samples.

Improvement of the cell structure has a significant influence on the impact strength. Michaeli et al. [18] also found that the absorbed energy during impact test increases up to three times when the foam morphology is finer. The finer cell structure can be made with breathing mold technology (refer to Chapter 8). It may need to be further verified by more data in the future.

Shimbo and his team examined the influence of cell size on tensile fracture strength of foamed part. With the small cell size and high cell density of microcellular foam in unfilled polyethyleneterephthalate (CPET), PP, and PC, the tensile fracture strength (only comparison among foamed parts) is increased [19]. If a foamed part is compared to a solid part, the strength decreases even for the fine cells. However, when the cell size is 3 μm or less, the tensile fracture strength is almost equal to that of the unfoamed one. Also, Shimbo pointed out that an increase in cell surface area causes molecular orientation, thereby affecting strength improvement [19]. The ratio of surface area of unfoamed and foamed plastics can be estimated with Equation (5.14) in Chapter 5 [19].

Shimbo also reported that the tensile fracture strength of foamed CPET is strongly related to the cell structure. With a cell size of 1 μm of morphology in CPET material, it maintains about 80% of unfoamed CPET with twice the foam magnification [19]. This is definitely a significant improvement with small cell size of foam with double the volume after foaming.

Park and co-workers [20] studied the effects of morphology of microcellular on the sound absorption behavior. Their statistical analyses of the acoustical curves of microcellular foams show that at lower frequency range (less than

1000 Hz), porosity and cell size effects are the dominant parameter. In medium frequency range (less than 2000 Hz), smaller cells have the major effects on the acoustic behavior. However, at high frequency (larger than 2000 Hz) the cell density plays the major role in sound absorption, and the porosity effects diminish. The modeling work results in the optimal performance of the morphology with higher porosity, lower cell density, and larger cell size [20].

3.4.2 Skin Thickness

Based on the modeling work, the skin thickness is the important factor for the flexural modulus [3]. Figure 3.3 shows a clear border that is typical for a PC microcellular part. Further experiment has been made to change the skin and core thickness in Figure 3.3. Once the mold temperature is set up higher than the one for the sample made in Figure 3.3, the skin becomes thinner and the big cells change to small ones. If the mold temperature in the cavity side is set up different from the mold temperature in the core side, the skin thickness will be different and may change the strength drop rate as well.

The skin thickness can be calculated by theoretical model. There are many mathematical models for estimation of skin thickness. A simple model for prediction of skin thickness has been used successfully for the microcellular foam part. It was a formula proposed by Lenk (21) in 1978, and was modified in this book for a microcellular sandwich structure part. It is defined as the following:

$$t = 2\alpha_p^{1/2}\tau^{1/3}\left(\frac{T_f-\theta}{T-\theta}\right) \tag{3.4}$$

where t is the skin thickness on the one side of the microcellular foam part; T is the polymer melt temperature; T_f is the polymer melt freezing temperature; θ is the mold temperature; τ is the time for melt touching the mold, which is approximately the hold time plus cooling time; and α_p is the thermal diffusivity of the polymer The skin thickness can be estimated by Equation (3.4). When the density reduction is equal to the weight reduction, the foamed core volume can be calculated. The simulation method will be discussed in Chapter 9; also see reference 22.

However, the skin thickness estimation may not be successful with Equation (3.4) for the morphology in Figure 3.10. There is no obvious skin thickness in this PC reinforced by carbon fibers with addition-filled Teflon material. There may be a need to find out the thermal diffusivity of this compound to do the correct modeling of skin thickness.

3.4.3 Fiber Orientation

Fiber orientation in the injection molding part is a disadvantage in glass-fiber-reinforced material. Microcellular injection molding provides an opportunity to improve the anisotropy caused by fiber orientation [6, 23]. Figure 3.16a

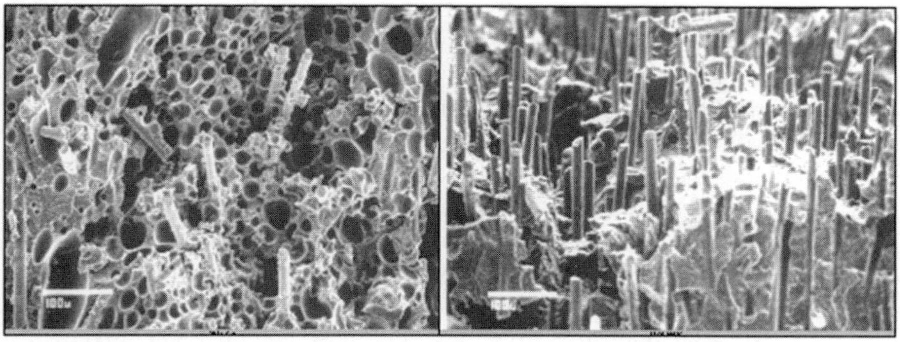

(a) (b)

Figure 3.16 Fiber disorientation in the foam and orientation in the solid, PBT with 20% GF, 15% weight reduction, N_2 gas, 4-in./sec injection speed with 40-mm plunger (white bar indicates 100 μm), [23]. (**a**) Microcellular foam. (**b**) Solid. (Courtesy of Trexel Inc.)

shows the morphology of fiber distribution in a microcellular part. It is obvious that the microcellular process results in some fiber disorientation in the center foamed core. Figure 3.16b displays the morphology of a solid part with strong fiber orientation in the mold flow direction. The material properties of solid and foam parts are obviously different. If we define a ratio of Ld/Td, Ld is the property measured parallel to the mold flow direction and Td is the property measured perpendicular to the mold flow direction. Then, the solid sample's tensile strength ratio for Ld/Td is about 2.2, and the microcellular sample's ratio for Ld/Td is about 1.74, which is a 21% reduction of the ratio Ld/Td in the microcellular sample. The ratio of the Young's Modulus in the flow direction to the transverse direction is 1.99 for the solid in Figure 3.16b. The ratio of Young's Modulus for the microcellular part in Figure 3.16a reduces to 1.52—that is, 23.6% reduction of anisotropy.

Many morphologies of glass-fiber-reinforced microcellular parts show obvious fiber disorientation because of significant cell growth around the glass fibers even if the weight reduction is only 5%. To verify this assumption, a glass-fiber-reinforced sample is made with big cells (25% weight reduction). The glass fiber disorientation is obvious because of the big cells.

For carbon fibers in PC microcellular foam the obvious fiber disorientation is shown in Figure 3.10. Even in the skin area, which should be a high shearing zone, the carbon fibers are not orientated as usually solid material processing. With many small cells around the carbon fibers, cell growth is the only reason that explains the driving force to disorient the fibers around the cells.

On the other hand, the processing conditions and gas percentage will influence the glass fiber disorientation significantly. The most important parameter of processing for fiber disorientation is injection speed. This is because a nice cell structure will usually be created at high injection speeds. The high gas percentage will help to generate more cells as well. Furthermore, fiber itself is a good additive for heterogeneous nucleation promotion. In conclusion, an excellent cell structure is the key factor to improve the fiber orientation in the microcellular part. More details regarding cell growth and fiber disorientation in injection molding processing will be discussed in Chapter 6.

There is one more phenomenon related to the shear-induced migration of filled material during injection [24]. The distribution of particles in the cross-sectional planes of injection-molded specimen present evidence of significant migration of filled materials from the surface to the interior. In microcellular injection molding for metal powder the heavy metal powder always moves from the high shear rate area to the low shearing area. The same trend seems to occur for glass fibers in the microcellular injection molding. However, the final result will be a balance between the migration of glass fibers moving into the center area while cell growth expels the glass fiber away from the center (see Figure 3.16).

3.5 OTHER SPECIAL CELL STRUCTURES

There are several special cell structures that represent either (a) the future research direction in industry or (b) the challenges to be faced for modifications of processing.

3.5.1 Ultra-microcellular Morphology

This is the real microcellular part if every part can be made with an ultra-microcell that has the cell size of less than 5 μm, or even the size of 0.05 μm to keep the transparency of the foam made by a transparent material. Figure 3.17 is the morphology of unfilled polypropylene (PP) that has the microcell

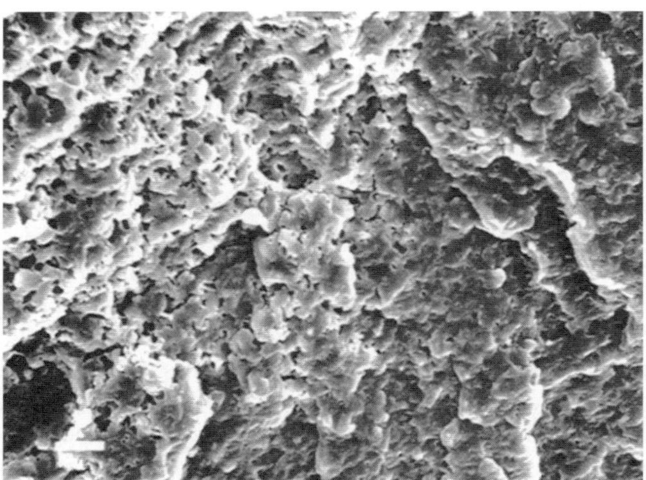

Figure 3.17 Ultra-microcellular foam (white bar indicates 10 μm), unfilled PP.

size smaller than 5 microns. This cell structure is also found in the 0.0005-m thin-wall microcellular container made by unfilled PP for the packaging part. The slim technology developed recently uses a microcellular part to get cycle time reduction and smaller cell sizes [25].

3.5.2 Bimodal Cell Structure

The bimodal cell structure was reported in the water- and butane-blown polystyrene (PS) [17]. It is basically the cell structure with greatly different sizes of cells together, where a big cell is 100 times bigger than the others adjacent around it. This bimodal cell structure with huge differences in cell sizes in the adjacent area is a special morphology that is not common in microcellular foam. Figure 3.18 shows another kind of bimodal cell structure that has one big cell spot of about 250 × 300, shaped like an oval surrounded by a solid. The inside of the cell spot has hundreds of small cells with the size ranging from 3 μm to 5 μm. A sample is made with the unfilled polypropylene (PP) foamed with 1 weight percent of N_2 gas at 20.7 MPa (3000 psi) back pressure during dosing in the screw. This kind of bimodal cell is different from the bimodal cell reported in reference 17 because the big cell itself is full of small cell structures while the wall around the big cell is the solid material. It is another complicated morphology found in the unfilled PP microcellular part. This bimodal cell structure is not a good microcellular structure and can be eliminated by either changing the blowing agent or modifying the processing conditions.

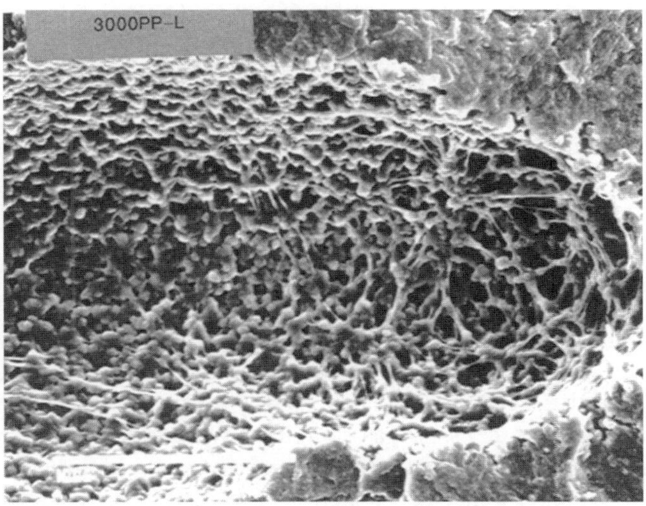

Figure 3.18 Bimodal microcellular foam (white bar indicates 100 μm), unfilled PP.

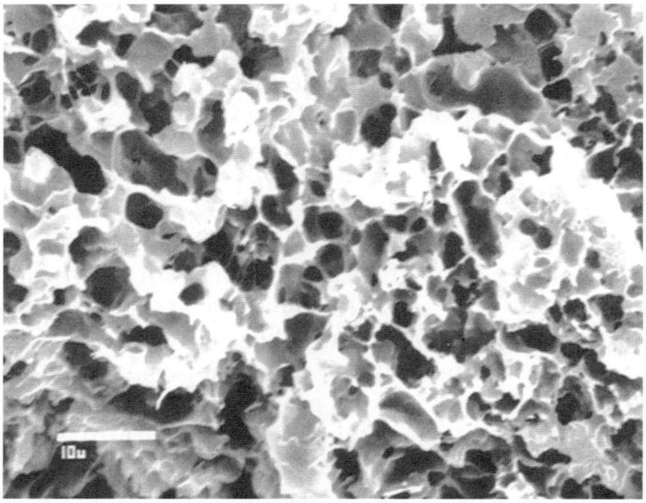

Figure 3.19 Open-cell microcellular foam (white bar indicates 10 μm), unfilled PP.

3.5.3 Open Cell Structure

The open cell structure is a mixture of open cells with adjacent closed cells in unfilled polypropylene (PP). Figure 3.19 shows the typical morphology of open cells mixed with closed cells in the PP microcellular part. The open cells may form on the straight interfacial boundary [1], or from the collapse of the cells because the wall thickness is too thin. Both open and closed cells are microcells because all cell sizes are 10 μm or less. This open microcellular structure may bring different applications from closed microcellular structure. The processing of the open microcellular foam basically involves (a) the high weight percentage of gas and high (b) nucleation. Then, the surface quality of open microcellular foam is even worse than the closed cells of the microcellular part.

There are potential applications for open cells of the microcellular injection molding part. One of the applications is the filter material that controls the open size of cells to filter small particles.

3.5.4 Cell Structure with Different Gases

The same material will have different cell architectures if the foam is made from different gases. Figure 3.20 shows the different morphologies of cell structures in HDPE. Figure 3.20a is an HDPE microcellular sample made by CO_2 gas that has large cells. N_2 gas samples have the finest cell structure, as shown in Figure 3.20b. Although argon is an inert gas seldom used for microcellular foam, it makes a nice cell structure in HDPE in Figure 3.20c, and may make the largest cell size among all three samples.

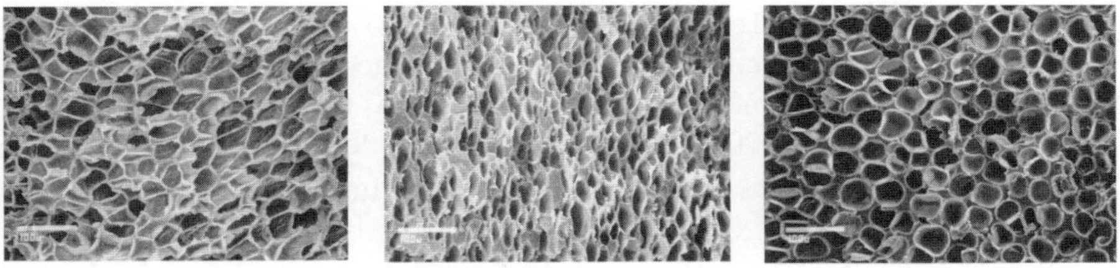

(a) (b) (c)

Figure 3.20 Morphology of HDPE with different gases. (**a**) CO_2, (**b**) N_2, (**c**) Ar. (Courtesy of Trexel Inc.)

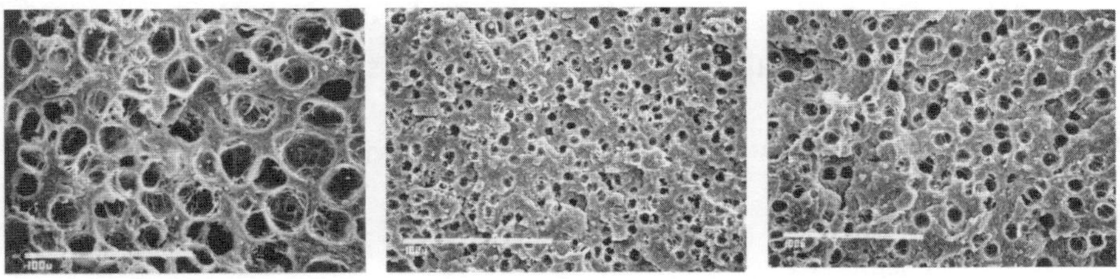

(a) (b) (c)

Figure 3.21 Morphology of RPVC with different gases. (**a**) CO_2, (**b**) N_2, (**c**) Ar (white bar indicates $10\,\mu m$). (Courtesy of Trexel Inc.)

Similar results of rigid PVC with three different gases were also tested. The sample made by N_2 gas shows the finest cell size in a rigid PVC microcellular sample, as shown in Figure 3.21b. Argon gas as blowing agent creates a fine cell size that is as small as the cell size made by N_2 gas. However, in Figure 3.21c the rigid PVC sample made by argon has a better small size cell structure than the CO_2 sample illustrated in Figure 3.21a.

3.6 CONCLUSIONS

The morphology of microcellular injection molding part explores many research topics. We need to characterize the morphologies of microcellular part further when more physics and chemistry principles are explored in the future.

Overall, the skin–core is the most common microcellular morphology model. The interface zone between skin and core is negligible in most cases unless the thickness is larger than 4 mm.

There are two possible core models: close-packed cell structure and non-close-packed cell structure. The possible factors influencing on the morphologies of microcellular foam are summarized as follows:

- Pressure drop rate *dp/dt* needs to be high enough for necessary nucleation.

- Fillers and glass fibers are good for morphology.
- A minimum gas percentage is necessary for the acceptable morphology of microcellular foam. Normally, insufficient gas percentage will result in larger cells (may be 200 μm and greater) and a fewer number of cells.
- CO_2 gas can create a close-packed cell structure more easily than can N_2 gas.
- N_2 gas can make the smallest cell size in most cases.
- Argon gas is also a good blowing agent, and the morphology proves its microcell structure.
- The first stage of gas mixing is critical for uniform cell distribution in the molding part.
- Amorphous material usually has better cell structure than crystalline material.
- Mold temperature is more important for the skin morphology of crystalline material.

REFERENCES

1. Suh, N. P. *Innovation in Polymer Processing*, edited by James F. Stevenson, Hanser/Gardner Publications, Cincinnati, 1996, Chapter 3, 100–105.
2. Xu, J., and Pierick, D. *J. Injection Molding Technol.* **5**, 152–159 (2001).
3. Xu, J., Kishbaugh, L. *J. Cell. Plastics* **39**, 29–47 (2003).
4. Xu, J. *SPE ANTEC Tech. Papers*, 2770–2774 (2006).
5. Doroudiani, S., Park, C. B., and Kortschot, M. T. *Polym. Eng. Sci.* **36**(21), 2645–2662 (1996).
6. Xu, J. *SPE ANTEC Tech. Papers*, 2158–2162 (2008).
7. Yuan, M., Turng, L. S., Spindler, R., Caulfield, D., and Hunt, C. *SPE ANTEC Tech. Papers*, 691–695 (2003).
8. Turng, L. S. *SPE ANTEC Tech. Papers*, 686–690 (2003).
9. Kharbas, P., Nelson, P., Yuan, M., Gong, S., Turng, L. S., and Spindler, R. *Polym. Compos*, **24**, No.6, 655–671 (2003).
10. Jo, C., and Naguib, H. E. *SPE ANTEC Tech. Papers*, 1902–1906 (2008).
11. Marrazzo, C., Maio, E. D., and Iannace, S. *J. Cell. Plastics*, **43**, 123–133 (2007).
12. Ray, S. S., and Okamoto M. *Progress Polym. Sci.* **28**(11), 1539–1641 (2003).
13. Semerdjiev, S. *Introduction to Structural Foam*, SPE Inc., Towanda, PA, 1982.
14. Strauss, W., Ranade, A., D'Souza, N. Anne, Redy, R. F., and Paceley, M. *SPE ANTEC Tech. Papers*, 1812–1816 (2003).
15. Muller, N., and Ehrenstein, G. W. *SPE ANTEC Tech. Papers*, 1830–1834 (2003).
16. Xu, J. *SPE ANTEC Tech. Papers*, 2089–2093 (2007).
17. Throne, J. L. *Thermoplastic Foams*, Sherwood Publishers, 1996, Hertford, UK, p. 154.

18. Michaeli, W., Florez, L., Krumpholz, T., Obeloer, D. *SPE ANTEC Tech. Papers*, 1024–1028 (2008).
19. Shimbo, M., Higashitani, I., and Miyano, Y. *J. Cell. Plastics* **43**, 157–167 (2003).
20. Serry Ahmed, M. Y., Atalla, N., and Park, C. B. *SPE ANTEC, Tech. Papers*, 1109–1112 (2008).
21. Lenk, R. S. *Polymer Rheology*, Applied Science Publishers, London, 1978.
22. Han, S., Kennedy, P., Zheng, R., Xu, J., and Kishbaugh, L. *J. Cell. Plastics* **39**, 475–485 (2003).
23. Kishbaugh, L., Levesque, K. J., Guillemette, A. H., Chen, L., Xu, J., and Okamoto, K. T. U.S. Patent No. 7,364,788 B2 (2008).
24. Hong, C. M., and Jana, S.C. *SPE ANTEC Tech. Papers*, 1841–1845 (2004).
25. Trexel Web Site, http://www.sabic-ip.com/.

4

MATERIALS FOR MICROCELLULAR INJECTION MOLDING

For injection-molded parts, the price of raw materials, particularly engineering plastics, contributes significantly to the total part cost. Various process methods, such as gas-assisted and structural foam, have been employed over the years to reduce cost by decreasing the material required to form the part. The microcellular foam technology is a recent advance in process technology that reduces part weight by introducing a gas in its supercritical state into the melt and then forming a great number of microcells in the part. While similar to structural foam, this technology with microcells has the added benefit of enabling foam parts to be molded at thin wall thickness. Significant weight and cycle time reductions can be achieved even in thin wall parts with microcell structure. Therefore, microcellular injection molding can be a replacement processing method not only for structural foam molding but also for regular injection molding.

The rapid expansion of the microcellular injection molding industry has resulted in a need to study the different materials suitable for microcellular injection molding. Many polymers are already successfully used for microcellular applications. More and more potential new materials or composites to be used for injection molding have been tested either by batch process or by extrusion process. The batch process is the first step to ensure that the material or composite is foamable with certain blowing agents at certain conditions. It may just give a direction for the injection molding process to follow and to guide it toward industry application. On the other hand, both extrusion and injection molding processes are required to make a single-phase solution in the plasticizing unit where is the screw and barrel with gas dosing or with

Microcellular Injection Molding, Edited by Jingyi Xu
Copyright © 2010 John Wiley & Sons, Inc.

chemical blowing agents. Then, most extrusion processing results can be quickly converted to the injection molding process since the process in the extruder is similar to the first stage of injection molding (see Chapter 2), except the viscosity is different. Therefore, some material test data have the potential for injection molding to be introduced with either batch process or extrusion process in this chapter.

From the processing point of view, an important classification of thermoplastics is that of crystalline and amorphous materials. Thermoset material is not popular for the microcellular process and will not be covered in this book. The filled material contents used in this industry result in unique characteristics of these composites, which dramatically influence their processing conditions of microcellular injection molding because of the benefits of heterogeneous nucleation. The homogeneous nucleation is an ideal case that is discussed for some cases as well.

Theoretically, every thermoplastic can be used for microcellular processing with different processing windows. To understand the material features as the good candidates for microcellular processing, this chapter will give a brief review of the major materials successfully used for microcellular injection molding. Some specific processing requirements of typical materials and related part property changes with microcellular foam will be discussed as special case studies since one can refer to them for similar materials. The discussions will include foamability, mechanical, rheological, thermal, shrinkage, and flammability properties of selected plastic materials. Some resins used more commonly for microcellular part are evaluated in the microcellular process with more detail. Others may be briefly introduced since either they can be referred to some typical material test data with detailed discussion in this chapter or they are still in a developing stage without enough reliable data available yet. Also, some processing data in the batch process are introduced if they give some principal guidelines and direction of development for real injection molding processing. In addition, the new developed materials, such as nanocomposites, metal powders, special compounds, and biopolymers, are discussed in this chapter for the current applications and future works.

4.1 STRUCTURE AND CHARACTERISTICS OF POLYMERS

A simple review of the structure and characteristics of polymers is necessary for understanding a more complex microcellular structure in polymers. At least, the basic type, the structures, and the terminologies of polymers will be discussed briefly in the following.

4.1.1 Types of Polymers

There are different ways to typify a carbon-based polymer. Two of them are based on the nature of the chemical reaction of polymer generation. They are additional polymer and condensation polymer, respectively. Three classifications of polymer types are related to injection molding processing; they are based on the arrangement of the atoms in the polymer, which will be discussed briefly in the following [1].

4.1.1.1 Straight-Chain or Linear Polymer. This is simply formed by connecting one carbon to another in a long continuous chain through two of the available bonding sites. The chains could be many thousands of units long. In fact, these various configurations will be constantly changing due to molecular motions such as vibration, stretching, or bending. In most polymers, the covalent bond between the individual monomer units has freedom of rotation so that the bond angle may be fixed while still permitting enough rotation to allow the chains to alter their configurations from coiled structure to stretched chains, as shown in Figure 4.1. However, if it is stretched to be straight, the chain is aligned in the straight line and is referred to as "straight-chain" or "linear" polymer.

4.1.1.2 Branched Chain Polymer. Some substitute groups may attach the remaining two bonds on each carbon atom, resulting in branches in the chain of linear polymer (see Figure 4.1). Since the branches are generally few in number and occur irregularly along the main chain, only such substituted structures are still defined as a linear polymer. The branched linear polymer is better material for the microcellular process than the linear polymer. It is because the branched material nucleates more cells.

4.1.1.3 Network Polymer. If the remaining bond sites on this linear chain are interconnected with other chains, a network polymer forms. This may occur as the growth of the chains proceeds from inception, or it may be induced via reaction of specific reactive sites associated with the chains. It is the well-known process of "cross-linking" and is important for some

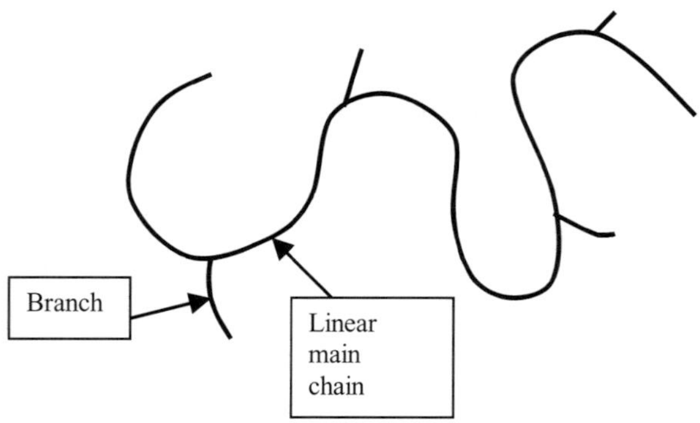

Figure 4.1 Branched linear polymer.

microcellular processes, such as extrusion and blowing molding. However, this "cross-linking" may not be critical in injection molding.

4.1.2 Structure of Polymers

A complex microcellular structure increases the difficulty of understanding the foamed structure in the basic structure of polymers. For further development of the microcellular technology with better cell structure, the polymer structure will be a basic requirement for microcellular processing.

4.1.2.1 Terminologies. There are some basic terminologies regarding to the polymer structure and characteristics. All the classifications of different polymers are defined by these basic terminologies, which will be discussed briefly in the following.

1. Monomer. There are some important terminologies for the polymer structures. The basic term is *monomer*, which is defined as the small reactant molecules from which the polymer is made. The simple example is an addition polymer, where both the monomer and the repeating unit within the polymer have the same composition as the following:

$$n\mathrm{CH_2{=}CH_2} \rightarrow -(-\mathrm{CH_2CH_2}-)_n-$$

The condensation polymer will have a composition similar to that of the monomer. The substances involved in the reaction are called monomers, and the reacting unit will contain the basic structure of both.

2. Homopolymer and Copolymer. A polymer with a single repeating structure is called a *homopolymer*. Thus, a polymer with more than one repeating structure is called a *copolymer*. There are different copolymer structures determined by the spaced position of the parts of the copolymer, such as a block copolymer, a graft copolymer, and a terpolymer. In microcellular injection molding, copolymers have been used more and more since their excellent properties. Usually, an alternating graft copolymer and a random block copolymer are used in microcellular processing.

3. Alloy. The term *alloying* is defined as the physical admixture of two polymers. The object of alloying is to combine resins with widely differing properties into a homogeneous mass and thereby overcome phase-separation problems associated with blending of polymers that differ significantly in molecular weight. It is simple to make an alloy by establishing sites of compatibility along molecular chains such that a degree of physical cross-linking occurs. Alloy, sometimes called blend, is the most popular material used in microcellular injection molding. There are so many different blends available on the market. However, only some typical alloys that have been used successfully for microcellular foam will be introduced in this chapter.

4. Thermoplastic and Thermoset. It is important to classify plastics as either thermoplastics or thermosets. Thermoplastics are those materials that undergo no permanent change on heating. Therefore, it can be heated again and molded repeatedly. Thermoplastic polymers contain linear molecules that are not cross-linked. On the other hand, thermoset materials are capable of a high degree of cross-linking. However, after the cross-linking, it can no longer be heated and cannot flow again. In the microcellular injection molding process, only thermoplastics are used widely. Thermoset is not a popular material for microcellular parts, and it will not be covered in this book.

4.1.2.2 Structure of Polymeric Materials. The structure of polymer and the behavior of the material are based on both the chemical nature of the individual monomer link and the way in which the chains fit together. Several well-known structures are discussed in the following.

1. Molecular Weight. One of the important features of polymers is the molecular weight, which is only for thermoplastics. Linear polymers are discrete, although tangled, molecules, and their size can be expressed in terms of molecular weight. The molecular weight is a function of the degree of polymerization, which is the number of structurally recurring units in the chain, each with its own specific molecular weight. There are different methods to determine molecular weight of polymers, and different methods result in different values.

2. Molecular Weight Distribution (MWD). There is also the concept of molecular weight distribution (MWD), which only represents an average value of the weights of the individual molecules. The generalized formula for a linear polymer is –(—*M*—)*n*–, where *M* is the repeating unit and *n* is the degree of polymerization. For example, in a sample with an average *n* value of 5000, there will be a percentage of molecules with *n* values of 50 and of 50,000.

3. Number Average Molecular Weight. The number and weight average molecular weight is the commonly accepted sense and is defined as follows:

$$\overline{M}_n = \sum N_i M_i \tag{4.1}$$

where $\overline{M}_n$ is the number average molecular weight, N_i is the mole fraction of the that has a molecular weight M_i, and M_i is the molecular weight.

4. Weight Average Molecular Weight. The weight average molecular weight is given as

$$\overline{M}_w = \sum W_i M_i \tag{4.2}$$

where $\overline{M}_w$ is the weight average molecular weight and W_i is the mole fraction of the polymer that has a molecular weight M_i. For the same size of molecules, the two methods give identical results. However, when there is a different size distribution, $\overline{M}_w$ is greater than $\overline{M}_n$. Then, $\overline{M}_n$ imparts greater importance to small molecules, while $\overline{M}_w$ stresses the effect of large molecules. Hence, the ratio of $\overline{M}_w/\overline{M}_n$ is a measure of the molecular weight distribution (MWD) [1].

5. *Crystallinity.* For solid materials consisting of small molecules, intermolecular forces are large enough to restrain the molecules in an ordered array that is crystalline structure. The molecular arrangement in crystallines is essentially regular throughout the mass, although there are possibly different structures of polymer crystals. It is important to know that this solid crystalline will melt at the exact temperature that corresponds to the thermal energy input needed to disrupt the intermolecular forces. The crystallinity will have significant influences on microcell structure for microcellular injection molding.

6. *Crystallites and Spherulites.* Polymer molecules may align themselves in an amorphous matrix such that select areas become more ordered than the surrounding environment. Such ordered regions are called crystallites. Then, crystallinity of a polymer indicates the degree of such ordered crystallite regions.

A typical crystallite has dimensions that are less than the length of the extended polymer molecules. This gives the possibility that a given molecule can traverse more than one crystallite region, as well as the amorphous regions in between. The transition from crystallite to amorphous area is not sharp and delineated, and semi-ordered transition regions exist as well. The relative percentages of these crystalline regions vary widely with the nature of the polymer.

There are no known commercial polymers that are 100% crystalline. However, a number of crystallites will order themselves further into larger groups. Such groups are named spherulites. Spherulites are observed in a polarizing microscope, and they rarely exceed 100 μm in diameter [1].

7. *Phase Transition.* There is no a marked change in such physical properties as specific volume, heat capacity, and viscosity at the melting point of polymer. However, a change will occur in a temperature range. Essentially, the melting point T_m of a polymer is that temperature at which the crystallite regions lose their order. Since there is a spectrum of crystallite regions that will disperse, the energy required for such dispersal is spread over a wide range. Once the large volume of gas adds into the molten polymer, the temperature of the polymer–gas solution can be lowered below its melting point without freezing. It is because the gas is dissolved into the molten polymer acting as a plasticizer. It is obvious that in semicrystalline material the dissolved gas delays the

formation of crystalline structure. Then, the gas will have phase separation from the polymer and freeze instantaneously the microcellular structure because of the prior supercooling of the matrix phase [2]. It is one of the reasons why microcellular cooling occurs so fast if the part thickness is not over a certain range (usually less than 2.5 mm).

Another important temperature range is T_g at which the polymer reaches its glass transition point. The polymer is stiff and dimensionally stable if the temperature is below T_g. However, once temperature is over T_g, the polymer becomes more rubberlike.

Glass transition effects are greatest with amorphous materials. Below the T_g, thermal energy is insufficient to allow chain atoms to do more than vibrate about their equilibrium positions, and the polymer is glasslike and brittle. However, if the temperature is beyond T_g, the polymer properties reflect this molecular motion by first appearing tougher and more flexible and then, finally, more rubbery in nature and even-flowing as a fluid. Adding enough gas, such as CO_2, into amorphous material will decrease the glass transition temperature significantly. For the PMMA material, its T_g decreases approximately near room temperature at a concentration of 35 weight percent (wt%), or higher of CO_2 gas. It means that the foaming is even possible at room temperature. However, to produce uniform homogeneous cell structure at room temperature, the gas concentration needs to be very high (even higher than 35 wt% for CO_2). In addition, the cell structure will be uniform if either the pressure is higher or foaming temperature is higher than room temperature [2].

4.2 CRYSTALLINE MATERIALS

Crystalline and semicrystalline materials become one of the major microcellular injection molding applications since they have excellent properties for wide applications in industries. The typical materials are polypropylene (PP), polyethylene terephthalate (PET), Polybutylene terephthalate (PBT), and polyamide (PA). The detailed technical review results are discussed for these typical semicrystalline materials below. The other semicrystalline materials are briefly introduced for their possible microcellular processing conditions.

4.2.1 General Characteristics of Crystalline Materials for the Microcellular Process

Every polymer has different chemical structure and some other parameters that may have amorphous and/or crystalline morphology [3]. The unique characteristic of crystalline materials is the crystallization during mold cooling. The crystallization during cooling will create several special issues for microcellular processing:

- The crystallization during cooling may expel the gas near crystalloid so that the cell structure of crystalline material may not be as uniform as the cell structure of amorphous material [4].
- The gas absorption and diffusion take place almost exclusively through the amorphous regions [5, 6]. The gas diffusivity decreases severely by increasing the crystallinity with slow cooling rate [4]. As a result, there is low gas solubility in the crystalline regions.
- It is difficult to foam near the melt temperature, so the proposal from Colton [7] is to process semicrystalline material above the melting point. The initial stages of foaming must occur before crystallization begins.
- During cooling, the crystallization will release the laden heat that is also not good for cell growth.
- Even under high pressure (34 MPa) for semicrystalline materials, there are still some crystalline regions unsaturated by gas at the supercritical state.

Kumar and Gebizlloglu [8, 9] observed the crystallization of PET induced by the dissolved CO_2. Baldwin et al. [10, 11] reported the significant differences in the microcellular processing characteristics of amorphous and crystalline polymers. Therefore, the suggestions are to use different strategies for the microcellular processing of amorphous and crystalline polymers based on their experiment on the effect of CO_2 on crystallinity of PET materials.

In addition, semicrystallines may have a very complex texture, which is strongly dependent on the processing conditions, specifically cooling rate and the presence of impurities [4]. The study result from Doroudiani et al. [4] shows the importance of crystallinity and a texture of crystallites. The common methods to improve the cell structure on the molding part of crystalline material are:

- Increase the cooling rate to decrease the crystallinity.
- Add fillers or other additives to create more heterogeneous nucleation.

However, high cooling rate will result in less crystallinity in the material, which may reduce the part's physical strength. Therefore, adding the fillers may be the realistic way to get both necessary crystallinity and nice cell structure in a microcellular part. In Doroudiani et al. [4], the nucleation of spherulites at the same sites during recrystallization is the evidence for the heterogeneous nucleation. This heterogeneous nucleation occurs on the insoluble particles and larger macromolecular chains that do not move because of the high viscosity of melt [4]. In fast cooling, the molten polymer goes through the maximum crystallization temperature (i.e., around 116 °C for HDPE from DSC experiments in reference 4) quickly. This leaves most of the molecular chains in amorphous form, and then possible finer cells are created in the amorphous region. Similar observations were made from PP, PBT, and PET samples [4].

Although all the results discussed above are from the batch process, the similar strategies are used in microcellular injection molding process. There are three methods used for the crystalline material processing with nice cell structure and nice skin surface without help from gas counterpressure and co-injection processes.

One method is the hot and cold mold technology. It is used to keep the mold temperature warm up to above the melting point of polymer. Therefore, this process not only irons the surface flat and smooth but also gives enough time for crystallization and gas diffusion back into skin.

Another method is the coated insulation film on the mold surface. Its function is similar to that of hot mold because it does not instantly cool the skin of the microcellular part. It also smoothes the skin surface so that the gas-laden material will not stick with a mold cold wall surface. Therefore, it significantly reduces the friction between material and mold surface and reduces the shearing near the skin as well.

Adding fillers into material is the efficient method used the most in current industry practices. The fillers added into the crystalline material promote the microcell structure. The major reason is the heterogeneous nucleation. On the other hand, the filler can help to hold the gas nearby and get more chances to share the limited gas for more cells that will be small since the number of fillers and gas are separated from big cells.

Michaeli and Bussmann [12] presented the morphological structure that shows the shear-induced crystallization occurring during injection. In future academic works, this crystallization from shearing needs to be considered in the microcellular foaming for both crystallizations during cooling and injection shearing.

4.2.2 Polypropylene for Microcellular Process

Polypropylene (PP) is a semicrystal material and is one of the most popular materials widely used in many areas since 1957. PP is also a material with relatively low cost and low density (about 0.9 for unmodified polymer), but with good mechanical and processability properties and high temperature resistance. The spiral molecule interlocks with other similar chains to form the crystallite structure that gives polypropylene its physical properties. The entwined helix holds the chains closely together, resulting in a polypropylene with better heat and creep resistance than polyethylene. However, polypropylenes exhibit many of the properties of polyethylene because of its long-chain aliphatic hydrocarbons. PP has moderate tensile and high-impact strengths and are resistant to many inorganic solvents. PP also has excellent electrical properties. The outstanding resistance to low-frequency flexing gives PP the living hinge usage, and oriented PP further aligns the molecule of PP for very good fatigue resistance. The high water resistance and lower absorption of moisture enable PP to be an extremely processable material because it remains dimensionally stable in humid environments.

However, PP is not a good barrier material for oxygen and other gases, and it is also the fastest burning of the olefins. The successfully flame-retardant treatments that benefit from heterogeneous nucleation for microcellular foam are possible. Perhaps the most outstanding features of PP are the low density and good rigidity. PP benefits more from microcellular foam since it is possible to reduce more weight of microcellular PP. Many new polypropylenes with improved engineering properties have been developed. The most promising materials are the heat resistance PP, glass-fiber-reinforced PP, and filled PP. All of these additional materials added into pure PP make microcellular PP processing much easier. In addition, PP has good un-notched and excellent notched impact strength. However, the notch-dependent impact strength of pure PP may need to be considered during microcellular processing. This notch-dependent impact strength of pure PP can be improved by adding glass reinforcement, blending with nitrile rubber, or copolymerizing with ethylene. By these modifications, the notched impact strength can be raised 20 times higher than unmodified PP. Furthermore, microcellular PP can improve the impact strength and resistance of fatigue load with very small cell size.

4.2.2.1 Unfilled Homo-PP and Co-PP. The unfilled homo-PP is usually not recommended for microcellular processing because of the difficulty to make uniform cell structure. It is well known that the unfilled homo-PP has weak melt strength and melt elasticity at elevated temperature [13]. However, there are some applications for unfilled PP if the strength drop is not the major issue for the unfilled PP microcellular part.

The temperature during foaming of PP should be maintained higher than the crystallization temperature. However, it cannot be too high that the gas will diffuse out of the cells due to cell wall rupture. The processing temperature for unfilled PP is about 204–288 °C (400–550 °F). With saturated CO_2 gas the processing temperature of PP may be reduced in the range of 20–30 °C at the same injection pressure without gas. Or at the same processing temperature as solid PP without gas, the cavity pressure can be reduced by 75–80% for the CO_2 gas-laden PP melt [14]. It is recommended that the weight reduction for unfilled PP should not be more than 10–15% since the strength drop is significant for tensile strength of unfilled PP with microcellular process. The major benefits of microcellular unfilled PP are the dimension stability including less warpage and almost zero sink mark. With the uniformly cooled mold and wall thickness not over 2.5 mm of the unfilled PP microcellular part, the cycle time reduction may reach 10–15%, which is the potential big cost savings.

Generally, unfilled PP is difficult to make good part of microcellular injection molding. Figure 3.4 is the typical morphology of unfilled PP microcellular made by injection molding with CO_2 gas as blowing agent. The cell size varies from 5 microns to 80 microns. Guo et al. [15] reported some observations for the cell nucleation and initial cell growth of PP with N_2 gas under various experimental conditions in a batch processing simulation system. The results from their studies show that the branched PP (WB 130) has higher cell density than the linear PP (HE351) with N_2 gas as blowing agent. Although this result is from the batch process, the conclusion is valuable for the injection molding as the reference data. This batch process result tells us that the diffusivity of N_2 in branched PP is lower than that in linear PP because the low diffusivity of N_2 gas in the branched PP will allow more gas retention in the sample to nucleate more cells after the gas pressure is released in the batch process chamber [15]. However, the injection mold is a closed system if the venting of the mold system is neglected. Therefore this conclusion may not be fully applied for microcellular injection molding. On the other hand, the experimental results from Guo et al. [15] show that the foamability of the PP/N_2 system is good at the initial stage of foaming. A high pressure drop rate and a large gas content have significant effects on high cell density for both branched PP and linear PP [15]. These are the major measurements to modify unfilled PP cell structure to be acceptable by the part strength requirement. A high talc content in PP and a high temperature of processing will also induce a high cell density [15] that also matches the trend for the effects from talc content in PP (to be discussed below) and processing temperature (high temperature needs less energy for cell nucleation) in microcellular injection molding. The other observations from Guo et al.'s experiments are useful to direct the processing of injection molding for these two different structure PP materials with N_2 gas as blowing agent:

- The nuclei density is slightly lower in branched PP foams than in linear PP foams.
- The cell coalescence is much less prevalent in branched PP foams than the one in linear PP. It means that more stable cell growth occurs in branched PP.
- The cells in branched PP foams are mostly closed cells, and the cells in linear PP are connected to each other.

For reasons similar to those for branched PP versus linear PP, the polypropylene–ethylene copolymers (co-PP) are found to have more gas around the crystalline than does the homo-PP. Due to a good elastic nature of co-PP, foamed PP copolymers are known to have high impact strength, high resilience, and good chemical resistance.

Park and Dealy [16] at McGill University also tested similar PP materials with carbon dioxide (CO_2) gas. The supercritical carbon dioxide has been used in their research. A high-pressure sliding-plate rheometer and a rotational rheometer were used to determine the combined effects of PP materials, gas contents percentage, pressure, and temperature. The effect of temperature could be described by the Arrhenius equation [17]. The pressure influence can be found by the Barus equation [18]. The effect of content percentage of CO_2 gas obeys the Fujita–Kishimoto equation [16, 19]. Intenerated conclusions for

CO_2 gas in these long-chain branching and linear PP materials from reference 16 are:

- Both long-chain branched PP and methyl side group in PP increase temperature and pressure sensitivity.
- The methyl groups in PP also increase the CO_2 gas content percentage sensitivity while long-chain branched PP has less effect on it.

The microcellular foam of PP is usually modified with different filled materials that are defined as fillers. The most common fillers are the nucleating agents and clarifying agents [20]. They will be discussed in a filler section below.

4.2.2.2 Impact-Modified PP. This is modified PP with a rubber phase in the PP matrix. The additional rubber is excluded as the second phase even though it can be distinguished from the PP phase microscopically because the material acts in most respects as a homogenous phase. The processing conditions are varied based on the percentage of rubber additives and the size of the rubber phase in the PP. It is always thought that the mixer needs to be designed to uniformly distribute the rubber phase, which is as important as the gas to be mixed uniformly in the molten polymer. The major change of the impact-modified PP is the promotion of the impact strength. Therefore, the cell size of impact-modified PP must be good to maintain the impact strength of the original solid PP.

The weight reduction leads to a change in mechanical properties that must be controlled in a specified small range for the load-bearing components or for the safety-relevant parts. Mueller and Ehrenstein [21] studied the constancy of properties from injection molding foams with this impact-modified PP, which is a copolymer with a melt volume rate of $16\,cm^3/10\,min$. The mold is a breathing mold (see Chapter 8 for the details on mold and molding technology) that can set up to 3 mm open stroke after injection. It is a good method to produce the constant properties of final microcellular foam with the special procedures [21]:

- The breathing mold can make relative constant properties of the foamed part. It is true for both physical and chemical foaming.
- The foamed part remains the tensile and the flexural modulus as well as yield strength in tension within a range of a few percent with breath molding technology.
- The lower residual stress in the foamed specimens contributes to high property constancy for this impact-modified PP microcellular part.
- By increasing the delay time for mold opening, the skin gets more chances to be solidified thicker and an improvement of the mechanical properties is possible.

- The constancy of the foamed part properties is equally as high as that of a solid molding part when an appropriate foam molding approach is in service.

4.2.3 PBT for Microcellular Process

Polybutylene terephthalate (PBT) is also known as polytetramethylene terephthalate (PTMT). It is a semicrystalline polymer and has been used in end-user industry since 1969 [22, 23]. Since there are only two functional groups on the precursor acid and alcohol, only linear polymers are formed. The number of possible polymers is great since all that is required for a polymer is a new pair of acids and alcohols. Therefore, PBT is one of the current commercial success members of linear polyesters. It is tough and durable for high-performance applications, and it is designed for molding and competes with other engineering materials. The same chemistry applies to the manufacture of PBT as PET. The glycol used is 1,4-butanedial, and the resulting polymer has the following characteristics [1].

PBT has a low T_g that facilitates rapid crystallization at typical mold temperature 50 °C, which will result in a short cycle. With high injection speed, glass-filled grades of PBT will produce good surface finishes. PBT can be foamed nicely with microstructure. It adapts well to printing, painting, gluing, hot-stamping, spin, and sonic welding. Typical molding shrinkage values for unfilled PBT range from 18 to 24 µm/mm. Glass-fiber-reinforced grade PBT has a shrinkage value of about 2–6 µm/mm, and it shrinks much less than unfilled PBT. The glass-fiber-reinforced PBT will undergo directional shrinkage; however, using microcellular structure helps for disorientation, which significantly improves the anisotropy problem that is common for glass-fiber-reinforced material [1, 24].

Like other crystalline thermoplastic resins, standard PBT grades are notch-sensitive. Therefore, the minimum radius of 0.76 mm or larger is recommended for the PBT part design. Fiber-reinforced PBT material improves the notch-sensitivity with much better toughness. In addition, reinforced PBT material significantly elevates the performance at high temperature. The coefficient of linear thermal expansion for most grades of PBT materials is below 2×10^{-5} mm/mm/°F. This low mean value of thermal expansion ensures excellent dimensional stability in a variable-temperature environment. Additional performance at high temperature is the low conductivity of PBT material [22]. All three performances are further promoted by microcellular foaming since the foaming further increases toughness, dimension stability, and thermal insulation.

There are many studies for the property of microcellular parts with PBT specifically focused on the glass-fiber-reinforced PBT, such as a 30% glass-fiber-reinforced PBT (PBT + 30%GF), Valox® 420 SE0-BK1066 [22–25]. Some selected results are discussed in a case study below with the comparisons for the property change between the microcellular part and the solid part. The equipment and standards of the testing for microcellular parts are the same as for solid parts. Some special concerns of testing and postprocessing for microcellular parts are discussed in Chapter 10. The excellent cell structures of glass-fiber-filled PBT is also discussed in detail in reference 24.

4.2.3.1 Processing. The processing results of a microcellular part are compared with the results of conventional injection a molding part for the same material. The processing conditions for conventional injection molding are: barrel temperature range 232–254 °C, mold temperature about 43 °C, and melt back pressure of 3.5 MPa. Microcellular processing conditions are different from conventional processing with a higher melt back pressure of 14 MPa, and the shot size is smaller than for a solid. The short size of injection weight is called weight reduction percentage in the microcellular foam industry.

It is noted that the parts made by conventional molding are severely warped with such thickness as a solid part. Attempts to troubleshoot the problem by reducing the injection speed and pack/hold pressures, and increasing mold temperature and cooling time fail to produce warpage-free conventional molding parts. However, the microcellular process produces warpage-free parts with 25% and 10% weight reductions with the same mold and equipment. The maximum injection pressure is reduced by 6% for the thick part (3.7 mm) and up to 23% for the thin wall part (2 mm) when using the microcellular process. A reduction in the cavity pressure as high as 55–64% is experienced when using the microcellular process (25% weight reduction trial). There will be more benefits from reducing the clamp tonnage for the thin wall microcellular process because of the low viscosity of a gas-laden molten polymer.

Microcellular processing at lower melt temperature is successfully tested because the gas-laden material has low viscosity. The barrel temperature in the zones that have gas–melt mixture is reduced by 28 °C, and the melt temperature is 220 °C as compared to 251 °C when using conventional injection molding.

4.2.3.2 Shrinkage. For the Valox® 420SEO resin, as the weight reduction increases, the shrinkage value decreases (26% average reduction). This indicates that the MuCell® process helps to reduce the shrinkage values of the filled Valox® resin. However, no significant difference is experienced for the shrinkage between the samples molded using different melt temperatures.

4.2.3.3 Flammability Property. All molded plaques made by Valox® 420SEO resin also passed the criteria of the 5-V device flammability test.

4.2.3.4 Thermal Property. For the Valox® 420SEO resin, the results of the thermal conductivity analysis indicate that the insulation properties of the material increase with the increase in the weight reduction. A drop as high as 34% is experienced at the 20% weight reduction level, indicating that the microcellular foaming process increases the thermal insulation properties of the material. No significant difference between the thermal conductivity of the samples molded at different barrel temperatures is observed.

4.2.3.5 Rheological Property. The rheology data indicate that the Valox® 420SEO resin may be significantly affected from both the molding process and the gas-laden material. This result of low viscosity from the molding process is expected for a highly reinforced material, since processing may significantly break the glass fibers at high back pressure in the barrel. On the other hand, the gas-laden material also significantly decreases the viscosity.

4.2.3.6 Dynamic Mechanical Analysis. Similarly to the tensile properties, an average significant percentage decrease in the elastic modulus of the Valox® 420SEO resin is observed when using the microcellular process. However, the cell structure will be the key factor for maintaining the mechanical properties of the microcellular part. Usually the elastic modulus change will be the same level as the tensile strength change. The microcellular process did not significantly affect the glass transition temperature of the material (solid, $T_g = 163\,°C$; microcellular process with 25% weight reduction, $T_g = 156\,°C$).

The PBT 30% GF is also tested with very low (4%) weight reduction. The result of a low weight reduction test still produces good cell structure and obvious fiber disorientation as well. With 10–15% of weight reduction the cell structure becomes much better with small cells and high cell density. Then, the glass fiber disorientation is significantly improved. As the result of fiber disorientation, the average elastic modulus in the flow direction and perpendicular to the flow direction will be close, although the maximum elastic modulus is dropped with the foamed part [24]. Overall, PBT with glass fiber is an easy material to be used for microcellular parts with good balance between cell structure and fiber disorientation.

4.2.3.7 Flexural Properties. The flexural test is not carried out on the solid parts for the Valox® 420SEO resin, since they are severely warped. However, the test performed on the foamed parts with this material indicated that the flexural properties decrease as the weight reduction increases.

4.2.3.8 Tensile Properties. The effect of the microcellular process on the tensile properties is less severe with the thin part, which results in about 27% reduction in the modulus at 10% weight reduction because the cell structure is usually better in the thin wall part; better retention of the tensile properties of the Valox® 420SEO resin can be achieved by lowering the melt temperature and producing a small cell size.

4.2.3.9 Impact Property. As the weight reduction increases, the total impact energy of the Valox® 420SEO resin decreases for the microcellular part. However, it will be improved significantly by the cell structure [24]. Reducing the melt temperature does not help in retaining the impact properties of the Valox®.

4.2.3.10 Blister on the Surface. The blister problem is not encountered during the trial using the Valox® resin. This indicates that it is easy to control the microcellular process with the Valox® resin. Furthermore, the glass reinforcement present in the Valox® 420SEO resin helps to produce smaller cell structures due to added heterogeneous nucleation. This may help eliminate the phase separation problem, which can cause blisters on the surface of the foamed parts.

4.2.4 PET for Microcellular Process

Polyethylene terephthalate (PET) was developed in the 1930s at the E.I. Du Pont de Nemours and Company Laboratories [22]. PET is a polymer commonly used to make bottles, synthetic fibers, films, and food containers. The oldest member of this linear polyester group (same group as PBT) is the reaction product of terephthalic acid and ethylene glycol. This polymer demonstrates to a high degree the property variations that come from the mechanical working of resins to produce changes in orientation and crystallinity. The resin melts at approximately 255 °C. Then, it is quickly cooled, or quenched, and the morphology is almost completely amorphous with a density of approximately 1.33. Crystallization starts at approximately 72–80 °C because of the slow mobilization of the molecules. At 130 °C, density will rise to 1.37. In addition, with proper orientation and heat setting, a density of 1.40 is obtained [1].

Unfilled PET material is widely used for blowing injection molding. It displays excellent clarity and gloss, as well as substantial weight savings over glass. The PET bottle has greater orientation in the hoop direction than in the axial direction, such that the burst and impact strength makes the bottle shatterproof. The thermoplastic polyester gains most of its strength from the stiff chains in the polymer plus the strong interaction between the molecules.

The rheology study indicates approximately 33 °C (310 °C to 277 °C) of melt temperature reduction for the 2% CO_2 gas-laden PET with the same viscosity as the molten polymer without gas. The pressure drop rate for unfilled PET is about a minimum of 0.4 GPa/sec, and the higher rate of pressure drop helps to reduce the cell size further. The highest CO_2 level reaches 5–7%, and melt temperature of processing can be as low as 240 °C. A rapid recrystallization of PET occurs at temperatures below 220 °C.

Figure 3.5 shows the sample of preform PET with microcellular structure. It has excellent cell structure with uniform average cell size of about 40 μm. The smallest cell size is achievable in the range of 5–20 microns. However, the reasonable range of cell size is in the range of 25–50 microns. Based on the result of the PBT microcellular part, there is no surprise for the good structure of microcellular parts for PET since they are in the same group of chemical structures.

When the cell size is 3 μm or less in CPET (crystallinity = 33.2%) material, the tensile fracture strength becomes almost equal to that of unfoamed plastics [26]. Shimbo et al. [27] further tested the CPET material with the batch process to verify that the strength of CPET foam will increase with a decrease in cell size. The tensile fracture strength of foamed CPET material maintains about 80% of that of unfoamed ones, though it is twice the foam magnification.

4.2.5 PA (Nylon) for Microcellular Process

The polyamides have the amide group as an integral part of the linear chain and have been given the generic name "nylon," or PA as the abbreviation. The PA first used in commercial products in the United States was PA 6/6, made from adipic acid and hexamethylenediamine. The first number in the PA 6/6 represents the number of carbon atoms in the diamine, and the second number stands for the number of carbon atoms in the diacid. Amines containing 2–10 carbon atoms and acids containing 2–18 carbon atoms have been used. This nomenclature is an agreed-upon form accepted throughout the industry and is applied to nylons of all kinds [1]. The commercially successful nylons are PA 6, PA 6/6, PA 6/10, PA 6/12, PA 10, PA 11, and PA 12. Except for these nylons, there are some modified nylons that include: copolymers; compositions containing additives to impact-specific properties; filled and reinforced materials; and chemically modified nylons.

PA is a relatively easier material to use for the microcellular injection molding process. Figure 3.6 shows the sample of unfilled PA 10 with microcellular structure. It has good cell structure with uniform average cell size of about 40 μm. However, the cell distribution in the PA 10 matrix is not truly uniform, nor is the wall thickness among the cells. The thickness of the thickest wall is about the same size of the cell. In addition, the thickness of the thinnest wall is so thin that the adjacent cells are almost touching each other to be an open cell. There are more results regarding the glass-fiber-reinforced PA 6, or PA 6/6. According to the test results available so far, the microstructure of glass-fiber-reinforced materials, generally, have much better cell structure than do unfilled same materials [24]. The detailed test result from Spindle shows the excellent properties of PA 6 with 13 wt% of glass fibers [25]. Kelvin reported that a cable tie made by PA 6/6 at a thickness of 1.45 mm has a uniform 20-μm cell size at 8–10% weight reduction with 30% injection pressure reduction, with 30% of clamp tonnage savings as well [28]. As a comparison, the same PA6/6 material with 33% glass fibers produced a uniform cell size of about 10 microns in an air intake manifold gasket part, which has 5–20% weight reduction, more than half of cycle time savings, and only 30% of clamp tonnage for microcellular processing compared to solid processing [28].

Rhodia North America introduced two special grades of Rhodia's Technyl Star polyamide 6 and also introduced 6/6 materials that have relative low viscosity of molten polymer approaching water. The material has a trade name "XCell™" and has low melt viscosity tailored specifically for microcellular processing. The higher flow rates create a laminar flow thought the mold, coating the cavity walls with a skin that freezes off, preventing any gas from breaking through the surface [29]. It is 12% less dense and can mold thin-walled parts with less injection pressure. The cycle time savings is about 20–30%. The microcellular part of XCell™ results in less molded-in stress, good rigidity, and the ability to withstand high temperatures and impacts. This high surface-aesthetic part, like an engine cover, benefits from the weight and warpage reduction provided by a microcellular part without any affect on appearance [30].

Adequate drying temperature and thorough drying are a necessity for making good cell structure of microcellular parts for all nylons. It is because the moisture in nylon material may generate big cells. The required drying time for microcellular processing is the same as that for the solid nylon material.

4.2.6 Polyoxymethylene (POM) for Microcellular Process

Polyoxymethylene (POM) is one of the acetal materials. There are two basic types of acetals in worldwide use today: (a) an acetal homopolymer that was first introduced by Du Pont with trade number Delrin® in 1960 and (b) an acetal copolymer that was from Celanese with name Celcon® in 1961. Both materials belong to the same generic family, but they differ significantly in chemical structure, in some properties, and in some modes of behavior [22]. When the acetal copolymer is subjected the degradation conditions, depolymerization stops short at the C–C link. In other words, copolymers cannot unzip and thus possess a high degree of stability. By contrast, once the ester end groups of the homopolymer succumb to attack, the molecular chain is free to unzip along its entire length. In addition, the C–C links in the copolymer are alkali-resistant, while the ester end groups in the homopolymer are not. Therefore, the microcellular POM processing will be more stable for acetal copolymer than the stability of processing for acetal homopolymer [22]. As a summary, the homopolymer has slightly greater short-term strength properties than the copolymer, while the copolymer has greater thermal stability and chemical resistance.

Acetal is truly the first plastic with the strength properties approaching those of the nonferrous metals. The high-crystalline structure accounts for many of the good properties of POM material, which include stiffness, fatigue endurance, durability, high temperature resistance, good solvent resistance, low coefficient of friction and stick–slip, good appearance, and creep resistance [1]. POM material is extremely rigid without being brittle and is both tough and resilient, much like spring steel. This gives more room for microcellular processing to reduce the weight of the part and still keep the remaining properties in the range of application.

This must be controlled to avoid overheating, which will produce formaldehyde gas and could create the risk of damage or injury. The excess gas from degraded POM material is confined and causes an excessive pressure buildup if the injection molding temperature does not set up and is not controlled correctly. However, the microcellular processing provides the possibility of low temperature processing for POM foam injection molding. The inert gases may help to slow the degrading of POM material as well.

The homopolymer is generally molded at melt temperature 10–20 °C higher than the copolymer that will be molded at a melt temperature of 195–235 °C. Using CO_2 gas may reduce the melt temperature significantly. However, it is only applied if it is necessary since the high percentage CO_2 is also difficult to control stably during process. The mold temperature is usually about 95 °C for POM material processing. However, the high mold temperature (about 15 °C higher than normal) may be necessary to improve the surface finish of the microcellular part of POM. On the other hand, if the surface appearance is of micro importance, to fully make use of the benefits of short cycle time for microcellular processing the mold temperature can be set up lower down to 27 °C and below without concerning the molded-in stresses.

The acetal materials, both copolymer and homopolymer, are adaptable to virtually all postmolding processes for plastics. The ultrasonic welding for acetal microcellular material produces excellent welded parts. It can accept inserts inserted ultrasonically. The other welding methods are also successfully used for acetal materials.

POM material usually uses N_2 gas or chemical blowing agent to make foamed parts. It is an excellent material for microcellular injection molding. Although it exhibits high crystalline content ranging from 77% to 80%, the nice cell structure has been made and is illustrated in Figure 3.9. Morphology of POM with 25% glass fiber [24], N_2 gas, $dp/dt = 2.6 \times 10^{10}$ Pa/sec, 15% weight reduction. The microcellular process for POM material is usually focused on weight reduction since it involves costly engineering material and cycle time reduction for the high efficiency of production.

4.2.7 Polyphenylene Sulfide (PPS) for Microcellular Process

Polyphenylene sulfide (PPS) is good semicrystalline engineering material for microcellular processing. It is a linear polymer of modest molecular weight. However, heating this virgin resin below its melting point of 285 °C in air increases its molecular weight, yielding products possessing substantially higher melt viscosity. The crystallization temperature is 125–135 °C, and a glass transition temperature T_g is about 85 °C. When cooled from the melt at 80 °C/min, PPS exhibits a melt crystallization exotherm at 220–240 °C. The temperature is a very important processing parameter for the PPS microcellular part because the thermal history determines the degree of crystallinity and, to some extent, the mechanical and thermal behavior of the molded part. When a part is molded at a mold temperature near or below T_g, it becomes substantially amorphous material (less than 5% of crystalline). However, when the mold temperature is raised to about 130 °C, a high surface crystallinity (about 60%) is obtained. In general, the optimum properties are realized and postmold dimensional stability is increased when the crystallinity is high [22].

PPS compounds have very low mold shrinkage, generally in the 0.1–0.5% range. Therefore, PPS compounds hold very close dimensional tolerances. Part-to-part variations are negligible as long as the molding conditions are held constant. The dimension stability of microcellular parts is not an obvious benefit for PPS molding. However, the good benefit of microcellular processing will be the weight savings for the low-cost molding parts. On the other hand, microcellular processing eliminates the high packing pressure for the lower shrinkage requirement of regular molding of PPS.

Venting is important for PPS molding, specifically with microcellular processing. An improperly vented mold can result in burning in the far end of the cavity opposite the gate. The general rules for the increased venting capability of microcellular processing can be found in Chapter 5.

Mold temperature for PPS microcellular parts could be in a wide range. Normally, the mold temperature for the PPS is 38–140 °C. Although microcellular parts prefer the low mold temperature for short cycle time, the PPS parts will have low crystallinity at low mold temperature. The low mold temperature also results in a rough and mottled surface of the microcellular part of PPS. High mold temperature of about 120–140 °C necessitates slightly longer cycle time, but it provides high crystallinity and produces a better surface. It is necessary to point out that the microcellular process itself does not need the high crystallinity of PPS material to keep the dimensional stability, and the surface quality is poor anyway. Thus, in most cases the microcellular PPS processing will use low mold temperature. On the other hand, the high mold temperature will decrease the flexural strength because of reduced elongation resulting from increased crystallinity. In addition, using very low mold temperature (less than 38 °C) will have low heat-distortion temperature of about 149 °C of PPS part. A hot mold (higher than 135 °C) will provide a heat distortion temperature as high as 246–260 °C.

PPS material can be injection-molded at a melt temperature of about 302–357 °C [22]. Any filled PPS should use the upper processing temperature to reduce the wearing in the barrel, screw, and screw tip. Sometimes the difficulty of mold filling can be solved easily by raising the melt temperature of PPS. On the other hand, the gas-laden PPS melt can make mold filling easier and reduce the wearing of equipment because of the low viscosity of gas-laden PPS melt.

A medical staple gun of PPS has been successfully produced with microcellular processing. It reduces 30% weight and up to 50% of cycle-time reduction. In addition, the clamp tonnage has been reduced from 120 tons to 15 tons.

Mold temperature can be reduced from 80 °C to 15 °C. For the other example of PPS microcellular part, the weight is reduced by 5%, and cycle time is decreased by 20%. In addition, the sink mark depth of the PPS microcellular part is reduced by 80% [14].

Precaution should be observed when molding temperature is at or above 371 °C due to off-gas products generated by decomposition. The gases sulfur dioxide and carbonyl sulfide, products of decomposition, are considered irritants to the mucous membranes [22].

4.2.8 Other Important Semicrystalline Materials for Microcellular Processing

There are many other semicrystalline materials used more or less for microcellular injection molding. Only some important semicrystalline materials that are used more than others are discussed in the following.

4.2.8.1 TPO for Microcellular Processing. TPO is the new developed material for automotive parts. The major component of TPO is the material of PP. Therefore, understanding the PP material above is sufficient to process TPO accordingly. However, TPO has some unique features, such as low temperature flexibility and ductility, excellent impact/stiffness/flow balance, excellent weatherability, and free-flow pellet form for easy processing, storage, and handling. The major differences among different TPO materials depend on the PP structure. A linear PP and branched PP are blended with an ethylene α-olefin copolymer as the toughening elastomeric compound. The branched PP-based TPO (B-TPO) exhibits more strain-hardening behavior compared to linear PP-based TPO (L-TPO). The results of foaming of TPO materials show that the foaming behavior of the TPO is controlled by the interfacial properties and the blend microstructure that includes the initial droplet size and numbers in TPO. Based on the foamability of PP materials, the B-TPO has better foamability than the L-TPO because the cell density of a branched PP is about two times larger than the cell density of a linear PP [31]. However, the total foam density also depends the population densities of the dispersed rubbery droplets that may become initial nuclei and may create cells. It was found that the volume average droplet size for L-TPO is about 0.47 µm and for the B-TPO it is 0.67 µm. Then, the increase in the droplet population density for L-TPO before foaming is threefold greater than that of B-TPO. Hence, the final cell structure of TPO is determined by both of the factors discussed above. For example, the final cell density in L-TPO is actually higher than the cell density in B-TPO [31].

Overall, TPO material performs better for the microcellular process than does any pure PP since the mixed compound provides a blend microstructure that is better for the foaming process. The heterogeneous nucleation is the major factor for the improved cell structure of TPO.

4.2.8.2 Liquid Crystal Polymer (LCP) for Microcellular Processing. Theoretically, liquid crystal polymer (LCP) is good material for microcellular processing [2]. LCP foam was successfully made with nice cell structure in the batch process. However, LCP microcellular foam has not yet been successfully made from industry because LCP changes the phase suddenly during melting and also results in a very low viscosity of melt. The viscosity change occurring suddenly in the screw will cause the gas dosing and pressure-maintaining difficulties during screw recovery. It is critical for LCP microcellular processing to control the melting position inside of the screw to match the position of gas dosing in the barrel. Therefore, the proper equipment designing must match the melting characteristics of LCP material to control the sudden viscosity drops once LCP melted and also control gas dosing in the right position along the screw accordingly. Once the equipment issue is solved, LCP material is possibly good material for microcellular processing.

4.2.8.3 PEEK for Microcellular Processing. Polyetheretherketone (PEEK) [22] is also good material to be used for microcellular processing. It was marketed under the trade name Victrex®. It is a wholly aromatic polymer suitable for prolonged use at high temperature. The partially crystalline nature of PEEK, however, differentiates it from the amorphous PES. PEEK has a high crystalline melting point and outstanding chemical resistance, thereby opening entirely new applications in some very aggressive environments.

PEEK is a linear thermoplastic that can be processed in the range of 350–420 °C. It needs to be dried for 3 hours at 150 °C before processing. It is usually reinforced with glass fibers and is then processed at both high temperature and high pressure. The microcellular processing can lower the injection pressure of PEEK molding. However, the high temperature processing is still necessary for PEEK microcellular foam. In most cases of PEEK microcellular processing, N_2 gas is used as a blowing agent since it is easily controlled for cell structure and surface finish. The N_2 gas can be added as high as 0.5–0.7% in PEEK microcellular foaming process. However, this dosing percentage of N_2 gas cannot lower the viscosity of glass-fiber-reinforced PEEK as much as CO_2 gas does. The target of PEEK microcellular part is definitely the material savings since it is expensive engineering material. Then, cycle time savings is another important goal as well.

4.2.8.4 Polyethylene (PE) for Microcellular Processing. Polyethylene (PE) has been used for the injection foaming industry for a long time. Both high-density polyethylene (HDPE) and low-density polyethylene (LDPE) are suitable for microcellular injection molding. However, unfilled HDPE and LDPE are similar to unfilled PP, with difficulty to make a good microcell structure. Usually, unfilled HDPE and LDPE need some fillers added to promote the nucleation, and then the cell structure can be improved significantly.

There are also many other ways to improve the microcellular cell structure of PE materials. One of them is to bend the different melt indices (from MI = 5, to MI = 30) of PE materials to improve the transient extensional viscosity [32]. Also, the cross-linking agent dicumylperoxide (DCP) is used for increasing the melt strength. The purpose of this modification of PE material is to reach a good balance between resin viscosity and branching of a compound PE. Then, this modified PE is the compound of PE materials not only with different melt indices but also with some extent of cross-linking inducing branching. The result of modified PE is similar to the branched PP material that enables us to maintain the cell expansion and prevent cell collapse during cell growth and shaping in the injection mold [32].

Similar to PP with filler as the modification with more heterogeneous nucleation, filled PE is another major improvement measure to make acceptable microcellular structure for all PE materials. The details of fillers will be discussed in a filler section below.

4.3 AMORPHOUS MATERIALS

The morphology of amorphous material consists of different structures from crystalline material. The unique structure in amorphous material is a random array of polymer chains with no regular molecular arrangement, the interaction of which determines the amorphous material properties. Amorphous materials show no sharp melting point and only a range of soft temperature to determine that they transfer from a solid to an ultrahigh-viscosity liquid. Typical amorphous materials for microcellular of injection molding are general-purpose polystyrene (GPPS), polycarbonate (PC), acrylonitrile/butadiene/styrene (ABS), and high-impact polystyrene (HIPS). Usually, amorphous material will have wide processing window for microcellular injection molding, along with excellent cell architecture. However, ABS and HIPS will have wide variations since the different rubber phases in the base material may significantly influence the results. Therefore, the cell architectures for different grades of ABS and HIPS may be different as well.

4.3.1 General Characteristics of Amorphous Materials for Microcellular Processing

Generally, the amorphous material has thick skin. It can be estimated to be about 15–20% of the whole thickness for thermoplastics. The skin thickness of amorphous material will be 1.5–2 times as thick as that of crystalline material. The gas at the supercritical state will dissolve the gas uniformly throughout the matrix of amorphous material. Usually, the cell structure of amorphous material has better uniform cell structure and smaller cell size compared to crystalline material.

4.3.2 GPPS for Microcellular Processing

The commercial general-purpose polystyrene (GPPS) is the most popular material for this initial equipment development of microcellular injection molding. It is normally a brittle and clear, colorless material without foaming and becomes white after foaming. A small amount of lubricant may be added to encourage molding flow and to aid in ejection from the mold. This polymer has low density (1.05–1.08). The GPPS part made by injection molded is rigid and water-resistant and has good electrical properties. The GPPS part can be produced in a wide range of colors. It will soften at just above 100 °C, and it becomes a viscous fluid at around 185 °C. It is resistant to acids, alkalis, alcohols, vegetable oils, fats, and waxes. GPPS is used unmodified for most applications, except for the occasional addition of pigments, dyes, and flow characteristics. A major application of GPPS is in the manufacturing of foams.

The unfilled GPPS morphology displayed in Figure 1.1 is the production part with 0.3 wt% of N_2 gas as blowing agent. It has an average size of 25 μm and has a cell density of about 8.1×10^7 cells/cm^3. It is a typical cell structure for a stable process and acceptable surface finish. However, the batch process can make the cell size as small as 1–20 μm. The cell density can be anywhere from 10^9 to 10^{12} cells/cm^3. With this kind of cell structure of microcellular PS material, the material will be tougher than the original materials by as much as a factor of six [2].

Usually, the cell structure in GPPS is closed-cell no matter what kind of gas is used. However, the open-cell microcellular GPPS can be made by nucleation and promoting the cell growth at a temperature higher than 150 °C, or high saturation pressure is used to maintain the postnucleation cell pressure high enough to break open the cells. Generally, the solubility of N_2 gas is considerably lower than that of CO_2, although the diffusivities of CO_2 and N_2 are nearly the same [2].

GPPS is also the easy material for excellent microcellular structure with extremely low density and high cell density. As a comparison of different cell structures with different gases, the GPPS microcellular foam with 16 wt% of CO_2 gas as blowing agent creates larger cells and thin wall thickness among the cells that is similar to that of the close-packed model proposed by Suh [2]. However, this structure is not popular for most of the injection molding parts and may be only good for some insulator or package applications.

4.3.3 PC for Microcellular Processing

Polycarbonate (PC) with commercial significance is derived from bisphenol A (4,4′-dihydroxy-2,2′-diphenylpropane). This monomer is made from the coreaction of phenol and acetone under acidic conditions. The other monomers used are phosgene and diphenyl carbonate [1]. The molecular weights produced range from 25,000 to 35,000 for molding grades. PC has a higher free volume at its T_g than normally observed with other polymers. The high free volume below T_g gives its damping capacity of the polymer over a wide temperature range, and a limited degree of crystallinity combines to produce PC's high level of toughness. This may be the reason why a large percentage of gas can be dissolved into PC's free volume at low temperature around T_g. The impact strength of PC depends on its thickness. PC also exhibits excellent creep resistance and excellent optical properties. The heat distortion temperature is in the range of 130–143 °C.

The PC injection mold must be prepared for high processing temperature and high melt viscosity of the material, even gas-laden, somehow reducing the viscosity. However, microcellular processing reduces the cavity pressure significantly (see the data in the case study below).

PC melts adhere strongly to metal and, if allowed to cool in a barrel, may pull pieces of metal away from the wall of barrel because of shrinkage. Therefore, PC usually needs to be purged out in the barrel if the machine is going to shut down for a while for microcellular injection molding process.

PC material also needs to be dried before processing. Any residual moisture in the resin not removed by proper drying will chemically react with the resin at the processing temperature, reducing the molecular weight, which can result in a loss of toughness and impact strength.

PC is, generally, an easy amorphous material for microcellular injection molding. There is little crystallization on cooling, and after-crystallization has not been observed [1]. This is an important feature of PC to make it easy to use for microcellular foaming. Figure 3.3 is typical of SEM with skin–core structure of a microcellular part, a cross-sectional view for unfilled PC (white bar is 1 mm) [33]. The blowing agent is N_2 gas, average cell size reaches 45 microns at sample thickness 3.7 mm with skin thickness 0.65 mm. It is made with mold temperature 71 °C, melt temperature 304 °C, pressure drop rate (dp/dt) 1.7×10^{11} Pa/sec, weight reduction 13%. It shows small cells in the center area and shows large cells near the skin (except one or two big voids as processing defects).

The study of mechanical properties of microcellular injection-molded PC with N_2 gas has been carried out by Hwang et al. [34]. The results show that the toughness increases for the microcellular PC part as compared to the solid PC part. The effects of processing conditions for microcellular PC are summarized as follows:

- As the melt temperature increases, both the tensile strength and the toughness increase.
- As the mold temperature increases, the tensile strength increases and the toughness decreases.
- As the melt plasticizing pressure increases, the tensile strength increases and the toughness decreases.
- As the shot size increases, the tensile strength increases as well.

The specific case studies for the material Lexan® resins with 20 weight percent of glass fibers and without glass fibers below represent the popular Lexan® resins used for microcellular parts. The selected results are discussed in the case studies below with the comparisons for the property changes between the microcellular part and the solid part. These case studies are typical reference data for high-viscosity amorphous materials, as well as reference data for the differences between filled and unfilled materials. The equipment and testing conditions are the same as the previous case study for PBT material.

4.3.3.1 Processing. The processing conditions for conventional injection molding of unfilled PC (Lexan® 940A-116) are: barrel temperature range 293–310 °C, mold temperature about 49 °C, and back pressure of the melt about 6 MPa. Microcellular processing conditions are different from conventional processing for a higher back pressure of melt 18 MPa, and the shot size is smaller than solid.

The processing conditions for conventional injection molding of filled PC (Lexan® 3412R-739) are almost the same as those for unfilled PC, so the results may be more comparable between them.

A reduction in the cavity pressure for unfilled PC material as high as 46% is experienced when using the microcellular process at 25% weight reduction trial. However, the highest achievable weight reduction for filled PC was about 15%, and the maximum cavity pressure was reduced by up to 45% with microcellular processing.

In addition, about 13% reduction in maximum injection pressure is reached in both unfilled and filled PC materials with microcellular processing.

Microcellular processing at lower melt temperature for this unfilled PC has been successfully tested. The barrel temperature in the last three zones is reduced by 28 °C, and the melt temperature is 220 °C as compared to 251 °C when using conventional injection molding.

The lowest melt temperature of unfilled PC microcellular processing that could be achieved without reaching the maximum torque of the machine was 283 °C as compared to 310 °C when using conventional solid processing.

4.3.3.2 Shrinkage. The shrinkage value of the unfilled PC resin slightly increases as the weight reduction increases. A 4% increase in the shrinkage values is experienced with the 25% foamed samples. It may be explained as the unfilled PC. However, for the filled PC resin, the shrinkage values decreased as the weight reduction increased. The shrinkage values of the filled PC samples molded using different melt temperatures showed no statistical difference of shrinkage.

4.3.3.3 Flammability Property. All molded plaques made by both unfilled and filled PC resins passed the criteria of the 5-V device flammability test.

There is no significantly difference between microcellular and solid parts for these PC resins.

4.3.3.4 Thermal Property. For both unfilled PC and filled PC resins, the thermal conductivity values drop as the weight reduction increases. A drop as high as 26% is experienced at the 25% weight reduction level, indicating that the foaming process increases the thermal insulation properties of the material. However, no significant difference between the thermal conductivity values of both the unfilled and filled PC materials molded using different barrel temperatures was observed.

4.3.3.5 Rheological Property. The rheological properties of the unfilled PC resin are affected by the injection molding process. This reduction in the viscosity value is intensified when using the microcellular process.

As expected for reinforced materials, the rheology data indicate that the filled PC resin is significantly affected from the molding process. Furthermore, the microcellular process accounted for an additional 12% average drop in the viscosity of the material. The screw design, back pressure, and residence time used in the microcellular process may be responsible for the significant drop in the rheological properties of the material.

4.3.3.6 Dynamic Mechanical Analysis. Similarly to the other mechanical tests, the DMA analysis indicates that the elastic modulus of the unfilled PC resin decreases (21% drop) when using the microcellular process. The elastic modulus of the filled PC resin decreases by 19% for using the microcellular process. On the other hand, the glass transition temperature of the material remained almost unchanged between foamed and solid parts.

4.3.3.7 Flexural Property. The flexural modulus of the unfilled PC resins decrease with the increase in weight reduction. For the parts molded with the microcellular process with a 25% weight reduction, the flexural modulus drops by 28% as compared to the solid parts.

For the filled PC resin, the flexural modulus was reduced by 17% when microcellular foaming the plaques with a 22% weight reduction. Reducing the melt temperature slightly improved the flexural modulus (7% drop) of the foamed samples.

4.3.3.8 Tensile Property. For the unfilled PC resins, a drop in Young's modulus of 38% is experienced with the parts molded using the MuCell process (25% weight reduction).

A drop Young's modulus of 31% was experienced with the filled PC microcellular foamed parts (22% weight reduction). Reducing the melt temperature did not have an effect on the tensile properties retention.

4.3.3.9 Impact Property. For the filled PC resin, the 3.7-mm-thick foamed parts with 22% weight reduction showed a 29% decrease in the total impact energy. Up to 10% weight reduction, the impact properties of filled PC are not affected. Reducing the melt temperature resulted in a slight decrease (6%) in the impact energy of the molded samples. The effect of the microcellular process on the impact properties was more severe with a 2-mm thin part, with a 33% reduction in the total impact energy at 10% weight reduction of microcellular foam.

4.3.3.10 Defect on the Surface. There are some blisters on the thin-wall part of unfilled PC, whereas there are no blisters on the same thin-wall part of filled PC. This indicates that the presence of glass reinforcement in the filled PC resin helps greatly the nucleation process and elimination of the phase separation problem that causes the blister on the surface.

4.3.4 Other Important Amorphous Materials for Microcellular Processing

There are more amorphous materials as successful candidates for microcellular injection molding. Several selected amorphous materials are introduced in the following sections since they are at least are tried with processing data and material property data available for industry applications.

4.3.4.1 High-Impact Polystyrene (HIPS) for Microcellular Process. It is obvious that the morphological cell structure of the high-impact polystyrene (HIPS) in Figure 3.2 is not as good as the morphological cell structure of general-purpose polystyrene (GPPS) in Figure 1.1. The size and shape of the rubber phase in HIPS will significantly influence the final morphological cell structure. It is similar to the cell structure of HIPS of the batch process shown in Figure 3.1, except cell size is large (up to 50 microns). However, the surface quality of the part in Figure 3.1 is not as good as the surface quality of the part in Figure 3.2, and the strength loss of the part in Figure 3.1 is significant as well.

4.3.4.2 Polymethyl Methacrylate (PMMA) for Microcellular Processing. Polymethyl methacrylate (PMMA) is a high-clarity and excellent light transmission material. It also has a good resistance to sunlight and has low density, which makes this resin ideally suited for optical parts. PMMA is also a good material for microcellular processing. Both N_2 and CO_2 gases are used successfully as blowing agents for the PMMA microcellular injection molding process. Typical N_2 gas weight percentage in PMMA is about 0.6–0.8%. The maximum of N_2 gas weight percent may be up to 1% for injection molding if the cell structure needs to be improved. However, the surface finish becomes very poor with high weight percentage of N_2 gas. On the other hand, the maximum CO_2 gas can be as high as 13% in PMMA for batch processing and 8% in PMMA for the injection molding process. It is obvious that the maximum gas percentage absorbed for PMMA in batch processing is higher than the maximum gas percentage that diffuses into PMMA in the injection molding process because of the limit of dosing time in the injection molding process.

There are obvious different appearances between foamed and unfoamed PMMA materials. The foamed one becomes white, which may only be good for the application without requirement of transparency.

4.3.4.3 Polysulfone (PSU) for Microcellular Processing. Polysulfone (PSU) has been successfully tested for injection molding with microcellular structure. There are four different basic sulfones: standard, polyaryl, polyether, and polyphenyl. One of the successful commercialized on the market is poly(aryl ether sulfone) based on 4,4′-dichlorodiphenylsufone and bisphenol A. This resin was introduced by Union Carbide Corporation as Bakelite polysulfone in 1965. The typical one has the trade name Udel®, which is a general injection molding grade resin. The most distinctive feature of the backbone chain is the diphenylene sulfone group that may be expected to have outstanding thermal and oxidative resistance. It keeps the rigidity even at elevated temperatures.

The flexibility in the backbone of the polymer is desired to impart toughness. It is from ether linkage and moderately augmented by the isopropylidene link. These ether linkages also contribute the thermal stability. Both ether and isopropylidene improve the processability of PSU material.

PSU absorbs small amounts of moisture. Although the presence of moisture does not chemically degrade polysulfone, it can cause foaming and results in some splay in the part. The recommended method for solving the moisture issue in PSU is to reduce the moisture content to below 0.05% before microcellular injection molding because the foaming caused by moisture results in macrocells and not microcells.

The increase in linear dimension of PSU is proportional to the amount of moisture absorbed and may be approximated as a 0.012% increase for every 0.1% of moisture absorbed [22].

Like all aromatic polymers, PSU exhibits poor resistance to ultraviolet light. However, pigmented and filled PSU resin help to improve the UV stability. For example, carbon black has been shown to improve the UV stability of PSU to the point where it has been used successfully as the absorber plate in solar collectors. On the other hand, carbon black color acts like an excellent nucleation agent in microcellular processing.

In many aspects, PSU has a rheology similar to that of PC because of its higher glass-transition temperature. Also both PSU and PC exhibit little shear sensitivity; therefore, the high processing temperature is an efficient way to lower the viscosity and to make smooth processing. It has been found that the microcellular process can lower the processing temperature, or lower the injection pressure with enough gas diffused in the molten resin of PSU.

4.3.4.4 Polyetherimide (PEI). Polyetherimide (PEI) is a high-performance, amorphous thermoplastic material. PEI material has regular repeating ether and imide linkages. The aromatic imide units provide stiffness, while the ether linkage allows for good melt-flow characteristics and processability. PEI material provides a good engineering material capable of meeting the difficult design requirements of many applications in the electrical/electronic, transportation, and industrial markets. This material has been commercialized with trade name Ultem® of General Electric (now Sabic Innovative Plastics) [22].

Ultem® material was used successfully for the microcellular process trials in Trexel in 1999. Since then, many companies have used it because it is a good engineering material used worldwide. It has a much wider processing window than most engineering plastics; this makes the microcellular processing easy as well, although the processing temperature is as high as 400 °C. The viscosity of Ultem® is also high and similar to viscosity of polycarbonate or polysulfone even if the processing temperature is high. One of the benefits of microcellular processing is the low viscosity of Ultem® material with gas during processing. Therefore, the microcellular injection pressure of Ultem® material is lower than the regular injection molding pressure. In addition, this material still needs to be dried 4 hours at 150 °C for microcellular processing. If cycle time is long, a hopper dryer is preferred during the microcellular injection molding process.

Another unique characteristic of Ultem® material is an excellent flame resistance and UV stability without the addition of flame retardant, or stabilizers. In other words, Ultem® material can be processed in a less corrosive environment without this harmful corrosive additive. The high heat deflection temperature and stable electrical properties are also special characteristics of Ultem® material. Generally, Ultem® material is expansive material so that microcellular processing results in another benefit of material savings with controllable cell structure. It can reduce the usage of Ultem® material for making microcellular foam without significantly reducing the material properties.

4.4 FILLER-FILLED MATERIALS

The filler is defined as any material added into a polymer matrix. Thus, is significantly different structurally or chemically from basic polymers [35]. Fillers are usually the inert substances added into polymers to reduce their cost and/or to improve the physical properties that include hardness, shrinkage, stiffness, and impact strength. Although most of fillers are inert, but there is considerable evidence that a strong attraction exists between fillers having chemically active sites and the functional groups in the polymer backbone [1]. Then, the fillers can be characterized by their interaction with the polymer matrix. The fillers can be adherent to the polymer matrix whether inherently or by special surface treatment. The fillers may absorb the polymer phase because of high surface area and inherent wettability.

The fillers can be grouped in several ways, based on (a) their structure, (b) the ways they interact with the polymer matrix, or (c) their composition. Structurally, the fillers may be aggregates with essentially round or polyhedral shapes such as clay or chalk, plates or flakes such as mica, or lamellar glass [36]. The shapes of fillers may affect the cell structure of microcellular parts and may also influence the final part properties.

Generally, fillers differ from reinforcements in two respects. One is that the fillers are generally small particles, and they do not markedly improve the tensile strength. The most commonly used general-purpose fillers are clays, silicates, talc, carbonates, and paper. Fillers act as pigments (e.g., chalk, titanium dioxide, and carbon black). Other act as an impart lubricity (such as molybdenum disulfide, PEFE, and graphite). On the other hand, there are some fillers increasing the electrical and thermal conductivity and magnetic properties (such as barium sulfate, powdered metals, and metal oxides). Most fillers increase the specific gravity of the matrix polymer [1]. Generally, the cell structure of filled material is simply better than the cell structure of unfilled material.

4.4.1 Organic Fillers

There are many organic fillers used for plastic industry. The typical organic fillers are from wood, shell, cellulose, synthetic organic materials, carbon-based fillers. Several typical successful organic fillers used in microcellular injection molding process are discussed as the following.

4.4.1.1 Wood Flour. Wood flour was first used by Dr. Baekeland to add them into phenolic resins in order to reduce the brittleness of the new phenol-formaldehyde materials [1]. This wood flour is continually used as a filler to reduce the mold shrinkage. It becomes more popular now for the wood compounds to satisfy some green-part requests made by environmentalists.

The growing trend to use wood fillers, such as wood flour, results in some interesting studies. One of the papers shows the influence of impact modification on the foamability of rigid PVC with CO_2 gas as the blowing agent, although it is made by the batch process [35]. The following results from reference 35 are helpful for the decision of whether or not to use microcellular injection molding for RPVC material. The key factors are summarized below:

- The weight percentage of CO_2 gas absorbed in RPVC with wood flour is in the range of 6.6–14.6. The impact modification will affect the CO_2 gas permeability of the materials. The amount of absorbed CO_2 gas is increased by either a cross-linked modifier (all acrylic) or an un-cross-linked modifier (CPE). However, the CO_2 solubility in RPVC relies on the degree of cross-linking in the polymer so that at a high level of gas sorption, a cross-linked modifier promotes more gas absorption than un-cross-linked modifier.
- The impact modifier with cross-links increases the diffusion of CO_2 in the RPVC. The uncross-linked modifier in the polymer promotes the rate of

gas diffusion because of the greater softening effect of uncross-linked modifier. It increases the acceleration of gas loss in the wood flour–RPVC compound. Therefore, only small portion of the gas in the composite has been utilized for the nucleated cell growth because of the accelerated gas loss.

- Consequently, impact modifiers are not a necessary ingredient in the RPVC/wood–flour composites because microcellular foam will increase the toughness of the RPVC material, which is similar to the effect of impact modifier.

4.4.1.2 Shell Flour. Shell flour can be obtained from grinding pecan, walnut, peanut shells, and so on. They ideally are called green fillers from the recycling the natural resources. It contains lignin and furfural that promote flow ability. In addition, shell flour has cutin, which is a wax with gloss, luster, and moisture resistance. The parts filled with shell flour have lower shear strength and lower resistance to impact than those containing wood flour [1]. Therefore, shell flour is not the popular filler for microcellular processing since the good flow property of shell flour is already one of benefits of microcellular processing. The other advantages and disadvantages of shell flour are also not encouraging with regard to its use as a filler.

4.4.1.3 Carbon Black. Although carbon black is an inorganic residue of the pyrolysis of organic materials, it is classified in the group of organic filler. Carbon black is the principal filler for elastomer and is also used as a pigment and fillers for various thermoplastics. The carbon black filler can reduce the sizes of crystallites, which is an important improvement because microcellular processing requires small cell sizes. The carbon black filler also improves resistance to stress-cracking and ultraviolet degradation of polyolefin. Therefore, carbon black is an excellent filler for the parts made by microcellular injection molding. One more issue needs attention: The black color is whitened somehow by foaming on the surface. It can be improved by raising the mold temperature. In addition, the initial spray during injection may leave this white color on the surface. It can be solved by slow initial injection speed until the runner system and gate filled the materials. This is a common approach to solve the surface issue of whitening locally in a microcellular part with any dark-colored material.

4.4.2 Inorganic Fillers

Compared to the typical organic fillers introduced above, inorganic fillers are more popular fillers used in thermoplastics with the microcellular injection molding process. The typical inorganic fillers are silica, talc, metals, and mineral fillers including wollastonite, barites, clay, nanoclay, calcium carbonate, altered novaculite, and so on.

4.4.2.1 Clays and Nanoclay. Clays, similar to silica and chalk, are fine polycrystalline mineral materials. The commercial clays, such as kaolins, have been used with a variety of forms and particle sizes. Water-classified clays offer close particles and control the color, brightness, PH value, and impurity content. Calcined clays are thermally treated to remove the water of hydration. Kaolin clays are used both as fillers and extenders for thermoplastics. In polyester premix, perform, and mat moldings the clay aids in consistency and moldability control. Generally, clay contributes to chemical resistance, low water absorption, and good electrical properties [1].

Wide application of recently developed nanoclay is growing fast. Organically modified nanoclay with dimethyl dehydrogenated tallow alkyl ammonium can be used as the layered silicate. The microcellular nanoclay composites represent the new research area. However, it is still not popular in the injection molding industry yet, and most activities are restricted in the university laboratory. On the other hand, there are a few practical applications of nanoclay for microcellular processing in industry, and they will be discussed in Section 4.6.

4.4.2.2 Mica. Mica and asbestos are structured mineral-type materials that are again basically crystalline in nature. Synthetic mica is a 1093 °C of mica for high temperature [1]. It has the density about 2850 kg/m^3, and the average particle size is about 5–20 microns. It is proved as good filler additives for creating good heterogeneous nucleation.

4.4.2.3 Talc. Talc is naturally as platelets delaminated mica flake that is in the range of 1–10 μm in thickness. It may be the mixture of such minerals with short glass or polyester fiber that is used for well balanced improvements of modular, flexural, and impact strength. It is widely used for PP material properties modification. This is properly the most well-known filler used for microcellular injection processing. Only 5–10% talc filler in the plastics may make great differences between poor foam with voids and microcellular foam with microcells. It is frequently used for PP resin modification. However, many research results show that talc content percentage is not necessary high when the gas weight percentage is high enough. On the other hand, when lower weight percentage of gas is used, then additional talc may help, but at certain level more talc does not play a significant role in cell nucleation anymore. For example, only 0.8 wt% of talc mixed into branched PP with 5 wt% of CO_2 gas can make fine-celled PP foam (the cell density is up to 10^7–10^9 cells/cm^3) at a tandem foaming extruder system [37]. Similarly, the block PP copolymer with 1 μm or less size of ethylene–propylene rubber (EPR) particles dispersed in PP simply does not need talc to make a nice cell structure. The cell structure of block PP copolymer without talc is as good as, or even better than, the result of a random PP copolymer with talc added as nucleation agent [38]. These results are good references for injection molding as well because the extruding process without considering extruding die is simply the same one in the first stage of injection molding.

4.4.2.4 Calcium Metasilicate ($CaSiO_3$). Wollasonite is the calcium metasilicate ($CaSiO_3$), which has a composition of 48.3% of CaO, and 51.7% SiO_2. It may contain trace to minor amounts of aluminum, iron, magnesium, manganese, potassium, and sodium. Wollasonite improves the tensile and flexural strengths, reduces the consumption of resins, and improves thermal and dimensional stability at elevated application temperature.

Asbestos is a first-rate heat-resistance improver, but it causes health issues and has been prohibited to use for many applications. Asbestos is likely to be replaced by a mineral–fiber mixture, such as wollastonite or $CaSO_4$ crystal microfiber.

4.4.3 Influence of Fillers for Microcellular Processing and Properties

The principal characteristics of the improvements from the incorporation of fillers result from the fillers reacting with, and are adherent to, the structure that is critical for the microcellular part with filler-filled materials. If nonadherent filler is added into plastics, it may still form micro voids that can help to form nuclei. However, the nonadherent fillers in the plastics generally make the filled material brittle and weak in tension and bending. Therefore the nonadherent fillers are not recommended for microcellular processing.

On the other hand, fillers are the great nucleation agents and largely decrease the barrier of cell growth. Lee et al. [38] reported the test for different sizes of fillers and different type of fillers with the following conditions:

Material: HDPE with 5 weight percent of fillers, $CaCO_3$ (3.5 μm versus 0.07 μm), and talc (5.0 μm) versus TiO_2 (0.3 μm), with blowing agent of CO_2 gas, processing temperature of 132 °C, and pressure drop rate of 0.15 GPa/sec.

Filler size should be chosen based on the nucleation conditions (gas level, pressure drop rate *dp*/*dt*). The fine filler size will have more influence of cell density at high *dp*/*dt* rate and high gas percentage. Assume that particle size distribution will have the following function:

$$f(x) = \frac{1}{a} e^{\frac{-x}{a}}, \qquad x \geq 0 \tag{4.3}$$

Then, assume that there is a critical filler size D_f that is a certain size of filler to be able to generate a critical size of nuclei where a cell can grow. Then, the number of particles whose size is larger than or equal to critical filler size D_f will be

$$N_{x \geq D_f} = \int_{D_f}^{\infty} \frac{f(x)}{\frac{4}{3}\pi x^3 \rho} dx \tag{4.4}$$

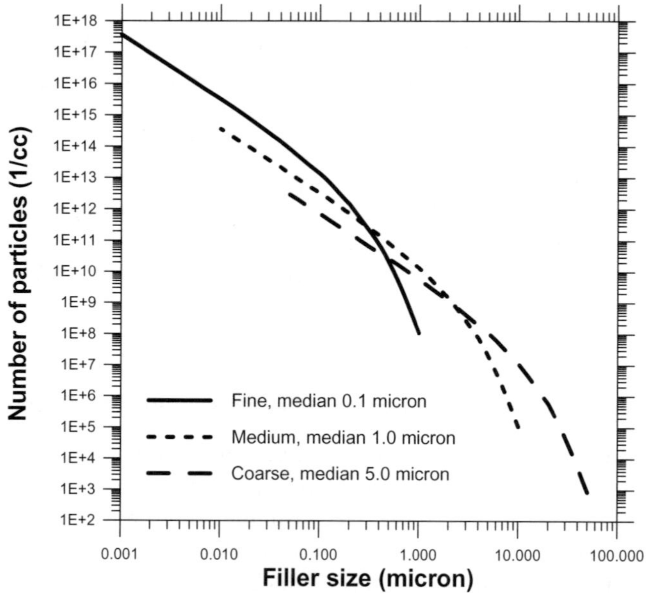

Figure 4.2 Filler size versus number of particles in unit volume [39]. (Reproduced with copyright permission of Society of Plastics Engineers.)

The results of particle size distribution are estimated in Figure 4.2. The fine fillers can have a huge number of particles, which is good for cell nuclei. However, the minimum cell size to become the nuclei needs a certain size of initial particles, or named as critical filler size.

The cell size in HDPE is affected by the filler size, as shown in Table 4.1. The fine filler will give a higher cell density at high saturation pressure. However, at low saturation pressure, a fine filler does not give the high cell density compared to a coarse filler size. The critical pressure is about 12 MPa, at which the fine filler size will have higher cell density than the coarse filler size at the same saturation pressure and gas percentage. Overall, fine filler size increases the cell density more quickly than does the coarse filler size with the saturation pressure increasing. The results can be explained by micropore theory and particle size analysis. The hypothesis is that gas is accumulated at the micropore. The size of micropore is proportional to the filler particle size. Based on the nucleation theory, only those particles that are larger than a certain size will become cells. This critical size of particle, called critical filler size, will be determined by many factors, such as gas pressure and gas percentage.

The cell structure approves the small cell sizes from fine fillers at high pressure with CO_2 gas, as shown in Figures 4.3a and 4.3b. Figure 4.3a shows the 3.5-micron coarse fillers that result in good cell structure. However, the 3.5-micron $CaCO_3$ filler is not as small as the cell size shown in Figure 4.3b made by 0.07-μm $CaCO_3$ fillers. On the other hand, at low pressure, coarse filler size will have better cell size in Figure 4.3c than the cell size made from fine filler size in Figure 4.3d. The cell size is larger, or cell density is lower, for the

TABLE 4.1 Cell Density (cells/cm^3) with Different Sizes of $CaCO_3$ Fillers in HDPE, Saturated with CO_2 Gas under Different Pressures

Gas Pressure (MPa)	Unfilled HDPE	3.5-Micron Filler Size	0.07-Microns Filler Size
6.9	1×10^6	1.1×10^7	1.9×10^6
10.3	3.9×10^6	2.3×10^7	1.4×10^7
13.8	2.8×10^7	4.8×10^7	3.4×10^8
20.7	4.9×10^7	3.1×10^8	1.5×10^9

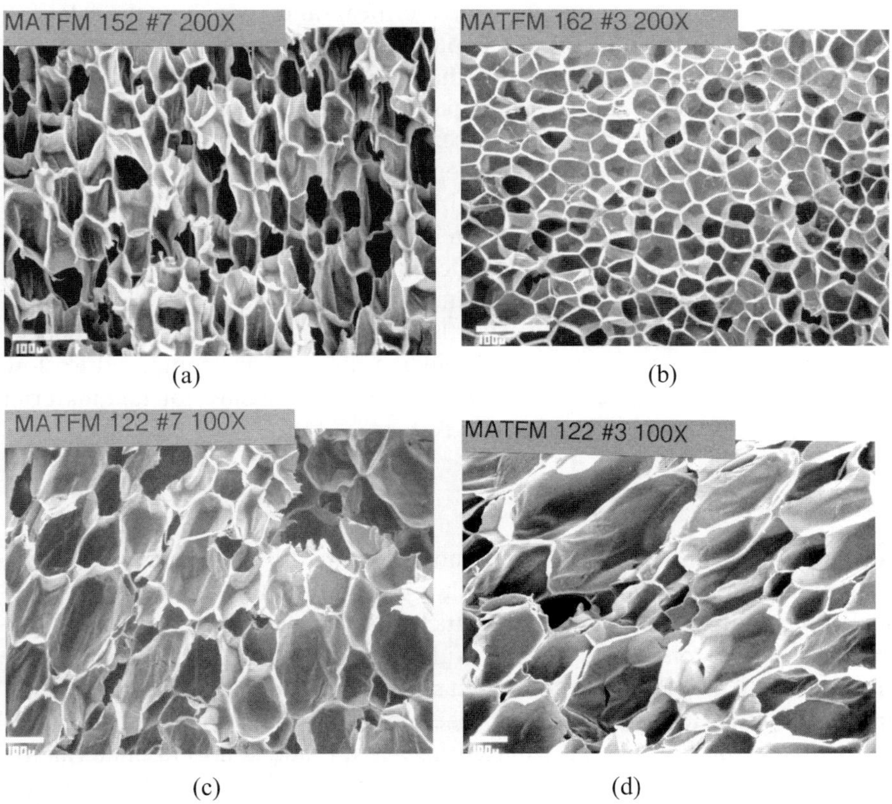

Figure 4.3 Cell structure with different sizes of $CaCO_3$ fillers in HDPE at different pressures with CO_2 gas. **(a)** Filler size of 3.5 μm at 20.7 MPa. **(b)** Filler size of 0.07 μm at 20.7 MPa. **(c)** Filler size of 3.5 μm at 3.5 MPa. **(d)** Filler size of 0.07 μm at 3.5 MPa [39]. (Reproduced with copyright permission of Society of Plastics Engineers.)

TABLE 4.2 Cell Density (cells/cm^3) with Different Fillers in HDPE, Saturated with CO_2 Gas under Different Pressures

Gas Pressure (MPa)	Unfilled HDPE	Talc-Filled HDPE Talc Size: 5 μm	TiO_2-Filled HDPE TiO_2 Size: 0.3 μm
6.9	3.8×10^5	5.6×10^6	9.0×10^5
10.3	1.1×10^6	1.8×10^7	8.1×10^6
13.8	6.1×10^6	3.7×10^7	4.2×10^7
20.7	3.0×10^7	1.2×10^8	7.6×10^8

fine filler size 0.07 μm at low pressure of 3.5 MPa compared to the result of coarse filler size 3.5 μm. The theory to explain the cell structure changes with filler sizes at certain saturation pressure is critical radius of cell, as the relationship shows in Equation (2.14) in Chapter 2. The cell growth is inversely proportional to the pressure. Therefore, the higher the pressure, the smaller the critical radius will be. Then, the small particles of fine fillers get more chances to become real cells at higher pressure. In other words, the critical filler size becomes smaller if the pressure is higher. In this case the theory really matches the experimental results and agrees with morphological results as well.

The results of filler type tests are listed in Table 4.2. The trend of cell density change with the different fillers, and talc versus TiO_2 varies with the saturation pressure of CO_2 gas. It is similar to the filler sizes trend shown in Table 4.1. In other words, the small filler size (0.3 μm) of T_iO_2 illustrates the quicker increase trend of cell density with the increasing pressure compared to slow response of increasing cell density with increasing pressure for talc material that has 5 μm of filler size.

4.4.4 Nucleating Agent in PP

A nucleating agent is the most common filler for PP processing. There are many types of nucleating agents, including conventional, advanced, and hyper-nucleating agents. The conventional nucleating agents are talc, pigment colorations, and so on. They are still widely used for not only microcellular processing but also the other foam industries. The advanced nucleating agents are phosphate ester salts, nanoclay, steel powders, and so on. The cell structure will be improved when the nucleating agent is added into PP material for the microcellular process. Figure 6.17b shows the morphology of the improved PP cell structure by adding 20% by weight of the nucleating agent of talc. The talc-filled PP has uniform cell size and reduces the largest cell size from 100 μm down to 50 μm.

4.4.5 Clarifying Agent in PP

This clarifying agent is used to promote the transparency of PP. It can increase the crystallization rate and crystallization temperature of the PP melt and thus increase the number and decrease the size of spherulites formed during processing. The decreased spherulite sizes are generally so small that the light will scatter less and will allow more transparency of PP. The small spherulites, and the uniformity of their distribution in the matrix of PP, definitely help for gas nucleation. The nucleating agents and clarifying agents may exist together in a matrix of PP. However, all clarifying agents will nucleate, but not all nucleating agents will clarify [16].

Seymour [40] summarized the general contributions to improve the plastic properties from some fillers and reinforcements. These trends are still true for microcellular processing. However, some special properties of microcellular parts that are not listed in Tables 4.1 and 4.2 are going to be found everyday by the practice of microcellular processing.

4.5 FIBER-REINFORCED MATERIALS

Reinforcements are strong, inert, fibrous materials incorporated in polymer matrix to improve physical properties. The strength of a reinforced plastic is related to its ability to transfer external stress on the resin matrix to the reinforcement [1]. Although the term *reinforcing* is used for even carbon black, named as reinforcing pigments, in most cases the reinforcing material means fibers. When the resin matrix is added to the filaments, the strength of the reinforced material is drastically altered. The fibers in the resin are bonded together far better than either the resin matrix or the fiber itself. It is obvious that the compression load and bending load benefits the most from reinforced materials with fibers. On the other hand, it is not necessary for the filaments to be continuous from one end to the other end of the part since the resin phase acts as a stress transfer medium to transfer the stress around discontinuities in the filaments. The thermal properties of the reinforced material are also substantially changed by the fibers. The thermal conductivity of the reinforced material will be increased to the level close to the thermal conductivity of the fillers. In addition, the thermal exposition of reinforced material will have only 10–20% that of the resin matrix. There are many different fibers to be used for reinforcing the plastics. They can be divided into two groups, organic material and inorganic material.

4.5.1 Organic Fibers

Organic fibers are not new in the plastic industry. They are not used as wide as inorganic fibers. However, organic fibers may be picked up by more and more applications that need to be considered for environmental issues. The typical organic fibers suitable for microcellular injection molding are discussed as follows.

4.5.1.1 Wood Cellulose Fiber. This is a newly developed organic or natural fiber in a composite, as the green production is required. It is a natural product of wood fibers, and the microcellular part with this kind of reinforcement may have a special market for home products. The wood-fiber-reinforced plastic is currently used for large-volume building applications, such as plastic lumber. Since the thermal stability of wood fiber is limited, wood fiber can only be used with thermoplastics whose melt temperature is below 200 °C [40].

The most common wood fiber is a very short wood fiber because of the higher bulk density and the free flowing nature. Low cost, familiarity, and availability are also contributing reasons why more people are using wood fibers in plastics applications [39, 41, 42]. For example, the hard wood fiber, Lignocel HBS 150-500, has the particle size of 150–500 μm.

The smallest coherent unit of cellular retaining characteristics of the bulk fiber is called a microcrystal. When added to liquid resins, these microcrystals produce thixotropic systems and thus may be used for viscosity control [1].

Jute fibers, yarn, and fabric have been used for reinforcements in phenolic composites and have low strength. Rayon may be used both as flock and as yarn. The impact resistance is a function of its fiber length.

Bledzki and Faruk [43] reported the result of microcellular-injection-molded wood–PP composites. They use (a) the hard wood fiber, Lignocel HBS 150–500, (b) PP with melt index 10.5, and (c) three different chemical blowing agents, exothermic (Hydrocerol 530), endothermic (Hydrocerol BIH20E), and endo/exothermic (Hydrocerol AB40E). A coupling agent maleic anhydride–polypropylene copolymer (MAH-PP) was used with 5% by weight relative to the wood fiber content. The pre-dried wood fiber was mixed with other materials before injection. The processing temperature is about 150–180 °C, the mold temperature is about 80–100 °C, and the injection pressure is below 20 kN/mm^2. They found that the microcellular structure of hard wood–PP composites were strongly affected by different chemical blowing agents in this injection molding process. The exothermic 2 wt% content shows the finer cellular structure. Also, the optimum chemical blowing agent content varied with the variation of wood fiber content. It is interesting that the composite produced by the microcellular process possesses a smoother surface than that of the nonfoamed composite. The surface roughness of microcellular composite is reduced nearly 70% compared to the nonfoamed composites. In addition, because of microfoaming, the odor concentration of wood-fiber-reinforced composites is less than that of the nonfoamed composite.

Yoon et al. [44] reported the feasibility of injection-molded wood-fiber/HDPE composite foam with N_2 as the blowing agent. Wood fiber (2020 American Wood Fibers) was used as filler with 50 wt% in HDPE (SCLAIR 2710 Nova Chemicals, with MI = 17). Maleated HDPE (MB265D, Dupont) was used with 3 wt% to modify the interface between wood fiber and the

HDPE matrix. The barrel operation temperature was 170 °C and N_2 gas delivery pressure was 19.3 MPa. The DOE result of shrinkage shows that the weight reduction, mold temperature, and gas amount are the significant factors for this HDPE wood composite microcellular foam. In addition, experimental results show that the shrinkage increases with the decrease of weight and with the increase of the gas amount. On the other hand, the warpage is related to the factors of mold temperature and injection speed. The warpage of HDPE wood composite increases with the increase of the mold temperature and injection speed, respectively. The tensile strength at yield and Young's modulus are strongly related to weight reduction, which decrease with the decrease of weight of the foam. However, compared to the HDPE the foamed HDPE wood composite possesses superior dimensional and mechanical properties. In particular, foamed HDPE wood composite will have 68% decrease of shrinkage, 91% decrease of warpage, 28% increase of Young's modulus, and 40% cost savings over the original neat HDPE [43].

4.5.1.2 Synthetic Organic Fibers. Aromatic polyamide fibers or aramids (e.g., Kevlar) form composites with excellent strength and heat resistance [1]. Du Pont refers to all of its aromatic nylons as "aramids." Three types of Kevlar aramids are available on the market. One is a high tensile strength, high modulus type for reinforcement of plastics, such as Kevlar 29 and 49. The other two are for ropes, cables, and rubber reinforcements. The aramid fibers of Kevlar 29 and 49 have 15% better tensile strength than E-type fiberglass. In addition, the modulus of aramid fibers is twice of that of E-glass. The thermal stability of aramid fibers is good, too. Therefore, the composites made with polyaramid fibers exhibit excellent dimensional stability. Aramid fibers resist most chemicals, except strong acids and bases. The main disadvantages of the aramid fibers are high moisture absorptivity (pre-drying is recommended), low compression strength, and difficult machining [1].

Polyester fibers (Dacron, Fortrel, Kodel, etc.) reinforce both thermoplastic and thermoset resins. The reinforced crystalline thermoplastic has unusually high specific strength since the polyester fiber serves as nucleating agents for crystallization [1]. Its usefulness in heterogeneous nucleation in the microcellular process is the same as its usefulness in crystallinity.

4.5.1.3 Carbon Fibers. Filaments of carbon were first made commercially in 1880 [1]. The true textile forms of carbon fibers were produced commercially in 1954. The high-modulus graphite yarns were introduced in 1965. The extremely fine filaments can be handled in much the same manner as glass fibers. There is complicated morphology for the carbon-fiber-reinforced PC compound that has filled with Teflon material. The sample is cut using scanning electron microscopy as shown in Figure 3.10. It still has excellent cell structure, with the average cell size about 15 μm. Some white ring around the cells is the Teflon.

4.5.2 Inorganic Fibers

Similar to inorganic fillers, inorganic fibers are far more popular fibers used in the microcellular injection molding process. However, glass fiber is the most widely acceptable inorganic fiber used in microcellular injection molding process. Therefore, only glass fiber as the typical inorganic fiber is discussed with detailed applications and final part properties in the following.

The reinforced material used for thermoplastics most is usually the glass fiber. It is comprised of lime–aluminum borosilicate glass that is relatively soda free. This kind of glass has great durability and is known as "E" glass because of its good electrical characteristics. E-glass fibers have greater mechanical strength, and it is the most frequently used for reinforcements. Glass surfacing and overlay mats are formed from a low soda, more chemically resistant glass called "C" glass that is also highly durable and resists strong acids. The next level of improvement is the "S" glass, which is developed with new a fiber-production method in which the fiber draws at significantly higher temperature. "S" glass is inherently strong, and its composite strength also dramatically increases when the optimized sizing for the composite is used. However, the principal reinforcement for plastics industry is continually "E" glass while "S" glass is usually for airspace with epoxy composites [1].

The sizing treatment of fibers protects the fibers from abrading each other. It also bonds the filaments together into a strand so they can be handled, and it may include a coupling agent that subsequently improves adhesion characteristics of glass fibers. Common to all these treatments are properties that allow one part of the chemical compound to adhere to the fiber surface while another part of the same molecule reacts with the resin matrix, thus affecting a chemical–mechanical molecular bridge. The coupling agents most used are those based on silane chemistry.

A fiber length of at least 100 μm is essential for significant reinforcement of thermoplastics. Although the long glass fiber is about 12 mm long in the long glass-fiber-reinforced pellets, the final fiber length is impossible to keep the original length since both screw recovery and mold filling will break most of long glass fibers. Such as E-glass fiber has the tensile strength about 3448 MPa. After the injection molding process, it reduces to 1931 MPa as a result of significant fiber breakage.

In fibrous reinforcement the stress becomes concentrated at the fiber ends. It allows the far larger area along the fiber length to continue to function as a stress-transfer surface and prevent interfacial crack formation.

The typical glass-fiber-filled material is PA, PBT, and so on. Only reinforced material morphology is discussed with SEM pictures because the glass fiber is visible among the cells. Similar to filled material, the cell structure of reinforced material is simply better than the cell structure of nonreinforced material.

Figure 3.14 is the morphology for a 30 weight percentage of glass-fiber-reinforced PA 6 material. It has uniform cell structure across the thickness direction and average cell size about 20 microns. Glass-fiber-reinforced PA 6 and PA 6/6 usually are good materials for the microcellular injection molding.

Figure 3.9 is the morphology of POM with 25% glass fiber. It shows excellent cell structure as well. Although there are several large cells, most cells are uniformly distributed around the fibers. The SEM picture of POM microcellular foam in Figure 3.9 shows excellent glass fiber disorientation, which is another advantage of microcellular foam.

Another ideal reinforced material is PBT. It is easy to make an excellent cell structure microcellular injection molding part no matter how much the percentage of fiber glass in it. Figure 3.14 is the morphology of the 30% glass-fiber-reinforced PBT with about 4% weight reduction; also N_2 gas is the blowing agent, and the average cell size is about 45 microns.

PC is also available in glass-fiber-reinforced grades that are a compromise with a small increase in modulus without substantial loss of toughness. However, at a high level of glass fiber range (20–40%), the toughness drops off markedly with increasing rigidity (promoting dimension stability) and decreasing shrinkage. Figure 3.10 illustrates a morphology of PC with 20% carbon fiber and 1% Teflon (white bar is 100 μm), with N_2 gas as the blowing agent. It is complicated cell structure with fibers and Teflon fillers mixed with PC base materials. However, the cell size is almost the same as the fiber diameter that is about 10 μm.

All glass-fiber- and carbon-fiber-reinforced microcellular parts made by injection molding show that the fiber always stays in the wall between cells. It means that the cells are only created around fiber, not in the same spot of fiber, so that fiber is still strongly held by solid material not in the cell or void of the part.

4.6 NANOCOMPOSITE-REINFORCED MATERIALS

Nanoclay-filled material is current growing quickly in the plastic industry. The polymer–silicate nanocomposites offer improved stiffness, heat resistance, barrier (to oxygen nitrogen carbon dioxide flavor and aroma), flame retardation, and dimensional stability with a small clay load that is less than 10%. Although it is still economically unfavorable, it has already become the new class of high-performance materials.

It was found that the nanoclay helps to facilitate the gas dissolution into the polymer melt. Gas in the melt also helps to further disperse the nanoclay platelets in the polymer–gas solution [44, 45].

However, the nanoclay is so small that the magnify ratio of scanning electron microscope (SEM) micrographs is too small to see the nanoclay. Therefore, most morphology studies do not show the nanoclay and cell structure together since there is a huge size difference between cell and nanoclay.

Dynamic rheological study is of special interests, because the bulk properties of the microcellular nanocomposites depend not only the bulk properties of the matrix, but also on the size and spacing of the cells as well as the physical and chemical properties of the nano-fillers in the matrix. The issues in microcellular nanocomposites are far more complicated because of the additional interfacial phenomena introduced with the clay/polymer/cell structure. Some other techniques may be used, such as (a) transmission electron microscopy (TEM) to take the picture of nanoclay dispersed in the polymer matrix, (b) dynamic mechanical analysis (DMA), and (c) X-ray diffraction (XRD) analysis. Several developed nanoclay composites are introduced in the following.

4.6.1 Nanoclay-Filled HDPE

The HDPE can be filled with nanoclay to become a nanocomposite compound. Jo and Naguib [47] reported the effects of nanoclay on the mechanical properties of injection molded HDPE nanocomposite microcellular foam. The clay only occupies a very small amount, in the range of 0.5–2 wt% [46, 47]. Both CO_2 gas and N_2 gas are successfully used for making good microcellular HDPE nanocomposites. The result from Jo and Naguib uses CO_2 gas as the physical blowing agent. The HDPE is SCLAIR 59A, Nova Chemicals, with MI = 0.72 g/10 min. The organically modified clay with dimenthyl dehydrogenated tallow alkyl ammonium (Cloisite 20A, Southern Clay Products) was employed as the layered silicate. In addition, the maleic anhydride-grafted HDPE (PE-g-Man, Fusabond MB100D, Dupont Canada) was used as the coupling agent. The nanoclay loading of 0.5, 1.0, and 2.0 wt% in the nanocomposite of HDPE were used. The result shows that nanoclay in both nanocomposite and nanocomposite microcellular foam plays a role to improve the mechanical properties, such as elastic modulus and tensile strength at yield. The mechanical properties of both nanocomposite and nanocomposite microcellular foams were improved most when 0.5 wt% nanoclay loading was used. It means that the nanoclay loading may not need much to optimize the nanocomposite of HDPE with improved mechanical properties.

Kuboki et al. [45] studied the effects of nanoclay on the mechanical properties of injection-molded HDPE nanocomposite microcellular foam with N_2 gas as the blowing agent. The clay occupies an even smaller amount than CO_2 gas as the blowing agent above, which is in the range of 0.1–1 weight percent. The result shows that the addition of nanoclay increased flexural strength and modulus of both solid and foam nanocomposites as the nanoclay contents are increased. However, the results indicate that the foam nanocomposites always exhibited lower flexural properties but higher notched Izod impact strength than solid nanocomposites at each given clay amount (up to 1 weight percent) [45].

The effect of crystallinity on the cell morphology and mechanical properties in HDPE/clay nanocomposite foam with CO_2 is also significant. Jo and Naguib

[48] found that the nanoclay loading in HDPE/clay nanocomposite foam contributes more to the foaming at a low crystallinity state. Cell size and distribution are more uniform as the crystallinity decreases. The toughness of HDPE/clay nanocomposite is markedly dropped as crystallinity increases. However, as clay content increases, the influence of crystallinity on the variation of cell size is diminished [48].

4.6.2 Nanoclay-Filled PA 6

Turng and his group did some systematic studies for nanoclay in the PA 6 material. The nanoclay dispersed into PA 6 nanocomposite is montmorillonite (MMT) with a typical platelet size 1 nm in thickness by several hundred nanometers across. N_2 gas at a supercritical state in the melt also helps to further disperse and facilitate the exfoliation of the nanoclay platelets in the polymer–gas solution, according to the XRD (X-ray diffraction analysis) data with PA 6 nanocomposites [46]. Yuan et al. [46] found that some clay decks or tactoids still exist and orientate in the melt flow direction. Those clay decks or tactoids behave as the normal fillers in terms of flow and orientation. The tensile modulus of microcellular nanocomposite part is usually higher than that of its corresponding base resin microcellular part [46]. The reason is the cell structure difference between them. The comparison of morphological cell structure (pictures from scanning electron microscopy) shows that the cell size of PA 6 nanocomposite is only 15 microns while the cell size of neat PA 6 is about 70 μm [49, 50]. This is because of the heterogeneous nucleation with nanoclay in the melt. It is the similar conclusion to the previous discussed results of the talc and mineral-filled materials. As the weight reduction of the PA 6 nanocomposite increases, the heterogeneous cell growth and formation of base resin becomes more obvious [46].

The other conclusions are that both nanoclay and processing conditions have strong influence on cell structures and nanocomposite of PA 6 part properties. In addition, the glass transition temperatures of microcellular PA 6 nanocomposite parts with and without nanoclay do not change too much. On the other hand, microcellular PA 6 nanocomposite exhibits a behavior more or less ductile.

4.6.3 Nanoclay-Filled PP

There is more interest for the nanoclay-filled PP because the PP can be used at higher service temperature, good stiffness, excellent chemical abrasion, impact resistance, and low material cost. Pathak and Jayaraman [51] investigated different approaches to build up the melt strength in linear PP with dispersed nanoclay (montmorillonite). They found that the surfactant used to treat the montmorillonite as well as the composition of the nanocomposite was critical to obtain significant strain hardening in uniaxial extensional flow of the nanocomposite melt. This conclusion is useful to control microcellular processing conditions accordingly, such as injection speed. Guo et al. [52] reported injection molding process results with linear and branched PP with nanoclay. The conclusions are as follows:

- There is no change in the extensional viscosity transient of the nanoclay composites based on the linear PP compared to the neat linear PP.
- The additional nanoclay to the PP matrix only provides additional nucleating sites. Therefore, nanoclay in PP microcellular foam will have increased cell density and decreased cell size compared to the cell structure of neat PP microcellular foam.

4.6.4 Nanoclay-Filled GPPS

Strauss et al. [53] use montmorillonite-layered silicate (MLS) to reinforce the GPPS and then make the microcellular foam successfully in the batch process at temperatures ranging from 75 °C to 85 °C and pressures ranging from 8 to 12 MPa. They found that the presence of clay in the sample results in a highly accelerated absorption rate for the supercritical CO_2. The key process parameters are the temperature and depressurization rate within the regime of heterogeneous nucleation. MLS concentration and dispersion observed in their study have a strong effect on cell density, shape, and size. The preferential nanocomposite orientation around the foam cells yield enhanced physical and mechanical properties [53].

4.6.5 Nanoclay-Filled LDPE

LDPE/silica nanocomposites improve both solid and foam properties. The microcellular foam made by this method shows that the average cell size is below 100 μm. An optimum weight percentage of silica content for the best cell structure of a microcellular part is about 6% in this experiment [54]. The nanosilica fillers have some effects as a nucleating agent in this system. Hence the nanosilica fillers help to make nice cell structures in microcellular processing of nanoclay-filled LDPE.

Hsu et al. [55] studied the effects of different montmorillonite (MMT) loadings in maleic anhydride (MA)-grafted LDPE (LDPEgMA) nanocomposites on the mechanical. The nanocomposites were made with 10 weight percent of maleic anhydride (MA). The modified montmorillonite clay has a density 0.2–0.6 g/cm^3, and a layer spacing of 2 nm. The conclusions from experiments are summarized below [55]:

- LDPEgMA nanocomposites (up to 5 weight percent MMT loading) are harder than the neat polymer, and they result in better wear resistance.
- LDPEgMA nanocomposites show better foaming quality including both cell size and cell density. The MMT clay and MA may act as nucleation agents.

- A small amount of clay can significantly increase the tensile strength.
- The thermal stability of LDPEgMA nanocomposites also improves from additional MMT. It may due to the better thermal barrier effects of the clay in the nanocomposites.
- Higher organoclay content or MA content appears to make the nanocomposites more brittle.

4.6.6 Nanoclay-Filled PBT

An organically modified montmorillonite (MMT) was successfully compounded with polybutylene terephthalate (PBT) and produced a good microcellular injection-molded part [56]. The effects of organoclay content, organoclay size (8–35 μm), and the screw speeds (80–100 rpm) on the mechanical and thermal properties are summarized as follows [56].

- The tensile strength, wearing resistance, and cell density of PBT nanocomposites reach the best results when the MMT content is about 1 wt%.
- The high screw speed gets the high tensile strength.
- MMT content helps to increase the thermal stability in PBT nanocomposites.
- Organoclay size 35 μm produces better cell structure than organoclay size 8 μm.
- Small content of MMT increases the tensile strength but decreases the ductility (elongation during tensile test). It is because the pure PBT is good ductile material before adding MMT.

4.7 BLEND MATERIALS AND COMPOUNDS

Blend material, or alloy, is simply an intermolecular compound. Two or more polymers in different proportions are compounded under defined temperature and shear conditions to processable granulates. They have two or more phases on a (sub) microscopic scale. As a rule, the polymer component in the lesser proportion forms the finely distributed disperse phase [53]. This is an excellent multiphase structure resulting in heterogeneous nucleation for microcellular processing.

Blend technology includes the coupling of the phase boundaries between poorly compatible polymers to prevent delaminating during the microcellular injection molding process. Too high a processing temperature or shear rate can effect a phase change that renders the multiphase system ineffective [53]. Therefore, the processing conditions of microcellular injection molding need to be set up according to this phase separation problem even if the viscosity of the microcellular process is lower than the same material without gas.

In the frequent combination of rigid, highly thermostable polymers with flexible elastic polymers to optimize the impact strength at low temperatures, the flexible elastic disperse phase, by stabilizing micro-deformation in the predominating hard matrix phase, dampens stresses that would cause fracture, in many cases (e.g., PC/ABS) synergetically down to temperatures below the cold impact strength of both components.

Heat-resistant advanced thermoplastic alloys with crystalline and amorphous phases are optimized to lower processing shrinkage and lower tendency to deform than the crystalline component, and they have better solvent resistance and environmental stress crack resistance than the amorphous component. The alloy itself again is the better combination for microcellular processing because of different phases in the same alloy material.

In thermoplastic elastomers, the predominant flexible component is the coherent matrix phase that determines the properties in the rubberlike state. Then, the major rubber phase may result in some nonspherical shape of cells and the shearing rate may be restricted at a certain level of speed of injection for microcellular processing. However, the gas-laden molten thermoplastic elastomer may help to lower the viscosity and make finer cells if all four steps of microcellular processing are controlled well.

There are many successful blends or alloys of plastics on the market. All of them are good for microcellular process since the heterogeneous nucleation contributes much better cell structure. Some of the most popular microcellular blends and alloys to be discussed in detail, along with some other important blends and alloys, are briefly mentioned in the following.

4.7.1 Noryl® for Microcellular Processing

Noryl® is the trade name of the material developed by GE Plastics (now Sabic Innovative Plastics). Basically, the material is the polyphenylene oxide (PPO)-based resin. GE introduced Noryl® that was defined as "modified PPOs" in 1966. Its key feature is that they are polymer blends of PS, or HIPS, with PPO. If polystyrene ($T_g \cong 90\,°C$) is blended with poly(2,6-dimethyl phenylene oxide) in equal quantities, a transparent polymer with a single $T_g \cong 150\,°C$ is obtained. Both PPO and PS have similar secondary transitions at approximately 116 °C, and the blends also show this transition. The viscosity of the blends is, however, lower than pure PPO, reflecting the influence of the lower-viscosity PS. Thus, these Noryl® resins have improved processability along with a reduced material cost.

The application of Noryl® resin is widely used for lighter weight, safer, and more durable products. Easily processed on conventional tooling, Noryl® injection moldable resins offer the fabricator a wide processing window with low shrinkage, inherent stability under processing extremes, and excellent flow. It features a range of heat deflection temperature from 82 °C to 193 °C (180–380 °F), coupled with high-impact strength, excellent mechanical and electric properties, and broad UL recognition. Although Noryl® resin has the lowest moisture absorption of any engineering plastic, it is good practice, particularly where surface appearance is critical, to dry Noryl® resin before molding. However, this resin should not be dried longer than 8 hours because excessive drying may result in a loss of physical properties.

The processing screw for Noryl® resin needs a conventional screw with a compression ratio between 2.3 and 2.5. It was found that the screw with compression ratio 2.5 might have a wide processing window compared to the screw with compression ratio 2.3. However, the compression should be accomplished over a gradual and constant taper since sharp transitions may result in excessive shear and material degradation. In addition, the depth of metering is deeper than that for a regular screw to reduce the shearing in the metering zone of the screw. Because of the gradual compression in the screw the ball valve is not recommended. A slide ring non-return valve with at least 80% of crosssection of flow area of the metering zone in the screw, is recommended. At no point should the clearance between screw and barrel be more than 0.0762 mm. The short open-bore nozzle with a minimum orifice of 4.76 mm is recommended for Noryl® resin without foaming. However, for better nucleation, the nozzle orifice for gas-laden molten Noryl® resin can be reduced to 3 mm. Also, the length of the orifice should be controlled as short as possible; a typical one is 4 mm or smaller. However, the glass-fiber-reinforced Noryl® resin may need to open the nozzle orifice diameter up to 7.9 mm to avoid premature wearing if the nucleation quality during injection is acceptable.

The temperature control is also important for Noryl® resin processing. Uniform mold cooling is also critical for maximization of cycle reduction of microcellular the Noryl® part, and control of the microcellular part has even better tolerance. It is generally advisable to maintain a small range differential in steel temperature over the mold cavity or core. Tighter controls in the mold cooling for Noryl® will provide greater processing latitude.

Noryl® resin is relatively easy to vent and does not exhibit the large volume of material gas typically of some plastic materials when thermally abused. However, the gas-laden molten Noryl® resin may have extra gas released during mold filling, and fast injection may trap air in as well. It may be necessary to increase the venting depth to 0.076–0.127 mm. The width will be 9.5 mm and spaced every 50 mm around the cavity if necessary. For the large part of Noryl® resin, there is a special venting channel designed named continuous vent in the mold. It will help to minimize the formation of residual flame retardant in the material and absorb the initial injection shock because of the fast injection required for microcellular injection molding [23]. This continuous venting channel is parallel to the flow direction of mold filling or along the parting line.

Noryl® resin molding is recommended to avoid the weld line at the point of high stress concentration. However, microcellular molding does not have this high stress level since the join weld line relies on the cell growth. On the other hand, it is the weak spot of the microcellular part so that it still needs to be carefully designed to avoid a weak weld line on the high stress spot for application.

There are two popular Noryl® resins that are widely used for microcellular parts. One is Noryl® MH230-GY2043 (UL2335, Pallet Material), which is a typical unfilled Noryl® material. Another one is Noryl® HM4025-701 (40% Glass/Mineral filled, High Modulus Grade, FR), which is a typical filled Noryl® material. Some selected results are discussed in a case study below with the comparisons for the property change between microcellular part and solid part.

4.7.1.1 Processing. The processing conditions for unfilled Noryl® with conventional injection molding are: barrel temperature range 204–266 °C, mold temperature about 38 °C, back pressure of melt 3.5 MPa. Microcellular processing conditions are different from conventional processing for a higher back pressure of melt 12 MPa, and the shot size is short. A higher barrel temperature range 260–288 °C is used for filled Noryl® material, and the rest of them are the same conditions as filled Noryl® material processing above.

The highest achievable weight reduction for the unfilled Noryl® material for acceptable quality of microcellular part is about 10%. Higher weight reduction caused blistering due to the phase separation in the SCF and molten polymer solution. When using the microcellular process with the highest weight reduction, the maximum injection and cavity pressures were reduced by approximately 15% and 30%, respectively.

The highest achievable weight reduction for the filled Noryl® material for acceptable quality of a microcellular part is 6–7%. When using the microcellular process with the highest weight reduction, the maximum injection and cavity pressures were reduced by approximately 5% and 17%, respectively. Compared to the results of unfilled Noryl® material, processing a filled material saves a lower percentage of tonnage and weight reduction since there is high percentage of glass fibers in the Noryl® material.

The microcellular process allows reducing the processing temperature significantly. The lowest melt temperature that could be achieved without reaching the maximum torque of the injection molding machine for unfilled Noryl® material was 249 °C as compared to 266 °C when using the higher barrel temperature settings.

The microcellular process allows reducing the processing temperature in glass-fiber-reinforced Noryl® material as well. However, it is not as significantly as unfilled material. The lowest melt temperature that could be achieved without reaching the maximum torque of the injection molding machine was 276 °C as compared to 288 °C. It is because the filled material needs more torque to turn the screw during recovery.

4.7.1.2 Shrinkage. The shrinkage values for the unfilled Noryl® MH230 resin is not clear since the cell structure may not be good to make good evaluation of shrinkage. However, the shrinkage values decreased (20% average reduction) when using the microcellular process for filled Noryl® material. This indicates that the microcellular process can improve the shrinkage properties of the Noryl® material when good cell structure is made since this fiber-filled material made much better cell structure than the cell structure of unfilled similar material.

4.7.1.3 Flammability Property. All molded plaques made by Noryl® resins, whether filled or unfilled, passed the criteria of the 5-V device flammability test. This indicates that the microcellular process does not affect the flammability properties at all weight reduction levels studied in this work. These conclusions are the same as for the previous case study for PBT material.

4.7.1.4 Thermal Property. The microcellular process had a minimal effect on the thermal conductivity values of the unfilled Noryl® resin. However, the thermal conductivity values for the filled Noryl® HM4025 resin was dropped significantly when using the microcellular foaming process. The thermal conductivity values of the samples molded at different barrel temperatures had no statistical difference for both filled and unfilled Noryl® resins.

4.7.1.5 Rheological Property. The rheological properties of the unfilled Noryl® resin were only changed with the microcellular process. It indicates that the gas-laden material will reduce the viscosity of unfilled Noryl® resin. The filled Noryl® resin is 40% glass-reinforced; a significant reduction in the material viscosity as compared to that of unfilled Noryl® resin may be from the comprehensive effects due to both gas-laden melt and glass fiber damage during processing.

4.7.1.6 Dynamic Mechanical Analysis. The elastic modulus of the unfilled Noryl® MH230 resin was affected (15% drop) when using microcellular processing as indicated by the DMA analysis. The microcellular process did not affect the glass transition temperature (solid, T_g = 139 °C; microcellular process—13% weight reduction, T_g = 140 °C). On the other hand, just like unfilled Noryl® resin, the glass transition temperature of the filled Noryl® material remained unchanged after using the microcellular process (solid, T_g = 148 °C; microcellular process—8% weight reduction, T_g = 150 °C).

4.7.1.7 Flexural Property. The flexural modulus of the unfilled Noryl® MH230 resin was not significantly affected by the microcellular process. A more significant reduction (16% drop) in the flexural properties was experienced with low melt samples. Similarly, the flexural modulus of the filled Noryl® HM4025 resin was reduced by 8% when achieving 8% weight reduction using the microcellular process.

4.7.1.8 Tensile Property. A drop in Young's modulus of 27% was experienced with the unfilled Noryl® MH230 foamed parts (13% weight reduction).

Reducing the melt temperature did not improve the retention of the tensile properties. With the thinner samples, this decrease in the modulus was not as severe as in the case of thick plaques (only 9% drop). However, the filled material has better results for retaining the tensile property with certain weight reduction. Young's modulus of the filled Noryl® HM4025 resin dropped by only 8% when reducing the part weight by 8% and about 6% for the 3.7-mm- and 2.5-mm-thick samples, respectively.

4.7.1.9 Impact Property. For the unfilled Noryl® HM4025 resin, a drop of 15% was experienced when reducing the part weight by only 6% using the microcellular process for the thick samples. However, for the heavily glass-fiber-reinforced material the fiber orientation may play an important role for final microcellular part properties. The microcellular processing actually helps to improve the fiber orientation, and the anisotropy problem usually improved significantly if the mold filling and cell structure was controlled for disorientation [24, 53].

4.7.1.10 Blister on the Surface. The blister problem is not encountered during the trial using both filled and unfilled Noryl® resins. This indicates that Noryl® resin is a good material for microcellular processing. It is another reason why Noryl® resin has been successfully used for microcellular injection molding for many years.

4.7.2 Acrylonitrile–Butadiene–Styrene (ABS) for Microcellular Processing

Acrylonitrile–butadiene–styrene (ABS) material is a mechanical blend of nitrile rubber and SAN copolymers. Improvement in weatherability and chemical resistance were achieved because of the acrylonitrile. The two most important types of ABS resins are blends of SAN copolymers with butadiene–acrylonitrile rubber (type I) and terpolymers of polybutadiene, styrene, and acrylonitrile (type II). The type I resin of ABS consists of 64% of a 70:30 SAN polymer with 36% of a 65:35 butadiene–acrylonitrile elastomer. There are many grades of ABS available on the market for injection molding. A typical type II resin contains a mixture of polybutadiene, polybutadiene grafted with acrylonitrile and styrene, and SAN copolymer. The graft polymer is essential for the development of optimum properties of ABS type II resin because blends of polybutadiene and SAN lack strength and toughness. Generally, the highest-impact material goes with the lowest tensile strength and modulus [1]. The typical properties of the final ABS polymer are:

- the amount of the elastomeric phase
- the molecular weight of the resin phase
- the styrene–acrylonitrile ratio in the resin phase

As amorphous material, ABS resins have low mold shrinkage and a wide melting range. ABS resins are high-viscosity material. It is characterized from ductile yielding down to low temperatures and up to high abuse strain rates. The dispersed rubber particles also provide good hot strength. The size of rubber particles is an important factor for microcellular processing.

A special thermoplastic, methacrylate–butadiene–styrene terpolymer (MABS), has been developed and combined good clarity with physical and chemical properties close to ABS. MABS polymers are two-phase materials, with the components only partially compatible. Processing characteristics, impact, and hardness are about the same for MABS and ABS.

Overall, ABS is an excellent material for microcellular parts since it is a blend of three different components. Each grade of ABS is tailored to provide a given property balance. It is great help with heterogeneous nucleation so that high density of cells becomes possible. Similar to the HIPS, the rubber phase in the ABS will be the factor to determine the final morphology of the ABS microcellular part. The SEM pictures in Figure 6.19 show the typical morphologies of the real part with great number of cells that have uniform cell sizes. The processing conditions are 127 rpm of screw (30 mm diameter, 26:1 L/D) speed, 0.8 weight percent of N_2 gas, 13.8 MPa of back pressure of screw recovery, and 249 °C of melt temperature [57].

CO_2 gas has been successfully used as a blowing agent in ABS microcellular processing as well. The 5 weight percent of CO_2 gas in ABS resin can make excellent microstructure for the ABS microcellular part.

4.7.3 PC/ABS Alloy

The typical alloy of ABS material used widely for microcellular processing is PC/ABS. Usually, PC is difficult to process and is an expensive resin. The PC/ABS alloy has excellent processing characteristics at relative low cost. The PC/ABS alloy has a higher heat resistance and impact strength compare to conventional ABS resins. Its resistance to flex fatigue is comparable to a straight PC. Gloss and surface appearance for PC/ABS is also excellent. PC/ABS alloy is actually good material to be used for microcellular processing because of help from heterogeneous nucleation. Figure 3.7a shows an excellent cell structure with average size of cell 10 µm that is distributed uniformly in the part.

With the same blend of material of PC/ABS in Figure 3.7a and the same other processing conditions, an air shot sample is made by injecting the gas-rich material into air, instead of into mold. The morphology of air shot in Figure 3.7b shows very fine 3-µm cells distributed uniformly in the whole sample. The real microcellular part can be made much better than the molding part if a constant pressure drop rate is kept during the whole injection period.

A brief summary of the PC/ABS alloy, Cycoloy® C6600-BK1005 (PC/ABS, FR), is introduced in the following.

4.7.3.1 Processing. The processing conditions for conventional injection molding are: barrel temperature range 221–254 °C and the mold temperature about 38 °C, back pressure of melt about 6 MPa. Microcellular processing conditions are different from conventional processing for a higher back pressure of melt 17 MPa, and the shot size is short. The solid part has a surface delaminating problem at low barrel and mold temperatures. The highest achievable weight reduction for this material for acceptable quality of microcellular parts is 20%. A reduction of 16% in the maximum injection pressure was experienced when using the microcellular process with 20% weight reduction as compared to conventional injection molding. The clamping tonnage was reduced from 200 tons to 100 tons when using the microcellular process. The maximum cavity pressure was reduced by 78% when reducing the weight by 20% using the microcellular process as compared to that for conventional injection molding [58].

The lowest melt temperature that could be achieved without reaching the maximum torque of the injection molding machine was 216 °C as compared to 251 °C when using conventional injection molding.

4.7.3.2 Shrinkage Property. The shrinkage values for the Cycoloy® C6600 resin are not changed significantly when using the microcellular process. Also, no statistical difference was observed between the samples molded at different barrel temperatures.

4.7.3.3 Flammability Property. For the 3.7-mm thickness, all Cycoloy® C6600 samples passed the criteria of the 5-V device flammability test. This indicates that the microcellular process does not affect flammability at the weight reduction levels in 20% or less. This conclusion is the same as the previous case studies for PBT and Noryl® resins.

4.7.3.4 Thermal Property. The thermal conductivity of the Cycoloy® C6600 resin dropped as the weight reduction increased. A drop as high as 34% was experienced at the 20% weight reduction level, indicating that the foaming process increases the thermal insulation properties of the material. This decrease in the thermal conductivity values was also observed with the samples molded using the low barrel temperature settings.

4.7.3.5 Rheological Property. For the Cycoloy® C6600 resin, when comparing the rheological properties of the virgin resin with the solid plaques, a 9% average drop in the viscosity was experienced. On the other hand, the viscosity dropped by 29% when using the microcellular process for the highest weight reduction. This indicates that this process significantly affected the rheological properties.

4.7.3.6 Dynamic Mechanical Analysis. The DMA analysis for the Cycoloy® C6600 resin indicates that the elastic modulus decreased (25% average decrease) when using the microcellular process. This result correlates with the data generated from the other mechanical tests. The glass transition temperatures of the solid and foamed parts were not changed much (T_g = 119 °C).

4.7.3.7 Flexural Properties. The flexural modulus of the Cycoloy® C6600 resin was reduced by 12% when foaming the plaques with a 20% weight reduction. A 16% drop in the flexural modulus was experienced with the parts molded using the low barrel temperature setting.

4.7.3.8 Tensile Properties. For the Cycoloy® C6600 resin, a drop in Young's modulus of 37% was experienced with the foamed parts (20% weight reduction). On the other hand, this drop in the modulus was reduced to 12% when molding the parts at lower barrel temperature settings. When using a lower melt temperature, the material degradation is less severe, which often results in a better retention of the material properties. The retention of the tensile properties (13% drop) was better with the thinner samples.

4.7.3.9 Impact Property. The impact strength test for this material is not clear. Usually, if the cell structure is controlled in the range of microcell sizes, the impact property should not drop too much. More experiments need to be carried out for the conclusions.

4.7.4 PC-PBT Alloy

PC/PBT is another successful material used for microcellular process. The case studies for an unfilled material Xenoy® 5220-BK1066 are briefly introduced below because most of the trends are similar to the previous case studies.

4.7.4.1 Processing. The processing results of a microcellular part are compared with the results of a conventional injection molding part for the same material. The processing conditions for conventional injection molding are: barrel temperature range 243–254 °C, mold temperature about 38 °C, and back pressure of melt 3.4 MPa. Microcellular processing conditions are different from conventional processing for a higher back pressure of melt 16 MPa, and the shot size is short depending on the weight reduction percentage targeted.

The solid part has surface delaminating problem at low barrel and mold temperatures. The highest achievable weight reduction for this material for an acceptable quality of microcellular part is 20%. The maximum injection pressure was reduced by approximately 8% when using the microcellular process with the highest weight reduction. The maximum cavity pressure was reduced by 43% when using the microcellular process.

4.7.4.2 Flammability Property. All samples passed the criteria of the 5-V device flammability test. This again indicates that the microcellular process does not affect flammability at the weight reduction levels evaluated in this investigation.

Furthermore, the microcellular foam process was generally shown to be able to eliminate part warpage while maintaining the flammability properties based several case studies above.

4.7.4.3 Thermal Property. The thermal conductivity values of Xenoy® 5220U decreased when using the microcellular foam technology. It is also a common characteristic of microcellular foam with excellent insulation property.

4.7.4.4 Dynamic Mechanical Analysis. As experienced with the tensile modulus, the elastic modulus of the Xenoy® 5220U resin decreased (14% drop) when using the microcellular process. However, the glass transition temperature of the material remained almost unchanged after using the microcellular process (solid, T_g = 135 °C; and MuCell® process—20% weight reduction, T_g = 138 °C).

4.7.4.5 Flexural Property. Similarly to the other mechanical properties, the flexural properties of the Xenoy® 5220U resin decreased when using the microcellular process. An 18% drop in the modulus was observed when reducing the weight by 20% using the microcellular foam process.

4.7.4.6 Tensile Property. The tensile plot for the Xenoy® 5220U resin indicates that Young's modulus decreased by 24% when using the microcellular process for the 20% weight reduction level. This property change is actually excellent to maintain the original strength with almost the same percentage of weight reduction.

4.7.5 Cross-Linked EVA for Microcellular Processing

Ethylene vinyl acetate (EVA) polymers are the thermoplastic materials widely used in medial, electrical, packaging, and so on. They are produced from copolymerization ethylene and vinyl acetate monomers. Therefore, it may belong to an alloy, or a blend from different materials. As the bulky acetoxy group content increases, the polymers become more amorphous and possess increased flexibility, rubberiness, low-temperature properties, tackiness, and heat sealability [59]. The vinyl acetate group can facilitate peroxide-catalyzed cross linking of the polymer chains and can couple with the inherent rubbery nature of the polymer that makes EVA suitable for a tough and abrasion-resistant foam.

Before injection, a palletized compound was fully formulated with a cross-linking agent, a blowing agent, and other ingredients. The barrel temperature is about 85 °C at the feed zone and is 90–95 °C at the tip of screw. Then, the mold temperature is up to 165–180 °C, at which the cross-linking and blowing agents reactions occur almost simultaneously under pressure, usually 300 atm. After molding cycle is complete the molds are opened at a rapid speed, and the nucleated bubbles grow instantaneously. The foam integrity is maintained by the high melt strength as a result of cross-linking reaction [59].

A closed microcellular structure is produced with the nice cell size from 5 to 30 μm, and the overall density of foam is 200 kg/m^3. It is a developing technology for the injection-molded EVA since the EVA part is usually made by compression molding. The applications of microcellular EVA are mainly for soles, sandals, tires, baby carriages, golf carts, and floating devices. It has more potential to be used for automotive parts, toys, and sports protective foams, and so on.

4.7.6 Kraton G-7722 for Microcellular Processing

The compound material is a growing application. The Kraton G-7722 is a typical TPE compound used for medical parts. A microcellular injection molding has been tried for the issue of high cost, poor autoclavability, and striction. The foamed morphology of Kraton G-7722 compound is shown in Figure 3.8. The cell size is not uniform, and cell shape is not spherical. However, the overall cell structure in Figure 3.8 is still in the range of microcellular processing. This sample is made with about 20% weight reduction for the cost savings. The processing conditions are: melt temperature 215 °C, mold temperature 38 °C, N_2 gas weight percentage ~0.8, and injection speed 0.076 m/sec (3 in./sec) with 40 mm diameter of plunger. Kelvin also reported the result of Santoprene TPV material in the microcellular process [28]. It has 23% weight reduction. The shore A hardness is reduced. In addition, the sealing capability and compression set have been improved by microcellular molding. Microcellular processing also removes the oil additives and no fogging issues.

4.7.7 Immiscible Blends for Microcellular Processing

The foaming of immiscible blends utilizes the blend properties as morphology, viscosity ratio, or constituent properties to make new cellular materials with a set of desired properties. A typical systematic research was published in reference 60. The methodology and results are summarized as the following.

The viscosity ratio of an immiscible (PPE/PS)/SAN–tenary blend system was varied by different PS contents in the PPE while the microstructure is changed via the SAN content. The PPE is poly(2.6-dimethyl-1,4-phenylene ether) from Mitsubishi Engineering Plastics, and PS is polystyrene from BASF. PPE and PS are miscible but immiscible to poly(styrene-co-acrylonitrile) (SAN from BASF) because of the acrylonitrile content of the SAN (19 wt%). The SAN content was systematically decreased while the PPE/PS ratio was fixed at 75/25 and 50/50. It results in a viscosity ratio of 0.37 and 0.99, respectively. Then, the influence of the microstructure of an immiscible blend on the foaming process in these two different viscosity ratios are listed below:

- By decreasing the content of the dispersed SAN phase, an increasing cell density and a decreasing cell size in final blend morphology for both viscosity ratio materials was achieved.
- The viscosity ratio affects the cell size. An elevated matrix viscosity restricts the cell growth and resulted in different foam densities. For example, the PPE/PS ratio of 50/50 has the matrix viscosity similar to the SAN viscosity and results in a higher expansion ratio and an increase in the density reduction.
- Regarding the strain hardening behavior of the blend system, the decrease of the SAN content and the change of the PPE/PS ratio will have no significant influence.

The important general conclusion is that the immiscible blend will help to improve cell structure and possibly promote the mechanical properties of microcellular foam.

4.8 METAL POWDER

The metal powder injection molding uses metallic powder that is atomized with small size in the range of 0.5–25 μm. The best particle size for molding corresponds to the ranges from 2 to 8 μm. However, the small, uniformly sized powder particles in the feedstock in order to achieve small, evenly distributed micropores are the key for microcellular processing to be used in MIM. It is because the small uniform metal powder particle is the possible nucleation spot. Overall, the powder particles are the excellent help for the heterogeneous nucleation. Then, the proprietary binder is usually a blend consisting of a polyolefin, a plasticizer, and a poly(ethylene oxide). For example, a simple binder may consist of 64% paraffin, 20% polypropylene, 15% carnauba wax, and 1% stearic acid. The weight percent of metal powder is about as high as 90%. The feedstock of the mixture between metal powder and binder can be injection-molded into a complicated shape with precise dimensions. In addition, injection molding allows a high volume of the metal powder parts at low costs. The typical processing temperature for metal injection molding is 150 °C, and the mold temperature is about 40 °C. Then, the binder must be fully removed without damaging the injection-molded part at a temperature below any sintering cycle used. The issue is always to reduce the distortion during the debinding. The debinding is a slow process to avoid any stress to distort the molded part. It is usually to remove the wax first and then remove the rest of the binder. The neat binder components decompose at 240–260 °C, and they end degrading at about 458 °C.

Although the metal powder injection molding is not new, the microporous process in metal injection molding is the new idea [61, 62]. In this process, similar to the microcellular plastic processing, the gas at supercritical state, such as carbon dioxide gas or nitrogen gas, is injected into the molten feedstock, which then dissolves into the binder that is similar to plastic with flow ability. As the feedstock with dissolved gas travels through the nozzle, the gas bubbles nucleate and then grow to form pores. This pore growth halts when the melt freezes as it comes into contact with the cold mold surface. The molded components are subsequently debinded and sintered with gas-generated pores preserved. The benefits of gas-laden feedstock of metal powder are to reduce the viscosity to make molding easier. Also, the microcellular process does not leave residual stress in the part so that the distortion during the debinding is reduced significantly as well.

The result is a metal part with a dense surface skin and pores ranging from 5 to 200 μm in diameter. The inventor Dwivedi believes that this dense surface is achieved upon molding because rapid surface cooling, when the feedstock hits the mold walls and releases gas, minimizes pore formation on the surface. During sintering, the small naturally occurring pores between metal particles are eliminated, leaving behind a large number of gas-generated pores in the microstructure. This process saves significant material in feedstock consumption.

It is true that the sample shows that a dense surface skin is formed over the porous interior of a molded microporous metal part. The microporous metal part may not care much about the surface finish. This technology has the potential to be used for microporous metal, ceramic, and intermetallic components with fully densified surfaces.

In addition to gas injection, the formation of microporous metal is similar to the microcellular process in that both processes use the same type of molding machine. The barrel and screw, however, are upgraded to accommodate the materials used in MIM, which results in one patent [63]. Another requirement involves the use of small, uniformly sized powder particles in the feedstock in order to achieve small, evenly distributed pores. The chemistry of the binder is also important: It should have high gas solubility when liquid and should have low solubility in solid state. It is suggested by the inventor that the process can be used for applications such as jewelry, sporting goods, lightweight structures, and heat insulating components.

The screw is designed with similar structure in Chapter 7. The compression ratio is low for MIM. The screw has five zones: feed, transition, metering, wiping, and mixing [63]. There is a trend that the heavy metal powder tends to migrate to the lower shearing area, which is the bottom of the screw channel. It may not be good for mixing the gas with moisture of powder and bond materials in the screw. Same concerning for mold filling that the metal powder will move from surface to the center of the part [24]. However, the cell growth is the strongest in the center area during mold filling that may expel the metal powder away from the center.

The recent growing application of magnesium molding may bring another opportunity for developing microporous structure in magnesium injection molding. The magnesium alloys attract interest for their creep-resistant properties for under the hood application The key advantage they offer over aluminum is the light weight; that is, it is 36% lighter than aluminum by volume and, when alloyed, has the highest strength-to-weight ratio of any structural metal. If the micropore is added into this alloy without obvious strength drop, it may make this strength-to-weight ratio even higher. The current technology is called Thixomolding. The processor and equipment supplier must be licensed by Thixomat to work with this technology. If this technology combines with microcellular technology, it will be a big help in dealing with the challenges in thin-wall molding for magnesium by a low-viscosity gas-laden mixture of metal powder and bond material.

Another potential application is powder injection molding. The powder means not only steel but also other materials, such as ceramic, copper, iron, nickel aluminide, cemented carbide, alumina, molybdenum, and tungsten. The powder may have a variety of shapes, sizes, and size distributions. The plastic material used for it is usually a blend of polymers, waxes, and other additives. It is common to have plastic as a binder. It holds the particles together at low temperature. Then, this mixture will be molded at high temperature. Similar to metal powder injection molding, all of the powder injection moldings can be made with microporous structure.

4.9 BIOPOLYMERS

Biopolymers are made from polylactic acid (PLA), polyglycolic acid (PGA), PLA/PGA copolymers polycaprolactone (PCL), PHA, polyhydroxybutyrate–valerate (PHBV), and starch-based resins. They are attracting growing market interest as green materials with no ties to petrochemical-based thermoplastics [64]. The major target of microcellular foam made with biomaterial is material saving since biomaterial can cost more than US$1000/lb. Then, some toughness and insulation properties of microcellular are the performance benefits that will become more important in the future development in the medical industry.

The biopolymer usually needs some special processing conditions that will be gentler than the regular polymer. Most of biopolymer does not like the extra heat and shear. The limit of hydrolytic stability must be controlled so that it will not exceed the processing limits. The differences between regular polymer and biopolymer are subtle, so the product design, tool design, and processing equipment modification must be considered for the changes for biopolymers. The narrow processing window may bring more challenges for the processing. For example, PHBV resin has the melting point about 154 °C, but it degrades at 182 °C. Too much heat in the biopolymer from either the heating system or from shearing can result in gels, black specs, or yellowing in the biopolymer part. Basically, in biopolymer processing we need to consider the following [64, 65]:

- Tight melt temperature control is necessary, so processing temperature should be controlled to within ±2 degrees.
- Shearing rate in screw for both plasticizing (screw rotation speed and back pressure) and injection (shearing rate controlled by injection linear speed) is important. Screw speed is about 50–200 rpm depending the diameter of screw. Injection speed is usually in the range of 0.012–0.05 m/sec. The low shear screw, like rigid PVC screw, will work for it. Also, the PET screw runs PLA successfully.
- Residence time in barrel needs to be limited in the range of biopolymers. A general rule of thumb is for shot volume to be 30–80% of barrel volume.
- Proper drying of material is necessary since biopolymers tend to be hygroscopic and moisture-sensitive. Or they will suffer a drop in molecular weight and melt viscosity, as well as increase potential for flashing and brittle parts. PLA and PHA are polyesters, and drying requirements are in the range of those for PET and PBT—that is, more strict than for ABS, PA, or PC.
- Feed-throat temperature should be about 21 °C, while recommended processing temperature will be 188–210 °C. The back pressure should be used with 0.35- to 0.69-MPa range. Metering zone temperature should be 188–204 °C.
- Mold temperature is about 24 °C, and expects part shrinkage 0.004 mm/mm. Mold should be run with water at around 24 °C to avoid plateout. However, mold temperature may be 49–60 °C if crystallization needs to be promoted.
- Most biopolymers are semicrystalline materials, but they tend to be relatively slow to crystallize or set up in the mold, even though they have relatively low melting temperature. Nucleation technology helps to improve both cycle times and heat resistance.
- Biopolymer with high level of amorphous (uncrystallized) reprocessed material may tend to stick with metal surfaces in processing. Adding mold release might help reduce chance of sticking to metal surface of mold.
- A minimized runner system is recommended to save the cost of waste material, specifically important for very costly medical grades of biopolymer.
- PLA retains heat more, so longer cooling time is required.
- PLA material does not flow well in thin wall over long distances. Increased injection pressure only increases the shear, which can cause breakdown and can cause the material to become brittle. However, adding gases can help to reduce the viscosity and make molding easier than solid.
- The noncorrosive components are needed for mold to resist the acidic properties of PLA and its tendency to plate out acid residuals on the walls of molding system.

- Special nozzle tip needs to be designed with low shear channels and thermal profile to counteract PLA's temperature hypersensitivity.
- Today's biopolymers are designed to process more easily. The processing window is widened by increasing primarily the temperature gap between crystallization and decomposition.
- If the biopolymer is exposed to ambient air, it can absorb enough moisture in 5 minutes to defeat most of the benefits of drying. On the other hand, if the drying temperature is too high, the material may soften and agglomerate in the drying hopper.
- A biopolymer is dried with starting moisture content of 2400 ppm down to 250 ppm in 5 hours at around 70 °C. The material should be dried to less than 400 ppm moisture, and best moisture content should be less than 100 ppm.
- Starch-based material must be purged out from the barrel by LDPE at the end of production to prevent excessive degradation.
- Molding a semicrystalline biopolymer can be up to 50% slower than more commonly used semicrystalline resins.
- The minimum gate size is 1 mm with full round shape. It is recommended to have 80% of part thickness.
- Most parts need 3–5 tons/in.^{2} of clamp force.
- PHBV is a biopolyester produced via bacterial fermentation of plant starches. It is a member of the PHA family. PHBV material is approved for food contact in Europe and is approved by the Biadegradable Products Institute (BPI), New York City, for composting. PHBV should be dried to 250 ppm moisture. Melt temperature is 170–175 °C, feed throat temperature is less than 135 °C, compression zone temperature is 145 °C, metering zone temperature is 155 °C, and adapter temperature is 161 °C. The shrinkage of PHBV is similar to that of PLA.

The market for biopolymer is growing fast. Biopolymers began from the film and sheet extrusion markets. Now biopolymers expand the market to rigid packaging, disposable cutlery, medical, and consumer parts. Also, PHA material is focusing on the agricultural applications, since PHA is biodegradable in water and soil. PHA targets the rigid application as well to replace ABS (such as cell phones and PDAs) and office equipment (such as printers). PLA biopolymer can be used to replace GPPS, HIPS, and ABS, and even PET and PP in some consumer goods. It shrinks like a styrenic and can be used in the same type of molds. PLA also is used for the injection stretch-blow molded bottles for water and dairy drinks. Other things that are produced using PLA include dental products, toys, tools, tableware, packaging, mugs, spoons, discs, sandwich picks, flying discs, drink stirrers, and so on.

However, biodegradable polymers, both synthetic and natural, often show poor foamability of fine cell structure. It is because the biodegradable polymer has poor rheological properties, poor thermal stability, poor solubility and inadequate diffusivity of the ordinary blowing agents, and insufficient setting mechanisms. Macromolecular design can be a way for improving the foamability of materials. For example, inducing some branching on linear molecules of synthetic polymers can enlarge the molecular weight distribution. It will improve the fundamental characteristics of material—that is, the strain-induced hardening behavior to allow the polymer to withstand the stretching forces at the latter stage of the cell growth. Several improvement criteria have been used to improve the foamability of different biodegradable polymers, both synthetic and natural [65]. Marrazzo et al. [65] did a systematic research for these criteria with batch processing and extrusion processing for the PCL foam. The results can be referred for the injection molding process of biopolymer foaming, and they will be discussed below.

4.9.1 Effect of Processing Parameters on Foamability of PCL

Three parameters have been controlled well to be able to process the PCL, gas concentration, foaming temperature, and pressure drop rate. For the biopolymer, foaming temperature can be considered the most important since it influences the crystallization/vetrification of the polymer and cell coalescence. Pressure drop rate is also a well-known important parameter—correlated to the thermodynamic instability necessary to generate as many nuclei as possible—that is mainly affecting the foam morphology. In addition, gas concentration is directly correlated to the availability of gas necessary to inflate the bubbles, which influences the final density as one of the important parameters for cell growth. One example from the batch expansion of the poly(ε-caprolactone)/N_2 system shows that the high pressure drop rate and high saturation pressure create the small cell size while the foam density is almost the same since the temperatures are set up the same for both experiments. However, the high temperature can reduce the cell size and increase the cell density simultaneously. Overall the processing window for the biopolymer is narrow, and it needs to be controlled carefully to tailor both foam density and cellular structure to the desired properties/applications.

4.9.2 Effect of Blowing Agent on the Foamability of PCL.

It is the same principle that the blowing agent induces plasticization of the biopolymer and reduces the viscosity and characteristic temperature of biopolymer. However, different blowing agents determine different processing windows. This effect on the foaming process of both poly (ε-caprolactone)/N_2 and poly (ε-caprolactone)/CO_2 is investigated with both batch and extrusion processes [66]. Similar to other materials, CO_2 gas plasticizes PCL more extensively, and it results in lower viscosity and foaming temperature. CO_2 gas has higher solubility in PCL than does N_2 gas. However, no matter how much more CO_2 gas can be solved in PCL and make processing temperature lower, the result of morphology of poly(-caprolactone)/CO_2 (about 250 μm mean cell diameter at 0.03 g/cm^3) is always showing the coarser cell than that of poly(ε-caprolactone)/N_2 (about 20-μm mean cell diameter with 0.2 g/cm^3). However, using a mixed blowing agent system such as 20%/80% of CO_2/N_2 in PCL results a density of 0.07 g/cm^3 and 50 μm mean cell diameter. In effect, the use the mixture of CO_2/N_2 led to the optimal use of the two blowing agents, with a synergic effect [66]. The further studies need to find if the ratio of this mixture of blowing system will make better results, and specifically to make the processing window wider.

4.9.3 Effect of Molecular Architecture Modification on Foamability of PLA

The rheological characteristics of the melt are important for the last two steps of microcellular foaming. It is because cell growth needs the melt strong enough to be elongated to expand cell diameter, and the strain-induced hardening will stabilize the cellular structure during the cooling period. For the PLA, two basic methods are used for tuning the rheological properties of the melt: to optimize the molecular weight and molecular weight distribution of polymer and/or to induce branching of the macromolecules [67].

As one example of the basic methods to improve the rheological properties of PLA, two chain extenders are used to treat PLA in melt stage: 1,4-butanedial (BD) and 1,4-butane diisocyanate (BDI) [68]. BD is used as the first coupling agent and acid value reducer to link carboxyl groups of PLA. Then, BDI is added to let it react with hydroxyl end groups of PLA to achieve chain-extended PLA [65]. It results in higher viscoelastic moduli and improves the cell structure of PLA. The morphologic results of chain-extended PLA shows the microcellular structure with 6.7×10^8 cells/cm^3 compared to neat PLA foam with 7.7×10^5 cells/cm^3 at the same processing conditions, consequently giving rise to smaller cell sizes in higher cell density chain-extended PLA, too [65].

4.9.4 Effect of Fillers on Foamability of PLA

The inorganic particles in a polymeric matrix of PLA will affect the thermal and rheological properties of PLA. It is more obvious to affect the PLA properties when the filler size has one nanometric dimension [68]. The aspect of ratio of nanometric particle is very high; as a consequence, the interaction between macromolecules and inorganic particles are enhanced to a great extent [65]. It is well known that any filler (with no exception for this nanometric filler in PLA) in a polymer melt will act as heterogeneous bubble nucleation sites, increasing largely the number of nuclei.

The effect of a modified montmorillonite, Cloisite 30B, causes a great increase in the rheological properties. With 5 wt% of nanoclay filler in PLA, the cell size of is about 30 μm, while neat PLA shows the larger cell size of about 250 μm [65]. However, this behavior may ascribe both to increased cell nucleation, due to the presence of the heterogeneous bubble nucleation sites, and to the intrinsically high viscosity and elasticity, caused by organoclay exfoliation and chain extension/branching. Finally, the fast crystallization rate for PLA nanocomposite is also helpful for stabilizing the cell structure during the foaming process [65].

4.9.5 Foams from Vegetal Proteins

These are typical natural biopolymers, such as zein. Zein is a protein that can be extracted from maize, is not water-soluble, and has potential applications in edible, and also biodegradable, packaging. It can extrude, for example, a 75%–25% weight mixture of zein–PEG400, at temperature 80 °C (80 rpm) with a lab-scale extruder [65]. The mechanical properties of these vegetal proteins are strongly related to primary and secondary structure of the proteins. In particular, one protein contains fewer β-sheets to be more deformable and fairly brittle [65]. This is also reported in the literature [69]. The zein material with fewer β-sheets was successfully fine-foamed with batch process at 70 °C after saturation at 180 bar of a 25–75% volume mixture of CO_2 and N_2 for 6 hours. However, another zein material with more β-sheets gave collapsed foam, not being able to withstand elongation deformation during foaming [65]. Therefore, the foaming process of a vegetal protein is possible and will be influenced by the hierarchical structure of the natural polymers. This is a very interesting biopolymer that is able to make microcellular foam for the future packaging applications.

4.9.6 Blend of Biopolymer

There are several blends of biopolymers on the market. Cereplast supplies a line composed of PLA blends and a new "hybrid" line of starch reactively blended with PP. A new blend material was just introduced into the market by Cheil recently. It is called GL-1401 and has better fatigue resistance, higher Izod impact strength, and higher flow rates than PC/ABS blend [70]. Bayer MaterialScience (Levekusen, Germany) also experimented with a PC/PLA blend creating Makroblend BC250 and BC400, with 25% and 40% PLA content, respectively [71]. Although it is a new material to be explored for the possible microcellular injection molding, both PC and PLA are good materials for microcellular processing; thus the blend of them shall be better for microcellular processing in the future. In addition, more and more materials will go greener in the future, and more blends with PLA or other biopolymers will be developed in the future.

REFERENCES

1. Schwartz, S. S., and Goodman, S. H. *Plastics Materials and Processes*, Van Nostrand Reinhold, New York, 1982.

2. Suh, N. P. *Innovation in Polymer Processing*, edited by James F. Stevenson, Hanser/Gardner Publications, Cincinnati, 1996, Chapter 3, pp. 93–149.
3. Doroudiani, S., Park, C. B., Kortschot, M. T., and Cheung, L. K. *SPE ANTEC, Tech. Papers*, 2183–2188 (1995).
4. Doroudiani, S., Park, C. B., and Kortschot, M. T. *Polym. Eng. Sci.* **36**(21), 2645–2662 (1996).
5. Michaels, A. S., and Bixler, H. J. *J. Polym. Sci.* **50**, 393, 413 (1961).
6. Sha, H., and Harrison, I. R. *J. Polym. Sci., Part B: Polym. Phys. Ed.* **30**, 915 (1992).
7. Colton, J. S. *Metals Manuf. Proc.* **4**, 253 (1989).
8. Kumar, V., and Gebizlloglu, O. S. *SPE ANTEC Tech. Papers*, 1297 (1991).
9. Kumar, V., and Gebizlloglu, O. S. *SPE ANTEC Tech. Papers*, 1536 (1992).
10. Baldwin, D. F., Park, C. B., and Suh, N. P. *Polym. Eng. Sci.* **36**, 1437 (1996).
11. Baldwin, D. F., Park, C. B., and Suh, N. P. *Polym. Eng. Sci.* **36**, 1446 (1996).
12. Michaeli, W., Bussmann, M. *SPE ANTEC Tech. Papers*, 796–800 (2005).
13. Park, C. B., and Cheung, L. K. *Polym. Eng. Sci.* **37**, 1 (1997).
14. Trexel Inc. Web site, http://www.trexel.com/.
15. Guo, Q., Wang, J., and Park, C. B. *SPE ANTEC Tech. Papers*, 2736–2740 (2006).
16. Park, H. E., and Dealy, J. M. *SPE ANTEC Tech. Papers*, 2534–2538 (2008).
17. Guzmán, J. de *An. Soc. Espanola Fis. Quim.* **11**, 353 (1913).
18. Barus, C. *Am. J. Sci.*, **45**, 87 (1883).
19. Fujita, H., Kishimoto, A. *J. Polym. Sci.*, **28**, 547 (1958).
20. Martinez, K., Turng, L. S., Kramschuster, A., and Lee, J. *SPE ANTEC Tech. Papers*, 2163–2167 (2008).
21. Mueller, N., and Ehrenstein, G. W. *SPE ANTEC Tech. Papers*, 593–597 (2005).
22. Margolis J. M., et al. *Engineering Thermoplastics—Properties and Applications*, Marcel Dekker, New York, 1985.
23. Sabic Innovative Plastics. Web site, http://www.sabic-ip.com/.
24. Xu, J. *SPE ANTEC Tech. Papers*, 2158–2162 (2008).
25. Spindler, R. Injection molding with MuCell microcellular technology. Advanced Injection Molding Processes, CFI Group, Ann Arbor, MI, August 21–22, 2001.
26. Shimbo, M., Baldwin, D. F., and Suh, N. P. *ATEM*, 309–313 (1993).
27. Shimbo, M., Higashitani, I., and Miyano, Y. *J. Cell. Plastics*, **43**, 157–167 (2007).
28. Okamoto, T. K. *Microcellular Processing*, Hanser Gardner Publications, Cincinnati, 2003, pp. 97–116.
29. Material Thoughts. Strong parts, smooth appearance. *Mod. Plastics Worldwide* **October**, 86 (2006).
30. Hoffman, J. M. Polyamides get a foam diet. *Machine Design* **September**, 118 (2006).
31. Maani, A., Heuzey, M. C., Carreau, P. J., and Khennache, O. *SPE ANTEC Tech. Papers*, 1391–1395 (2009).
32. Gendron, R., and Vachon, C. *J. Cell. Plastics*, **39**, 71–85 (2003).
33. Xu, J., and Kishbaugh, L. *J. Cell. Plastics* **39**, 29–47 (2003).

34. Hwang, S. S., Chen, S. C., and Chung, M. H. *SPE ANTEC Tech. Papers*, 776–780 (2005).
35. Matuana, M. L., and Mengeloglu, F. *J. Vinyl Additive Technol.* **7**(2), 67–75 (2001).
36. Levy, S., and DuBois, J. H. *Plastics Product Design Engineering Handbook*. Van Nostrand Reinhold, New York, 1977.
37. Kaewmesri, W., Lee, P. C., Park, C. B., and Pumchusak, J. *J. Cell. Plastics* **42**, 405–428 (2006).
38. Lee, E. K., Lee, K., Jung, P. U., Naguib, H. E., and Park, C. B. *SPE ANTEC Tech. Papers*, 925–931 (2009).
39. Chen, L., Straff, R., and Wang, X. *SPE ANTEC Tech. Papers*, 1732–1736 (2001).
40. Seymour, S. S. *Modern Plastics Encyclopedia*, McGraw-Hill, New York, 1968–1969.
41. Clemons, C. M. *Forest Products J.* **52**(6), 10–18 (2002).
42. Winata, H., Turng, L. S., Caulfield, D. F., Kuster, T., Spindler, R., and Jacobson, R. *SPE ANTEC Tech. Papers*, 701–705 (2003).
43. Bledzki, A. K., and Faruk, O. *SPE ANTEC Tech. Papers*, 1316–1320 (2005).
44. Yoon, J. D., Kuboki, T., Jung, P. U., Wang, J., and Park, C. B. *Polymer Processing Society, 24th Annual Meeting, June* 15–19, 2008.
45. Kuboki, T., Lee, Y. H., Lee, J. W. S., Zhu, W., Park, C. B., and Sain, M. *SPE ANTEC Tech. Papers*, 1997–2001 (2008).
46. Yuan, M., Turng, L. S., Spindler, R., Caulfield, D., and Hunt, C. *SPE ANTEC Tech. Papers*, 691–695 (2003).
47. Jo, C., and Naguib, H. E. *J. Cell. Plastics* **43**, 111–121 (2007).
48. Jo, C., and Naguib, H. E. *SPE ANTEC Tech. Papers*, 1902–1906 (2008).
49. Turng, L. S., Microcellular injection molding. *SPE ANTEC Tech. Papers*, 686–690 (2003).
50. Kharbas, H., Nelson, P., Yuan, M., Gong, S., Turng, L. S., and Spindler, R. *Polym. Compos.* **24**(6), 655–671, (2003).
51. Pathak, T., and Jayaraman, K. *SPE ANTEC Tech. Papers*, 103 (2007).
52. Guo, M., Heuzey, M., and Carreau, P. J. *Polym. Eng. Sci.* **47**, 1070 (2007).
53. Strauss, W., Ranade, A., D'Souza, N. A., Reidy R. F., and Paceley, M. *SPE ANTEC Tech. Papers*, 1812–1816 (2003).
54. Rodriguez-Perez, M. A., Garcia de Acilu Laa, P., Arevalo-Guiterrez, J., Saiz-Arroyo, C., Solorzano, E., and de Saja J. A. *SPE ANTEC Tech. Papers*, 939–943 (2009).
55. Hsu, P., Yang, J. P., Hwang, S. S., and Lai, Y. Z. *SPE ANTEC Tech. Papers*, 1652–1656 (2009).
56. Hwang, S. S., Hu, C. H., Sung H. J., and Hsu, P. *SPE ANTEC Tech. Papers*, 1657–1661 (2009).
57. Xu, J. *SPE ANTEC Tech. Papers*, 2770–2774 (2006).
58. Kishbaugh, L. A., Levesque, K. J., Guillemette, A. H., Chen, L., Xu, J., and Okamoto, K. T., U.S. Patent 7,364,788 B2 (2008).
59. Lee, J. *SPE ANTEC Tech. Papers*, 2060–2064 (1997).

60. Gutmann, P., Bangarusampath, D. S., Altsaedt, V., Ruckdaeshel, H., Schmalz, H., and Mueller, A. H. E. *SPE ANTEC Tech. Papers*, 950–954 (2009).
61. Chitwood, A., Micropores in MIM provide material savings. *Injection Molding* **April**, 92 (2001).
62. Dwivedi, R. K. U. S. Patent No. 6,759,004 (2004).
63. Anderson, G., and Xu, J. U.S. Patent No. 7,172,333 (2007).
64. Knights, M. Injection molding biopolymers. *Plastics Technol.* **April**, 39–44, 48 (2009).
65. Marrazzo, C., Maio, E. D., and Iannace, S. *J. Cell. Plastics* **43**, 123–133 (2007).
66. Di Maio E., Mensitieri, G., Iannace, S., Nicolais, S. L., Li, W., and Flumerfelt, R. W. *Polym. Eng. Sci.* **45**(7), 432–441 (2005).
67. Xanthos, M., Young, M. W., Karayannidis, G. P., and Bikiaris, D. N. *Polym. Eng. Sci.* **41**(4), 643–655 (2001).
68. Di, Y., Iannace, S., Di Maio, E., and Nicolais, L. *Macromolecular Mater. Eng.* **290**(11), 1083–1090 (2005).
69. Ray, S. S., and Okamoto M. *Prog. Polym. Sci.* **28**(11), 1539–1641 (2003).
70. Gao, C., Taylor, J., Wellner, N., Byaruhanga, Y. B., Parker, P. L., Mills, E. N. C., and Belton, P. S. *J. Agric. Food Chem.* **53**(2), 306–312 (2005).
71. Deligio, T. *Mod. Plastics Worldwide* **May**, 86 (2009).

5

DESIGN OF MICROCELLULAR INJECTION MOLDING

If microcellular polymers can replace solid polymers with 10% or more material reductions without significantly compromising material properties required, it might be a revolutionized way to save materials and protect the environment [1]. This original idea was motivated by Eastman Kodak in an attempt to search for a way to reduce the cost and improve manufacturing efficiency. Then, Professor Nam P. Suh of Massachusetts Institute of Technology soon developed microcellular technology to approach the solution for Eastman Kodak's request. A new microcellular injection molding machine that can make a uniform cell distribution part with the small cell size in the range of 5–100 microns for a wide range of materials was developed in 1998 [2] by Trexel, Inc. The microcellular foam generally offers improved properties compared to the ones from conventional foams [3]. The questions are, How do the process, part designing, and mold designing affect physical and mechanical properties? and What is the predictable quantity relationship between cell structure and mechanical properties?

The design of microcellular injection molding includes part design, material selection, mold design, and cost analysis. It determines the molding efficiency, mold structure designing, final weight reduction of the part, and part geometry designing. Mold and part designs will focus on the special requirements for microcellular injection molding. Quality of the microcellular part usually includes dimension stability, overall property change, and surface quality of skin. This chapter also introduces several approaches to improve—or even to make perfect—smooth skin. The property changes will be considered with

Microcellular Injection Molding, Edited by Jingyi Xu
Copyright © 2010 John Wiley & Sons, Inc.

different materials, different densities of microcellular parts, and different skin–core ratios. A useful index of weight reduction is used in this chapter to estimate the property changes of the part of microcellular injection molding. The detailed part and mold design guidelines of microcellular injection molding are summarized in this chapter.

5.1 PART DESIGN

The design procedures for microcellular parts are similar to the solid parts. They are function, geometric sizing, material selection, cost consideration, design life, performance evaluation, and test. It is necessary to review all of the design procedures briefly before the discussion of special issues for microcellular part design. First of all, it needs a careful definition of the function. It will simplify the design and will permit the widest latitude of alternatives possible in the design without compromising the function of the part. Then, select materials will include the physical properties, tensile and compression strength, impact properties, temperature resistance, environmental resistances, stiffness, differential expansion, and the dynamic properties. However, to select the material for making a functional part with both economical and practical senses, all other factors such as the cost, geometric size and lifetime need to be reviewed before the decision is made. The next step is the performance evaluation of the part. The part must be completely designed in this step. The final step is to make sample tooling and part for the test. More details of all of these design procedures for solid parts can be found in references 4 and 5. Some designing differences of wall thickness and ribs between the solid part and the microcellular part are listed briefly in reference 6.

Although solid part design may have some fundamental differences from microcellular part design, most of the design procedures above and the design rules for plastic parts are still the same for both microcellular and solid parts. Only some special rules related to the gas-laden melt and cells in the part need to be considered for the new rules of microcellular part design [7–13].

5.1.1 Part Geometry Design

Any geometric shape for a solid plastic part is an acceptable design for a microcellular part because (a) the single-phase solution has low viscosity and (b) cell growth during mold filling can fill any difficult filling area more easily than can regular injection molding. However, several issues regarding the microcellular part design are considerable to make full use of the advantages of the microcellular process. The factors to be considered are as follows:

- Wall thickness
- Symmetry for even mold filling that is critical for microcellular processing

- Parting lines
- Draft to be increased for microcellular processing
- Radii and fillets not as critical for microcellular process as for solid process because of microcellular molding filling the details in the mold easily
- Ribs restricted for the height for ease of demolding
- Holes to be considered for forming welding lines
- Bosses to deal with cooling issue of microcellular processing
- Undercuts not to be the issues to fill the corner, but cooling is always the concern
- Threads to be checked with the strength of the microcellular part
- Use of molded-in inserts with special attention for the strength of foam
- Tolerances usually no major issue anymore for the microcellular part
- Surface finish to always be a disadvantage of microcellular parts unless special measures are taken
- Sink mark to be no issue anymore with microcellular processing
- Warpage to be an advantage using microcellular processing in most cases

The details of the microcellular part designing are discussed in the following.

5.1.1.1 Uniform Part Wall Thickness. Wall thickness is always an important consideration when designing plastic parts. In microcellular molding, wall thickness may have a lesser affect on dimensional stability and a greater affect on cycle time compared to conventional injection molding. On the other hand, it is a well-known rule of the solid part design to make uniform cooling and possible non-warpage part. It is a more important issue that the microcellular part is designed with uniform wall thickness throughout the part. There is an advantage that the microcellular part removes the sink mark no matter how much is the variation of the thickness in the part because of the cell expansion. However, there are several issues that must be kept in mind to design the wall thickness of the microcellular part.

- Uniform wall thickness will make the uniform cell size distribution throughout the part. The sudden thickness change may lead to some void or big cells locally in the part. These void and big cells will significantly decrease the strength of the part and unnecessarily lengthen cooling time. The dimension stability may be the problem from the locally unbalanced shrinkage, and warpage may occur in the part.
- If the local thick section cannot be avoided from the structure design of the part, the gradually blended wall thickness from thin section to thick section is necessary to create nice cell structure in the thick section, along with less geometric dimension distortion. In any event, the wall thickness

should not vary by more than a ratio of 3:1, and the maximum thickness in the thick section should not be larger than 4 mm.

- The local thick spot may cause the post blow if the cooling is not long enough. It is common that the gate is located in the thick section of the solid part. If the same mold with this gate location is in the thick section of the microcellular part, a blowout will occur after the part is ejected from mold. From the mold filling point of view, the gate area is the position where the last material is injected into mold where both highest melt temperature and highest melt pressure exist in this gate area. Then, it will cause the longest cooling time, and most likely there will be a blowout area from the residual pressure in the last cooled cells.

On the other hand, the microcellular part is capable of having thin-wall thickness as thin as 0.3 mm with possible 4–8% weight reduction [6]. Wang et al. [13] test a 1-mm-thickness HDPE microcellular part with the cold sprue and fan gate; with 5% of CO_2, the weight reduction with fine cell structure achieves up to 20%. The cell size is in the range of 20–100 μm. The thin-wall microcellular part is actually defined with the thickness as 1 mm or less. This thin-wall part is a unique feature of the microcellular part because it creates very small cell size about 5 μm or less with valve gate. This is the major reason why the thin-wall microcellular part is still able to create some weight reduction. Furthermore, this small uniform cell size keeps the strength better than the big cell size in thick microcellular part.

5.1.1.2 Weight Reduction Related to Wall Thickness. The weight reduction percentage is also related to the part thickness if the flow ratio and molding conditions are the same. It is because the core thickness in microcellular foam will be increased with the overall part thickness. However, the skin thickness does not change much with the change of part overall thickness if the processing conditions are the same. Therefore, the overall weight reduction will be increased with a thicker part.

It is well known that the strength of a microcellular part will be decreased with an increase in the weight reduction percentage. Therefore, in this case, the microcellular part may need to be redesigned with extra thickness to compensate for the strength loss from weight reduction.

Overall thickness of a microcellular part is not recommended to be over 4 mm. It is because the thicker microcellular part may cause much longer cooling time. Consequently, it may lose a short cycle-time benefit and make the microcellular process more expensive. The critical thickness to have the insulation effect of the microcellular part is determined by the processing conditions. If thickness of 4 mm is necessary, then only high injection volume rate and short flow ratio (less than 100) of the mold filling may be possible to maintain the cooling time shorter than solid molding. However, if the flow ratio is very short, such as 50:1 or less, the uniform thickness of a thick micro-

TABLE 5.1 Suggested Wall Thickness for a Thermoplastic Microcellular Part

Materials	Minimum (mm)	Average (mm)	Comments
Acetal	0.38	1.58	
ABS	0.76	2.29	
Acrylic	0.64	2.36	
Cellulosics	0.64	1.9	
FEP fluoroplastic	0.25	0.89	
Nylon	0.38	1.58	
Polycarbonate	1.06	2.36	
LDPE	0.51	1.58	
HDPE	0.89	1.58	
Ethylene vinyl acetate	0.51	1.58	
PP	0.64	2.03	
Polysulfone	1.06	2.54	
Noryl (modified PPO)	0.76	2.03	
GPPS	0.76	1.58	
SAN	0.76	1.58	
PVC-rigid	1.06	2.36	
Surlyn (ionomer)	0.64	1.58	

cellular part is still possible. Another fact is that the thick part can get great help from microcellular injection molding if warpage is the major issue for some thick part without foaming. In this case the microcellular part is still the better choice to improve the warpage of the part. Microcellular processing definitely removes any sink mark from the thick part as well. Then, the injection speed may be required to be increased as much as possible, and mold temperature may be required to be lower than normal mold temperature. A further study needs to find out the relationship between thickness and processing condition for effective cooling simulation. As the general suggestion of wall thickness for a microcellular injection molding part, the recommended wall minimum thickness and average thickness are listed in Table 5.1.

5.1.1.3 Rib Design. Rib is the most common structure in the part design to increase strength and rigidity without increasing the wall thickness of the part throughout and the weight of the part. It is still the best way to increase the stiffness of a microcellular part to provide reinforcing ribs or gussets. The risk of wrong position of rib resulting in a surface appearance defect, the so-called sink mark, is not the issue anymore for a microcellular part. However, the wrong position of rib causing the warpage may still exist for microcellular part because of less strength of microcellular ribs compared to solid ribs. Overall, microcellular part needs to focus on the short ribs with larger draft angle even for the larger part. The "T" section formed by rib or gusset provides stiffness equivalent to a thick rectangular section, with less material and therefore less

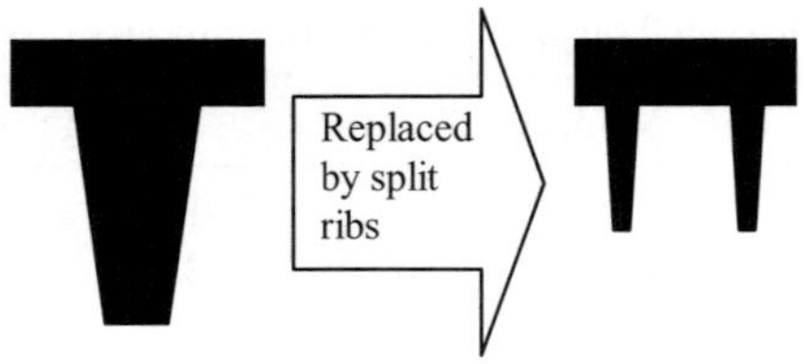

Figure 5.1 Split short ribs to replace deeper and thicker rib.

weight. Theoretically, as Okamoto indicated in his book [6], the rib designs for microcellular is no longer limited by the inability to pack out sink marks underneath the ribs. It is because the cell expansion is usually large enough to take care the shrinkage in the joining point of the rib and major part. In addition, the microcellular molding can fill the deeper and thin rib easily because of the low viscosity of gas-laden melt and cell growth in the mold to fill the last corner. The last corner mold filling easily from cell growth is usually taken care of by high packing pressure.

However, more real problems also occur because of (a) of lack enough venting in the rib area during mold filling (to be discussed in mold design below) and (b) ejection of the part with deeper ribs from the mold. Therefore, several new issues related to ribs in the microcellular part need to be addressed below.

- Rib thickness shall be the same as that of the solid—that is, about 40–80% of the thickness of the adjacent wall to which it is attached. Some material, such as acetal material, prefers the even lower percentage of maximum 50% only of the adjacent wall thickness.
- To increase the stiffness of the rib or the gusset, it is usually by increasing the height of the rib rather than increasing the thickness of rib. When the structural requirements demand rib thickness greater than the thickness of the adjacent wall, it is better to add additional ribs with decreased height of ribs that is named split ribs as shown in Figure 5.1. The spacing between two ribs should not be less than two times the thickness of the wall to which the ribs are attached.
- The cell growth in the rib may cause the ejecting difficulty of ribs. The solution is to increase the draft angle about twice of the draft angle for solid part.
- The foamed core in the rib will decrease the tensile strength that may cause the breakage of the ribs during ejection. Therefore, the radius at the base of the ribs is necessary not only to solve the problem of rib breakage during ejection but also to improve the part performance as the rib reinforcement.
- On the other hand, the rib minimum thickness can be much less than the solid as a result of the flow ability of gas-laden melt. However, it is also determined by stiffness requirement.

5.1.1.4 *Bosses Design.* Bosses are protruding pads, or raised sections, for enough mounting surfaces of the molded part and reinforcing the areas around holes or slots on the parts. They are used to distribute bearing loads and to transfer the loads to the main structure of the part. When the attachment loads are sufficiently high, the bosses may be designed with extra thickness. However, the microcellular part has the limit of thickness no more than 4 mm. Therefore, the solution will be to add the ribs to connect the bosses to the adjacent wall. For the same reason above, we may need to remove any solid boss from the microcellular part. With a solid boss the localized thickening of the wall section at the base of solid bosses, where it joins the wall, is too thick to microcellular part because specifically cooling problem will occur for solid bosses.

On the other hand, similar to the ribs, the height of boss needs to be designed as short as possible. The solid part has the rule of thumb that the height of bosses should not be more than twice the diameter. However, the microcellular part may require this height of bosses are no more than the diameter of the pad or boss. Otherwise, the draft angle needs to be increased, and the bosses will have a similar problem of ejecting and cooling in a short period.

In addition, long bosses can create molding problems unless extra vents are provided accordingly in the position of last mold filling, which is usually the bottom of the bosses around a hole in the cavity.

5.1.1.5 *Taper or Draft.* Taper or draft is necessary for a solid part in order to assure the easy removal of the part from the mold. Generally, it is required on the sides of the part and on cored holes. However, the microcellular part needs almost double the draft angle as does the solid part since the microcellular part expands more than shrinks during mold cooling. Sometimes this expansion is so strong that the microcellular part may stick on both male (core) and female (cavity) sections of a mold. The recommended draft angle for microcellular is a minimum of 2° to as high as 4° if the design of the microcellular part will permit it. In other words, the taper 1° to 2° per side will provide trouble-free operation of microcellular molding. If the short rib is designed as discussed in the solution above, then 2° to 3° of drafting angle, or 1° to 1.5° of taper on each side, will generally be satisfactory for microcellular molding.

5.1.1.6 *Radii and Fillets.* Radii and fillets in the microcellular part have the same functions as in the solid part. Radii and fillets promote the even flow of material at the corner during mold filling. It means that low pressure is required to fill the corner. Also the solid skin of the microcellular part does need radii and fillets to relieve the stress concentration associated with (a) change in wall thickness and (b) change in direction of internal loads. Radii and fillet also add the general strength of both the part and mold cavity, and they have beneficial effects on the aesthetics and "feel" of the molded part.

Wherever possible, the minimum radius of 0.5 mm should be designed at both external and internal corners of the microcellular part.

Radii and fillets are still an important consideration in the design of the crystalline material part. It is because the crystalline material is more notch-sensitive, such as acetal materials. The rule of thumb is that the fillet (inside of corner) radii should be in the range of 25–75% of the adjacent wall thickness, and matching outside corner radii should be sized to maintain uniform wall thickness at the bend.

5.1.1.7 Tolerances. The rule of thumb to specify the tolerance of the microcellular part is to use the tightest tolerance of the solid since the microcellular part can be designed with much tighter tolerance than the corresponding solid part without increasing the cost of manufacturing. In addition, when the tolerance is related to the flash, the microcellular part must have tighter tolerance anyway since the low viscosity of gas-laden melt can easily cause the flash unless the mold design has been prepared with tight tolerance on the parting line.

There is a special topic of tolerance of the microcellular part for the materials absorbing moisture. For example, the PA 6/6 can change as much as 8% in volume going from "bone dry" to a 70–80% relative humidity. With microcellular parts, this PA 6/6 part may have more than 8% of volume increase at the same relative humidity.

At least now most of the tolerances of microcellular parts are designed with the solid standard tolerances because of lack of the test data for microcellular parts. On the other hand, there are few issues of tolerance of microcellular parts. Therefore, the usage of solid part tolerance will definitely work for microcellular parts.

5.1.1.8 Holes. Through holes in the regular injection molding are definitely not a problem for microcellular injection molding either. The blind holes that are usually supported at one end only may, if the pin is relatively long, be distorted due to the deflection of the pin during regular injection mold filling at high injection pressure. For microcellular injection molding the injection pressure is low so that usually it is not the issue for blind holes, unless it is extremely long. Generally, the rules of some hole design for regular injection molding are still good for microcellular parts:

- The depth of the blind pin should be limited to twice its diameter.
- The distance between successive holes, or between a hole and a side wall, in a microcellular part should be a minimum of one hole diameter.
- If the hole is threaded, this minimum distance should be increased to about three times of the diameter of this threaded hole to avoid the problems due to stress concentration.

5.1.1.9 Undercuts. Undercuts can be molded equally as well in either a microcellular part or a solid part. However, shallow, simple undercuts on round thin-walled moldings that can sometimes be snapped or stripped out of the mold cavities during molding operation are the best choices. The microcellular part increases the toughness of the material so that any rule of undercut for a normal part will be good for a microcellular part as well. On the other hand, one rule of microcellular processing that must be kept in mind is that the foamed part not only shrinks but also expands during mold cooling so that it is possible that the microcellular part not to sticks with pin or core in a mold but also stays with cavity in a mold. Then, it becomes a special requirement for microcellular processing to either (a) determine where the part should stay or (b) design the ejector system accordingly.

A very special measure taken with the microcellular part for ejection is that some undercuts are made on the surface of the major core if the regular ejection system requires that the part stays with the core during mold opening. Otherwise, the microcellular part may easily stay with the cavity of mold that will cause the problem of demolding. It is because in most cases there is no ejection system in the cavity of the injection mold. If the undercuts are not allowed in the core surface, then the draft angle of the microcellular part must be increased until the part stays with the core during mold opening.

5.1.1.10 Threads. The microcellular processing can easily fill the detailed teeth, or sharp corners, that can be difficult to make with regular injection molding. The stress concentration in the nonfoamed threads is not the issue for the microcellular threads because of stress-free microcellular processing. On the other hand, some regular plastic thread designing rules are still applicable for microcellular thread as well but with different focus. For example, the roots and crests of all threads should be round with a 0.127-mm to 0.25-mm radius and are still required for microcellular threads. However, the purpose for the radius in the microcellular part is for the better strength of the thread teeth tip not, for mold filling difficulty in the sharp tip. The small diameter of threads may not be recommended for the microcellular part either. Usually, the thread size of 6 mm or below is better to select the machine threads after molding. Also regular injection molding will have some restriction of making very fine threads that are finer than class 2 and 28 pitch. However, the teeth strength of microcellular parts will have 20–30% drops depending on the cell structure and skin thickness. Microcellular threads will have similar rules for these fine threads to be used in the microcellular part because of the strength changes on the microcellular thread teeth.

5.1.1.11 Use of Molded-in Inserts. There are lots of successes of using molded-in inserts in engineering plastics. It will be similar to the threads design for microcellular above many rules of solid part will be useful for microcellular as well as long as the special attention paid for weak microcellular material and the expanding of the microcellular part during mold cooling. Some useful rules are summarized as follows:

- The insert should be free of sharp corners usually for a solid part. It is also required for a microcellular part for the strength issue.
- The end of the insert should have radius, preferably with a full spherical radius.
- Female threaded inserts should have blind internal holes rather than through holes.
- Wall thickness around the insert needs to be equal to the outside diameter of the insert, and in no case should it be less than one-half the outside diameter of the insert.
- Heating the insert to 200–210 °F before placement in the mold is helpful but not necessary for microcellular part anymore since it is stress-free processing.

5.1.1.12 Symmetry. For solid part design the symmetrical design is preferred because asymmetrical parts may create problems with dimensional control during or after molding. For microcellular part design it is more critical since the presence of balanced cavities is the key of uniform cells through uniform mold filling. The uneven filling paths will cause overfilling locally in the short filling path, while it results in underfilling locally in the long path to fill the mold. The overpacked cell structure will have too high residual stress and may cause post blow. On the other hand, it causes the problems of part ejection and construction as well. Overall, asymmetrical microcellular parts are the costly problems of part design, and it should avoid this kind design if it is possible.

5.1.1.13 Parting Lines. It is always true for the part design to have a simple parting line, which means single straight parting line. Primary considerations should be ease of part removal from the cavity, along with ease and feasibility in tool construction. Step parting lines also make a complicated mold structure and increase the tooling manufacturing and mold maintenance costs. It is also important to select the parting line location on the part to prevent any oversight on the functionally of the surface on which the parting line will appear.

5.1.2 Part Property

Similar to the structural foam, the microcellular part usually decreases the measured physical properties. However, like the discussion in Chapter 3, the microcellular part has unique cell architecture that will increase some of the physical properties, such as flexural modulus, elongation at break, and toughness. It is found that the glass-fiber-reinforced material may get more benefits from microcellular processing since it disorientates the fibers around the cells in the part. Overall, the uniform small cells in microcellular parts improve physical properties due to low residual stresses. The physical property drop of the microcellular part is more linear with weight reduction compared to physical property drop of structural foam with square relationship to the weight reduction. The details of the different property changes related to the weight reduction and models for prediction of these changes are presented in the following.

5.1.2.1 Models for Property Calculation. Many efforts have been made to predict mechanical properties based on the cell morphology with mathematical models [3, 14–20]. A common model based on the integral skin foams (ISF) comprises a low-density cellular core surrounded by a high-density skin of the same material [14, 15, 17]. It has been used with the models for the structure–property relation of homogeneous foam. The real structural foam part has an ISF structure, and the typical density distribution profile across the thickness section of the foamed part shows a bell shape. Between the solid skin and big cells in the center of the part, a transition zone is defined with small cells near the skin. This transition zone is a very complicated structure, and it is difficult to predict because it will vary by processing conditions, materials, and so on. There are lots of models of simplifying the theoretical model by neglecting the transition zone or even the core material [14, 16, 17].

Although the microcellular part is virtually stress-free without sink mark, bowing, and warpage, the mechanical properties of the microcellular part are generally less than the properties of the solid part except the toughness. The final properties of the microcellular part are not only on the base polymer but, more importantly, on overall structure of the part, which includes overall part density, density distribution, skin thickness, cell shape, and size. However, the microcellular part shall have the uniform cell structure in the foamed core and will clearly have a boundary between foamed core and solid skin (see Figure 3.3, Chapter 3) [4]. Usually, amorphous material has a clearer boundary between foamed core and solid skin than does crystalline material. The PC/ABS sample also displays a microcellular part with clear skin–core architecture. This cell structure is much different from the well-known bell-shape density profile of structural foam [14], and it is then possible to simplify the microcellular part property estimation by simple skin–core–skin model without transition zone [9]. With this simplified model, any traditional foam strength formula can be used for the mechanical property calculation for a microcellular injection molding part. It will be a combination result of mechanical properties of skin and core, respectively. There will be no more mechanical property calculation of transition zone for the microcellular part. If the cell size distribution is not uniform in some thick microcellular part, the average cell size can be used for using this model in Figure 5.2, and the result is still acceptable since the cell size difference in a microcellular part usually is not as big as the cell size difference in structural foam.

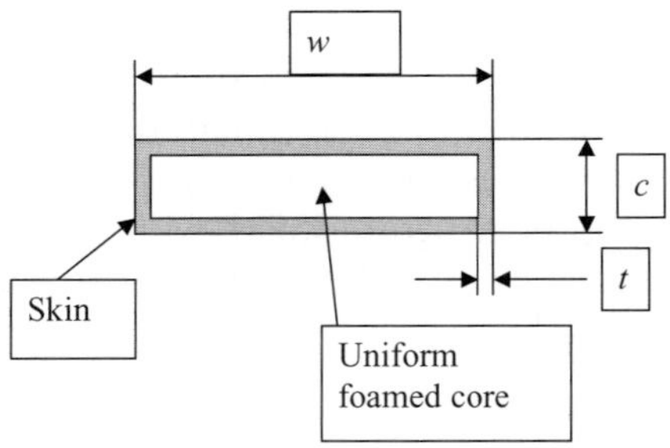

Figure 5.2 Sandwich model of a microcellular part [19]. (Reproduced with copyright permission of Society of Plastics Engineers.)

Bledzki and others proposed the similar ideal sandwich structure model as in Figure 5.2 [20], which is thick skin layers or pseudo skin layers and then foamed core. However, the vertical side of skin is neglected in the sandwich model. For the load calculation this sandwich model was simplified as Model U (density course of the cross section looks like a U). In addition, a model V was proposed by Bledzki that is distinguished by a small skin layer with larger cell sizes in the center of the core. The model U is usually made with amorphous material with cold mold temperature. Bledzki pointed out that the foamed part with morphology model U must use low gas concentration without gas counterpressure, and the weight reduction varies in the range of 5–10%. On the other hand, the part to be made with morphology model V needs to run high gas counterpressure and high gas concentration at high mold temperature, so the samples from production of the microcellular process may not be necessary because the high mold temperature increases cycle time.

On the other hand, as with most morphologies of microcellular parts illustrated in Chapter 3, the cell characteristics are simplified for microcellular parts because of uniform cell size, shape, and distribution in a good microcellular part. The cell shall have spherical shape with average cell size across the thickness direction. If the uniformity of the cell architecture is reached, the cells model can be further simplified to "square-in-square" as a two-dimensional model.

5.1.2.2 Mechanical Properties of Microcellular Part. It is well known that the cell structure may determine the final properties of a foamed part. It may not be an important factor for flexural strength, but it is a critical factor for tensile strength and impact strength. If cell sizes are not uniform, they will not distort uniformly when they are under load. In reality, certain weak cells that include thin-wall cells and big cells will distort to a great extent before stronger cells with thick walls and small spherical cell shape show any signs of distortion. This is the reason why the big cells and voids in the microcellular part will cause premature failure during application. As the traditional mechanical calculation method, to estimate the mechanical properties of structural foam

TABLE 5.2 Retaining Strength Percentage of Foam Material to the Solid Material, PBT + 30% Glass Fibers

Property	4% Weight Reduction	10% Weight Reduction	15% Weight Reduction
Flexural modulus	101	95	89
Flexural strength at yield	91	85	79
Young's modulus	93	84	74
Tensile strength at break	91	84	75
Elongation at break	100	106	106
Izod Impact	100	94	89

from cell structure is very difficult because the cell architecture of structural foam is classified as irregular [16]. Microcellular foam has much uniform cell structure across the part, and it may provide the possibility to estimate the mechanical properties with assumption of uniform cell structure. However, it is important to inspect the cell structure before drawing any conclusions about any physical property test result for a microcellular part because the microstructure of core material can make big differences in the same material with the same percentage weight reduction. On the other hand, the skin structure will be an effective factor specifically for tensile strength and flexural strength of a microcellular part. However, the main influence factor is truly the overall microcellular foam density if there are no big voids in the part. The higher the density, the better the mechanical strengths, including Young's modulus, tensile strength, strength at yield, flexural modulus, and flexural strength [20]. On the other hand, impact strength is a little bit different, and the conclusions are different from papers under different conditions.

All the results shown in Table 5.2 are discussed with detailed SEM pictures in Chapter 3 (see Figures 3.12–3.14). The morphology of each sample can explain well the reasons for the physical property changes in Table 5.2 because of their cell structure differences. Although all three samples are good microcellular parts, the cell structure in 15% weight reduction sample is the best with more small cells uniformly distributed throughout the part.

The mechanical properties of PP with 20 wt% of talc in Figure 5.3 show the similar trend as the PBT data in Figures 3.12–3.14. The flexural strength is more related to skin thickness than is tensile strength. Therefore, the foam quality or weight reduction will change the flexural strength but not as significantly as the change for the Izod impact strength and tensile strength. An is interesting result of elongation at yield shows that the 15% weight reduction sample has a higher elongation than does the 10% weight reduction sample. This is because the 15% weight reduction sample has more toughness than the 10% weight reduction sample as a result of high cell density.

The comparisons of the mechanical property changes between PP with 20 weight percent talc and unfilled PP are illustrated in Figure 5.4. Overall, the

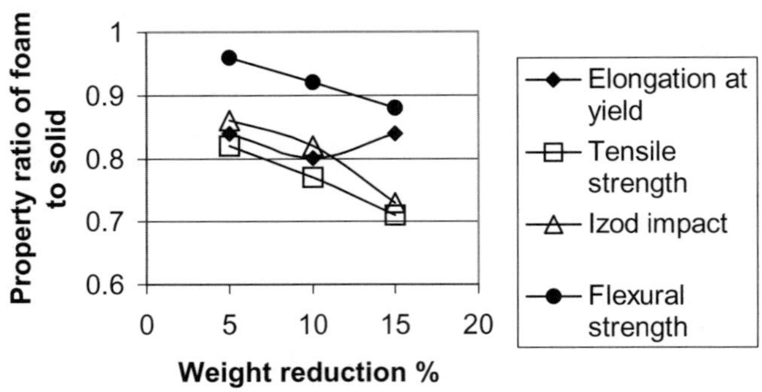

Figure 5.3 Mechanical property changes with the weight reduction percentage for PP with 20 weight percent of talc.

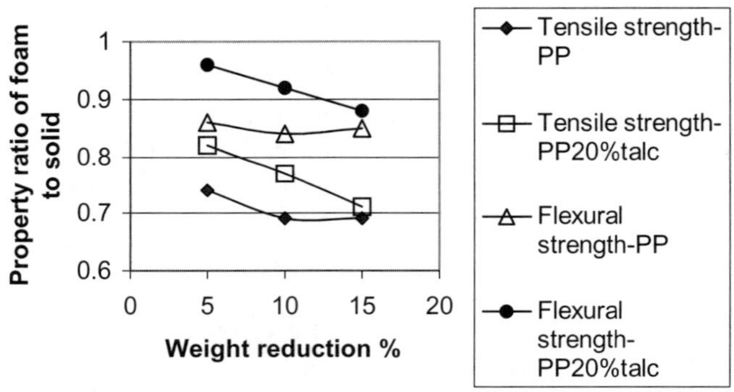

Figure 5.4 Comparison of mechanical property changes with the weight reduction percentage for PP + 20% talc and unfilled PP.

filled PP shows less mechanical property drops with the increasing weight reduction than the unfilled PP. However, the difference between two materials is obvious at the low weight reduction percentage. The unfilled PP property loss is going to be leveled at high weight reduction or at high foaming percentage. On the other hand, the filled PP does have less property loss at low weight reduction. This result of filled PP may be from the better nucleation and cell structure compared to unfilled PP at low weight reduction. At high weight percent reduction the foam quality of unfilled PP and filled PP are more likely close so that the property retention percentage for both materials are close as well. Therefore, it is recommended to use filled PP in the less weight reduction case to get the minimum property drop with better cell structure.

In Figure 5.5, the tensile strength is the lowest retention percentage with weight reduction for PA material Zytel® (73G33 HSLBKB031) with 33 weight percent of glass fiber. The Izod impact strength and flexural strength are in the same range of property drop except that the Izod impact was in the same percentage between 5% and 10% weight reduction.

The toughness of PC has an obvious improvement over microcellular foam if it breaks brittle with the solid part [20]. It is proved by the notched impact strength of unfilled PC actually from both morphology models U and V [20]. In addition, the weight reduction percentage must be about 8–18% to improve the toughness of unfilled PC [20].

The properties of the glass-fiber-reinforced parts are strongly related to fiber orientation and cell structure. Overall the glass-fiber-reinforced materials have better elongation, flexural strength, and Izod impact than unfilled material. However, the tensile strength loss is significant for both glass-fiber-reinforced material and unfilled same material. More studies need to be carried out with a careful review of the fiber orientation, cell architecture, and physical tests, with careful sampling for both the position and mold flow direction of samples.

A special ratio R_{sw} is defined as the ratio of strength loss percentage to the weight reduction percentage in Chapter 3. A detailed discussion of R_{sw} related to the cell structure can be referred to Table 3.1. From a part design point of view, the analysis of a relationship between cell structure and the R_{sw} values is given as the following.

It is obvious that the 10% and 15% weight reduction samples shown in Figures 3.12 and 3.13 have almost the same value of R_{sw} because the cell sizes of both 10% and 15% weight reduction samples are the same. Although the cell size of 45 μm in the 4% weight reduction sample is in the range of microcellular foam made in current industry equipment, the cell size of 15 μm in 10% and 15% weight reduction samples shows a significantly lower R_{sw} compared to the cell size of the 45-μm sample. The conclusion is that the small cell size is the key factor for the low R_{sw} value that is critical for the strength drop rate at certain weight reduction. This is a recommended design rule of microcellular part to have this ratio R_{sw} as low as possible through finest cell structure, such as 15 μm of cell size. The lowest ratio R_{sw} of the microcellular part will have the smallest strength drop for maximum weight reduction that is truly material saving without significantly losing the unfoamed material strength.

Knowing the ratio of R_{sw}, the strength drop may be estimated with the weight reduction that can be known instantly after the injection molding part is made. For example, the ratio in Table 3.1 is about 1.6 of tensile strength drop for the 10% weight reduction sample. It means that the tensile strength drop percentage for this PBT material is 1.6 times as high as the weight reduction percentage for the same morphology of 10% weight reduction sample. Therefore, if the same morphology as the 10% weight reduction sample in Figure 3.13 is used but 20% weight reduction is made, then the tensile strength of the 20% weight reduction sample will be predicted with the reduction about 32% of the solid tensile strength ($1.6 \times 20 = 32$). Similarly, a sample with the same cell structure as the 4% weight reduction sample will have R_{sw} about 2.25 for both tensile and flexural strength. It means that the strength loss percentage is 2.25 times as high as the weight reduction percentage with the cell structure in the sample shown in Figure 3.14. Therefore, if the sample of 20% weight reduction is made with the same cell structure as the 4% weight reduction sample in Figure 3.14, then the strength drop prediction of this sample is about 45% from solid strength ($2.25 \times 20 = 45$).

There are still some obvious differences shown in the morphologies of 10% and 15% weight reduction samples. Since the cell size is almost the same for both 10% and 15% weight reduction samples, the comparison is focused on the cell density difference. It seems like the larger cell density is helpful for less flexural strength loss. The highest cell density of 15% weight reduction sample shows the lowest flexural strength ratio. On the other hand, the weight reduction percentage may dominate the tensile strength drop percentage. Therefore, the 15% high weight reduction sample has a slightly higher R_{sw} value than the one for the 10% weight reduction sample. However, fine cell size may play an important role to keep this ratio R_{sw} low that is proved by both 10% and 15% weight reduction samples.

The results of more specific mechanical properties varied with the different cell structures and weight reduction percentages for PBT +30% glass fibers samples are listed in Table 5.2 and are analyzed as the following.

- Only 4% weight reduction keeps the same original flexural modulus as the solid material. It is because the cell structure in 4% weight reduction sample has only about 30% of foamed thickness in the center of part, as shown in Figure 3.14 in Chapter 3. Solid skin may have 30% of total thickness on each side encapsulated in this foamed center layer. It is easy to understand that the flexural modulus is largely determined by the skin thickness as the bending feature of the testing for measuring of flexural test.
- The test results of Young's modulus and tensile strength are normal as the general trend of other test results without checking the cell structure. However, the changing percentage of the strength drop rate versus the increase of weight reduction percentage is in the range of the linear to the square-law relationships of density-dependent tensile strength for foams.
- The elongation at break is great evidence that the microcellular foam makes tougher material than the solid. The elongation at break keeps the same for 4% weight reduction sample (with 45 μm of cell sizes) as the solid sample. However, even better elongation at break performance for both 10% and 15% weight reduction samples (both with 15-micron cell sizes) gives a conclusion that small cells truly promote the toughness of material.
- The data of Izod impact strength in Table 5.2 also show great improvement of toughness that is reflected by less impact strength drop with the increase of weight reduction of the foam as shown in Table 5.2. The 4% weight reduction sample again keeps the same impact strength as solid.

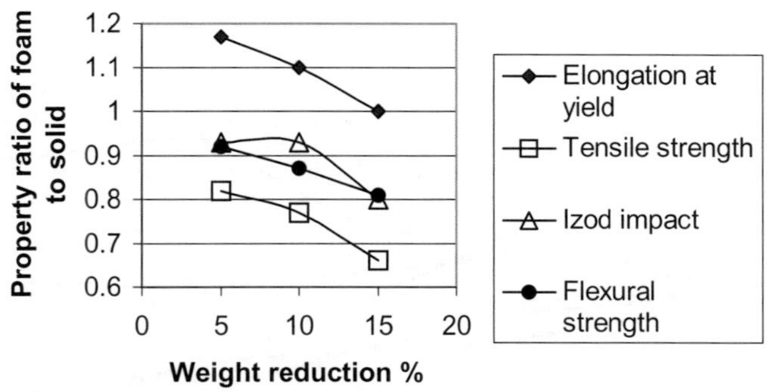

Figure 5.5 Mechanical property changes with the weight reduction percentage for Zytel™ 73G33 HSLBKB031(PA) with 33 weight percent of glass fiber.

In addition, both 10% and 15% weight reduction samples have the relationship of impact strength reduction versus weight reduction even less than the linear relationship. The only reasonable explanation for this improvement is the results of morphologies shown in Figures 3.12–3.14. However, the impact strength test with or without notches is the most complicated test that is very difficult to predict because it will include many factors including the fiber orientation in the final part [21].

The morphology is the only visible picture to explain the property changing data of the part in Figures 5.5–5.8. It is the recommendation that for any material test data to be published, it must be verified by morphology inspection that is displayed in Chapter 3. Otherwise, there may be more and more different mechanical property data published with large variation for the same materials with same weight reduction. The major reason of the variation is from the cell architecture differences among the samples made by microcellular injection molding.

As in the results calculated above, the relationship between weight reduction percentage and strength drop percentage is not a completely linear relationship. This relationship is unique for each mechanical property. (See Appendixes C, D, and E.)

Overall, the relationship between flexural modulus and weight reduction percentage of the microcellular part for all data in Table 5.2 seems like a linear relationship. This conclusion is similar to the result for structural foam [14]. It verifies that the flexural modulus primary depends on the skin and not on the foamed core. Furthermore, the linear relationship between flexural modulus and weight reduction can be written as the formula of the ratio of foam to solid versus weight reduction percentage:

$$R_{fs} = a_{fs}x + b_{fs} \tag{5.1}$$

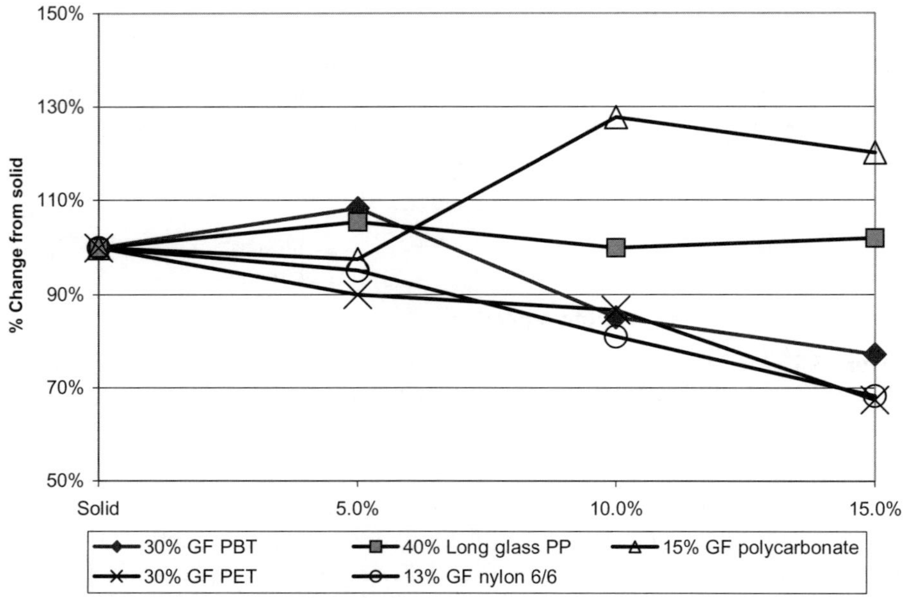

Figure 5.6 Mechanical property changes with the weight reduction percentage for glass-fiber-reinforced materials [22].

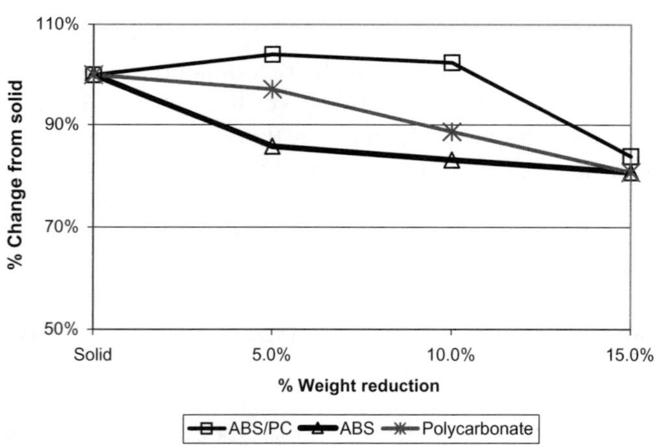

Figure 5.7 Mechanical property changes with the weight reduction percentage for unfilled materials [22].

where R_{fs} is the strength ratio of foam to solid, a_{fs} is the coefficient in linear equation of flexural modulus (see Table 5.3), and b_{fs} is the constant in linear equation of flexural modulus (see Table 5.3); x the weight reduction percentage and is given by

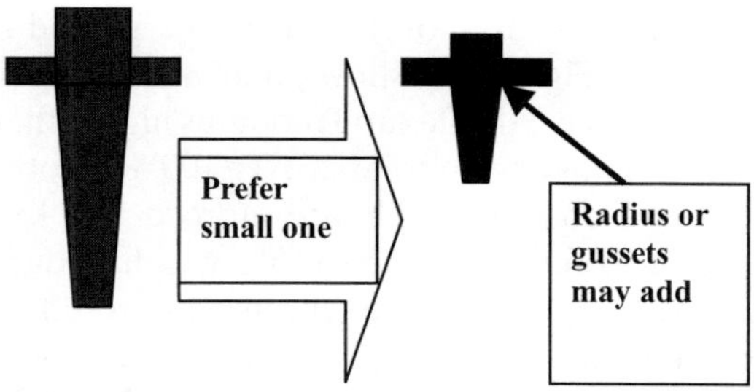

Figure 5.8 Modification for the cold sprue with short and small size.

TABLE 5.3 Coefficient and Constant for the Linear Equation of Flexural Modulus Calculation for Different Materials

Materials	a_{fs}	b_{fs}	Comments
PBT + 30% GF	–0.0109	0.9552	
PA + 33% GF	–0.011	0.9767	
Noryl + 20% GF	–0.014	0.9467	
Noryl	–0.014	1.01	
PP + 20% talc	–0.008	1	
PP	–0.001	0.86	

$$x = 1 - \frac{V_g}{100} \tag{5.2}$$

where V_g is the percent volume fraction of the cells.

Theoretically, the additional term of the modulus of gas within the cells is also the part of the function of foam flexural modulus. However, the gas modulus is extremely small in relation to the solid modulus of plastic, so it is safely neglected in the flexural modulus calculation [14].

One also needs to keep in mind that the flexural modulus measured above is for a short period of time. The creep occurs when a load is applied on the microcellular part over a long time span. This is common property for all thermoplastics and foams as well. They need to be measured separately. Generally, the following measures can be taken to increase the flexural modulus:

- Increase overall density.
- Increase skin thickness on the outside of bending to prevent the tensile rupture but decrease the skin thickness on the inside of bending for less wrinkling form compression [18].
- Improve cell structure that is not as important as skin but is good for flexural modulus.
- Add ribs at right positions.

The glass-fiber-reinforced material overall shows very good flexural modulus with microcellular foam. Figure 5.6 shows that at least three microcellular materials with glass fibers have the flexural modulus higher than the solid material (the percentage of change from solid is over 100%). Both 40% long glass-fiber-reinforced PP and 15% glass-fiber-reinforced PC keep the superior modulus property of the solid even at up to 15% weight reduction, as shown in Figure 5.6. This can be explained by the fact that the results of the fiber glass orientation inside of the plastics are improved by the microcellular foaming.

The similar result of the microcellular part test of flexural strength of PBT with 30% glass fiber from Spindler [22] in Figure 5.6 is close to the data shown in Table 5.2. However, one sample with 5% weight reduction is different from the result in Table 5.2 since Spindler found that the flexural modulus at 5% weight reduction is even higher than that of the solid.

However, Spindler also reported that the modulus of flexural strength for unfilled material ABS/PC with microcellular structure has a higher value than that of the same solid material, as shown in Figure 5.7. This may be related to the skin thickness since PC has thick skin formability. In addition, ABS/PC blend helps for high nucleation so that the cell architecture of the sample is very good.

The property of microcellular foam in tension gives the poorest performance in all figures of the properties of tests above. The tensile strength is usually determined by the cell structure. In other words, if there are some voids in the part, the tensile damage will begin from the weakest voids existed in the part and will result in very low tensile strength during the test. The measurements for increasing tensile strength are listed below:

- Increase overall density.
- Increase skin thickness.
- Improve cell structure.

Again, impact strength of a microcellular part is the most difficult property to be estimated correctly. It was suggested that for structural foam the Izod impact test results should be used to compare polymer grade only, not different polymers [14]. It was found that the impact strength may be reduced with an increase in melt index. However, there are several factors to increase the impact strength of microcellular part:

- Increase overall part density.
- Increase skin thickness.
- Use high-grade with higher-impact-strength polymer.
- Improve the cell structure.

"Improve the cell structure" has the significant influence on the impact strength. Michaeli and others also found that the absorbed energy during an impact test can be increased up to three times when the foam morphology is finer [23]. The finer cell structure can be made with breathing mold technology (see Chapter 8).

A glass-fiber (GF)-reinforced microcellular part has a considerable improvement effect on creep resistance, tensile strength, and flexural modulus. In addition, carbon fiber (CF) has been used for reinforced microcellular parts as well. CF provides much better properties than does GF. There are several reasons for CF material's superiority over GF material:

- The elastic modulus of CF is about three times higher than that of GF.
- The tensile strength of CF is not as strong as that of GF. But the real molding part with CF has much stronger tensile strength compared to GF part. It is because the fraction of load transmitted to the CF is much higher than the one to the GF. In other words, the tensile load on the CF part is taken by carbon fiber more than by resin matrix itself.
- There are much big different CTE values for CF compared to GF. It is also an important reason for the differential shrinkage performance of CF compared to GF.
- CF has less elongation (or stain) at the same load compared to GF.

Fiber orientation in the injection molding part is a disadvantage of glass-fiber-reinforced material. Microcellular injection molding provides an opportunity to improve the anisotropy caused by fiber orientation. Figure 3.16 shows the morphology of fiber distribution in a microcellular part. It is obvious that the microcellular results in some fiber disorientation in the center foamed core. The microcellular processing improves the fiber orientation, and it makes more uniform properties distribution in the microcellular glass-fiber-reinforced part. This can be considered as the advantage of a microcellular part with fiber reinforcement.

The rule of thumb of overall mechanical properties of a microcellular part can be summarized as the following:

- A linear relationship of flexural strength to the weight reduction (or density reduction) can be used safely for microcellular part design.
- A square-law relationship is still good for tensile strength calculation of a microcellular part, although cell structure can improve it significantly with fine cell structure.
- Impact strength design is difficult to specify without checking the cell structure and fiber orientation (if filled).
- The elongation at break is the toughness index. It is generally increased with microcellular foam. However, it may depend on the original material toughness.

5.1.2.3 Shrinkage of Microcellular Part. The shrinkage of solid material, particularly of semicrystalline materials, is the major source of dimension issues of the nonfoam part. However, microcellular foam almost eliminates the shrinkage problem even for crystalline material because of the cell growth during mold cooling. Some laboratory experiments reported the details of shrinkage tests for microcellular and made a comparison to the shrinkage of the same solid materials [8, 22]. For the filled materials, the shrinkage values decreased (16% average reduction) when using the microcellular process. It is true especially for glass fiber reinforcement in the material that enhances nucleation and the formation of evenly distributed microcells. The uniform microcell structures can help reduce part shrinkage significantly. Kelvin presents the similar result about the shrinkage values for four different grades of PP. The average shrinkage values are 0.0173 mm/mm for unfilled copolymer PP, 0.013 mm/mm for 20% talc-filled PP, and 0.0012 mm/mm for both 30% and 40% glass-fiber-filled PP, respectively [6]. The shrinkage values for filled and unfilled PP are so much different because the cell structure may not be good in the unfilled PP. It will be much better actually for unfilled PP if the cell structure is improved to a microcellular structure. Therefore, the shrinkage value for unfilled PP needs to be verified with morphology, and it cannot be used in real application without checking the morphology to be sure it is a uniform microcellular architecture.

For some single-phase impact modified amorphous material, such as PPO/HIPS, the shrinkage value of microcellular part does not change with the weight reduction [6]. However, for PC-based materials, such as PC/ABS and IMPC, the shrinkage rate increases as the weight is reduced [6]. For ABS material the shrinkage gradually increases as the weight reduction increases [6]. The multiphase materials mentioned above are more affected by microcellular process because the percentages of shrinkage increase at 15% weight reduction level are 30% for ABS and PC/ABS and 58% for IMPC, respectively [6].

Experiments for microcellular processing also presented some similar unnormal results for unfilled materials. The explanation for this increasing shrinkage in some unfilled materials is that larger cells may be present within the polymer structure. These cells can collapse during the cooling process of the parts, which may result in higher shrinkage values. When using the microcellular process, a 6% average increase in the shrinkage values was experienced with the unfilled samples [8, 22].

Anyone associated with injection molding knows that shrink rates are as much a byproduct of part design, gate location and type, temperature, cavity pressure, and son on, as is the actual rate of shrink of the resin. Studies are being performed on numerous resins in an effort to better understand the affect that microcellular processing has on shrink rate. These studies are conducted on test bars measuring approximately 165 mm long by 19 mm wide by 3.2 mm thick. Table 5.4 shows the results of three test results [22].

TABLE 5.4 Shrink Rate (mm/mm) [22]

Sample	Solid	Solid	2.5%	2.5%	5%	5%	10%	10%	15%	15%
Direction	Lin.[a]	Trav.[b]	Lin.	Trav.	Lin.	Trav.	Lin.	Trav.	Lin.	Trav.
PC/ABS	.0064	.0041	.0058	.0033	.0066	.0049	.0067	.0055	.0071	.0060
30% GF PBT	.0037	.0166	.0039	.0136	.0040	.0135	.0037	.0178	.0040	.0179
13% GF Nylon 6/6	.0084	.0085	.0065	.0049	.0070	.0121	.0078	.0127	.0078	.0151

[a]Lin.: Means linearly parallel to mold flow direction.
[b]Trav.: traverse the flow direction.

There are two patterns that develop in these tests. The first is that microcellular processing seems to have little affect on the shrink rate in the direction of mold flow in these test specimens [22]. It is the fact that all results fall within the manufacturer's suggested range for each material.

The second pattern is that the shrink rate traverse to flow appears to be much more influenced by microcellular processing especially with glass-filled polymers [22]. This conclusion is similar to the result above. At a 2.5% weight reduction, the shrink rate falls slightly from that of the specimens molded in a solid. As the weight is further reduced, the shrink rate increases considerably. It is likely that the faster fill rates associated with microcellular processing may result in higher molecular and glass fiber orientation and therefore increase the shrink rate in the direction traverse to flow. This would be most evident in a long test bar.

Initial test results seem to indicate that the "foaming action" of the supercritical fluids has little affect on the shrink rate of a resin. Molecular and fiber orientation may be a larger consideration [22]. The actual shrink rates in any given application can vary considerably with part design, material selection, and processing conditions.

Since shrink rates vary not only from part to part, but also within the same part, microcellular processing's affect on shrinkage will need to be taken into account as one more variable when designing a mold for the process. However, the conclusion is that the shrinkage and warpage can be reduced with microcellular processing [22, 24]. Kramschuster et al. [24] reported the quantitative study of shrinkage and warpage behavior for microcellular processing and found that the weight percentage of supercritical fluid (SCF) and injection speed affect the shrinkage and warpage of a microcellular part most significantly.

Trexel also published some useful information regarding the dimensional variations between solid parts and microcellular MuCell® parts [8]. Table 5.5 shows that a typical microcellular part has a superior dimensional stability when compared to a solid part. Even if the microcellular part has more weight

TABLE 5.5 Dimension Variations for Microcellular and Solid Parts [8]

	Solid Part (g)	MuCell® Part with 5.2 Weight Reduction (g)	MuCell® Part with 9.2 Weight Reduction (g)
Weight	0.019	0.029	0.032
Length	0.019	0.012	0.011
Width	0.017	0.009	0.009

Source: With permission from reference 8 to reproduce as a table (POM adapter with 71.5 × 105.9 mm, flat part, wall thickness 2 mm, gate size 1.75 × 0.75 mm, solid part weight 23.85 g).

variation than that of the solid part, the dimensional variations for both 5.2% and 9.2% weight reduction microcellular parts is 18–25% less than that of solid parts in both directions.

5.1.2.4 Thermal Property. Based on the report from some microcellular applications, the insulation properties improved when foaming the parts using the microcellular process [8]. As expected, the thermal conductivity decreased as the weight reduction increased, which can be explained by the presence of a larger number of gas microcells in the part structure. For the Noryl® MH230 case, the values of thermal conductivity will not be affected by the foaming process.

5.1.2.5 Acoustical Property. A similar principle of thermal insulation from microcellular structure, the sound absorption behavior of microcellular foam material, is also a unique advantage with porosity structure of microcellular foam. Microcellular foam provides better acoustical material than does conventional foam. The studies carried out by Park and co-workers [25] show that the effects of porosity and cell density are not cross-correlated and that porosity is not a sufficient parameter to account for the interconnectivity of the pores in the microcellular foam material. The sound absorption behavior of microcellular foams has been divided into three distinct frequency ranges in which different effects of morphology were studied. Their statistical analyses of the acoustical curves of microcellular foams show that at lower frequency range (less than 1000 Hz), porosity and bigger cell size effects are dominant parameters. In this range of frequency the flow resistivity of microcellular material along with the thermal dissipation dominates the sound absorption performance (thermal characteristic length can be estimated from the behavior in this range). The sound absorption coefficient grows linearly in this low frequency range until the maximum sound absorption reached. Therefore, the microcellular foam process may focus on the number of cells no matter what sizes are in this low frequency of vibration range. In medium frequency range (less than 2000 Hz), smaller cells have the major effects on the acoustic behavior (peak frequency). Therefore, microcellular foam processing needs to focus on the cell size control for the best acoustic behavior of the foam in this medium frequency of vibration range. At high frequency range (larger than 2000 Hz), porosity effects diminish and cell density effects result in another smaller peak [25]. The viscous characteristic length and tortuosity of microcellular material are dominated in this range. Therefore, at high frequency the cell density plays the major role in sound absorption, and the porosity effects diminish. Cell size effects are still in action in this range but are not the major effects anymore. The modeling work results in the optimal performance of the morphology with higher porosity, lower cell density, and larger cell size [25].

The test method ASTM C-1050 was employed to determine the normal acoustical absorption coefficient of microcellular foam.

5.1.2.6 Electric Property. Spindler reported the test result for electric properties of microcellular parts [22]. Dielectric strength testing was performed on 30% GF PET, PC/ABS, and 30% GF PBT. The lab found that there was no change in dielectric strength when comparing solid samples to MuCell® samples at up to 20% weight reductions in all three resins.

All three resins were also tested for volume resistivity, as discussed in Spindler's report [22]. The PC/ABS blend and the 30% GF PET exhibited the same volume resistivity results up to 20% weight reduction when compared to a solid. The results of 30% GF PBT were similar to that of a solid up to approximately 10% weight reduction.

5.1.2.7 Other Properties. The microcellular process had no significant effect on the flammability properties for all materials PET, PC/ABS, and PBT [22]. Some solid sample had even a worse result than that of the microcellular sample.

Weight reductions of 5% or less should not require additional flammability testing by UL because foamed materials are already covered under UL 746D. Weight reductions higher than 5% will require additional testing by UL.

Chemical resistance should not change much from the microcellular structure. However, the microcellular may change the surface tension and roughness that may have some influences for some chemical resistance.

Volume resistivity of the microcellular part was also measured and was reported in reference 22. The PC/ABS blend and the 30% GF PET exhibited the same volume resistivity results up to 20% weight reduction when compared to a solid. The volume resistivity of 30% GF PBT microcellular foam is similar to a solid up to approximately 10% weight reduction.

5.1.3 Assembly for Microcellular Parts

There are quite diverse possibilities for joining plastic parts together. Most of methods of unfoamed plastic parts joining methods are good enough to be applied for microcellular plastic parts. However, there are some exceptions that need to be addressed for microcellular part assembly.

- The mechanical strength may drop about 15–20% for a microcellular part, depending on the weight reduction and cell structure. Therefore, the snap-fit design may need to be made either thicker or structurally stiffer for a microcellular part. However, the toughness of the microcellular part will help to be able to withstand more deformation during snap-fit.
- The threaded closures of a microcellular part offer a stress-free part but need to increase the engagement of teeth to compensate for the strength reduction of foamed teeth.
- For ultrasonic welded part there is no significant difference between solid and microcellular materials to transmit ultrasonic energy during the welding process. However, the cell structure and skin thickness play an important in for the quality of welding with microcellular parts. The glass-fiber-reinforced material will maintain the same quality of the welding of the solid part, and even better sometimes because of the disorientation of glass fibers in the microcellular part [26].

The microcellular parts assembly process selections may refer to the same rules as solid in Table 10.1 [27]. The amorphous material is good for every method of welding. However, semicrystalline thermoplastic and olefin materials have some application limit for ultrasonic welding. Most composite materials are only good for an electromagnetic bonding method to join them together. The part itself will influence the selection of welding as well. The thin wall is only good for ultrasonic and hot plate welding methods. The large and complex geometry parts are not good for an ultrasonic welding method, either. The details of welding methods are discussed in Chapter 10.

Finally, the big issue for the microcellular part design is the poor surface finish. However, there has been much progress made to improve the surface quality of microcellular part. It will be discussed in a special section for proper mold and part designing for the better surface finish of the microcellular part in this chapter.

5.2 MOLD DESIGN

The microcellular mold determines the final shape of the part. It also determines the raw material usage and the final finish of the part. The optimum mold design may reduce the waste of the raw material. Based on the unique performance of the molding with microcellular foaming, the mold design shall be improved as the following to meet the special requirements of microcellular injection molding.

5.2.1 Mold Materials

The microcellular injection molds are basically the same as the regular injection molds, except for high thermal conductivity and low strength requirements for microcellular molds. Several factors that determine the materials for cavity and core in microcellular mold materials are discussed below.

5.2.1.1 Steel. The microcellular molds will use mostly the same material as the regular injection molding molds. It is because thin wall molds still dominate the microcellular application while thin-wall molds require a high-strength steel mold even when microcellular injection pressure is low. The most common steels used for microcellular mold are 1045, 4140, 4150, P-20, H13, 420M, and so on.

P-20 is the standard steel for unfilled polymer processing. It has a medium mold lifetime, that is, 1 million parts. P20 steel is good for texturing on the surface. It is fair to have a nitride surface for an extra surface harness of P20 steel mold, but in most cases the mold surfaces are made without nitriding. The surface hardness of mold made by P20 steel without nitriding is about 29–36 Rockwell C. If it is necessary, this steel can be flame-hardened with care to avoid too much dimension changes. Another advantage of P-20 steel is its good polishability. In addition, it has fair weldability.

The hardened steel for the longer lifetime of mold may use H13 tool steel. It is primary steel for the cavity blocks where hardness is required with some ductility. It is good for high volume in unfilled materials and intermediate volume in loadings of less than 30% mineral or glass. It requires finish machining after the heat treat, which adds cost but provides longer lifetime. The typical hardness after heat treatment of H13 tool steel is about 43–65 Rockwell C. It has good polishability as well. It is very good material if nitriding is necessary to get the hardness over 60 Rockwell C. However, the H13 steel has poor weldability.

In addition, 420M steel is good mold material as well. It is fair for texturing. It is very good steel for the nitriding if a hardened surface is required. The machinability is similar to H13 and is not as good as P-20. However, the polishability is better than both P-20 and H13 steels. It can be hardened by a flame-hardening method, which is the cheapest heat treatment.

Just recently a family of prehardened mold steels from France became notable for mold makers as the better option than the standard P20-grade steels for the injection mold. It has higher thermal conductivity that is about 18% higher than the one of P-20 steel, which can lead to faster molding cycles. On the other hand, this steel is easier and faster to machine and has 15–20% higher yielding strength than P-20 steel. The available grade of this steel is Superplast® 300, 350, and 400 from Industeel USA. However, the cost of Superplast® is about 10% more than the P-20 steel. Overall, the cost of steel is only about 10–15% of a total cost of a mold. This steel has some advantages, such as (a) faster machining with less downtime and (b) short cycle time from better heat conduction property, which makes up for the difference of the cost of steels [28].

Also, a new line of stainless steel ejector is developed for the medical, food contact, or other sensitive uses of clean environment. It is the Z413 ejector from Hasco America Inc. The steel has high chrome content (16–18%), which prevents corrosion and makes the ejector pin extremely long-wearing [28].

There are more steel materials used for molds of microcellular processing. They can follow the same rules as for regular injection molding and should have longer lifetime since the microcellular process has low viscosity, low injection pressure, and low tonnage of clamp force [29].

5.2.1.2 Carbide Material. For a high percentage glass-fiber-reinforced material to be processed (above 40 weight percent), carbide material is necessary for a reasonable mold life. This is the best material for the applications with good wear resistance and higher durability. It may crack if improperly machined. The hardness of carbide material can reach 90–95 Rockwell C.

5.2.1.3 Aluminum Material. Although an aluminum alloy could be used for good thermal conductivity, it may be only used for prototype mold and some insert that needs high-efficiency cooling. However, the heavy-duty aluminum alloys are possible to be used for microcellular mold, such as 7075-T6 (thermal conductivity 140 W/(kg m) and ultimate tensile strength 503 MPa with Brinell hardness 150 [29]). Another common aluminum material for the quick and low-volume (about less than 250,000 part life) mold is QC-7. It can be easily machined but easily damaged, too. It is difficult to weld and has only 5 Rockwell C hardness.

Therefore, the precise injection mold will use steel for a long service life and for reliability. Aluminum may be successful to be used for microcellular processing with the combination of aluminum and steel. This method may provide the advantage of high thermal conductivity aluminum for cooling and more wearing resistant steels as the high wearing area in the molds. Also, the mold base shall use the steel as well.

5.2.1.4 Sintered Metal Material. Sintered metal is also used for microcellular processing, which may have the advantage of venting through the porous surface of the sintered metal. However, this mold may only produce a total of 100,000 parts during its lifetime. It is only recommended for unfilled, or lightly filled, polymers processing. Maximum part size is limited to 25–65 cm^2, with hardness <50 Rockwell C.

In addition, the most attractive advantage to use sintered metal for the microcellular mold is the porous structure of sintered metal itself. It may solve the extra venting requirement of microcellular processing since the sintered material has porous structure through out the entire mold. Any gas in the part can be released from the porous sintered metal in any position of mold. It also partially solves the problem of surface roughness caused by trapped gas for the skin because the gas in the skin can simply be released through the porous wall surface of mold. However, a porous material can be easily filled from degraded materials if the wrong material temperature control or local mechanical shearing heat causes degrading. The other filler or color particles may stay on the porous surface to finally block the venting path through porous material. Therefore, the necessary venting areas are still needed even if porous material is used for the microcellular mold.

5.2.1.5 Others. For the microcellular part there is some insert or core where one needs to intensify cooling material that is preferred for using a beryllium–copper alloy. This alloy offers both hardness (up to 42 Rockwell C) and very good thermal conductivity. Generally, the thermal conductivity of beryllium copper can be as much as six times greater than that of steel. Thus, it offers a special advantage of part designs with the requirement for better cooling. The 1.7% beryllium has prevailed in regular mold-making. It provides tensile strength of up to 1200 MPa and can be hardened up to 440 Brinell. For a microcellular mold, beryllium material will be used for cores, inserts, sprues, and hot spots such as bosses that are in the most common positions of post blow after the microcellular part is ejected out from the mold. The range of hardness of Brinell 330–360 is generally sufficient for the practical purpose. More details of the mold materials and the rules on how to select them can be obtained from a mold-making book [29].

There is a brief summary of mold materials given below based on different requirements. Microcellular process for the special part may need all of them, or just part of them:

- The mold material should be free machining to reduce the tool cost
- Capable of a number one finish with five minutes of number 120 stoning
- Hard enough to be resistant to wear, usually over 50 Rockwell C
- Capable of flexing 30 degrees for five million cycles without fracturing
- Good thermal conductivity of water
- Low cost

5.2.2 Mold Surface Coating and Structuring

Mold coating is usually used for corrosive protection for mold surfaces. The chemically aggressive substances during processing are mostly hydrochloric or acetic acid. The coating to protect the mold surface is the electrolytic plating with chromium (more frequently usage) or nickel (more corrosion protect coating). The thickness of chromium plating is about 5–200 μm, in special cases between 0.5 mm and 1 mm [29]. Chrome plating provides not only corrosion resistance but also the wearing resistance since it can reach a hardness of 900 HV. The nickel plating is very good for corrosion resistance, but nickel plating is relatively soft compared to chrome plating.

The mold surface must be well polished before the coating process since the buildup surface quality of the coating depends the original quality of the base material surface. In addition, the coating on the mold surface above is effective only if the thickness of the deposit is uniform, and the sharp edge in the mold must be avoided. Nonuniform thickness and sharp corner causes stress in the protective layer. It, then, can lead to peeling off the plating layer under the loads. The hazard of a nonuniform deposit is particularly great in the molds that have intricate contours (undercuts). In addition, the bending stress may easily pose the danger of cracking.

The new ceramic coating on the microcellular mold greatly reduces the surface roughness of the microcellular part. However, it is still in the trial stage. There is also a heat insulating coating on the mold surface that improves the microcellular part surface quality by a minor increase of cycle time [29].

Some tests with a polyester film on the cavity surface of microcellular mold were carried out in Trexel. The test result of the microcellular part made with this polyester film in the mold shows a smooth surface without any swirl mark. The reason for the advantage of film on the mold surface is that the film reduces the cooling rate of the skin of the part and the flow friction. Therefore, it results in less internal stress, less orientation, and less flow marks on the surface of the microcellular part. The gas may be slid on the smooth-coated mold surface without any breaking. Then, the severely sheared gas bubble on the surface shown in Figure 6.25 no longer exists on this film-coated mold surface. The surface bubble on the coated mold surface is either (a) compressed to flat since there is no immediately freezing skin or (b) diffused back in the polymer melt with a long cycle time and a relatively higher mold surface temperature than a cold metal mold surface [30].

As a similar approach like film on the surface of mold, a special coating of polyimide on the mold surface will achieve the class "A" surface as well. The polyimide coating on the mold has been developed from Asahi, Kasei Corp. [31], named Clear Surface Molding (CSM); it provides a high-gloss finish devoid of weld lines straight from the mold. It is also good for microcellular processing with a nice surface. The cavity surface of CSM mold incorporates a 0.1-mm-thick polyimide insulation layer covered by a 30-μm-thick protective epoxy-silicone hardcoat. The protective coat is polished to a mirror finish (103% gloss at 60°). The hardcoat has a hardness of 3H; and tests have shown that it does not degrade, even after 30,000 shots.

The end result of polyimide coating on the mold not only creates a high-gloss finish even at lower mold temperature but also promotes the crystallinity of crystalline material. It is because the layer of polyimide coating acts as an insulator to maintain the temperature in the surface region of the part. However, this temperature is only maintained for the instant immediately after melt injection. The coated surface temperature is typically 70°C higher at 0.1 sec after injection is completed.

The mold surface can be structured with textures. Overall, it results in a significant improvement by reducing the visibility of the swirl. There are different explanations for the principles of bright reflections of textured surface. It may provide better venting of the escaping gas and trapped air and a concealing of the surface defects caused by foam process. Another explanation is that textured surface reduces the shear strain of growing gas bubbles on the cavity wall by the mechanical anchoring of the polymer in the structured mold surface [30]. The most common explain is that the appearance of silver streaks is greatly reduced by a diffuse backscatter [11, 30]. The textured mold surface is a very effective method for the light color microcellular part. Xu [11] reported some bubble stretching on the microcellular surfaces with smooth mold surface, but he also reported different depths of texture on the mold surface (see Figure 6.25). The result of different depth of textures on the mold surface shows the deep structured mold surface making the bubble stretching even worse than the shallow depth of structure on the mold surface. However, the overall vision of the surface quality is not much different by observation, and both are acceptable by the real microcellular part. There are several applications are used for the real parts. One of them is to add texture to conceal the gas swirl on the microcellular part surface only on the exposed area. It can save the cost of texturing on the local surface wherever it is necessary only, instead of on the whole mold surface. The different textures may need to be studied in the future to find better textures to improve more of the surface quality of microcellular foam.

There are two roughness-forming mechanisms for microcellular processing: interface roughness and free flow front roughness [11]. Both roughness-forming mechanisms are fully discussed in Chapter 6 from a processing point of view. Interface roughness dominates most of the microcellular foaming process. Although small and uniform cell structure is the important factor for improving the surface quality of microcellular foam, mold surface technologies can make a better surface of microcellular foam. The typical mold surface technologies are summarized as follows:

- Hot and coated mold can eliminate the interface roughness for (a) amorphous material and (b) crystalline material, with careful control of the crystallization on the surface. The free-flow front roughness may be considered at the end of mold filling for hot or coated molds if the injection time is too long. The free-flow front roughness can be used as the control factor for the limit of maximum injection time.
- Texture surface mold may benefit from the optical effect of reducing the appearance of silver streak but making the bubble shearing on the surface worse. However, the texture surface with carefully designed grooves to match the mold flow direction may help to vent the surface gas or air that will reduce the breaking gas bubble on the surface of microcellular foam.
- The crystallization may expel the gas, and it may make the surface quality even worse with expelled gas on the surface. Therefore, a balance between percentage of crystallinity and small cell size is a way to improve the surface quality of crystal material.

There are some special discussions in the final section of this chapter. The detailed designing and researching are presented to solve more issues of surface quality of microcellular part.

5.2.3 Mold Venting

Mold venting is important for successful microcellular mold filling. It is initially assumed that the venting gap for microcellular molding needs less than regular molding because the gas-laden melt is a low-viscosity material that needs a tight seal for preventing flash. It needs some corrections based on the facts below:

- The injection speed is so fast that the gas estimated from the free-flow front and the trapped air in the difficult venting area will cause some regular venting channel size too restrictive to release gas and air in time. An insufficiently vented cavity may cause trapped air and gas, localized burning, and subsequent tool damage. It may also result in short shot on some corner that is the final difficult venting area.
- Although the gas-laden melt has low viscosity that may cause easy leaking and result in more flash, the real case is that the mold cavity pressure is much lower than that of the regular injection molding. Thus, the force to push low-viscosity material leaking through vending channel is very low compared to high cavity pressure of regular injection molding. Therefore, the vent gap depth may be kept the same as for the solid injection molding since it needs the large gas and air volume rate released for the venting area.

The critical gap depth for venting of microcellular molding is kept the same as for the regular molding, and it is suggested that we follow the high end of the material manufacturing recommendations:

- For crystalline thermoplastics PP, PA GR-PA, POM, and PE, the critical depth is 0.015 mm for solid.
- For amorphous thermoplastics PS, ABS, PC, and PMMA, the critical depth of venting is 0.03 mm for solid material and about 0.05–0.08 mm for microcellular material.
- For extremely fluid materials with high melt index, the critical depth of venting is 0.003 mm for both solid material and microcellular material.
- For polyolefin material the venting depth may be in the range 0.025–0.05 mm.
- The venting depth is more sensitive to be changed. Generally, if the rectangular venting channel is designed, the depth of venting channel will use the data recommended above. The width can be increased up to 12 mm if the mold structure is allowed.
- The last area to be filled in the mold may have the largest volume of gas and air to be released, and the pressure there may be the lowest in the path of mold filling. Therefore, it may be safe to increase the venting width 100% and depth 50–75% more [6] to effectively increase the venting area.

- The first mold trial with short shot may help to find the critical areas to be vented for microcellular injection molding. It needs to add the new venting channel in the critical areas that are detected by the mold trial.
- Venting land length should be as short as possible. It is recommended not to be over 1 mm long if it is possible.
- Venting relief channels are about 1–2 mm and should be connected to the atmosphere.
- The sprue puller is another location to add venting outlets. Usually the sprue puller needs about 0.04-mm clearance in diameter for venting. Or the flat may be cut with 0.05-mm gap in puller pin for venting in sprue.
- Bosses and deep ribs may need vent to be designed if the depth of rib or boss is 4 times higher than thickness of the microcellular part.
- The venting will occur in the ejector pins. For microcellular processing the ejector pins should have the ground flats from 0.05 to 0.1 mm, and the typical pin clearance of 0.025 to 0.04 mm may not enough to vent gas and air together at high injection speed of mold filling.

The permissible gap depth recommended above can be exceeded by several tenths of a millimeter if the pressure at the gap is appropriately low (<5 MPa) [29]. For microcellular parts, this low pressure is always there unless the processing is changed to fill the difficult thin-wall mold or super-microcellular parts with extremely high injection speed.

The flow rate of air and gas to be vented from the mold can be given from a modified formula with gas volume [29]:

$$\dot{V}_{\text{vent}} = \frac{V_{\text{mold}} + V_r + V_{\text{gas}}}{t_j} \tag{5.3}$$

where $\dot{V}_{\text{vent}}$ is the total venting flow rate in the mold, V_{mold} is the volume of molding, V_r is the volume of runner system, and V_{gas} is the volume of released gas from the flow front.

Generous venting is absolutely paramount when utilizing microcellular technology. Since the viscosity of the resin is reduced by as much as 50% with single-phase solution, the mold may fill considerably faster than when molding the same part as a solid. Keep in mind that the "pack" portion of the process is provided by the "foaming action" as the supercritical fluid returns to a gas phase. This "pack" is low foaming pressure and also increases the need for sufficient venting.

Recommendations are summarized below:

- Mold vents should normally be cut to a depth that is 50–75% deeper than a solid. Starting steel safe is the smart approach, especially with complex contoured parts, but be prepared to increase the depth as required.

- Increase the amount of venting. Peripheral venting is a must. Consider the use of self-cleaning ejector pins on ribs, bosses, and hard-to-fill part geometries. While this suggestion will hardly seem a revelation to some, most molds are simply not vented properly. In conventional injection molding, these inadequacies are often compensated for in the molding process by reducing fill speed, increasing melt temperatures, and high-pressure packing. These traditional fixes will counter the affects of the MuCell® process and inhibit uniform cell structure, weight reduction, and improved dimensional stability.

One more thing that needs to be kept in mind that in most cases less clamp tonnage is required when processing with microcellular foam. When clamp pressure is reduced, the actual vent depth under pressure may already be increased due to less steel deflection.

Finally, there is a special requirement of tighter mold for any flame retardant in the material to have a microcellular molding. It is because the moderate cavity pressure built up in the mold cavity will help to minimize the formation of residual flame retardant condensation by keeping it in the material. A large part of Noryl® resin has a special continuous vent in the mold. A 6.4-mm-wide channel is cut parallel to the flow direction of mold filling, and it acts as a reservoir that absorbs initial injection shock, uniformly distributes compressive heat buildup, and adequately discharges cavity air across natural parting line imperfections while maintaining sufficient parting line integrity to prevent flashing [21].

5.2.4 Mold Runner and Gate Design

Microcellular mold may use either a runner system or a gate system. Some simple mold may use direct cold sprue. General rules for microcellular mold gate and runner systems are balanced gates and runner systems. All hot runner and gate systems must be valve-gated. Similarly, the cold sprue must use the shut-off nozzle on the machine. It is because prior to injection, the single-phase solution of resin and supercritical fluid must stay under pressure. It is necessary in order to keep the supercritical fluid in solution. Due to this requirement, hot manifold systems with conventional hot drops cannot be used with the microcellular injection molding process. Runners must be conventional cold runners or hot manifolds with valve gates. The exit end of the manifold for the hot runner system must blend in with the nozzle smoothly without areas of hang-up. The details of microcellular mold gate and runner systems are discussed in the following.

5.2.4.1 Valve Gate. The valve gates now are the most popular and the best methods used in microcellular injection molds. The advantages of valve gates for microcellular injection molds are summarized below:

- The valve gate usually has the smallest size of opening in the entire flow channels in the runner system and nozzle so that it is an ideal nucleation spot before the single-phase solution from the barrel is injected into the mold. This is the best approach to move the nucleation spot as close as possible to the mold cavity to avoid the duplicated nucleation. Therefore, the nozzle orifice can be selected as big as possible to truly move the nucleation spot to the valve gate.
- The valve gate can be designed to delay the opening action of the valve gate until the injection movement truly occurs. It can avoid the wrong sequencing that would cause the valve gate to open too early before injection action truly takes off and gets full speed. It is why the spring-loaded valve gate from hot runner systems of Fisa has been used successfully in lots of microcellular processing (refer to Figure 7.5).
- Another advantage of the valve gate with either pneumatic or hydraulic actuator (see Figure 7.6) is that the valve gate is controllable individually to change the valve opening differently from gate to gate to make even mold filling for the multicavity mold.
- The valve gate usually seals at high pressure and needs to be designed for sealing up to 34.5-MPa melt pressure to avoid prefoaming before injection and some extra pressure requirement for more SCF (gas at super critical solution) percentage in the solution.
- The valve gate system will be easily set up for gate close default of safety recommendation from SPI to protect the person who is working in the mold area for sudden power loss [12, 32]. Then, the pressure in the valve gate may be released safely by retracting the nozzle tip away from the mold and releasing the gas-pressurized hot material between the valve gate system and nozzle where a purge guard will protect any hot material spray.

5.2.4.2 Cold Sprue. Cold sprue is the simplest gate system if it works for some application of microcellular mold. The big concern is the cooling of the regular injection molding size of cold sprue and the strength of sprue to be pulled out from the mold for demolding. The root of sprue is usually much thicker than the thickness of the part itself, so it takes extra time to be cooled down. It is also weaker since it is the last one to be cooled, and the foam in the thick joining point of sprue significantly reduces the strength that is not strong enough to be pulled out from sprue bushing. It makes the sprue ejection even worse because the cell grows in the sprue so that it foams out against to the mold and sprue bushing as opposed to shrinking away from sprue bushing. This will result in sprue expanding during the cooling stage, and shrink is not enough to control overexpansion. Then the sprue stays in the sprue bushing and breaks with the runner system or part during ejection. Some suggestions may help, such as changing the sprue size to short and small with big draft angle (see Figure 5.8). The other method to modify the sprue design is to add gussets and radius around the root of sprue and then add extra cooling channels around the sprue while even adding a cooling pin into the cold well [6, 8]. However, it may not be worth doing a complicated sprue design since the reason to use the cold sprue just is because it is a simple solution for the possible simple molding portion of a microcellular part. Hence, if the mold is so simple for using sprue, some easy suggestions are summarized below:

- Reduce the sprue diameter since the low-viscosity material does not need big-size sprue for packing and high injection pressure anymore.
- Keep the shortest cold sprue length, which may be successful if it is as short as 50 mm. It can be designed with deep sprue bushing to extend the nozzle deep into the mold.
- Increase the sprue draft angle up to 5–7 degrees included angle if the sprue can be made short. It only works with short sprue since the large draft angle increases the root diameter of sprue at the junction of sprue to the part or runners significantly with regular sprue length.

If the cold sprue can be used successfully without the cooling and sticking in the sprue bushing, it will be the economic solution for the simple microcellular mold. The shut-off nozzle is usually used for the cold sprue mold that needs to review the opening diameter of nozzle orifice to ensure enough nucleation through the nozzle tip. However, the cold sprue also is a possible solution for the three-plate mold where the needle gates may be used. In this case the nucleation spot will be in the needle gate, not in the nozzle orifice, so that the shut-off nozzle tip should have a larger opening orifice diameter related to valve gate size.

The cold sprue in the three-plate mold will stick in the sprue bushing more often than any other mold. The three-plate mold typically designs for the larger sprue and runners to minimize the pressure drop through the runner system for solid injection molding. Therefore, with microcellular processing the expansion of cells and soft foamed material in the runner and sprue cause both sprue and runner system to stick within the mold. It is a big limitation to reduce the cycle time if cold sprue is used in three-plate mold. There are different methods to improve it.

One method to improve sprue design is to reduce the size of sprue to small and short, as shown in Figure 5.8. Now the sprue size is much smaller and may be comparable to the size of the runner in Figure 5.8. The draft angle for a small sprue is possible to be increased because the sprue length is much shorter than the original one. If it is necessary, the radius and gussets can be added in to the junction between sprue and runner. However, adding radius and gussets may increase the junction size between sprue and runner that needs to be checked for the junction size to avoid significant cooling time increase.

In addition, the sprue puller (also named sucker pin) needs to be redesigned with much more force to pull out the sprue from the sprue bushing. It can be reached by (a) changing the undercut design to grip the sprue better and (b) changing the material of puller to beryllium–copper alloy for quick cooling for the sprue junction spot with runner.

Another effective design for the sprue demolding is to use spring-loaded sprue bushing [29]. It uses a strong spring around the sprue bushing that pushes the sprue bushing away from the sprue when the sprue puller holds the sprue. There is a mechanical stop to stop the sprue bushing from moving further; this is to hold the sprue bushing in the position where the nozzle will be push it back. This needs a sprue break action from machine. After the part cooled, the nozzle will be retracted with certain distance to allow the sprue bushing to move away from the sprue. It will significantly reduce the load on the sprue puller when it just needs to hold the sprue on the position, and the whole runner system still be held by the whole runner plate in the mold.

5.2.4.3 Runner System. The runner system for a microcellular mold is basically the same as the solid mold. However, the size of the runner system can be reduced since the low viscosity of gas-laden material is not necessary with packing stage through the runner system anymore for microcellular foam. The small-size runner system can reduce the cooling time as well. It may bring the sprue size down since the biggest diameter of the sprue is also determined by runner cross-section area.

The runner system for a microcellular mold has a more critical requirement for the balanced runner system because the microcellular mold will have more processing issues for the unbalanced runner system. The first filled cavity will have extra material filled in that will cause overpacking and cause post blow or flash. On the other hand, the restricted cavity may have (a) a short shot that possibly causes voids in the last filled area (b) or a short shot on the difficult-to-fill corners. There are several major possible design issues causing an unbalanced runner system, and the related solutions are summarized as follows:

- Runner length may not be equal, and it may be corrected by redesign of the runner system.
- Runner cross section may not be facilitated with the same area and same quality of surface finish. It can be corrected by reworking the runner system.
- Runner system may not have equal cooling around because the cooling system has poorly designed cooling water loops. This kind of thermal imbalance may be improved by decreasing the water temperature between inlet and outlet. In addition, rearrange the inlet and outlet of cooling channel to get close thermo-balanced cooling.
- The best natural balanced runner system is the radial pattern. The “H” pattern of runner system may still have a little problem because the inner-corner flow length is shorter than the external-corner flow length.

It can be solved by MeltFlipper™ runner design developed by John Beaumont [33]. One of experiments shows the MeltFlipper™ technology improving the shear-induced imbalance in the runner system of microcellular injection molding, and it provides a more robust process window.

- Runner or mold venting may cause a similar issue as would an unbalanced runner system that needs to be detected to have the mold closed loosely, and it may make a short shot to find the venting restricted area. It shall be corrected to open the width of vending channels (first choice since width is not as sensitive as depth of venting channel). If it is not enough or width restriction limits the opening, then depth of venting channel can be reworked with the careful control not to cause a flash problem from the deep venting channel.
- Sometimes, the final runner system is reached by the gate size balance analysis and is modified accordingly. It is because the gate size is so small compared to runner size that is much more effective than the runner size for the dimension changes.
- Hot runner system uses about 60% of full wattage for extended heater life. Usually, 3 watts per square centimeter of manifold is used as the rule of thumb to estimate the necessary heating power for the hot runner manifold.
- Insulation is necessary for the hot runner manifold body to be installed against the main mold body and machine platen. Utilize either air gap or insulator to minimize the heat transfer to the mold properly. *Note*: Doubling the air gap results in an increase of 8 times in insulation value. The insulator must have a careful layout to be sure that the insulation is uniform as well for the whole runner manifold.
- The best runner cross section is the full round shape because it permits the greatest amount of material to flow. However, it may not be an economic geometric shape for manufacturing. Therefore, two more popular cross-section shapes are used for the runner system. One is a trapezoidal shape, which is the most popular one used in the runner system. Another one is a half-round runner, which permits the least amount of flow or produces the greatest chilling effect.

5.2.4.4 Mold Gate Design. Mold gate design is more flexible for microcellular molds since the low-viscosity material is tolerated for different gate sizes and shapes. However, the gate location is critical for balanced microcellular mold filling pattern in the same mold. The detailed discussion following will be focus on the gate size and location.

Mold gate size is related to nucleation rate and shearing rate in the gate during injection. The gate size always needs to check with the nozzle tip orifice diameter as well to be sure that the nucleation is focused on one location only. Similar to runner size design, the gate size design for the microcellular process does not need to consider the packing stage. On the other hand, the microcellular process of injection molding usually has high injection volume rate so that the shearing in the small gate needs to be checked for maximum shearing rate allowed by material manufacturing. This maximum shearing limit for microcellular processing can be increased to about 30–40% because the gas-laden melt provides lower viscosity than that of the original solid material. The mold gate size is determined by the nucleation rate required by different materials. The theoretical nucleation rate can be estimated from Equations (7.2), and (7.3). If the nozzle orifice is the nucleation position, the nucleation rate can be calculated from Equation (7.1). The general rule is that the minimum pressure drop rate of 10^9 Pa/sec needs to be achieved in either nozzle orifice, or valve gate, whichever is the nucleation position. However, in most practical cases the heterogeneous nucleation will help significantly for nucleation. Therefore, the real pressure drop rate 10^9 Pa/sec may not be necessary for the filled material. In addition, the sizes of multi-gates must be uniform for all gate sizes unless different gate sizes are used to balance the mold filling. As a general rule, changing the gate size will be the last measure to improve mold filling balance.

Mold gate location is important to make successful microcellular processing. First of all, it must be balanced design for the mold filling even if the part geometry is not symmetric. The reason for it is similar to the runner system, but it is now the even filling pattern in the same part.

If the part thickness is different, the gate must be located in the thin section, instead of located in the thick section as a rule of regular injection molding. It is because the pack stage is not necessary for microcellular molding so that the gate does not need to be prepared for packing the mold to remove the sink mark and to compensate the shrinkage during mold cooling stage. For microcellular processing, the gate in the thick part will cause several issues like post blow in the thick gate area since it is the last area to be cooled and the pressure is also the highest. Then, both temperature and pressure will provide the condition for post blow. This post blow issue caused by the gate in the thick session can be solved only by extra cooling time. Therefore, this is the special rule that does not locate the gate of the microcellular mold in the thick section.

Similar to the problem of the gate in the thick session, the gate is always the last part of the microcellular part to be cold down. Therefore, the microcellular process is more likely to cause the blemish around the gate area. This blemish defect can be eliminated with better cooling around the gate. In addition, the blemish issue can be improved with venting more and adding texture in the gate area [8].

Gate location also influences the weight reduction percentage and possible cell architecture. Although the microcellular process can be used for large flow ratio (the ratio of longest flow length L_{flow} during mold filling to part thickness c), it is recommended to design the gate location to minimum flow length to the thickness ratio that may result in a maximum weight reduction and a nice

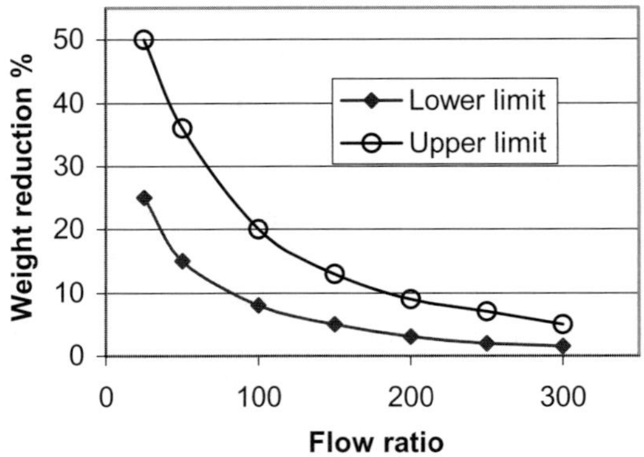

Figure 5.9 Typical relationship of weight reduction with flow ratio.

cell architecture. The short flow ratio will have low cavity pressure, which means more energy savings from both injection pressure and clamp tonnage. The gate location keeps short, and even flow pattern may help to improve the surface appearance. In addition, the low flow ratio keeps the injection time short so that the cell growth in free flow front of mold filling will not become too big. Although the weight reduction is driven primarily by flow ratio the injection volume rate, there are many more factors to influence the relationship between weight reduction and flow ratio, such as mold temperature, flow balance of the part, the overall thickness of the part, filler type and weight percentage, total injection time, venting system, and surface finish requirement. A reference chart in Figure 5.9 may be referred to as the primary design guidance to determine the initial weight reduction percentage at certain flow ratio that will be used to determine the mold runner system and number of gates in the mold. In most microcellular injection molding, the flow ratio is about 60:1 to 125:1 without special high injection volume rate (with a standard injection molding equipment). It can reach up to 350 for the thin-wall application with high injection volume rate and hot runner system with valve gate. Although the weight reduction can reach up to 50% for some extremely short flow ratio and high injection speed capability, the over foamed part is not a structural part with enough strength because the wall thickness among cells is too thin. Therefore, the good rule of thumb for the weight reduction is 5–15% for most application of microcellular parts. On the other hand, the zero or just 0.5–1% weight reduction is a special application for the part that needs good surface finish and no strength drop while the mold filling needs low viscosity of gas-laden melt to fill the mold without any dimension problems (see the details in Chapter 8). This becomes another well-known practice for microcellular injection molding, and it has more potential applications in regular injection molding industry.

In addition, the following factors are emphasized in order to obtain the most benefits of microcellular processing:

- Design in short, thin runner, and sprue with increased cooling.
- Don't size the runner for packing.
- Pursue optimum balanced flow (both volumetric and thermal).
- Design the runner system for faster fill rates with considering cooling requirement.
- Consider that lower flow factors yield the highest weight reductions, along with nice uniform cells.
- Utilize valve gates whenever possible. Valve gates are preferred since the pressure drop and subsequent nucleation takes place in the part (not the runner) and nozzle.
- The location of gates is critical for uniform fill rates, cooling efficiency, weight reduction, and pressure drops. Multiple gate locations may not be required to fill the part, but may aid in attaining uniform cell structure within a part with proper welding line positions.
- The land of the gate can cause excessive pressure loss and premature freezing of the gate. It should be designed as short as possible, such as 0.5–1 mm.

5.2.5 Mold Cooling System

Mold cooling may be the safe issue for microcellular foaming since the smaller the cells, the higher the residual pressure in cells before demolding. The foamed part must be cooled enough, not only reaching the ejection temperature but also forming the strong solid skin to hold the cells expanding. The post blow in the demolded foam part is dangerous for hot gas and hot plastic popping out. The post blow sometimes occurs after the part is demolded for even a long time. It causes some blister in the surface of the foamed part. A broken sprue is also a common problem occurring in a not properly designed mold. The sprue is always the hottest spot because it is the thick section in the mold and the last part of mold filling with pressure still maintained there. It is a misunderstanding that the weight reduction will create pressure free zone near the gate or sprue. Even if the weight reduction is more than 20%, there is some residual pressure in the gate or sprue area because the pressure-free area from weight reduction is true only in the free flow front since the mold does not fill in full there. A broken sprue stays in the mold and immediately stops the automatic cycling. Unless the similar safety system is designed for this case [12, 32], it is not safe for the operator to take the broken sprue from the mold because the safety guard is opened and single-phase solution is still under the pressure.

Since the microcellular process can yield cycle time much shorter than traditional conventional molding, cooling paradigms will need to be challenged. As discussed earlier, sufficient cooling is extremely critical in obtaining maximum cycle time reduction and optimum uniformity of cell structure.

Philosophies pertaining to cooling in injection molds vary considerably in the industry.

In addition, microcellular processing has the cooling benefit since the cell growth pushes the skin of the part touching the cold mold wall tighter than the solid part. The foamed part also benefits by quickly releasing the pressure and a larger number of cells, which results in a dramatically cooling time reduction. The physics behind this phenomenon is the quick energy release during injection that results instantly in a gas pressure reduction from high to almost zero. But any thick foam part may have longer cooling time than the solid because the heat transfer is not efficient for the foamed part when the insulation dominates the cooling stage. However, thickness is not the only restriction for the fast uniform cooling. The uniform thickness is a critical factor if the thick part is necessary. Longer cooling time may also be related to poor single-phase solution and wrong processing conditions or mold design because any big cell or void in the part is the air insulation spot with is the slowest cooling position.

Uniform mold cooling is critical for (a) maximization of the cycle reduction of the microcellular part and (b) control of the part tolerances. It is generally advisable to maintain less than a range from 5 °C (for small mold) to a 20 °C (for large mold) differential in steel temperature over the mold cavity or core. Tighter controls in the mold cooling for microcellular processing will provide greater processing latitude.

Typical mold cooling channel layout would incorporate 12.7- to 15-mm diameter of cooling channels 38–50 mm apart and 12.7 mm below the cavity and core surfaces. It will be determined by geometry of mold and configuration of cooling circulation. Most rules to design the cooling channels in regular molds are still useful for microcellular mold design—except the gate and sprue area, which are the places usually not to be paid much attention with regard to regular mold design.

Spot temperature control in the mold is always a challenge for injection molding. The bubbler (cascading) should be used for larger standing cores when conventional channeling is not possible. In some series standing cores the bubbler-baffles may be used to guide the water cooling to go through the necessary paths to cool the corners, which are difficult to be cold with regular water channels in the mold. Heat pipes can be used in smaller standing sections as well as chrome-plated beryllium copper inserts and pins.

The chiller capacity for the common resins to be processed with solid injection molding is listed in Table 5.6 [34]. Although it is for solid processing, the microcellular mold usually has extra benefit with quick cooling from gas-laden material within 2.5-mm thickness. For example, if 135 kg/hr of ABS is processed, divide the 135 by 22.7 (~6), which means that a 6 US ton chiller is required to cool this mold only.

General recommendations for cooling system of microcellular mold are summarized below:

TABLE 5.6 Chiller Capacity for Common Resins [34]

Resin	Kilogram/hour/ton
HDPE	13.6
LDPE	15.9
Acrylic	15.9
PP	15.9
Nylon	18.2
PPE	18.2
Urethane	18.2
PET	20.4
ABS	22.7
PS	22.7
Acetal	22.7
PVC	34.1

- Maximize the amount of cooling that will have ample turbulent water flow throughout the tool. The recommended Reynolds number is 5000–10,000 in the mold-cooling passages. Extra emphasis should be put on having a cooling circuit adjacent to all molding surfaces and placed as close to the molding surface as possible.
- Eliminate hot spots in cavities, cores, and runners through either turbulent flow or heat pipes.
- Add cooling to the sprue bushing and runner blocks if utilizing a cold runner.
- Utilize heat pipes, or bubblers, in hot spots that are typically associated with relatively long exposed sections of steel such as core pins. The high heat transfer material such as beryllium copper can be used for mold material for these hot spots.
- Utilize thermally conductive metals in mold construction. It may be a combination of steel as the base metal and aluminum as the mold cavity and core. It is possible for microcellular molding since the low injection pressure of microcellular process.
- If any standing steel in the mold with height/width is greater than 3:1, then we need to add extra cooling.
- Generally, running the cooling system with treated water (and a corrosion inhibitor) at 18°C (50°F) will provide better heat transfer and higher productivity than a system running at lower temperatures (such as ethylene glycol). Be sure that the freeze protection circuit in the chiller is set to shut off the chiller if the cooling fluid temperature drops below 7.2°C (45°F).

5.2.6 Mold Ejecting System

The microcellular foam processing may result in a demolding problem since the gas pressure in the cells will not be released immediately like in a gas-assist process. The regular mold design makes use of the plastic shrinkage during cooling to let the part stay on the core side, which is usually the moving half of the mold with ejection system for demolding. If the expansion of cells is so strong, the foamed part may stick on the cavity of the mold and cause a demolding difficulty.

The ejector pin used for microcellular foam may need to increase the size of the contact area. For the thin-wall microcellular part the solution will be to increase the number of same size of ejector pins. The demolding of the microcellular part will have the issues below for demolding:

- A soft surface of microcellular part compared to solid part
- High ejecting force required because of cell growth
- Rough surface of microcellular part

The first issue requires the ejector pin contact area increase to reduce the contact stress on the ejecting position on the part. Otherwise, either a visible ejector mark or a surface damage will be left on the microcellular part surface. Then, the cell growth of the part during cooling makes the ejecting design even more challenging because it touches both the core and the cavity very well and increases the ejecting force requirement. The last issue above may not be important, but the rough surface of the microcellular part may contribute some more friction force during ejection.

If the modifications of ejection system are not possible, the ejector force and speed need to be reduced. The longer cooling time may be necessary to increase the hardness of the surface of the microcellular part to be strong enough to withstand the ejector contact stress during ejection. Also, slowing down the ejection speed at the beginning of ejection helps to reduce the damage on the surface during ejection because it reduces the impact during ejection.

Overall, the most efficient way to improve the ejecting process of microcellular parts is the part design with larger draft angle, or taper. Similarly, the taper required for glass-reinforced material that significantly reduced the shrinkage value of the material during the cooling stage will create more difficulty for ejection. Therefore, the tapers of microcellular part need at least double the regular tapers of solid part.

5.3 MATERIAL PROPERTIES VERSUS WEIGHT REDUCTION

Microcellular structure brings a new structure–property relationship for homogeneous foam. Kumar and Vander Wel [35] studied the PC microcellular structure–property relation and proposed a good model for the tensile modulus prediction of PC microcellular foam without skin or ISF structure. However, the easy parameter of the microcellular part is the weight reduction percentage that can be calculated either from shot size reduction on the controller of injection molding machine or by just weighing the samples and comparing the results between solid and microcellular parts. There was a simple model for prediction of mechanical properties related to weight reduction of microcellular foamed injection molding parts presented by Xu et al. [19]. To use this model easily for the strength prediction of the real microcellular foam part during the injection molding process, it is the objective of the discussion in this chapter to propose this model with more mathematical and engineering details.

It is important to deal with the transition zone between skin and core for a conventional foamed part because the property in the transition zone is difficult to be calculated correctly without the empirical relation data for this zone. This transition zone is varied also with different materials. For a traditional foamed part the amorphous polymer such as PC usually yields broader transition zones than does a crystalline polymer [35]. But a good microcellular foamed part can provide a clear skin–core structure. Figure 3.3 shows a typical SEM of this section view from the unfilled PC (a typical amorphous material) sample of an injection molding microcellular foam part. A similar morphology of this kind of microcellular foam is shown in Figure 3.14 (4% weight reduction PBT foam with 30% glass fiber reinforcement, a typical semicrystalline material). Theoretically, every microcellular part should have this kind cell architecture to make full use of the benefit of microcellular technology. Otherwise, there are more improvements to be made to achieve this excellent true microcellular cell structure. It has unique characteristics for a microcellular plastic injection molding part with a uniform density profile across the foamed section except the skin. It is obviously different from the bell-shape density profile across the thickness of the conventional foamed part. There is no significant transition zone shown in the PC microcellular sample in Figure 3.3, and it is important to simplify the model of a microcellular foam part without this transition zone so that a simple mathematical model of mechanical property can be developed easily and realistically. Based on this fact, a sandwich structure model is described as a uniformly foamed core encased by skin frame without a transition zone between them (see Figure 5.2). The cell morphology effect for the tensile and flexural strength calculations is neglected in less than 10% error of prediction if the cell size is equal to or less than 100 μm and cells distribute uniformly. Since microcellular injection molding is controllable weight reduction technology, we introduce this easy prediction model that uses only weight reduction percent and skin thickness as the input data to calculate the final mechanical properties of tensile and flexural strength for a microcellular part. In addition to the input data above, the cell morphology must be considered for the impact strength modeling.

Skin thickness can be estimated by experimental data or by mathematical modeling. Simple experimental data show that skin thickness can be estimated to be about 10% of the whole thickness for crystalline thermoplastics and 15–20% for amorphous thermoplastics. Then, for gas counterpressure molding the skin thickness can be up to 30% of the whole thickness. On the other hand, the skin thickness of the co-injection process will comprise about 40–45% of the whole thickness.

Then, the core thickness t_c is defined as the following based on the estimation of skin thickness (refer to Figure 5.2)

$$t_c = h - 4\alpha_p^{1/2}\tau^{1/3}\left(\frac{T_f - \theta}{T - \theta}\right) \tag{5.4}$$

where t_c is the core thickness on the microcellular foam part, T is the polymer melt temperature, T_f is the polymer melt freezing temperature (melt point for crystal polymer, soft point for amorphous polymer), θ is the mold temperature, τ is time of melt touch the mold, that is approximately the hold time plus cooling time, and α_p is the thermal diffusivity of the polymer.

The measured average thickness of skin for the PC microcellular foamed part based on the SEM photo in Figure 3.3 is about 0.65 mm. The skin thickness can be estimated from the second term in Equation (5.4). The calculated skin thickness for the PC sample in Figure 3.3 with Equation (5.4) is 0.70 mm; and it is good enough (7.7% error) for the estimation of the skin thickness, which will be a key parameter to be used for calculating the strength for the real part.

The modeling procedure for the mechanical property versus weight reduction is illustrated in the following sessions. The real weight reduction in the core needs to be found to estimate the real core strength without skin effect. For filled material the weight of filler and glass fiber will be deducted from the weight calculation of core material. It is because the filler and glass fibers do not get the weight reduction, and only base material in the core reduces the weight by the foam. Finally, the property calculation as the whole part is estimated with the models proposed in the following sections.

5.3.1 Model for Tensile Strength (See Appendix C)

The tensile strength is the most important mechanical property, and it follows a general square-law relationship to predict the tensile strength for closed-cell structural foam for a long time. However, the microcellular foam test results show some different relationships between filled and unfilled materials. Therefore, the different models for unfilled material and filled material are proposed for investigating the relationships between tensile strength and weight reductions.

5.3.1.1 Unfilled Materials. To simplify the mathematical model reasonably, some of assumptions are given as follows:

- A sandwich model is proposed for the microcellular part. It means that a uniform foamed core is encased with a uniform solid skin (see the model shown in Figure 5.2).
- Average cell size is equal to or less than 100 μm.
- In tension, the entire cross section of the sample is subjected to a uniform stress, regardless of whether it involves skin or core sections.
- The elastic modulus of unfilled material microcellular foam is proportional to the square of density of the foamed core that obeys a traditional square-law relationship [14, 16].

Then, a tensile strength ratio R_{tu} is defined as [19]

$$R_{tu} = \frac{wh-(h-2t)(w-2t)+(1-R_{\text{core}})^2\,(h-2t)(w-2t)}{wh} \tag{5.5}$$

where R_{tu} is the ratio of tensile strength of unfilled foam to unfilled solid, w is the width of the sample, h is the whole thickness of the sample, t is the average skin thickness of the sample, and R_{core} is the ratio of weight reduction of foamed core.

A ratio of R_w is defined as the weight reduction ratio of the whole microcellular part including skin and core. Generally, R_w is known from the molded microcellular part by the method introduced above. Then, R_{core} can be calculated [19] as

$$R_{\text{core}} = \frac{wh(1-R_w)}{(h-2t)(w-2t)} \tag{5.6}$$

5.3.1.2 Filled Materials. Filled material includes fillers and glass fiber that will not change the specific gravity because filler and glass fibers will not foam. Therefore, when a calculation of the weight reduction of the core material with filler or fibers is performed, the filled material weight needs to be deducted from the real weight reduction calculation of the foamed core. Some extra assumptions (all assumptions above for unfilled material are also good here) are as follows:

- Filled materials are distributed in the foamed core uniformly.
- Glass fiber distributes uniformly in all directions.
- After foaming the weight percentage of filled material in the core decreases with the same ratio R_{core}.
- The elastic modulus of microcellular foam made by filled material is linearly proportional to the density of the foamed core.

Then, the real weight reduction ratio R_{gf} (see Appendix D) of core material without filled material will be [19]

$$R_{gf} = \frac{\rho_s - (1 - R_{\text{core}})\rho_{sf}(1 - W_{gf} R_{\text{core}})}{\rho_s} \tag{5.7}$$

where R_{gf} is the real weight reduction ratio of the core material without filled material, W_{gf} is the weight percentage of filled material before foaming, and ρ_{sf} is the density of filled solid material. In Equation (5.7), the same formula of R_{gf} is used for both filled materials because of the same method of separating the weight percentage of filled materials from unfilled base materials. R_{core} is the real weight reduction ratio of the core material with filled material, and it can be calculated by Equation (5.6).

A similar equation can be derived for the strength ratio of foamed part R_{tf} with filled material to the nonfoamed part with filled material [19]

$$R_{tf} = \frac{wh - (h - 2t)(w - 2t) + (1 - R_{gf})^m (h - 2t)(w - 2t)}{wh} \tag{5.8}$$

where m is different for different filled materials. For glass fiber, m is about 1.2; and for fillers, m is approximately 2.0 that is once again the same as the square-law relationship. The input data required for the calculation here are only the whole part weight reduction R_w and skin thickness t.

5.3.2 Model for Flexural Strength (See Appendix E)

It is similar to the tensile strength for a microcellular part, and two formulae will be introduced for both unfilled and filled materials. In addition to the all assumptions above for tensile strength calculation, the skin thickness for microcellular foam with injection molding is assumed to be so thin that the wrinkling of the upper skin under compression during flexural strength test will not be considered. Also, the effect of skin on the side (vertical effect) is not important for flexural strength since the microcellular foam part is usually 3 mm or less as a thin wall part compared to a structural foam part (4 mm above), and it can be neglected. Therefore, only the skin in thickness direction and tensile failure of the lower skin during flexural strength test are considered. If the part is thick, then the wrinkling of the upper skin under compression must be considered a mathematical model proposed by Vaidya and Khakhar [18] is recommended.

5.3.2.1 Unfilled Materials. Similar to the uniform-density structural foam [16], the flexural modulus of unfilled materials of microcellular foam shows proportionality to the square of the density. A flexural strength ratio of unfilled foam to unfilled solid R_{fu} is defined as [19]

$$R_{fu} = \frac{2t^3}{h^3} + \frac{6t(h - t)^2}{h^3} + (1 - R_{\text{core}})^2 \left(\frac{h - 2t}{h}\right)^3 \tag{5.9}$$

Then, the real weight reduction for flexural strength model in the foamed core is $R_{\text{core–f}}$, which can be calculated with the following equation [19]

$$R_{\text{core-f}} = \frac{h(1-R_w)}{h-2t} \tag{5.10}$$

The difference between the real weight reduction in flexural strength calculation and that in the tensile strength calculation is that the side skin effect is neglected in bending deformation.

5.3.2.2 Filled Materials. With the assumptions here being the same as those in the tensile strength calculation, the microcellular foam of filled material can be calculated with a simple model of flexural strength. It has been found that the flexural modulus of filled material microcellular foam decreases linearly as the density decreases. A similar equation can be derived for the strength ratio of the foamed part with filled material (both fillers and fiber glass) to the nonfoamed part with filled material [19]

$$R_{ff} = \frac{2t^3}{h^3} + \frac{6t(h-t)^2}{h^3} + (1-R_{gf})\left(\frac{h-2t}{h}\right)^3 \tag{5.11}$$

The input data required for this calculation are once again the whole part weight reduction R_w, the thickness of the whole part h, and skin thickness t.

5.3.3 Model for Impact (Izod) Strength

Izod impact (machined notch) strength modeling is very complicated work since it is strongly related to a structure–property–density relationship. There is no good report so far for Izod impact strength model related to weight reduction. Only falling-weight impact properties of microcellular foam for PVC without skin is reported from Juntunen et al. [36]. It shows a linear relationship between falling-weight impact strength and density reduction of the PVC microcellular foam. Juntunen et al. [36] also reported that the falling-weight impact strength of CPE microcellular foam retains a much higher percentage of virgin polymer impact strength than does the PVC microcellular foam. It means that there will be a complicated relationship between morphology and impact strength, and it is not a linear one in most cases. Izod impact strength prediction has been studied in the structure foam industry, and it is still a good reference for the microcellular foam. Refer to the results of Izod impact strength study for structure foam [16]; and based on the empirical data a special model tentatively with weight reduction, skin-to-core ratio, and cell density, it can be written as [19]

$$R_{iz} = \sqrt[3]{\frac{N_j}{N_{sf}}}(1-R_{\text{core}})^4\left(1-\frac{2t}{h}\right)^{-1} \tag{5.12}$$

where R_{iz} is the Izod impact strength ratio of foamed part to solid part, N_{sf} is reference cell density for standard foam, and it will be calculated with average cell size 100 μm, and N_j is cell density [1] for microcellular part witch is calculated from the foamed core section, and it is given as

$$N_j = \frac{1 - R_{\text{core}}}{0.0001 d_{\text{cell}}} \tag{5.13}$$

where d_{cell} is the average diameter of the cells with the millimeter unit. The impact test itself is sensitive for surface roughness. The break tough and break brittle materials will also have different results.

5.3.4 Results for Tensile Strength Model Application

The comparison result between experimental data and prediction of model verifies the tensile strength model with an acceptable accuracy of prediction. Also, the experimental data show that the filled and unfilled materials do have different results of tensile strength changes with weight reduction percentages.

5.3.4.1 Filled Materials. The filler-filled PP data are listed in Table 5.7 [19]. The results of predicted tensile strength loss versus weight reduction are acceptable with the error percentage range of prediction from 3.38% to 5.98%. The structure of PP sample shows that an average cell size is less than 100 μm, and average skin thickness is 0.38 mm.

The prediction error percentage of fiber-filled PBT is less than 6.93% (see Table 5.8 [19]). The PBT samples with 30% glass fibers have the uniform cell structure and small cell sizes in the range of 10–50 μm, and the average skin thickness is 0.25 mm. The cell growth may help to improve the glass fiber disorientation during mold filling and improve the mechanical properties of microcellular foamed glass-fiber-reinforced part.

5.3.4.2 Unfilled Materials. The predicted data in Table 5.9 [19] matches the measuring data reasonable well for the unfilled PC samples at 8% and 13% weight reductions with the error percentages only 1.2% and 2.6%,

TABLE 5.7 Tensile Strength Ratio Comparisons, 20% Talc-Filled PP [19]

Measured R_f	0.820	0.770	0.710
Predicted R_f	0.869	0.801	0.734
Weight reduction %	5	10	15
Error % of prediction	5.98	4.02	3.38

Source: Reproduced with copyright permission of Society of Plastics Engineers.

TABLE 5.8 Tensile Strength Ratio Comparisons, 30% Glass-Fiber-Filled PBT [19]

Measured R_f	0.910	0.840	0.750
Predicted R_f	0.879	0.838	0.802
Weight reduction %	4	10	15
Error % of prediction	–3.41	–0.24	6.93

Source: Reproduced with copyright permission of Society of Plastics Engineers.

TABLE 5.9 Tensile Strength Ratio Comparisons, Unfilled PC [19]

Measured R_f	0.850	0.770
Predicted R_f	0.840	0.750
Weight reduction %	8	13
Error % of prediction	–1.2	–2.6

Source: Reproduced with copyright permission of Society of Plastics Engineers.

respectively. The average skin thickness is about 0.5 mm and the average cell size is 45–100 μm, which is typical of PC microcellular architecture.

For tensile strength loss the skin thickness may not be a very important factor if a certain whole weight reduction percentage must be maintained. For example, the same 10% of whole weight reduction is maintained, but skin thickness and core weight reduction are changed accordingly. Then, even if the skin thickness increases about fourfold, the tensile strength of the foamed part with thick skin increases only 6% while the foamed core weight reduction must be changed from 11.7% to 28.3% to maintain the same 10% weight reduction for the whole part.

On the other hand, if the skin thickness increases about four times with the same 10% weight reduction of core, the tensile strength loss of foamed part will decrease 10.8%. But the whole weight reduction in this case is also decreased to about half of the original weight reduction. It is well known that the larger number of smaller cells can increase the tensile strength. The economic way to increase the tensile strength without compromising the weight reduction and cycle time is to make the cells smaller and the cell density higher.

5.3.4.3 Result Comparisons Between Filled and Unfilled Materials. A relationship in Figure 5.10 [19] is the calculation result of prediction of tensile strength change versus whole part weight reduction percentage from the tensile strength models proposed above for both filled and unfilled materials. The total skin thickness in Figure 5.10 is selected to be the same about 12% of whole part thickness for both unfilled and filled materials. The results in

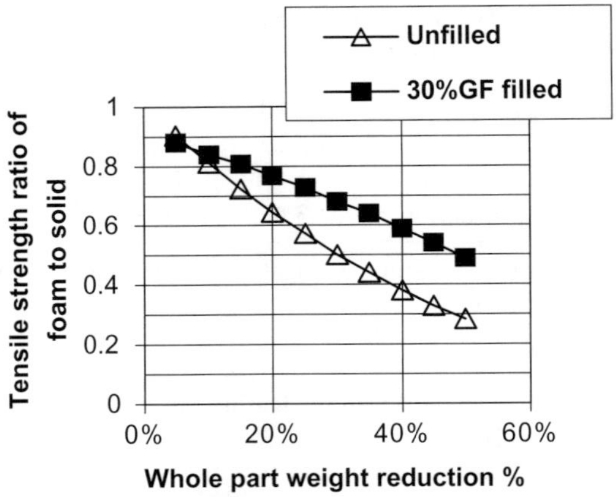

Figure 5.10 Tensile strength change versus whole part weight reduction % for filled and unfilled materials [19]. (Reproduced with copyright permission of Society of Plastics Engineers.)

Figure 5.10 show obvious less tensile strength loss of the glass-fiber-reinforced material compared to the tensile strength loss of unfilled material at the same 12% of whole part weight reduction. Generally, the weight reduction percentage suggested to be controlled in 10% for unfilled material and 20% for glass-fiber-reinforced material if less than 20% tensile strength loss is the target of part design limit.

Xu reported a comparison between measured and predicted tensile strength among different materials, such as PP with 20% talc, PBT with 30% glass fiber, and unfilled PP with same 10% weight reduction. The results constantly show that the filled material has less strength loss than unfilled material. These results verify that the tensile model is reasonable as a tool for prediction of the strength versus weight reduction. In addition, the prediction of strength loss versus weight reduction for filled material has less prediction errors compared to unfilled material because the filled material always has better cell structure than the unfilled same material. On the other hand, a conclusion from morphology study is that the tensile strength of microcellular foam is sensitive for the cell structure [19].

5.3.5 Results for Flexural Strength Model Application

The flexural strength verification for the model prediction is very good as well. The analysis shows that the real data sometimes show better results than predicted for filled material, which is an error from neglecting the fiber disorientation effect from the microcellular foaming process.

5.3.5.1 Filled Materials. The flexural strength of filler-filled PP data is listed in Table 5.10 [19]. The results of predicted flexural strength loss versus

TABLE 5.10 Flexural Strength Ratio Comparisons, 20% Talc-Filled PP [19]

Measured R_f	0.960	0.920	0.880
Predicted R_f	0.944	0.914	0.884
Weight reduction %	5	10	15
Error % of prediction	−1.67	−0.65	0.45

Source: Reproduced with copyright permission of Society of Plastics Engineers.

TABLE 5.11 Flexural Strength Ratio Comparisons, 30% Glass-Fiber-Filled PBT [19]

Measured R_f	0.910	0.850	0.790
Predicted R_f	0.912	0.883	0.857
Weight reduction %	4	10	15
Error % of prediction	0.22	3.88	8.48

Source: Reproduced with copyright permission of Society of Plastics Engineers.

Table 5.12 Flexural Strength Ratio Comparisons for Unfilled PC [19]

Measured R_f	0.980	0.910
Predicted R_f	0.940	0.905
Weight reduction %	8	13
Error % of prediction	−4.08	−0.55

Source: Reproduced with copyright permission of Society of Plastics Engineers.

weight reduction percentage are acceptable with the error percentage range of prediction from 1.04% to 2.27%. The fiber-filled PBT data are also good enough with the error percentage less than 8.5% (see Table 5.11 [19]). It has been found several times that the flexural strength of microcellular foam part with filled material is even higher than the one of the solid part, and it is also shown in the paper presented by Spindler [22]. It may be true that the glass-fiber-filled material part made with microcellular structure and fine cells help for disorientation of glass fiber. To predict this kind of data, the model must be developed with cell structure factor and it will be a further study in the future.

5.3.5.2 Unfilled Materials. The predicted data in Table 5.12 [19] agree well with the experimental data of unfilled PC samples at 8% and 13% weight

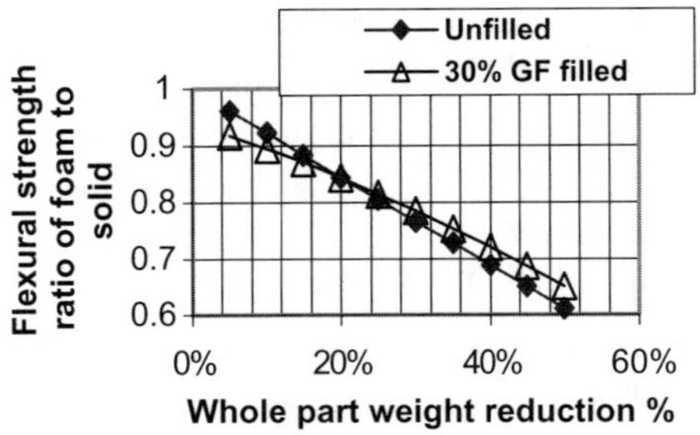

Figure 5.11 Flexural strength loss versus whole part weight reduction % for filled and unfilled materials [19]. (Reproduced with copyright permission of Society of Plastics Engineers.)

reductions, with the prediction error percentage only 4.08% and 0.55%, respectively.

For flexural strength skin thickness is a very important factor if a certain weight percentage of the whole foamed part must be maintained. With the same example of tensile strength if the skin thickness increases about fourfold, the flexural strength loss of foamed part decreases significantly about 11.5%, which is double the rate of tensile strength data, with the lower density foamed core but same whole weight reduction. It is because the maximum stress of the bending test during three-point flexure loading is on the skin.

5.3.5.3 Result Comparisons between Filled and Unfilled Materials. Figure 5.11 shows the results of modeling for flexural strength loss versus whole part weight reduction with the same materials and conditions as the ones in Figure 5.10. Theoretically, the weight reduction percentage should be controlled in less than 25% for unfilled material and less than 30% for glass-fiber-reinforced material if 20% flexural strength loss is allowed.

The similar comparison between predicted and measured flexural strength for different materials with 10% weight reduction are reported from Xu et al. [19]. Since it is the result of the same samples as the ones in the tensile test, it shows no significant difference among unfilled and filled materials so that the flexural strength related to cell structure is not as sensitive as the tensile strength. It can also be explained by the stress distribution in the thickness direction during the flexural strength test which shows zero stress near the center, neutral axis, maximum stress at surface, bottom and top surface [19].

5.3.6 Results for Impact Strength Model Application

Figure 5.12 shows the impact strength prediction results for 20 wt% talc-filled PP and 30 wt% glass-fiber-filled PBT. The results of modeling for 30 wt% fiber-glass-filled PBT are agreed with the experimental data very well in all

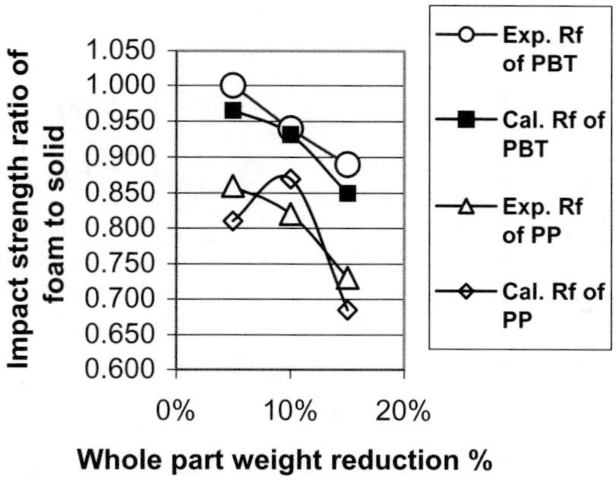

Figure 5.12 Impact strength ratio of solid to foam versus whole part weight reduction % for filled materials. PP, 20 wt% talc; PBT, 30 wt% glass fiber; Exp., experimental data; Cal., calculation results [19]. (Reproduced with copyright permission of Society of Plastics Engineers.)

ranges of weight reduction samples. But the 20 wt% talc-filled PP data present an acceptable prediction because 4% and 15% weight reduction samples have underpredicted impact strength and 10% weight reduction sample has overpredicted impact strength. The overpredicted impact strength is a result of fine cell size in the 10% weight reduction sample.

5.3.7 Promotion of Mechanical Properties of Microcellular Foamed Part

As a general trend, it is true that small cell architecture to promotes impact strength [9, 16, 23]. It is not only proved for microcellular injection molding, but also the sample of extruding and sample from batch process verify the same conclusion that a small cell size will result in better impact strength. Lee and others provide a data of impact strength of a rigid PVC sample with 30% weight reduction. The result shows that impact strength increases with the reducing cell sizes, as shown in Figure 5.13 [8]. Although the sample is made by a batch process, the result verifies better impact strength resulting from small cell size. On the other hand, the data shown in Figure 5.13 are truly the test results reflecting the sole effect of cell size of the sample without effect of skin or interface between skin and core. Shimbo and his team also show that the small cell size and high cell density of microcellular foam for unfilled polyethyleneterephthalate (CPET), PP, and PC can result in the increases of the tensile fracture strength (only comparison among foamed parts) [37]. However, if the foamed part is compared to the solid part, the overall trend is that the strength decreases even if the foam is made with fine cells. Furthermore, when the cell size is 3 microns or less, the tensile fracture strength is almost equal to that of the unfoamed one [38]. In addition, an increase in cell surface area causes molecular orientation, thereby affecting

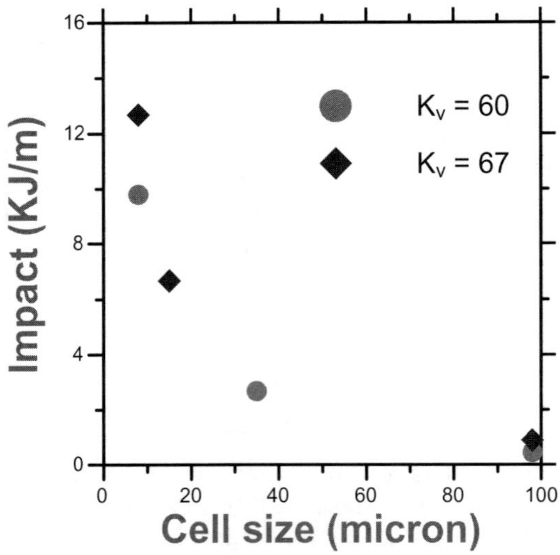

Figure 5.13 Effect of cell size in rigid PVC microcellular foam on the impact strength. (Courtesy of Trexel.)

strength improvement [37]. The ratio of surface area of unfoamed and foamed plastics can be estimated by [37]

$$\frac{S_F}{S_S} = m^{2/3} + (m-1)\frac{a_{\text{cube}}}{b_{\text{cell}}} \tag{5.14}$$

where S_s is the surface area of an unfoamed cube, S_F is the surface area of a foamed cube, m is the magnification from foaming, d_{cell} is the average diameter of the cells, and a_{cube} is the length of the cube to be measured for calculation of surface area of a foamed part.

According to Equation (5.14), the ratio of surface area becomes $S_F/S_S = 10^3$ in the case of $a_{\text{cell}} = 10\,\text{mm}$, and $m = 2$, and $d_{\text{cell}} = 10\,\mu\text{m}$. It means that the surface area of foamed plastics increases 1000 times compared to that of original unfoamed plastics [37].

Shimbo also reported that the tensile fracture strength of foamed CPET having a cell size of 1 μm maintains about 80% of that unfoamed CPET with twice the foam magnification [37]. This is a significant improvement with small cell size in microcellular foam.

It is possible to make the microcellular foamed part with reinforced materials such as glass fiber have higher flexural strength and impact strength under certain conditions of morphology and processing conditions. For example, a 30% glass-fiber-reinforced PBT has been made with 105% flexural strength ratio of foamed part to solid part. Kumar reported that the CPET microcellular foam, up to 40% weight reduction, is still maintaining 90% or more of the virgin CPET impact strength [38]. Kumar's data can be verified by Equation

(5.12) since the cells density can significantly increase the impact strength. As a visual evidence, the SEM pictures for either PBT with 30 wt% glass fiber made in microcellular injection molding or CPET sample in Kumar's paper show the extremely small cells and large number of cells [39]. The other factor may be that the fiber orientation has been improved compared to the solid part, and it may be the most significant evidence of the reason for the property promotion with microcellular foam for reinforced materials. To predict this kind property promotion, a morphology model needs to be considered in the mathematical model in the future study.

5.3.8 Conclusions for Mechanical Property Modeling and Improvement Methods

The conclusions for mechanical property modeling and improvement methods are summarized as follows:

- All models presented in this chapter can be used for general prediction of the mechanical property loss versus weight reduction percentage for any material. The prediction error percentage is less than 10% if the foamed part has uniform microcellular structure where the cell size is equal to 100 μm or less.
- Generally, filled material will have less tensile, impact, and flexural strength loss than the unfilled material at the same weight reduction percentage and skin thickness. The input data of weight reduction and skin thickness are good enough to predict the tensile and flexural strength ratio of foamed part to solid part.
- The skin thickness is not an important factor for the tensile strength loss versus weight reduction. But it is a critical factor for the promotion of the flexural strength of the microcellular part. For reasonable tensile and flexural strength loss, such as less than 20%, the recommended weight reduction is about 10–25%.
- The impact strength is strongly related to cell morphology in addition to the weight reduction percentage and skin thickness. The fine cells and uniform distribution are critical factors for high impact strength, and they are very helpful for increasing tensile and flexural strength.
- The toughness of PC has no improvement from foam unless it breaks brittle with the solid part. It is also true for notched impact strength of unfilled PC [20].
- The strength of foamed plastics generally increases from a decrease in cell size. This improvement of strength is more significant for crystalline plastics compared to amorphous plastics.
- It is an interesting conclusion that the microcellular foamed crystalline plastics increase the strength due to the molecular orientation. The explanation from Shimbo is that the cell density increases when the cell size

becomes small, and the surface area of cells increases greatly and, then, molecular orientation is caused with an increase in surface area [37].

- The strength reduction due to foaming may be controlled by using molecular orientation according to the theory from Shimbo et al. [37].

5.4 SURFACE QUALITY IMPROVEMENT FROM MOLD AND PART DESIGNS

The surface finish of a microcellular part must be considered as the important factor to design the part including the selections for mold, materials, process, and molding methods. The surface improvement through processing is fully discussed in Chapter 6, and it is well discussed in Chapter 8 with more details of special processing for better surface finish of microcellular foam. It is well known that in conventional microcellular molding the highly polished molds will not result in a high gloss finish with the microcellular process [8]. The special topics of part design, material, and mold selections are discussed in this chapter since the surface quality of a microcellular part excluded it as a visually exposed part.

5.4.1 Material

The aesthetic quality of microcellular molded parts can vary considerably. Most microcellular parts exhibit a splay-like appearance that is generally consistent from shot to shot. While many factors contribute to the differences in appearance of a microcellular part versus a part molded in a solid, the material itself is the principal variable in this equation.

5.4.1.1 Filled Material. As a rule, filled resins vary the least cosmetically. If the SCF (gas) percentage is to be controlled as minimum, the filled material may make some acceptable microcellular part with the similar surface quality of a solid part. One reason is that the gas bubbles that are spread over the surface of the parts oftentimes look quite similar to that of a glass-rich surface. Another reason is that less supercritical fluid dosage is usually required to attain a uniform microcellular structure with reinforced resins. Therefore, reducing the gas percentage in the microcellular part will improve the appearance of the foamed part.

5.4.1.2 Low Viscosity and Adhesion Material. A special developed material with low viscosity, or low adhesion to metal, tends to allow the sliding on the interface. Therefore, it is also an interface roughness improvement. The spherical grade PA 6/6 with 30% glass fiber reinforcement material developed by Rhodia has been successfully used for the automotive industry to make a microcellular part with excellent appearance. The material has the trade name Xcell™ and is registered as Technyl®. This material is optimized for microcellular processing, and it offers the part performance with no trade-off in surface finish. On the other hand, this special material may not be the solution for free-flow front roughness. Therefore, the injection time still needs to be controlled within 3 sec or less.

5.4.1.3 Solution for Unfilled Material. Unfilled polyolefin can oftentimes come out of solution so quickly that surface bubbles or "orange peel" form on the outer skin of the part. The appearance can be improved somewhat by the combination of material modification, processing, mold, and part design, such as the measurements used in current industry:

- Using a texture on the cosmetic areas of the part
- Switching to lighter colors
- Raising the mold temperature
- Using low-viscosity material
- Increasing the nucleation rate

5.4.2 Part Design for the Surface Quality Issues

Microcellular parts are designed with some criteria for the possible smooth surface quality through easy processing. Although most of them were already discussed more or less in the part design section above, the important design details are emphasized in the following.

A uniform thickness throughout the part will avoid the big cells, and prevent the pressure suddenly drop during mold filling.

The new overlapped smooth surface on the foamed bottom layer has been developed. It is similar to co-injection, but only one side has smooth surface. This technology is named Dolphin Skin processing and will be discussed in Chapter 8.

The texture on the microcellular part needs to be designed for the depth and pattern to cover the foamed surface without obvious observation of swirl. A suited and effective part design to improve the visibility of the surface roughness of the microcellular part is to use texture on the part surface where it is allowed. In addition, different textures can conceal the possible surface defects. A careful designed texture on the part surface will superpose the rough surfaces of microcellular part so that the appearance of silver streaks can be concealed by a diffuse backscatter [30].

5.4.3 Mold Design for the Surface Quality Issues

The hot mold surface has less friction or adhesion between mold metal surface and melt, and most importantly it has no cold skin on the mold surface so that the surface of a part slides on such surface acting as a plug flow (the theoretical velocity distribution is equal in the thickness direction). Then, ideally, there is no shearing between flow layers in depth direction. In addition, the breaking bubble from a free-flow surface will be easily ironed or repaired by a hot mold surface under the cavity pressure.

There is a new joint development between Trexel and a Japanese molding company Ono Sangyo, and it is named MuCell Gloss™ [8]. It finally opens the way to a high-surface-quality microcellular part. The microcellular technology from Trexel is combined with the rapid heat cycle molding (RHCM) from Ono Sangyo. The RHCM molding process involves adding two fully independent temperature control circuits in the mold so that the surface can be maintained above the heat distortion temperature of the polymer during the mold cooling stage and can then be rapidly cooled once mold filling is complete. Although there must be a small cycle time penalty for using this technology, perhaps two or three seconds, an important gain is the improvement in surface finish and big improvement to remove the weld lines. While molding onto a warm mold surface, the microcellular part maintains the maximum surface gloss. This technology is also particularly suitable for glass- or mineral-filled resins, where it prevents the reinforcement from making the surface. It has an additional benefit with RHCM technology that provides slow cooling at the mold surface. It allows a high crystalline content layer to develop, giving significant improvements in flexural modulus and surface hardness of the microcellular part. The RHCM process also enables excellent replication of microscopic surface features to be achieved, prompting Sabic Innovative Plastics (previous GE Plastics) to explore the microcellular plus RHCM for optical disc production. Key target applications are high-quality complex molding where weld lines currently make the painting necessary, or where a high gloss surface must be achieved with low levels of molded-in stress. While a traditional microcellular process is used most often as a cost reduction, MuCell Gloss™ will be used predominantly where a higher-quality surface can displace a secondary process and reduce the overall system cost. It has been tried for a highly complex automotive center console fascia that is achieved class A surface with 6–10% weight reduction. Also, special mold alloy steels must be used to work on this highly thermo-stress environment.

On the other hand, hot mold may not always work for the smooth surface of crystalline material foam. This is because more crystallization on the hot mold surface will create high nonuniformity of cell structure and, then, rough surface. A balance between percentage of crystallinity and small cell structure with warm mold is still possible to result in better surface quality of crystal material microcellular foam.

The special coating on the mold surface will achieve the class "A" surface as well. The details have been discussed above [30, 31]. However, there is no commercial coating really used for microcellular production mold yet.

Turng and his group presented the results of inserting PTFE film on the mold. Different thickness of PTFE film on the mold has been studied, and the conclusion is that 175 μm thickness is enough to eliminate the surface swirl marks of a microcellular injection molding part [40].

The potential benefits of the microcellular molding technology are genuine. The microcellular process can lower part costs and improve dimensional stability. Maximum savings with microcellular processing can only be obtained by designing the part and mold for the process while giving careful consideration to material selection. However, the microcellular molding is not only a process but also a technological breakthrough that has, in a sense, created literally thousands of new materials that will benefit the part design of injection molding. Understanding how to harness this process, design the part and mold accordingly, and take advantage of these new materials is the challenge that the plastics industry will face over the next several years.

REFERENCES

1. Shu, N. P. *Innovation in Polymer Processing*, edited by James F. Stevenson, Hanser/Gardner Publications, Cincinnati, 1996, Chapter 3, pp. 93–149.
2. Xu, J., and Pierick, D. *J. Injection Molding Technol.* **5**, 152–159 (2001).
3. Kumar V., and Vander Wel, M. M. *SPE ANTEC, Tech. Papers*, 1406–1410 (1991).
4. Levy, S., and DuBois, J. H. *Plastics Product Design Engineering Handbook*, Van Nostrand Reinhold, New York, 1977, pp. 224–231.
5. Malloy R. A. *Plastic Part Design for Injection Molding*, Hanser/Gardner Publications, Cincinnati, 1994, pp. 47–75.
6. Okamoto, T. K. *Microcellular Processing*, Hanser/Gardner Publications, Cincinnati, 2003, pp. 123–135.
7. Xu J. *SPE ANTEC, Tech. Papers*, 2770–2774 (2006).
8. Trexel Inc. Web site, http://www.trexel.com/.
9. Xu, J., and Kishbaugh, L. *J. Cell. Plastics* **39**, 29–47 (2003).
10. Xu, J. *SPE ANTEC, Tech. Papers*, 2158–2162 (2008).
11. Xu, J. *SPE ANTEC, Tech. Papers*, 2089–2093 (2007).
12. Xu, J. *SPE ANTEC, Tech. Papers*, 594–598 (2004).
13. Wang, J., Lee., J. W. S., Yoon, J. D., Park, C. B. *SPE ANTEC, Tech. Papers*, 2168–2172 (2008).
14. Semerdjiev, S. *Introduction to Structural Foam*, SPE, Inc., Towanda, PA, 1982, pp. 5–6.
15. Shutov, F. A. *Handbook of Polymer Foams and Foam Technology*, edited by D. Klempner, and K. C. Frisch, Oxford University Press, New York, Chapter 3, 1991.
16. Thron, J. L. *Thermoplastic Foams*, Sherwood Publishers, Hertford, UK, 1996.
17. Thron, J. L., and Shutov, F. A. Structural foams, in *Encyclopedia of Polymer Science and Engineering*, Vol. **15**, John Wiley & Sons, New York, 1989, pp. 771–797.
18. Vaidya, N. Y., and Khakhar, D. V. *J. Cell. Plastics* **33**, 587–605 (1997).

19. Xu, J., Kishbaugh, L., and Casale, M. *SPE ANTEC, Tech. Papers*, 1905–1909 (2002).
20. Bledzke, A. K., Rohleder, M., and Kirschling, H. *SPE ANTEC, Tech. Papers*, 944–949 (2009).
21. Sabic Innovative Plastics. Web site, http://www.sabic-ip.com/.
22. Spindler, R. Injection molding with mucell microcellular technology. Advanced Injection Molding Processes, CFI Group, Ann Arbor, MI, August 21–22, 2001.
23. Michaeli, W., Florez, L., Krumpholz, T., and Obeloer, D., *SPE ANTEC, Tech. Papers*, 1024–1028 (2008).
24. Kramschuster, A., Cavitt, R., Ermer, D., Chen, Z., and Turng, L. S. *Polym. Eng. Sci.* **45**, 1408–1418 (2005).
25. Serry Ahmed, M. Y., Atalla, N., and Park, C. B. *SPE ANTEC, Tech. Papers*, 1109–1112 (2008).
26. Kishbaugh, L., Kolshorn, U., and Bradley, G. Vibration and ultrasonic welding conditions and performance for glass fibre filled PA 6 and PA 6.6 injection moulded using the MuCell® microcellular foaming process. Smithers Rapra Blowing Agents and Foaming Processes Conference, RAPRA, May 2007.
27. Witzler, S. Assembly Methods, *Injection Molding Mag.* **April**, 29 (2009).
28. Robin, K. *Plastics Technol.* **April**, 27 (2009).
29. Menges, G., and Mohren, P. *How to Make Injection Molds*, Hanser Publishers, Munich, 1993.
30. Michaeli, W., and Cramer, A. *SPE ANTEC, Tech. Papers*, 1210–1214 (2006).
31. Mapleston, P. *Modern Plastic*, **October**, p. 29 (2000).
32. SPI Machinery Division. Recommended Guideline for Entrained Gas Processing in Horizontal Injection Molding Machines (EGPHIMM), May 2003.
33. Beaumont, J. *Plastics Technol.* **April**, 64–69 (2001).
34. Witzler, S. *Injection Molding Mag.* **April**, 18 (2000).
35. Kumar V., and Vander Wel, M. M. *SPE ANTEC, Tech. Papers*, 1406–1410 (1991).
36. Juntunen, R. P., Kumar, V., Weller, J. E., and Bezubic, W. P. *J. Vinyl Additive Technol.* **6**, 93–99 (2000).
37. Shimbo, M., Higashitani, I., and Miyano, Y. *J. Cell. Plastics* **43**, 157–167 (2003).
38. Shimbo, M., Baldwin, D. F., and Suh, N. P. *SPE ANTEC, Tech. Papers*, 309–313 (1993).
39. Kumar, V. *SPE ANTEC, Tech. Papers*, 1892–1896 (2002).
40. Lee, J., and Turng, L. S. *SPE ANTEC, Tech. Papers*, 1662–1666 (2009).

6

PROCESS FOR MICROCELLULAR INJECTION MOLDING

It is well known that the microcellular foam process follows four basic steps [1]: gas dissolution, nucleation, cell growth, and shaping. All four steps must occur in a microcellular injection molding machine, such as in (a) a most popular reciprocating screw (RS) of a injection molding machine and (b) a proper mold to successfully produce microcellular foam. Furthermore, all four steps rely on the right processing technologies once the equipment and material are selected.

Compared to regular molded nonfoaming parts, microcellular foams of injection molding have been developed for a short cycle and have unique stable dimension benefits. However, the wrong process parameters of injection molding may give less benefit to microcellular parts and may cause some premature failure of the parts. The most popular method for making a single-phase solution is to add blowing agent in the plasticizing unit with the reciprocating screw (RS). Unless specified, it is the process unit to be discussed in this chapter. There are already some publishers who have given some technical details of real RS injection molding for microcellular foams [1–6]. The microcellular process parameters from molding and plasticizing were mixed in most previous studies so that the results were complicated for the real application. In this chapter the analyses of processing for microcellular foam will be divided into two stages: plasticizing and molding. With the methodology of two-stage analysis the complexity of microcellular injection molding is simplified, and it also helps to quickly to find out the optimized processing conditions for the best quality of a microcellular part [7]. The first-stage

Microcellular Injection Molding, Edited by Jingyi Xu
Copyright © 2010 John Wiley & Sons, Inc.

process includes material plasticizing in the normal three-zone screw, gas dosing into barrel, gas mixing, and diffusing in the molten plastics. It is obvious that the first stage occurs in the plasticizing unit of the injection molding machine plus gas dosing system including chemical blowing agent option. The second stage is the molding stage, which includes the nucleation, cell growth, and mold filling and shaping. Therefore, the second stage is a system with injection unit and mold unit. If the processing parameters are set up with these two stages separately, there will be several topics focus on the specific issues that are related to the microcellular molding only, such as gas dosing; gas mixing and diffusing into melt; nucleation and cell growth; and shaping in the mold. The guidelines and empirical processing data of microcellular injection molding are summarized for each topic as follows.

6.1 GAS DOSING

The essential requirement of microcellular foaming is the ability to create a single-phase solution of gas and molten polymer. Further requirement is to continually hold the gas under certain pressure in single-phase solution throughout the cycle of the machine until the single-phase solution enters the mold. However, plasticizing the plastic in the barrel must be well done first before adding gas. If the single-phase solution is poor in the first stage, the molding process cannot make acceptable microcellular foams in the second stage. Therefore, a complete gas dosing analysis is introduced below for every well-known important parameter, such as: gas percentage, back pressure of recovery, melt temperature, output rate of screw, pressure difference between gas and melt at the gas injector position, and residence time of shearing for a gas–melt mixture in the screw.

6.1.1 Output Rate of Microcellular Screw

The output rate of microcellular screw needs to be either (a) predicted when the screw is designed or (b) measured during the new screw test. It is important to calculate the settings of a gas dosing unit to reach (a) the adequate gas dosing weight or (b) volume percentage in molten plastic. The output is also the key parameter to see how fast the molten plastic moves through the gas dosing spot in the position of barrel and cleans up the previous gas package with fresh molten plastic for the next gas droplets from gas injector. For quick reference empirical data the output rate of microcellular screw is about 10–20% less than that of the conventional screw. The reason is the higher leak flow through the flights of screw and inside of screw channels caused by the low viscosity of gas-rich material and high back pressure in the screw required to maintain a single-phase solution in front of screw tip. However, the low torque (about 10% lower with gas rich material in the screw) gives a possible increased screw speed if the microcellular screw is necessary to match up the original output of material without gas in the same size screw. At least the test

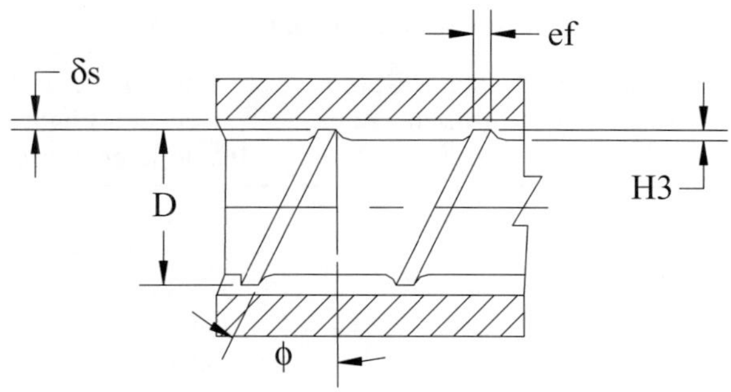

Figure 6.1 Screw geometry for calculation the output rate of recovery.

was successful for polyolefin material where the screw circumferential velocity is increased up to 1.25 m/sec without burning the molten plastics with inert gas in the melt. It is because inert gases, such as N_2 and CO_2, may protect the plastic burning inside of barrel at such high shear rate.

The geometry of a single screw is illustrated in Figure 6.1 with necessary dimensions to be used for calculating the output rate of screw recovery. The metering model is used for estimating the output rate [8, 9]. For quick estimation of the metering rate to decide the gas dosing rate, the Isothermal Newtonian Fluid Model is good enough. It is a modified one-dimensional flow between parallel plates:

$$Q_v = \frac{\pi^2 D^2 N_s H_3 \cos\phi \sin\phi}{120} - \frac{\pi D H_3^3 \sin^2\phi}{12\eta_s}\frac{(P_t - P_s)}{L_3} - \frac{\pi^2 D^2 \delta_s^3 \tan\phi}{10\eta_c e_f}\frac{(P_t - P_s)}{L_3} \tag{6.1}$$

where Q_v is the volume rate of screw recovery, D is the outside diameter of screw, N_s is the screw rotation speed (rpm), H_3 is the depth of metering zone in the screw, ϕ is the helix angle of flight in the screw, L_3 is the length of metering zone in the screw, P_s is the pressure in molten plastic near the gas injector, P_t is the pressure in the single-phase solution accumulated in front of screw tip, η_s is the modified viscosity of molten plastic in screw channel, η_c is the modified viscosity of molten plastic in the clearance between top of flight and barrel, δ_s is the clearance between outside diameter of screw and inside diameter of barrel, and e_f is the axial width of flight.

The viscosity in the screw channel and clearance between top of flight and inside diameter of barrel will be determined by the average shear rates in the screw channel and clearance between top surface of flights and inside diameter of barrel:

$$\dot{\gamma}_s = \frac{\pi D N_s}{30 H_3} \tag{6.2}$$

where $\dot{\gamma}_s$ is the representative shear rate in the screw channel, and

TABLE 6.1 Shear Rate in Different Screw Diameters at Maximum Screw Speed[a]

Screw Diameter (mm)	Shear Rate in Metering Zone (1/sec)	Shear Rate in Mixing Zone (1/sec)	Shear Rate in the Clearance Between Flight of Screw and Inside Diameter of Barrel (1/sec)
30	897	448	11,278
40	673	336	9,577
50	563	282	8,573
60	497	248	7,898
70	453	226	7,404
80	420	210	7,024
90	395	197	6,720
105	375	187	6,469
120	344	172	6,077
135	332	166	5,919
150	313	156	5,656
160	305	153	5,546
170	298	149	5,446
180	292	146	5,354

[a]Maximum screw speed is calculated with 1(m/sec) circumferential velocity at the outside diameter of the screw.

$$\dot{\gamma}_c = \frac{\pi D N_s}{60\delta_s} \tag{6.3}$$

where $\dot{\gamma}_c$ is the average shear rate in the clearance between top of flight and barrel.

The details of viscosity calculations are discussed in Chapter 9. It can be determined by representative shear rates in Equations (6.2) and (6.3), at certain processing temperature. The maximum shear rates in both metering zone and clearance between flight and inside diameter of barrel have been estimated in Table 6.1. All the calculations are based on the 1 m/sec as the maximum circumferential velocity at the outside diameter of screw. P_s is the reading pressure from the pressure transducer near the gas injector, and P_t is the setting value for the back pressure of screw recovery. The back pressure P_t in front of screw tip and the melt pressure in the gas dosing position is very close and they are almost constant at different screw speeds (refer to Figure 7.9 in Chapter 7). Therefore, using P_t to represent the melt pressure near the gas dosing position is approximately close enough for the real pressure distribution, as shown in Figure 7.9. Then, the output rate can be calculated quickly with Equation (6.1).

It is important to estimate the output rate for microcellular screw with all three different flows (see Figure 6.2). To show clearly the flow in the clearance δ_s, the clearance in Figure 6.2 is exaggerated with unscaled dimensions. It is just schematically showing the distribution of every flow pattern in the screw

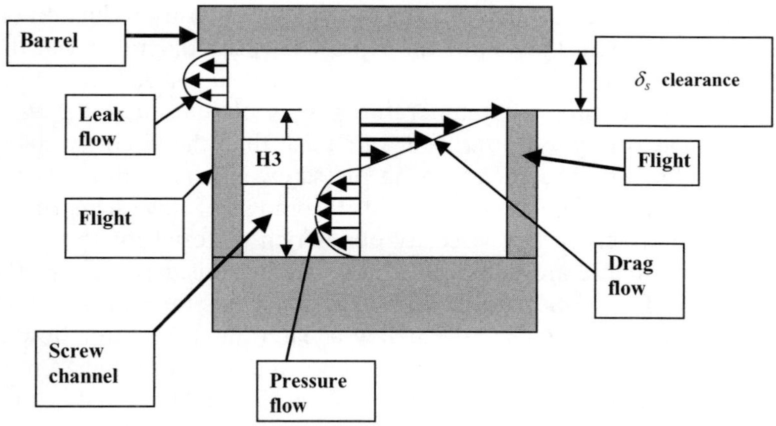

Figure 6.2 Three different flows during screw recovery.

and barrel assembly. The first term in Equation (6.1) is the drag flow rate that is the volume rate to convey material forward. It is proportional to the channel depth H_3 and the screw rotation speed N_s. The second term in Equation (6.1) is the pressure flow rate, or the back flow rate. It is the volume flow rate caused by pressure gradient without effect of side of flights. Pressure flow rate is usually inversely proportional to the viscosity of the material in screw channel. It is also proportional to the pressure gradient in the channel that is related to the back pressure P_t directly. It is important to pay attention to the effect from depth of channel H_3 because the pressure flow rate is proportional to the third power of this depth. Although the volume rate is mainly the difference between the drag flow and pressure flow, the leak flow rate of the third term in Equation (6.1) is not negligible. It is because the back pressure of microcellular injection molding is usually 5–10 times higher than the one for regular injection molding. It is similar to the pressure flow rate and greatly depends on the clearance δ_s by the third power. On the other hand, the leak flow is the same as the drag flow rate that is proportional to the second power of the diameter of screw.

To estimate the output rate of a microcellular screw is more complicated than it would be for a conventional screw because of gas-rich material in the downstream of a metering zone. This viscosity reduction may have three effects for the screw recovery. One is that the torque requirement for the screw may be reduced. Also, the positive effect from low viscosity is the low resistance to the metering pumping action. However, the low viscosity may cause easier back flow of any gas-rich material to the metering zone. Sometimes the screw design fails because the gas-rich material leaking back in metering zone can reduce the positive pumping capability in the metering zone. Therefore, it is essential to design a screw where the melt pressure in the metering zone is always higher than the melt pressure in the downstream where the gas dosing spot and mixing zone for the single-phase solution is located.

Refer to Figures 9.3, 9.4, and 9.5 for the unique viscosity reduction of gas-laden molten polymer; they show the typical viscosity diagrams that display the effect viscosity in GPPS as a function of nitrogen gas concentration and shear rate at different melt temperatures. It is obvious that the viscosity change with gas is larger at low shear rate zone than the viscosity change at high shear rate zone. The shear rate in the screw channel can be calculated from Equation (6.2), and it is usually in the range of low shear rate zone. Therefore, the effects of low viscosity of gas-rich material for the effects to the screw performance are significant. They shall be considered during the gas dosing process. The ABS material with CO_2 gas shows the similar trend that the viscosity reduces significantly at low shear rates no matter how much percentage of CO_2 gas is added into molten polymer [6]. The effect of nitrogen gas dissolution on molten ABS polymer viscosity is similar to that of the CO_2 gas. Generally, this trend of gas-laden melt of polymer with low viscosity especially pronounced at low viscosity rate is quantitatively similar to that of most resins [6].

6.1.2 Gas Dosing Percentage

Based on the gas solubility in different polymers (refer to Chapter 2) from available publications, the gas weight percentage in the current microcellular injection molding machine can be estimated as the half of the published data [10–12]. The current microcellular equipment can simply not handle that much high solubility data made by experimental of batch process. On the other hand, the highest gas solubility in the injection molding process does not need to be as high as the gas solubility in each material tested in the batch process. It is because high shearing rate and high processing pressure are available in the injection molding process that helps to speed up the gas diffusion process significantly compared to batch process. Therefore, high shearing rate and high pressure are actually the key advantages of the injection molding process compared to the batch process. However, the gas dosing time in an injection screw is only one minute or less, instead of hours of gas dosing time required in the batch process. Such a short dosing time is obviously a disadvantage of the injection molding process. From many years' practices, the recommended gas weight percentage is summarized in Table 6.2. As the data in Table 6.2 show, a much lower gas weight percentage is required for real injection molding, and the characteristics of injection molding gas dosing is small gas dosage at high pressure, high temperature, and short dosing time. Then, for the discontinuous injection molding process the required gas flow rate can be given by

$$Q_g = \frac{R_{scf} Q_{\text{part}} \rho_{\text{poly}}}{t_{gv} \rho_g} \tag{6.4}$$

where Q_g is the flow rate of gas in gas injector, ρ_g is the density of gas, ρ_{poly} = density of unfoamed polymer, Q_{part} is the volume of the part, R_{scf} is the

TABLE 6.2 Recommended Gases Level in Different Molten Plastics

Plastics	Nitrogen Weight %	Carbon Dioxide Weight %	Comments
PE	1–2	5–8	Highly susceptible to voids
Unfilled PP	1–2	4–8	Can actually get to 2% in high L/t applications
Filled PP	0.2–0.8	2–8	
Glass-filled PP	0.3–0.5	1–5	Glass much more effective than talc, then the gas % can be reduced
GPPS	0.3–0.7	2–4	Very good foamability
PC	0.4–0.6		Very good foamability
HIPS	0.5–1		Impact modifiers may make cell structure more difficult; however, if the size is small, there may be excellent cell structure.
ABS	0.4–0.8	5–6	Impact modifiers may make cell structure more difficult; however, if the size is small, there may be excellent cell structure.
Impact modified PC	0.5–0.8		Impact modifiers may make cell structure more difficult; however, if the size is small, there may be excellent cell structure.
Glass-filled PC	0.3–0.5		Independent of impact modifiers
Unfilled POM	0.3–0.5		
Glass-filled POM	0.2–0.4		
Unfilled PA	0.4–0.8		
Glass-filled PA	0.2–0.5		Excellent cell structure control
Glass-filled PBT	0.2–0.4		Excellent cell structure control
Glass-filled PET	0.3–0.5		Excellent cell structure control
Unfilled PSU	0.5–0.7		
Unfilled PEEK	0.4–0.7		
Glass-filled PSU	0.3–0.4		Excellent cell structure control
Glass-filled PEEK	0.3–0.5		Excellent cell structure control

overall ratio of weight of gas to the weight of the part, and t_{gv} is the opening time of gas injector during dosing.

R_{scf} is just an overall ratio of weight of gas to the weight of the part, and it is recommended in Table 6.2. The real gas dosing process is much more complicated. The velocity analysis of reciprocating screw is provided for correct calculation for gas percentage in the molten polymer before it is mixed uniformly [5]. This analysis result may explain the reasons for some failures of dosing processes.

The reciprocating-screw (RS)-type injection molding machine is the most popular injection molding equipment used worldwide. The reciprocating screw rotates like an extruding screw to plasticize the plastic material, and it conveys the molten plastic forward until it is accumulated in front of the screw tip. This accumulation of molten material (shot size) occupies space in front of the screw tip and pushes the screw moving back axially at velocity U_t, which is either calculated from extrusion theory [Equation (6.1)], or measured from output rate of a real machine. On the other hand, the molten material inside of the screw flows along the spiral channel of the screw, which is the melt velocity related to the screw itself. At the same time, the molten plastics inside of the screw move backward axially with the screw together related to the barrel. Usually this axial movement of the screw has been neglected in the screw performance study for the regular injection molding screw because it is relatively small compared to the velocity U_c, which is the circumferential linear velocity of rotation on the outside diameter of the screw [13]. The calculation of velocity ratio of U_t/U_c is only about 0.018–0.054 for the general-purpose (GP) screw with diameters from 25 mm up to 180 mm (see Figure 6.3). It means that U_t is only about 1.8–5.4% of the U_c. It is also the obvious trend that the velocity ratio of U_t/U_c is decreased with the screw diameter increasing. Usually the screw output rate is determined by the metering section so that the melt velocity related to the screw can be calculated with metering section geometry. If the screw melt channel in the metering section is simplified as an annual channel (neglecting the flights) with inside diameter (ID) as the root diameter of screw and outside diameter (OD) as the screw outside diameter the melt velocity related to screw in this annual channel in axial direction is defined as U_a. Then, all GP screws with the diameters from 25 mm to 180 mm have been studied with a velocity ratio U_t/U_a. The result is about 0.13–0.29 and is shown in Figure 6.3. It means that U_t is about 13–29% of the U_a. Therefore, the axial velocity of the screw moving back related to the barrel is comparable to the axial melt velocity related to the screw itself, and it cannot be neglected when some processing needs to know how much the influence of screw axial movement is related to the barrel for the correct calculation of real gas dosage. A gas dosing process to add gas through the barrel is in a fixed position for the dosing port in the barrel while the screw is moving axially related to the barrel [14, 15]. It causes the dosing position related to the screw to change with the screw moving backwards. In addition, most special foaming processes use a two-stage reciprocating injection screw with a decompression zone, where the gas or liquid color is added. If we assume that the depth of the decompression zone in the screw is double the depth of the metering zone in the screw, a melt velocity related to a screw at the decompression zone in axial direction U_d can be recalculated. Then, a ratio of U_t/U_d is shown in

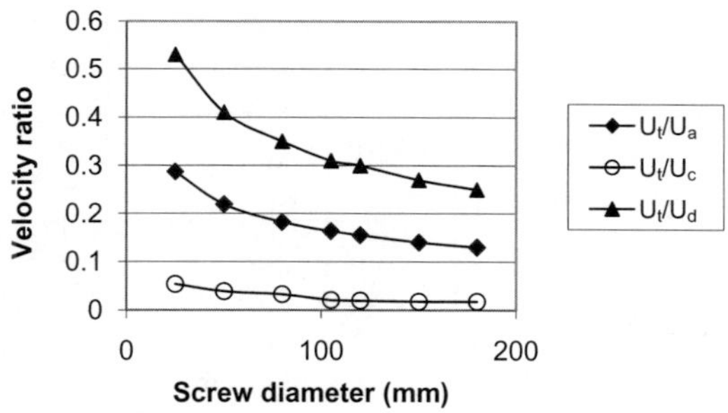

Figure 6.3 Velocity ratio versus screw diameter.

Figure 6.3. It is in the range from 0.25 to 0.53. It means that U_t is about 25–53% of theU_d. The ratio of U_t/U_d is significantly important to calculate the real velocity of melt related to the barrel during two-stage screw recovery. The ratio of U_t/U_d has been ignored in the injection molding industry for this quantitative relationship among these velocities, and the influence on the real processing such as liquid agent dosing or venting processing has also been ignored. With this velocity analysis a few simple formulae of the relationship among these velocities versus the screw geometric parameters have been derived. A detailed application of the velocity analysis for liquid agent dosing is discussed with focus on the gas dosing process. It shows an important effect for dosing percentage changed with the screw geometry. It is also useful to estimate the mixing variation with the different mixing history and time related to the velocity differences to be calculated with this model. It provides a useful mathematical tool for necessary velocity analysis of all the processes similar to the liquid agent dosing process.

If we chose the barrel as a fixed static reference that is the same reference as the earth, then an absolute velocity can be defined as a velocity related to the barrel reference only. To simplify the mathematical model, some assumptions are made for this model:

- Know the screw output volume rate Q_v that can be measured or estimated by mathematical model above [see Equation (6.1)].
- The absolute axial velocity U_t of the screw moving back during recovery and the melt velocity related to the screw itself are constant during the whole recovery stroke.
- The absolute axial velocity U_m of the plastic melt related to barrel is constant during the whole recovery stroke.
- Neglect the total volume of flights in the screw when the flow channel area is calculated.

From Figure 6.4, it is easy to understand a simple case of continuum concept that a cylindrical volume of accumulated shot size (a) in a unit time is equal

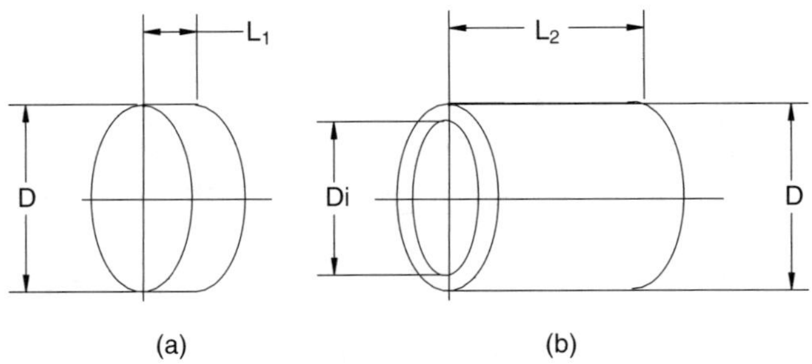

Figure 6.4 Continuum concept of shot size in front of screw tip and inside of screw channel.

to the annular volume inside of screw melt channel (b) in the same unit time. In addition, both metering section and dosing section are assumed with full channel of melt. The recovery process is usually determined by a metering section. Then, an absolute axial velocity formula for plastic melt related to a barrel can be derived (see Appendix F):

$$U_m = \left[\left(\frac{D^2}{D^2 - D_i^2}\right) - 1\right] U_t \tag{6.5}$$

where D is the outside diameter of screw and D_i is the root diameter of a screw located in the port of barrel for liquid agent injection.

U_t can be given:

$$U_t = \frac{4Q_v}{\pi D^2} \tag{6.6}$$

If the depth of dosing zone is shallow (less than twice of the metering depth) and the pitch is small (less than outside diameter of screw), the flight volume may need to be considered. Then, the U_m formula becomes

$$U_m = \left(\frac{D^2}{D^2 - D_i^2 - \dfrac{4H_e N_e W_e}{\pi}} - 1\right) U_t \tag{6.7}$$

where H_e is the depth of a mixing channel in a dosing zone, N_e is the number of flights in a dosing or venting section, and W_e is the width of flights.

We can further define a ratio of R_u:

$$R_u = \frac{U_m}{U_t} \tag{6.8}$$

The R_u ratio in Equation (6.8) represents an axial velocity difference between the plastic melt forward movement related to the fixed position in the barrel and to the screw backup movement during screw recovery. From Equations (6.5), (6.6) and (6.7), it is clear that the real velocity relationship between U_m and U_t is only determined by the screw geometry in the metering and dosing zones. It will be a fixed value for the specific screw geometry. Therefore, analysis of microcellular processing will be easier with known screw geometry.

A typical process that is necessary to analyze the processing parameters related to the velocity ratio in Equation (6.8) is the liquid dosing process. A liquid dosing process is to feed fluid agent, such as gas at supercritical state, liquid blowing agent, liquid color, and other liquid additives, into the melt in the middle of the barrel. The model of velocity analysis above has been successfully used to develop some relationship among those velocities to the local real dosing rate. Once the gas is at a supercritical state, it has liquid-like performance. Therefore, supercritical fluid (SCF) for all gases used in the microcellular process is a special kind of liquid-like agent so that the name of liquid agent will be used in the following analysis. The dosing ratio of the liquid agent to plastic melt is calculated by the agent weight flow rate divided by the plastic weight output rate. It is an overall average ratio for a consistent process, and it works well for the extrusion dosing process since the extrusion screw does not move in the axial direction. But it is confusing to calculate correctly about the real dosing ratio of injection molding for reciprocating screw with the liquid dosing port fixed in the barrel. The velocity analysis makes this calculation easy and results in an intensify ratio of local dosage to overall average dosage.

Similar to the velocity derivation, a real local output Q_r that is observed at the barrel reference can be calculated by the measured or calculated Q_v of plastics output during the whole screw stroke:

$$Q_r = Q_v\left(\frac{D_i^2}{D^2}\right) \tag{6.9}$$

Therefore, the real output Q_r past through the fixed position of barrel is always lower than the overall output Q_v from the screw itself. In other words, the liquid agent dosing will be intensified locally because the material that had been added with a certain rate of liquid agent will be partially added again since the screw is moving back during recovery. It results in the fact that the local dosage is always larger than the overall average dosage. This intensified dosing may cause difficulty for mixing of liquid agent with melt downstream after the gas injection area. Specifically, it is also sensitive for the gaseous dosing in the injection screw because of the extremely low viscosity of all gases. Normally, the plastic melt exhibits viscosities in the range of 10–1,000,000 poise (g/cm sec). The liquid agent such as blowing agent, whether gaseous (e.g., N_2, Ne, Ar, He, CO_2) at supercritical state or liquefied gas (e.g., pentane), when in the feed inlet of line of the injector, will normally exhibit viscosities in the range of 0.00005–0.05 poise (g/cm sec) [16]. It is important to check if the local gas dosing condition is beyond the limit of recommended gas saturation level for specific material and processing conditions [12]. A local instant dosing percentage R_f is given as

$$R_f = R_{scf}\left(1+\frac{1}{R_u}\right) \tag{6.10}$$

In Equation (6.10), R_{scf} is the overall average weight percentage of fluid agent in molten plastics. Now, the possible intensify ratio of dosing can be defined as

$$R_i = \frac{R_f}{R_{scf}} = 1+\frac{1}{R_u} \tag{6.11}$$

R_i in the equation above is the intensify ratio of dosing that will show how much the average dosage of liquid agent is to be increased in the specific geometry of screw. This intensify ratio is determined only by the screw geometry and not by the processing conditions.

A popular question is, How much and how long can a gas at supercritical state be diffused into the plastic melt without phase separation or gas oversaturation in the plastic melt? Now with velocity analysis a local dosing rate can be calculated correctly, and then a criteria of gas dosing level can be found in Table 6.2. The data in Table 6.2 provide typical operating levels for nitrogen and carbon dioxide in various materials. It is recommended to use the minimum amount of gases necessary to achieve the desired results regarding the existed gas intensify issue in the gas dosing process, specifically for small screws. Unfortunately, the targets of weight reductions and cycle-time reductions are much easier to be achieved compared to control cell structure. The weight reductions and cycle-time reductions are normally achieved at as much as 50% lower SCF levels than the fine cell structure is achieved at higher SCF levels.

Tables 6.3 and 6.4 show that the results of different depths of dosing zones will have different intensify ratios of dosing. A 20-mm-diameter screw is the smallest one studied in this chapter to find out the relationship between depth ratios of metering zone to dosing zone and intensify ratio of dosing. The depth ratio is defined as the dosing depth H_d to the metering depth H_3. This ratio is usually larger than one since the decompression in the gas injection port is required for easily adding the gas into a lower pressure zone. In addition, this deeper mixing zone creates a pressure difference between mixing zone and metering zone to prevent the gas-rich material coming back to metering section. Therefore, the possible depth ratios have been selected as 1, 2, and 3. It simply means that the dosing depth used in Table 6.3 will be equal to the depth of metering zone, double the depth of metering zone, and three times

TABLE 6.3 Intensify Ratio of Dosing for Different Depth in Dosing Zone of 20-mm Diameter Ccrew, Metering Depth H_m = 1.8 mm

Depth of Dosing H_d to Depth of Metering H_m (H_d/H_m)	Root Diameter of Screw in Dosing Zone, D_i (mm)	R_u	R_i
1	16.4	2.05	1.49
2	12.8	0.69	2.44
3	9.2	0.27	4.73

TABLE 6.4 Intensify Ratio of Dosing for Different Depth in Dosing Zone of 180-mm-Diameter Screw, Metering Depth H_m = 7.5 mm

Depth of Dosing H_d to Depth of Metering H_m (H_d/H_m)	Root Diameter of Screw in Dosing Zone, D_i (mm)	R_u	R_i
1	165	5.26	1.19
2	150	2.27	1.44
3	135	1.29	1.78

TABLE 6.5 Intensify Ratio R_i for Different Diameter Screws with H_d/H_3 = 3

Screw Diameter (mm)	20	30	60	120	180
Ri	4.73	2.60	2.37	1.93	1.78

as deep as the depth of metering zone. It has a strength limit in the feed zone because the root diameter is too small with the depth ratio of more than 3 for a small screw. Then, Table 6.3 shows that the intensify ratios of dosing will be 1.49 for depth ratio 1, 2.44 for depth ratio 2, and 4.73 for depth ratio 3, respectively. It is really a concern for depth ratio 3 because the intensify ratio is as high as 4.73. In other words, if the average dosing is 1 wt%, the local dosing really is 4.73 wt%. On the other hand, Table 6.3 shows that the result of a 180-mm screw is much better to have a small intensify ratio. At the same depth ratio of 3, the intensify ratio of a 180-mm screw is 1.78, which is much less than the intensify ratio of a 20-mm screw. It is true that the intensify ratio of dosing in a big screw is not as serious as the one in a small screw, so only special precise gas dosing equipment is necessary for the small screws. If the same depth ratio 3 is used for all different-diameter screws from 20 mm up to 180 mm, the intensify ratios for five different-diameter screws are calculated and listed in Table 6.5. The 20-mm screw is the worst case with the highest intensify ratio 4.73. A 30-mm screw is still at high intensify ratio 2.6 but is much lower than a 20-mm screw. Then, the intensify ratio decreases slightly as the screw diameter increases. The intensify ratio is kept about 2 after the screw diameter is larger than 30 mm. It means that the local dosing weight percentage is about double compared to average weight percentage of dosing.

As an application example, this result partially explains why the real gas dosing weight percentage in plastic actually used in current reciprocating screw injection foaming machine [6] is only about the half of the gas weight percentage saturation limit in specific plastic recommended in reference 12. It is useful information for the correct calculation of real dosing weight percentage to match the screw dosing capability. Also, the mixing issue can be studied below with this velocity analysis since it is a factor affecting the variation of both a mixing length and a mixing time for the whole shot size of material.

Regardless the intensify ratio of screw, as a general guideline, a nitrogen level of 0.1% is a good starting point when running a filled material. Unfilled materials typically require a higher level; 0.3% nitrogen is a good starting level. While these are starting points, they should not be considered optimum levels.

6.1.3 Gas Dosing Pressure Setting

Before the gas pressure setting, an adequate back pressure in front of the screw tip must be selected to make single-phase solution. The minimum pressure must be higher than the critical pressure of gas at a supercritical state. There are two major gases for microcellular injection molding: nitrogen and carbon dioxide. Their critical pressures are 3.4 MPa (–147 °C as the critical temperature) for nitrogen gas and 7.22 MPa (31.1 °C as the critical temperature) for carbon dioxide gas, respectively. Although the necessary pressure only needs to be equal to the critical pressure for supercritical fluid, the real processing back pressure is much higher than this critical pressure. It is because the real injection molding processing must be finished in such a short time with a short mixing length of screw. The easiest way to promote the gas diffusion rate in the molten plastics is the high pressure setting in the microcellular molding machine. For example, for nitrogen gas the minimum back pressure is usually 7 MPa or higher for fiber-glass-filled material and is 13.8 MPa or higher for unfilled material. These pressure settings for different materials can be found in Chapter 4. In addition, the gas dosing must have a pressure difference between gas pressure and molten plastic in the same position of gas injector. However, this pressure difference must be controlled as small as possible. The adequate range of this pressure difference is from 0.35 MPa (50 psi) to 0.69 MPa (100 psi). From Equation (6.1) it is well known that the drag flow in a small screw is not as strong as the drag flow in a big-diameter screw. Therefore, the small screw will take a low limit of pressure difference. A simple analysis in the following helps to understand the pressure difference between gas and melt playing the role for gas dosing inside of a screw.

The small pressure difference between gas and melt at the gas injector position helps to make a small gas droplet size. The big pressure difference may create initial gas surging (possible big gas package). A gas pressure difference test was carried on in a 30-mm-diameter small microcellular screw with 26:1 *L*/*D*. It uses both N_2 gas and GPPS melt in this experimental screw. The processing conditions are listed below:

- 127 rpm screw speed (52 1/sec shearing rate)
- 13.8 MPa back pressure with the melt pressure 16.2 MPa at gas injector position
- 0.5% of N_2 gas at constant 17.7 MPa
- 440 °F melt temperature

The processing pressure difference of gas and melt has been tested with 1.5 MPa and 0.34 MPa, respectively. With a cycle time of 35 sec, the 1.5-MPa pressure difference ran less than a half-hour, and the stable automatic molding lost control gradually. However, the 0.34-MPa pressure difference ran continually for more than 4 h without any problems. It is recommended that the pressure difference between gas and melt is about 0.34 MPa or less in small screw diameters and is about 0.67 MPa or less for big screw diameters, which are 100 mm and up. As a general rule, the pressure difference between gas and melt at the beginning of a gas opening should be as small as possible. The minimum pressure in the gas injector is only required to overcome the initial melt pressure and to clean up the gas outlet as it opens.

The overall pressure profile in the whole cycle is shown in Figure 6.5 for polypropylene (PP) processing. The top fine line represents the supercritical fluid (SCF) dosing pressure, and the bottom one is the melt pressure at the same position of SCF dosing injector. Once the screw begins to rotate, the melt pressure quickly builds up to the set up value. Then, the SCF injector opens after the screw moves axially. To see the details of the pressure changes between SCF dosing and melt, a special test was carried out with the similar conditions to the settings in Figure 6.5, but with a longer cycle time at slow screw speed (see Figure 6.6). Then, a little higher pressure was built up in the middle of the screw. The local view of pressure profile in Figure 6.6 clearly shows the pressure relationship between SCF dosing and melt. The initial SCF pressure decreases from the setup value when the SCF injector opens. The melt pressure instantly rises to meet the SCF pressure somewhere in the

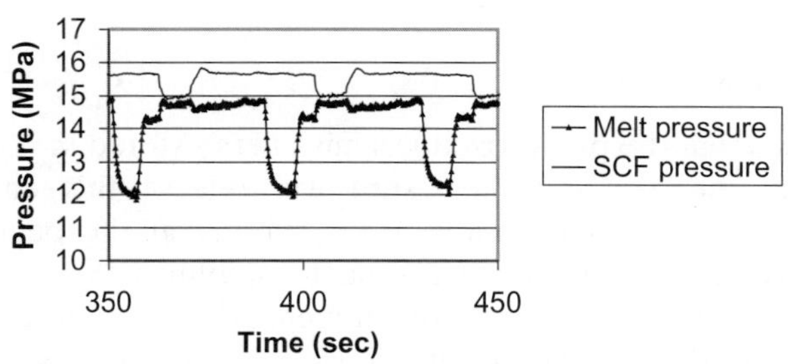

Figure 6.5 Pressure profile in the whole cycle of reversal screw.

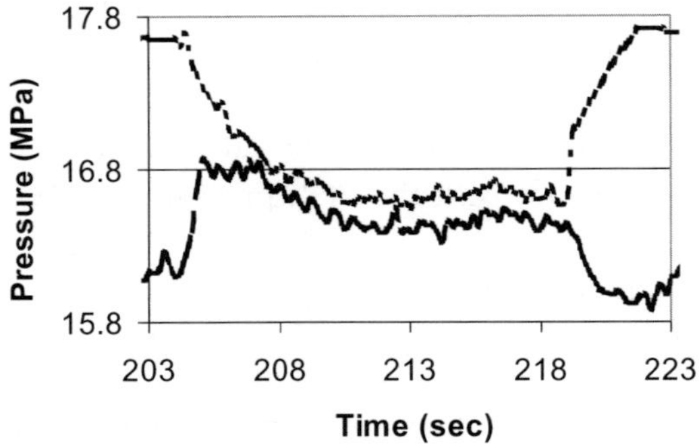

Figure 6.6 Local view of pressure profile in the reversal screw.

middle, which is between SCF setting pressure and the existing melt pressure in the barrel. The simultaneous increase in melt pressure is helpful to reduce the sudden gas pressure surge at the beginning of the gas injector opening. The actual melt pressure keeps reducing gradually until the gas dosing finishes. It is an approved design with reversal channel design of screw because all cycles shown in Figure 6.5 are repeatable pressure profiles. There are more details of equipment design to be discussed in Chapter 7.

On the other hand, if the gas droplet is small, it is easily carried away from a gas dosing spot by the plastic melt. In this way, the gas will not stay in the same area and will join with additional gas added later again. The viscosity of N_2 gas is only about 0.0001% of the plastic melt. It forms a low-viscosity material layer on the interface between gas and melt. If gas is added through the barrel, the upper layer of the material in the screw channel will have a low-viscosity gas layer. This gas layer creates a lubrication layer as the interface between the barrel and screw flights. If this layer becomes too large to move out the gas dosing area, it will influence the highest shearing area in the screw. In other words, the upper layer in the screw channel is the critical pumping area for the drag flow that determines the pumping rate of the screw. Consequently, the gas dosing process will lose the control, which is the test result of larger pressure difference 1.5 MPa. Therefore, the small gas droplet size will help to decrease the distance between gas drops for quicker gas diffusion and mixing.

6.1.4 Gas Dosing Time

It is well known that the pressurized gas has energy stored in it. Once the gas injector opens, the gas under pressure will release this energy quickly. Therefore, the gas dosing is simply a gas surging at the beginning of gas injection opening. It requires more time to stablilize the gas dosing at a certain value (see Figure 6.6). It is recommended to have a reasonable long recovery time if the mixing result allows a slow screw speed with low shear, such as 5 seconds, or a longer screw recovery time. Otherwise, a pulse opening of the gas injector may be the solution for the very short screw recovery time. It usually controls the frequency of pulse to match the short recovery time.

Finally, the gas dosing opening and closing time must be controlled inside of the screw recovery period. The beginning of gas dosing, or gas injector opening, is determined by the beginning of the screw stroke and not the screw rotation. It is because the beginning of screw rotation only begins to build up the pressure to match the setup value of back pressure. It takes time for the action of screw rotation to build this pressure. Then, the screw begins to move with setting speed in axial direction after the back pressure has been built up fully (note that the pressure sharply rises to the setup value once the screw starts rotation in Figure 6.5). This axial movement also means that the positive drag flow inside of the screw is large enough for gas dosing. The experience of gas dosing suggests setting up 3 mm or more as the minimum of screw initial stroke before adding gas. Another option is to make use of the reading of the middle pressure near the gas injector. If this pressure reaches the setting value, the gas injector can safely open. On the other hand, the gas dosing must be stopped ahead of the end of screw recovery. This is important for small screws because the intensify ratio of local dosing is much more severe than that for the big screw. Closing the gas injector about the position where the screw has 20% of the screw stroke left will guarantee to clean up the gas dosing spot, thereby preparing the easy start next cycle. The pressure profile in Figure 6.5 displays the relationship between gas dosing and screw recovery. Once the screw begins to rotate, the gas dosing delays about 2 sec. Then, there is 4 sec of gas dosing time, and the gas injector closes before the end of screw recovery. It leaves an additional 4 sec of screw recovery stroke without gas dosing. This is a special gas dosing approach with almost half the time to close the gas injector ahead of the end of recovery. It is necessary for a small screw with only 30-mm diameter or less. It is so sensitive for the gas local intensify dosage that it may mess up the normal gas dosing if there too high an overdosing gas percentage during recovery.

6.1.5 Conclusions for Gas Dosing

The gas dosing needs to consider the pressure difference, opening time, and possible intensified local overdosage. The guidelines for the reasonable dosing settings are summarized as follows:

- The pressure difference between gas dosing pressure and the melt pressure must be kept as small as possible. It is recommended in the range from 0.35 MPa (50 psi) to 0.69 MPa (100 psi). Small screw diameter must take the low value of this pressure difference. However, the big screw diameter may allow a high value of pressure difference. The very small screw diameter, such as 20 mm or less, may allow it to be as small as only 0.138 MPa (20 psi).

- The gas opening time must be controlled within the screw recovery period. It is important to use the screw axial movement signal to control the gas injector opening, not the screw rotation signal. It is also recommended that the gas dosing must be ended ahead of the end of screw recovery stroke.
- There is an intensify ratio to be considered to calculate the local gas dosing percentage. Specifically for a small-diameter screw the intensified local gas percentage may cause the failure of gas dosing in the screw, and even the overall dosing percentage is still in the range of gas dosing limit.
- The gas dosing is simply a gas surging at the beginning. It requires more time to stablize the gas dosing. It is recommended to have a reasonable long recovery time, such as 5 seconds or longer. Otherwise a pulse opening of gas injector may be the solution for the very short screw recovery time.

6.2 GAS MIXING AND DIFFUSING

After gas enters into the molten plastic at the supercritical fluid state, it needs to be mixed and dissolved with molten plastic inside of screw mixing zone. The theory in Chapter 2 is required for a gas to be completely diffused into molten plastic to make a single-phase solution. It seems impossible with such short residence time in the screw based on the theory discussed in Chapter 2. However, the real process with special mixing elements does complete the gas mixing and diffusing well to create a single-phase solution in the short mixing length with the overall residence time as short as 6 sec. The only explanation for this fact is the excellent mixing element design and the elevated processing conditions. The mixing elements at adequate processing conditions create a unique environment to accelerate the gas mixing and diffusing process. The practical processing conditions of the equipment are not even comparable with any laboratory processing conditions related to the published theoretical data. The details of effects from mixing and processing conditions are discussed one by one below. All well-planned experimental results verify the approaches of the microcellular process. Finally, some concise conclusions are summarized as the guidelines for the processing of gas mixing and diffusing.

6.2.1 Mixing Effect

Although the smallest drop size pursued above by dosing design is in the wiping section, it does not guarantee the final quality of mixing and diffusing. The mixing in the downstream of wiping is critical for making a uniform single-phase solution. The viscosity of molten plastic is so high that the laminar flow dominates the mixing process. Several specific mixing functions can be divided into two basic mixing functions: dispersive mixing and distributive mixing. For the direct result of gas diffusion, or so-called single-phase solution processing, every mixing effort is focused on the gas diffusion quickly and uniformly.

6.2.1.1 Distributive Mixing. Distributive mixing refers to the uniform distribution of gas droplet in the molten plastics. It does not require high stress. The mixing design includes the dividing elements, such as mixing pin the screw channel, discontinuous flights, multichannels or multiholes as inlets and outlets alternatively in different sections, and so on. The result shall be the gas distributed uniformly in space as illustrated in Figure 6.7. Theoretically, distributive mixing moves the gas droplet from place to place, and it has no size change of droplet during this movement. Therefore, all gas droplets shown in Figure 6.7a are all located on the top layer of the wiping section when they enter into the molten plastic from the gas injector. After ideal distributive mixing, they move uniformly in space as shown in Figure 6.7b; for example, some droplets are now in the bottom layer of the molten plastic. However, the number and size of droplets do not change. The existing distributive mixer designs are typically designed with splitting reorientation effects. There are several slotted flight designs that satisfy this requirement. The best one is the Saxton mixer [17]. It is a modification from the Dulmage mixer [18]. The simple change in the Saxton mixer is the helical grooves separating the multiflight sections, instead of tangential grooves. Then, this mixer has the good distributive mixing ability with a positive wiping ability as well. Therefore, it has low pressure drop and still has possible pumping capability in the mixing section because of the continuous wiping action of this kind of mixer.

It is possible to do some quantity calculation for the distributive mixing. For example, the multichannel mixer may have several sections. Then, the mixer may be defined by the possible number of striations as follows:

$$S_n = (N_{\text{mix}})^{Sf} \tag{6.12}$$

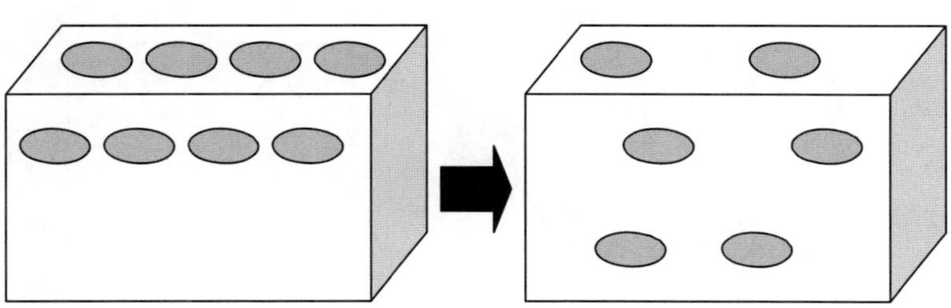

(a) (b)

Figure 6.7 Distributive mixing for the gas droplets. **(a)** Wiping section. **(b)** Mixing section with distributive mixing.

where S_n is the number of striations divided by distributive mixing element with multichannel, S_f is the number of multichannel sections, and N_{mix} is number of the mixing channels in one section.

There are two contributions for the gas diffusion process from distributive mixing. One is that distributive mixing movement uniformly moves the gas droplets in the whole space of the screw channel. In addition, it will separate the gas droplets from each other and will then decrease the risk of gas concentrated in one area to form a big gas pocket. In other words, it actually makes the distance between adjacent gas droplets even bigger to avoid the coalescence of gas droplets. Another good feature is the distributive mixing of the gas droplets to make a uniform gas diffusion in the whole molten plastics. It will be very important mixing for the mold filling stage with uniform cells in the whole microcellular part. More details of structural designing for this mixer are introduced in Chapter 7.

6.2.1.2 Dispersive Mixing. Dispersive mixing is stress-related mixing [19]. It refers to the reduction of the droplet size when the stress of the droplet surface exceeds the coherent strength of droplet. As the concept shows in Figure 6.8, the gas droplets initially formed from the gas injector in the wiping section of the screw are distributed with the pattern as illustrated in Figure 6.8a. After dispersive mixing, the number of droplets are increased by splitting the original big droplet into two or more small droplets. Therefore, it will have more droplets with smaller sizes. The small droplet size and more droplets are the most important contributions from the dispersive mixing. Both more droplets and small size of droplet will increase the contact surface area of gas droplets to the molten plastic. It is the most efficient way to accelerate the gas diffusion rate. However, as shown in Figure 6.8b, all small droplets are still lying in the same position or nearby (top layer only since it is the original droplet comes from) without distributive mixing involved. This is still not an acceptable result because the uniform gas diffusion is required in whole space, not in a local spot.

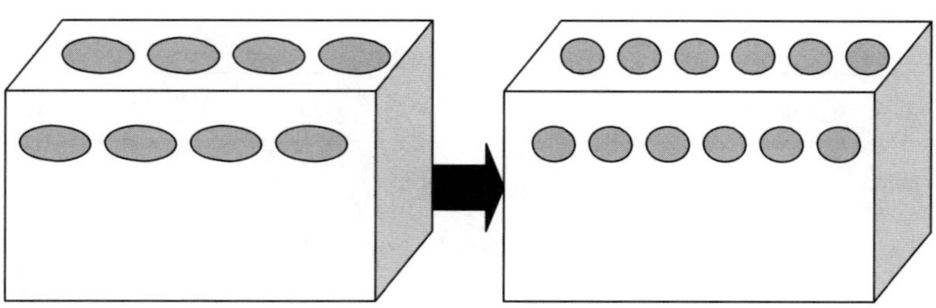

(a) (b)

Figure 6.8 Dispersive mixing for the gas droplets. **(a)** Wiping section. **(b)** Mixing section with dispersive mixing.

6.2.1.3 Comprehensive Mixing. The real mixing in the microcellular screw will have a comprehensive mixing that includes both dispersive and distributive mixings. The final result is the huge number of small size gas droplets uniformly distributed in the molten plastic, the so-called single-phase solution. The shearing field in the mixing section is determined by the combination of the drag flow and pressure flow as illustrated in Figure 6.9a. Assume that the final shearing field height is h_1 (h_1 may not equal to the depth of mixing channel because the pressure flow), and the distance between the two cylindrical elements including the gas before shearing is l_1. Then, an initial cylindrical gas element before shearing is defined as a cylinder with diameter d_1, height h_1. After shearing, it is stretched as two elongated cylindrical elements (simplify the elongated oval shape to cylindrical shape) that have diameter d_2 and height h_2. These two elongated cylinders are no longer in the vertical position, while they lie in the position of an angle θ_{ld}. Hence, the new distance between two elongated cylinders becomes l_2. It is obvious that l_2 is shorter than l_1. In other words, the new distance between elongated cylinders is reduced and can be calculated as

$$l_2 = l_1 \sin(\theta_{ld}) \tag{6.13}$$

When the shear field is sufficiently strong that the angle θ_{ld} becomes small enough, then sign (θ_{ld}) = tan (θ_{ld}) and Equation (6.13) is approximately rewritten as

$$l_2 = l_1 \tan(\theta_{ld}) = l_1 \frac{h_1}{L_r} = \frac{l_1}{\dot{\gamma}_{\text{mix}}} \tag{6.14}$$

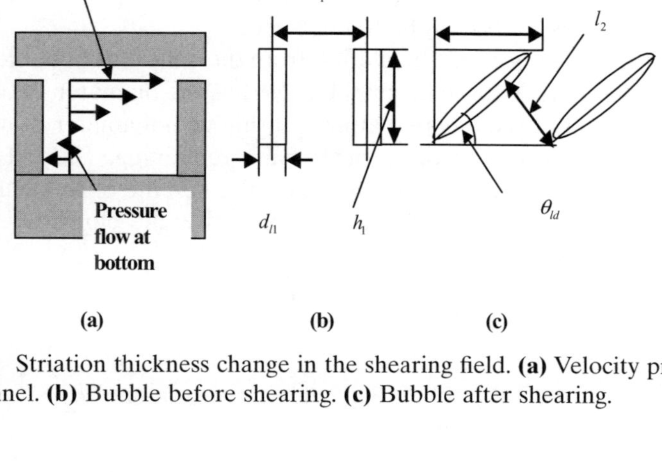

Figure 6.9 Striation thickness change in the shearing field. **(a)** Velocity profile in the screw channel. **(b)** Bubble before shearing. **(c)** Bubble after shearing.

where $\dot{\gamma}_{max}$ is the shear rate in mixing channel, h_1 is the height of drag flow in shearing field of mixing section, l_1 is the distance between two bubbles before shearing (sometime it is known as thickness of plastic for gas diffusing), l_2 is the distance between two elongated bubbles, L_r is the distance moved for the shearing that is related to zero shearing position, and θ_{ld} is an angle of deformed bubble with the horizontal line (original angle is 90°).

It is recommended to use Equation (6.13) for a small-diameter screw or for slow screw recovery speed. For high screw speed and large-diameter screw, Equation (6.14) is the simple one to do the estimation. It is because the shear rate is easier to be calculated, similar to Equations (6.2), and (6.3):

$$\dot{\gamma}_{mix} = \frac{\pi D N_s}{30 H_{mix}} \tag{6.15}$$

where H_{mix} is the depth of the shearing zone in the mixing section, and it equals approximately the depth of the mixing zone.

It is obvious that the high shear rate $\dot{\gamma}_{max}$ is helpful to reduce the plastic thickness between two bubbles. It is an efficient way to speed up the gas diffusion. Assume that the shearing field creates the angle θ_{ld} of 30°; then, the plastic thickness l_2 reduces to half of the initial plastic thickness l_1. This is significant reduction of plastic thickness between bubbles. Assume that the average depth of mixing section is about twice as deep as the depth of metering section; the estimated $\dot{\gamma}_{max}$ values at maximum screw speed are listed in Table 6.1.

The effect of $\dot{\gamma}_{max}$ to reduce the thickness of plastic between bubbles is extremely significant. The biggest value of $\dot{\gamma}_{max}$ is from a 30-mm screw, and it is as high as 448. Hence, the plastic thickness l_2 is given as $l_2 = l_1/\dot{\gamma}_{mix} = l_1/448$. It means that theoretically the thickness of plastic for gas diffusion is only 1/448 of original thickness of plastic in the wiping section. It may be explain well about why the high screw speed can make a better single-phase solution; even the residence time is so short at high screw speed.

There is another important calculation from the concept of bubble shearing in Figure 6.9. Assume that the original bubble has the diameter d_{l1} and length h_0, and suppose that it is elongated after shearing with diameter d_{l2} and length h_2 at the assumption of the cylindrical shape remaining. Then, if the local pressure and temperature are kept the same, the volume of this bubble must be kept the same before and after shearing. Hence, the elongated length of bubble h_2 is given by

$$h_2 = h_0/\sin(\theta_{ld}) \tag{6.16}$$

where h_2 is the stretched bubble length after shearing and h_0 is the original bubble length before shearing.

With the same simplified approaches as for Equation (6.14), the final formula of stretched bubble length h_2 can be rewritten as the function of shear rate $\dot{\gamma}_{\max}$:

$$h_2 = h_0 \dot{\gamma}_{\text{mix}} \tag{6.17}$$

With the assumption of constant volume of the bubble before and after shearing, the diameter of stretched bubble d_{l2} is expressed as

$$d_{l2} = d_{l1}\sqrt{\frac{h_0}{h_2}} = d_{l1}\Big/\sqrt{\dot{\gamma}_{\text{mix}}} \tag{6.18}$$

where d_{l1} is the bubble diameter before stretching in the shearing field and d_{l2} is the bubble diameter after stretching in the shearing field.

Equations (6.14), (6.17) and (6.18) are ready for the calculations for all droplet geometric dimensions before and after shearing. They are related directly to the shearing in the mixing zone. Generally, it is reasonable to use the shearing rate in Equation (6.15) because the real shearing deformation is larger than the simplified linear shear rate whenever the pressure flow exists in the mixing zone.

Practically, the shearing deformation will elongate the bubble long enough to be broken into small droplets before it can be stretched so long. There is a ratio of shear force to the surface force, known as the Weber number W_e [see Equations (7.11), (7.12), and (7.13)]. The critical W_e is about 300 for the simple shear field [20, 21]. When the stretching deformation in the bubbles exceeds the critical value of the Weber number, the bubbles break into small droplets. This is another important factor that the gas diffusion will be fast with the disintegration of bubbles in the strong shearing field.

6.2.2 Temperature Effect

The temperature effect to the gas diffusion may have two general trends to be discussed. One of them is that the gas solubility in the plastic decreases with the increasing temperature. Another general trend is that the gas diffusion rate in the polymer will increase with higher temperature.

6.2.2.1 Gas Solubility at High Temperature. The barrel temperature for microcellular injection molding is usually the same as the regular injection molding. However, the high back-pressure requirement may bring more heat in the screw. Based on the theory in Chapter 2, the high melt temperature will generally reduce the solubility of gas in the molten plastic. In terms of the difference in the specific volume change between gas and molten plastic with the temperature change, it may be one of the reasons to explain this phenomena. Therefore, the result of high melt temperature is to reduce the gas dosing percentage to avoid the extra gas out of solution. On the other hand, the low gas dosage increases the thickness of plastic between two adjacent bubbles so that it seems like an increase in the gas diffusion difficulty because the plastic between gas bubbles becomes thicker.

Another processing condition change that needs to be addressed is that the intensify dosing rate must be checked again to avoid overdosing locally. It is because the low gas solubility in the polymer brings more sensitive issue of possible overdosing, and then the gas is more likely to be out of solution in these polymers.

6.2.2.2 Gas Diffusion Rate at High Temperature. The gas diffusion time is very slow at the room temperature. The estimated diffusivity of CO_2 in most of thermoplastics is in the range of 5×10^{-8} cm^2/sec [12]. The diffusivity of N_2 is nearly the same as that of CO_2 at room temperature. However, at the elevated temperature up to 200 °C the diffusivities of both gases are three to four orders of magnitude greater than those at room temperature. For both gases, the estimated diffusivity coefficient in polymer is in the range of 10^{-4} cm^2/sec to 10^{-6} cm^2/sec at 188 °C to 200 °C without any shearing and at atmospheric pressure [12, 20]. It may be used for a qualitative estimation or comparison to select the processing conditions. However, it is not ready to be the useful data used for quantitative calculation of the practical injection molding process.

Therefore, the general trend is that the diffusion rate of gases in the molten plastics will increase with the temperature rise. That is the opposite trend with the solubility in the plastics. These factors must be considered to set up correct parameters of processing. On the other hand, the data in reference 12 show a decrease of gas diffusivity for HDPE with increasing temperature. For CO_2 gas it is about 5.7×10^{-5} cm^2/sec at 188 °C, while it is about 2.4×10^{-5} cm^2/sec at 200 °C. The explanation is the complex of crystalline material that may have a significant effect for the gas diffusion process from crystalline change in addition to the temperature change.

6.2.3 Pressure Effect

Similar to the temperature effect on a gas diffusion rate, the pressure effect will be focused on two issues as well. One trend is that the gas solubility in the plastic will increase with increasing pressure. Another trend is that the gas diffusion rate in the polymer also increases with higher pressure. One can also learn more about this effect from the basic theory in Chapter 2.

6.2.3.1 Gas Solubility at High Pressure. It is well known that the high pressure in the processing will help to increase the solubility in the molten plastic. At low pressure the amount of gas dissolved into the plastic is very low. However, the amount of gas dissolved into plastic is much higher at high pressure. This high solubility at high pressure can help to increase the gas dosage during screw recovery. Then, the gas droplets will be closer to each other so that the distance between plastic and bubbles is small. It means the initial l_1 in Figure 6.9 is small. Therefore, the initial gas diffuse task is not as difficult as the big l_1 to leave a thicker plastic layer to be diffused by gas.

6.2.3.2 Gas Diffusion Rate at High Pressure. The gas diffusion rate is also a function of pressure. The amount of gas dissolved into plastic is determined by the saturation pressure at a certain temperature. It is well known that the gases need to be pressurized into the supercritical region at a certain temperature. Thus, the supercritical state can enhance the solubility and diffusion rate.

The pressure is the most important parameter to be controlled for better microcellular processing. It is because high pressure is positive for both gas solubility and gas diffusion rate. Therefore, increasing the pressure will definitely accelerate the gas diffusion process. However, one must be aware that the increase in pressure will have several more parameter changes associated with it. For example, as the mechanical energy increases the melt temperature, the melt viscosity will decrease, and the output rate of plasticizing unit may be decreased as well. In addition, a decrease in output rate will result in overall residence time increase, which will be discussed below.

The high pressure requirement in the foaming industry may cause a common misperception that the lower the pressure, the safer the operation. An experiment has been carried out to verify both safety and process ability related to the single-phase solution by injection into air with constant injection speed [4]. The adequate pressure must be set correctly to make a single-phase solution. Otherwise, the poor single-phase solution in the barrel will never allow the second stage of the molding process to go smoothly. In addition, the poor single-phase solution causes some safety issues that will be discussed in the molding stage.

6.2.4 Residence Time Effect

There are three possible different residence times to be considered in the injection molding process. The first one is the static residence time, which is simply the total time of the gas-rich material that statically sits in the screw and barrel in the screw idle period. The second resident time is dynamic residence time during screw rotation. The dynamic residence time is defined as the total time of the gas melt to be sheared in the screw rotation. The third one is the residence time of material during injection. When the injection occurs, the material trapped in the screw stays in the space where is between the screw tip and the middle of check valve. It can be considered as static residence time since most materials in the channels of mixing and wiping are restricted between two check valves, and there is only possible movement for little material on the top layer of channels that contacts the barrel and may be dragged backwards during injection. Therefore, only two residence times—static residence time and dynamic residence time—are considered in this book.

The static resident time can be estimated from the method introduced by Park and Suh [20]. The GPPS material is processed during a 60-sec cycle time. The gas diffusion time is given by

$$t_d = l^2 / D_a \tag{6.19}$$

where t_d is the SCF diffusion time, l is the thickness of plastic for gas diffusion, and D_a is the gas (SCF) diffusivity coefficient.

Then, the thickness of plastic for gas diffusion is assumed to be 1 mm; and the SCF gas is CO_2, which has the D_a of 1.3×10^{-5} cm^2/sec at 200 °C. It results in the static residence time t_d of about 769 sec. For the total cycle time of less than 60 sec, the static residence time is almost 13 times longer than the total cycle time; therefore, the static residence time can be neglected during the calculation.

There is no dynamic gas diffusivity coefficient in real processing conditions available now. One experiment was carried out to verify that short dynamic gas diffusion is possible. The screw runs as slow as possible to eliminate the static residence time at every screw idle time. It still results in good microcell structure of the part. In addition, the experimental results show further that the dynamic residence time can be as short as 6 sec at high back pressure and high screw speed to make a perfect single-phase solution.

It has been verified by industry practices that the effect of static residence time of gas is not as important as the dynamic residence time for gas diffusion. On the other hand, the most efficient way to increase the mixing effect is to increase the shear rate. It means that the screw speed must be raised as high as possible. However, the high screw speed results in very short dynamic residence time. It presents a processing dilemma that needs to be decided: Which one is the priority to be guaranteed? All well-planned experiments show that the shear rate is the most important parameter to pursue for the best microcellular quality of injection molding part at the condition of a stable gas dosing process. In other words, as long as a minimum stable gas dosing time is satisfied, the screw speed shall be raised as fast as possible to reach the high shear rate.

6.2.5 Material Effect

Polyolefin resins typically require significantly higher nitrogen levels to achieve good cell structure than do most other materials. These materials are also more likely to have significant cell structure variation from the gate to the end of fill. This situation will be aggravated by increased wall thickness, greater than 3.0 mm. It should be expected that the final nitrogen levels when running unfilled HDPE or unfilled PP will be 1% or higher. In fact, nitrogen levels as high as 2% can be run with these unfilled materials.

As with all materials, the addition of fillers improves the efficiency of the nitrogen added to the polymer. The most common filler with polypropylene is talc. As talc levels approach 20% or more, the nitrogen level will be in the range of 0.5–0.75%. Glass fiber is the more efficient filler than talc, and the nitrogen level can be decreased to 0.5%.

Amorphous resins can be split into groups such as (a) polystyrene, polycarbonate, acrylic, and SAN, which do not contain an impact modifier, and (b) ABS, HIPS, and impact-modified PC, which contain an impact modifier. For those materials without impact modifiers, nitrogen blowing agent levels will be about 0.4%. These materials typically achieve excellent cell structure at relatively low levels of supercritical fluid. Cell structure will be essentially uniform from gate to end of fill and will be microcellular. Adding an impact modifier has the effect of increasing cell size at equivalent SCF levels. In order to achieve a cell structure that is microcellular or close to microcellular, nitrogen blowing agent levels need to be closer to 0.7%.

Fillers can significantly improve the cell structure in amorphous resins regardless of whether the materials are impact-modified. This is particularly true because most amorphous resins are filled with glass fibers that are highly effective as a nucleating agent and in controlling cell structure. The addition of as little as 10% glass fibers will allow the nitrogen level to be decreased to 0.3–0.5% while still maintaining a microcellular structure. The low level of gas dosing will make the microcellular processing more stable and will enable it to be easily set up even for a new mold and new material trial.

Semicrystalline engineering resins show behavior similar to that of polyolefin resins. Unfilled resins show larger cell structure variation in the part from the gate position to the end of flow. Unfilled resins also require higher nitrogen levels to achieve good cell structure, which is in the range of 0.5–0.7%. The addition of 20% or more of glass fiber will allow the supercritical fluid level to be dropped to in a range of 0.25–0.3%. Other fillers types such as mineral will also act as a nucleating agent and allow for low SCF levels but will inhibit weight reduction. An example of these is the reinforced PA with glass fiber that achieves 20% weight reduction but may only get 15% weight reduction with an equivalent amount of mineral filler.

In addition to the specific information above, there are some general trends that can be applied across all materials. Fillers act as nucleating agents, and they have the effect of improving cell structure and increasing the efficiency of the given level of supercritical fluid. Glass fiber is the most beneficial filler in terms of controlling cell structure and achieving weight reduction. Talc and mineral are less effective in terms of both cycle time reduction and weight reduction.

Amorphous resins always require lower nitrogen levels than do semicrystalline resins, although the presence of impact modifiers will require a higher SCF level. This trend is true regardless of whether it is an impact-modified amorphous resin or a semicrystalline resin such as a TPO or toughened PA.

The biopolymer microcellular usually requires special processing conditions that are more gentle than the regular polymer. The biopolymer simply does not like the extra heating and shearing during processing. However, the inert gas as blowing agent may help to protect the quick degrading of biopolymer in a barrel during recovery and gas dosing in the first stage. On the other hand, the limit of hydrolytic stability must be controlled so that it will not exceed the processing limits. Overall, the biopolymer microcellular processing is similar to the PVC microcellular processing that needs to control the heat and residence time in the barrel. The differences between regular polymer and biopolymer are subtle, and we need to understand the principle that is discussed in Chapter 4. The product design, tool design, and processing equipment modification must be considered for the biopolymers. The narrow processing window of biopolymer may bring more challenges for the microcellular injection molding processing. For example, PHBV resin has a melting point of about 154 °C, but it degrades at 182 °C. Therefore, over this temperature range the shearing heat in the screw cannot melt the biopolymer and cannot mix it with gas in time without burning it in the barrel. Too much heat in the biopolymer from the heating system or from shearing can result in gels, black specs, or yellowing in the biopolymer part.

The guidelines above apply to general injection molding applications, with wall thickness of 1.5–4.0 mm. Cell structure control becomes easier in thinner parts due to higher cavity pressure. The trends discussed above still hold, but cell structure variation from gate to end of fill will increase for the thick part. The thickness restriction of microcellular structure is also determined by the flow ratio of mold. The small flow ratio allows having thicker parts of microcellular injection molding.

6.2.6 Experimental Result for the First Stage of Microcellular Processing

There is still a big gap existing in theory and practical process of microcellular injection molding. For example, the gas diffusivity coefficient D_a for the dynamic processing is not available yet. The calculation in this book just uses the published D_a for static processing to do the comparisons between different processing conditions. Another confusing concept is the shortest residence time required for the minimum value of gas diffusing. Therefore, some well-planned experiments were carried out to verify how short a residence time is really needed to make a single-phase solution. In addition, the relationship between resident time and shearing strength was studied for the setup of an economic processing.

6.2.6.1 Amorphous Material. Making the single-phase solution is the target of this experiment with either longer dynamic residence time with weak shear rate or strong shear rate with short dynamic residence time. It was implemented in a 30-mm microcellular screw with 26:1 L/D and a gas injector in the wiping zone of screw. By using a back pressure of 13.8 MPa, N_2 gas weight of 0.3%, pressure difference (between gas and melt) of 0.34 MPa, injection speed of 0.051 m/sec, and different screw rotation speeds of 127 rpm and 390 rpm, both ABS and GPPS materials were tested in the plasticizing process.

TABLE 6.6 Results of Different Screw Speeds

Screw Speed (rpm)	390	127
Shear rate (1/sec), ABS	161	52
Shear rate (1/sec), GPPS	161	52
Diffusion time (sec), ABS	3	18
Diffusion time (sec), GPPS	2.3	14
Residence time (sec), ABS	6	24
Residence time (sec), GPPS	7	26

Then, related quantity analyses were made for the shear rate in the screw, the pressure difference between gas and melt, gas theoretical diffusion time, and residence time of shear for the gas and melt mixture.

In addition to the well-known important parameters, such as gas percentage, back pressure, melt temperature, the other key parameters studied here are the shear rate, pressure difference between gas and melt at the gas injector position, and residence time of shear for a gas–melt mixture in the screw.

The shear rate is the major factor to shorten the gas diffusion time. Increasing screw speed is the effective way to raise the shear rate. Both ABS and GPPS get six times shorter gas diffusion time with the screw speed increasing three times higher (see Table 6.6). The small gas droplet size also guarantees a consistent screw recovery. It also creates a higher number of potential nuclei for nucleation in the molding stage.

Residence time of shear may not be as important as the shear rate during gas dosing. The test result shows that the shear rate effect is obvious for GPPS gas dosing in the screw at the different average shear rates of 52 1/sec and 161 1/sec, respectively. With the velocity analysis method [5] the correct residence time of a gas–melt mixture in the screw can be calculated. A longer shear residence time of 26 sec is related to the low shear rate of 52 1/sec. This processing condition results in nonuniform cell structures, where the cell sizes range from 50 μm to 100 μm with even larger voids. However, the higher shear rate of 161 1/sec is associated with short residence time about 7 sec. This result of high shear rate but short residence time creates much better cell structure than low shear rate but long residence time. The success of the gas dissolution process strongly relies on the breaking gas droplet size. This also establishes new rules for the minimum pressure difference between gas dosing pressure and melt pressure. It helps for initial small gas droplet size and the long stable microcellular foam processing.

The results above show that the shear rate is the most effective parameter to make single-phase solution compared to the effects from residence time of shear. In addition, the pressure difference between gas and melt at the dosing plays an important role to make a stable process.

The shear residence time can be as short as 6 sec. This processing condition generates uniform cell structures from 30 μm to 50 μm without voids. The more obvious differences are shown with the ABS material. The longer residence

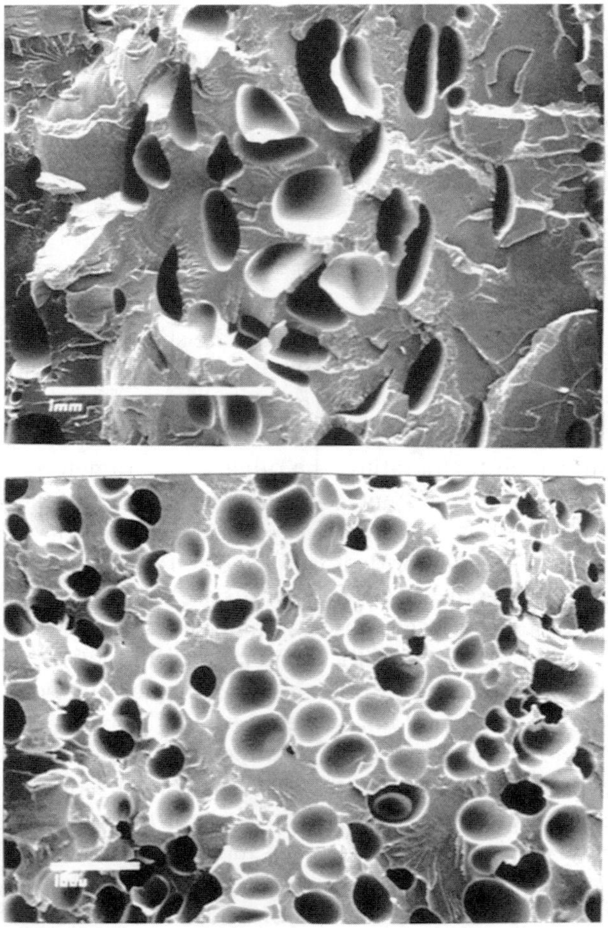

(a)

(b)

Figure 6.10 Cell structure for the ABS material [white bar indicates 1 mm of (a) and 100 μm of (b)]. **(a)** Shear rate 52 1/sec [7]. **(b)** Shear rate 161 1/sec [7]. (Reproduced with copyright permission of Society of Plastics Engineers.)

time of 24 sec and the low shear rate of 52 1/sec makes an average 300-μm cell size as shown in Figure 6.10a. However, the shorter residence time of 6 sec and the higher shear rate at 161 1/sec improve the cell structure with average 60 μm of cell size shown in Figure 6.10b. This result in Figure 6.10 and test result of GPPS may give the conclusion that the shear rate is the key factor to be controlled first to optimize the gas dosing process. The shearing residence time may help for gas dosing, but it is not a critical factor for gas dosing process. This conclusion may well explain the success of gas dosing through nozzle sleeve during injection from the new innovation in IKV [22].

To explain the results above, the new approach is to focus on the estimation of gas droplet size. If the gas dosing percentage is the same, then the smaller size of the gas droplet corresponds to a larger number of gas droplets. Therefore, it is important to make the initial gas droplet as small as possible. There are two effective ways to make such small gas droplets. The first method is the shear method. It stretches the gas droplet long and thin, and it breaks when the critical Weber number (ratio of shear force to surface force) reaches. The second method is the mechanical whipping by multiflights of screw in the gas dosing zone [6]. Material viscosity, shear rate, Weber number, screw whipping frequency, and flow rate of gas injection finally determine the breaking number and sizes of gas droplets. Using a similar method recommended in the reference paper [12, 20], the average breakage number of gas droplets can be theoretically calculated. For both ABS and GPPS, the breakage number of gas droplets is increased about five times when the screw speed is three times higher. This is why the shearing process is the key to speeding up the gas diffusion process in the melt.

Therefore, the small gas droplet size will help to decrease the distance between gas drops for quicker gas diffusion. The shear rate is the major factor to shorten the gas diffusion time. Increasing screw speed is the effective way to raise the shear rate. Both ABS and GPPS get six times shorter gas diffusion time with the screw speed increasing three times higher (see Table 6.6). The small gas droplet size also guarantees a consistent screw recovery. In addition, small gas droplet size and large number of droplets will help to create a higher number of potential nuclei for nucleation in the molding stage.

6.2.6.2 Crystalline Material. The results above show that the short residence time works for amorphous material. More experiments have been carried out for crystalline materials, such as PA, PP, and so on. The screw is the same screw as the one used for amorphous materials above. It runs at 390 rpm (161 1/sec average shear rate in the mixing section) at 20.7 MPa (3000 psi) back pressure with 0.6 w% of N_2 gas to make the microcellular structure of unfilled PP. It has about 100-μm cell size and acceptable uniformity of cell distribution. However, this crystalline material must be made with higher back pressure up to 20.7 MPa, along with a high N_2 gas dosage for microcellular processing. The CO_2 gas is a good blowing agent for crystalline material as well. Figure 3.4 shows the microstructure of unfilled PP with CO_2 gas as blowing agent.

A different microcellular screw is used to run the similar experiments above with the unfilled PP. It is the 60-mm-diameter screw with 28:1 *L/D*. It runs at 49 rpm, 103 rpm, and 166 rpm, respectively. The back pressure is 13.8 MPa (2000 psi). The 60-mm-diameter screw has much more positive drag flow than 30-mm-diameter screw so that the gas dosage allows adding weight percentage as high as 1% of N_2 gas. Figure 6.11 shows the result of microstructure of the samples of unfilled PP runs at three different screw speeds. The slowest screw speed is 49 rpm, and it has the average shear rate of 38 1/sec in the mixing section (with average 8-mm depth of mixing channel). The cell size is small but not uniformly distributed in the part. Figure 6.11a shows that the cells only

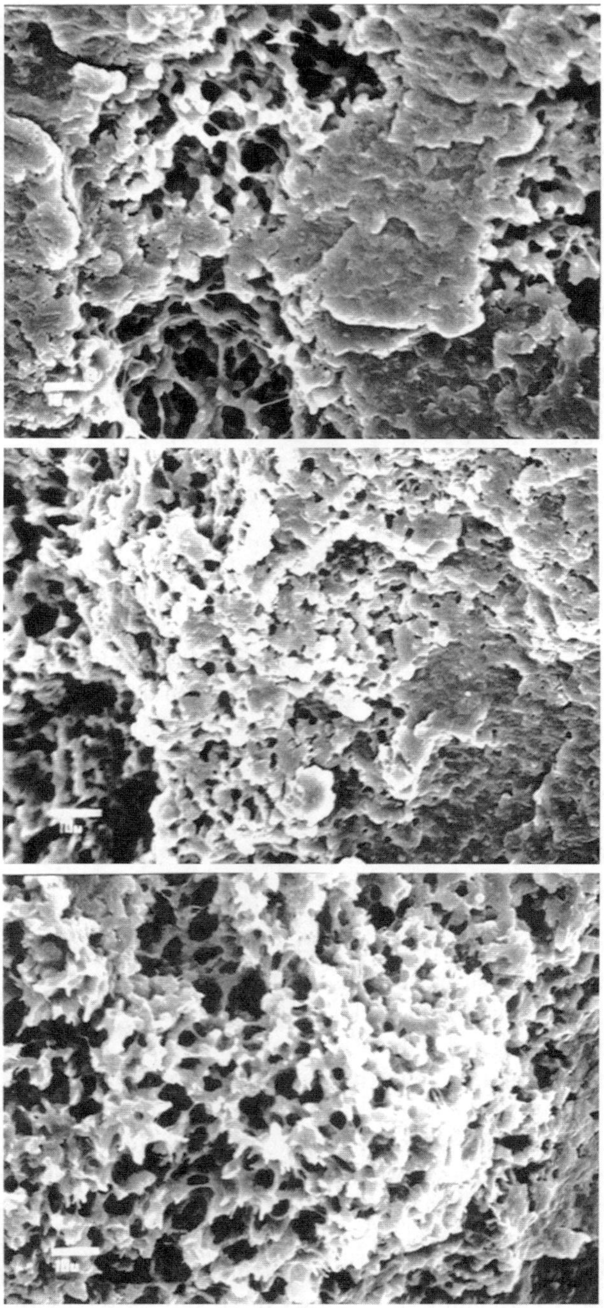

(a)

(b)

(c)

Figure 6.11 SEM pictures for unfilled PP samples (white bar indicates 10 μm). **(a)** Sample of shear rate 38 1/sec. **(b)** Sample of shear rate 81 1/sec. **(c)** Sample of shear rate 130 1/sec.

exist locally in the gas-rich area. Then, the screw speed increases to 103 rpm and the shearing rate is 81 1/sec. It is almost double the shear rate of previous slow speed 49 rpm. The result of cell structure is improved by smaller size of cells in Figure 6.11b. However, the cell distribution is similar to the result in Figure 6.11a. Finally, the screw speed increases to 166 rpm and related shear rate is up to 130 1/sec. The other conditions are kept the same as the previous two running conditions, and the result shown in Figure 6.11c is much better with both uniform cell distribution and small cell sizes as well.

From the results above, there are several technical differences to be summarized below:

- A big-diameter screw has more positive drag flow that allows running higher gas dosage. The combination of high shear rate and high gas dosage result in better cell structure in unfilled PP.
- The gas dosing in a big screw is easier than that in a small screw, and the back pressure of microcellular processing can be reduced to get the same cell structure made with a small screw. The benefit of low back pressure is that the output rate of screw increases.
- Shear rate is always the positive help for speeding up the gas diffusion process because the cell structure at high screw speed is always better than the cell structure at slow screw speed. On the other hand, the big screw can get the higher shear rate than the small screw at the same screw speed because the big screw has higher circumferential velocity at the outside diameter of screw than does the small screw.

6.2.6.3 Biopolymer. Generally, biodegradable polymers, both synthetic and natural, often show poor foamability of fine cell structure. It is because the biodegradable polymer has poor rheological properties, poor thermal stability, poor solubility and inadequate diffusivity of the ordinary blowing agents, and insufficient setting mechanisms. Usually, a biopolymer has a very narrow processing window, specifically, the thermal degradation temperature is too close to the melting temperature of all biopolymers. Three parameters have been controlled well to be able to process the PCL, gas concentration, foaming temperature, and pressure drop rate. For the biopolymer, foaming temperature can be considered the most important since it influences the crystallization/vetrification of the polymer and cell coalescence. Pressure drop rate is also a well-known important parameter correlated to the thermodynamic instability necessary to generate as many nuclei as possible that is mainly affecting the foam morphology. In addition, gas concentration is directly correlated to the availability of gas necessary to inflate the bubbles, which influences the final density as one of the important parameters for cell growth.

The screw for plasticizing, as well as other elements related to mold filling, must be gentle enough without creating too much shearing heat to burn the biopolymer. No dead corner is allowed in the screw, nozzle, and mold for processing biopolymers. Both CO_2 and N_2 gases are successfully used for biopolymer injection molding process. Here is the summary of processing criteria of microcellular biopolymers in the first stage.

- Tight melt temperature control in the screw and in the mold including all valve gates, nozzle, and so on, are necessary.
- Mechanical energy from the screw for both plasticizing (screw rotation speed and back pressure) and injection (shearing rate controlled by injection linear speed) need to be controlled precisely.
- Residence time in the barrel needs to be limited to the range of vendor's recommendation for each biopolymer.
- Proper drying of a biopolymer is required since the biopolymer tends to be hygroscopic and moisture-sensitive. Or they will suffer a drop in molecular weight and melt viscosity, as well as increased potential for flashing and brittle parts. PLA and PHA are polyesters, and drying requirements are in the range of those for PET and PBT—that is, more strict than for ABS, nylon, or PC.
- Biopolymers will need a low shear geometry designed screw; a rigid PVC screw will work in this case. Also, the PET screw runs PLA successfully, as well. It should have less than 3:1 (2.1–2.8) compression ratio of screw with 20:1 L/D. Feed-throat temperature should be about 21 °C, while recommended processing temperature will be 188–210 °C. Screw speed is about 50–200 rpm, depending on the diameter of screw. The back pressure of oil should be used with 0.345–0.690 MPa (50–100 psi) range. Metering zone temperature should be 188–204 °C. Mold temperature is about 24 °C, and the expected part shrinkage is 0.004 mm/mm. Injection speed is usually in the range of 0.013–0.051 m/sec (0.5–2 in./sec).
- The processing temperature should be controlled in ±2 degrees.
- Biopolymers with a high level of amorphous reprocessed material may tend to stick with metal surfaces in processing so that high-finish metal surface in the mold is required.
- A general rule of thumb is for shot volume to be 30–80% of barrel volume.
- If the biopolymer is exposed to ambient air, it can absorb enough moisture in 5 min to defeat most of the benefits of drying. On the other hand, if the drying temperature is too high, the material may soften and agglomerate in the drying hopper.
- A biopolymer is dried with starting moisture content of 2400 ppm down to 250 ppm in 5 h at around 70 °C. The material should be dried to less than 400 ppm moisture, and best moisture content should be less than 100 ppm.

- Starch-based material must be purged out from the barrel by LDPE at the end of production to prevent excessive degradation.
- For example, PHBV (polyhydroxybutyrate-valerate) is a biopolyester produced via bacterial fermentation of plant starches. It is a member of the PHA family. PHBV material is approved for food contact in Europe and is approved by the Biodegradable Products Institute (BPI), New York City, for composting. PHBV should be dried to 250 ppm moisture. Melt temperature is 170–175 °C. Feed-throat temperature is less than 135 °C, compression zone temperature is 145 °C, metering zone temperature is 155 °C, and adapter temperature is 161 °C. The shrinkage of PHBV is similar to that of PLA.

6.2.6.4 Short Screw Versus Long Screw. The screw L/D is reduced from 26:1 to 22:1, and the results are still good for making a single-phase solution. The four test materials are as follows:

Material 1: 20% talc-filled PP, MI = 12
Barrel temperature (°C): 27, 205, 221, 221, 221, 215, 215 (nozzle)
Back pressure: 13.8 MPa (2000 psi), 20.7 MPa (3000 psi)
Gas (N_2) weight %: 0.6, 1.0

Material 2: Unfilled PC
Barrel temperature (°C): 50, 298, 310, 295, 295, 295, 290 (nozzle)
Back pressure: 13.8 MPa (2000 psi), 20.7 MPa (3000 psi)
Gas (N_2) weight %: 0.6, 1.0

Material 3: 30% GF-filled PA6/6
Barrel temperature (°C): 50, 282, 288, 293, 293, 293, 302 (nozzle)
Back pressure: 13.8 MPa (2000 psi), 20.7 MPa (3000 psi)
Gas (N_2) %: 0.6, 1.0

Material 4: GPPS
Barrel temperature (°C): 27, 220, 240, 249, 249, 249, 238 (nozzle)
Back pressure: 13.8 MPa (2000 psi)
Gas (N_2) weight percentage: 0.6

The short screw with 22:1 L/D has a modified mixing section that creates some upside-down mixing flow. The total residence time is decreased 30% compared to the long screw with 26:1 L/D.

The quality of gas diffusion and mixing was checked through the SEM pictures for three screws tested in this project. PC samples with 13.8 MPa

(2000 psi) back pressure at 162 rpm were selected as the comparison samples to get some conclusions about the quality of gas diffusion and mixing performance of these three screws. The SEM micrographs of PC samples made from a 22:1, 30-mm screw and a 26:1, 30-mm screw do not have significant differences in cell structure. Both of them have uniform cell distribution and similar cell size variations at the same processing conditions and different dynamic residence time in the mixing section.

The 30% GF-filled PA6/6 material microcellular process in this short screw is successful as well. The cell structure of the sample made in the short screw (22:1 *L/D*) is as good as the cell structure of the sample made in the long screw (26:1 *L/D*). Both samples are processed with the same processing conditions at 20.7 MPa with 0.6 weight percentage (wt%) of N_2 gas.

The talc-filled PP does show good structure with only 13.8 MPa (2000 psi) in this screw with 22:1 *L/D*. The cell size is almost the same as the unfilled PP in a long screw with 26:1 *L/D* at 20.7-MPa back pressure. It is because the 20% filled talc in the PP helps to reduce the back-pressure requirement.

An additional test was focused on the gas problem for GPPS materials. It seems that this short screw with 22:1 L/D may need either higher back pressure up to 20.7 MPa, or higher weight percentage gas for making the same quality of the GPPS part at the long screw with 26:1 *L/D*. The problem for the short screw was that it could not get rid of the problems of gas popped out from the part after ejection with low back pressure of 13.8 MPa. The SEM picture in Figure 6.12a shows the cell structure in a short *L/D* screw with varied cell size from 20 μm to 300 μm. However, it can be run successfully at high back pressure 20.7 MPa (3000 psi) with 0.7% of N_2 without popping. After the back pressure is decreased down to 17.24 MPa (2500 psi), there was the same problem as the gas popped out from the ejected part. On the other hand, the long 26:1 *L/D* screw creates a much better cell structure shown in Figure 6.12b. It not only has small cell size from 10–40 μm but also has uniform cell distribution in the part. Then, this experiment runs well without any gas post pop in the part.

The conclusions from the screw *L/D* test are that only GPPS processing shows the significant difference result and that the rest of the materials, PC, PP, and PA 6/6, do not have much morphology differences in short and long screws. The reason for it may be from the shearing sensitivity of GPPS material.

6.2.6.5 Shear Rate Versus SCF Dosage. When the low SCF dosage is used to make some good surface finish of microcellular parts, a phenomenon of shear rate (screw speed) improving the quality of cells occurs. When the N_2 SCF is reduced below 0.2%, it results in a hollow channel in the part that is similar to the gas assist process. However, this hollow channel is removed when the screw rotation speed increases while the SCF dosage remains the same. The experiment about the shear rate versus SCF dosage was implemented in a 30-mm-diameter screw with 26 *L/D* GPPS material; different

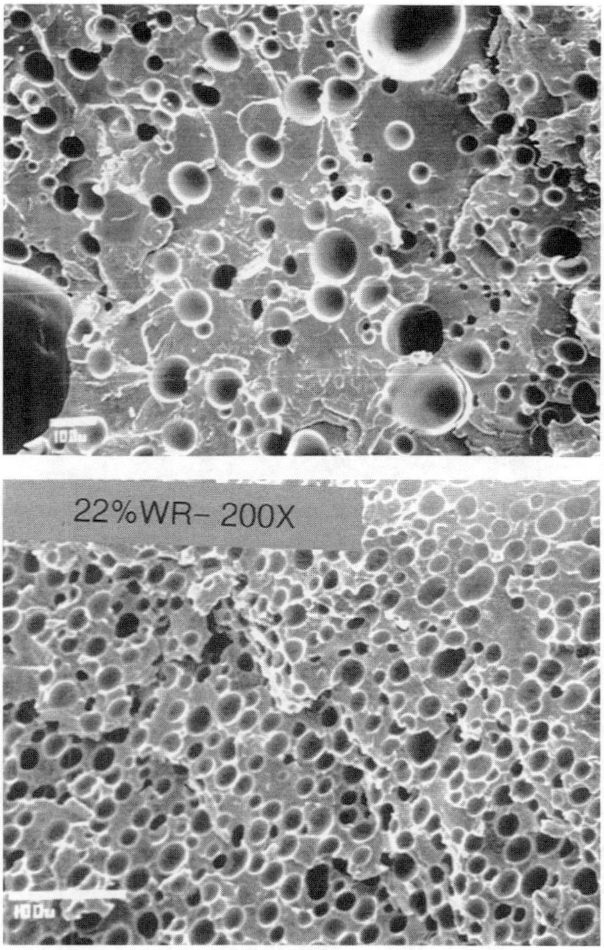

(a)

(b)

Figure 6.12 Cell structure for the GPPS material (white bar indicates 100 μm). **(a)** Sample in 22:1 *L*/*D* screw. **(b)** Sample in 26:1 *L*/*D* screw.

combinations between screw speed and SCF dosage are selected to find if it is possible to make the same quality of cell structure with some combinations. It is true that the low dosage of 0.3% of N_2 at high screw speed 520 rpm with shear rate 213 1/sec creates almost the same quality of cell structure as the high dosage of 1% of N_2 at low screw speed 127 rpm with shear rate 52 1/sec; other parameters are the same, such as back pressure 13.8 MPa (2000 psi). The result in Figure 6.13a shows the microstructure of the sample made with high SCF dosage at low screw speed. It has nice cell size and uniform distribution of cells with a few big cells more than 100 μm. However, the low dosage at high screw speed shows an excellent cell structure in Figure 6.13b as well. This result provides some processing window about the SCF dosage versus screw speed.

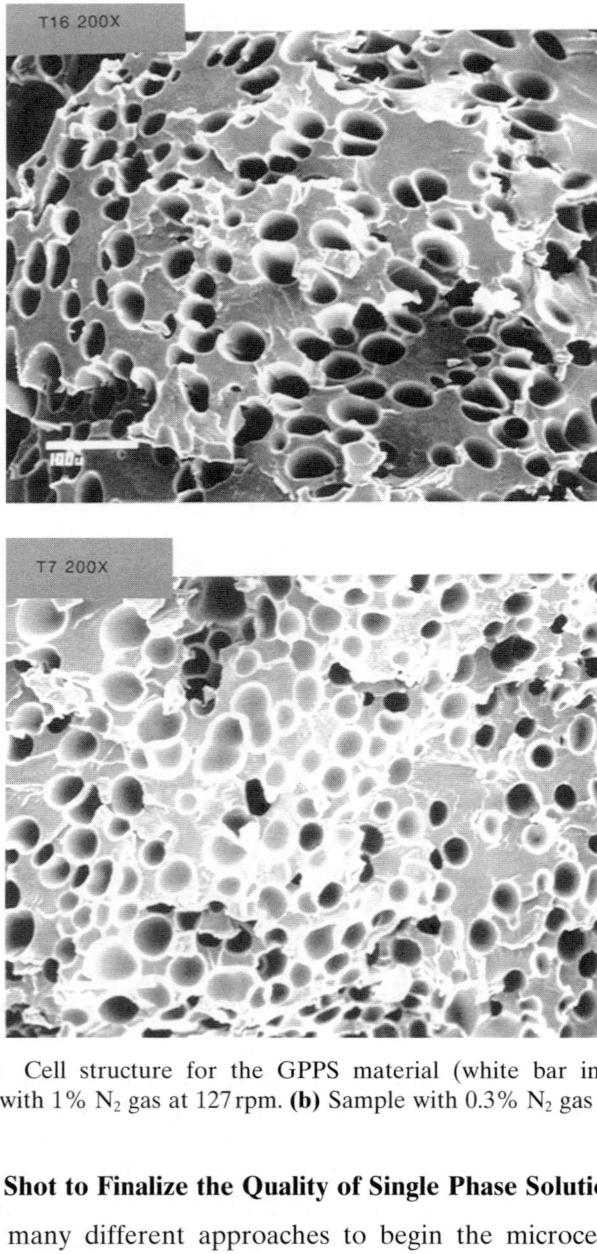

(a)

(b)

Figure 6.13 Cell structure for the GPPS material (white bar indicates 100 μm). **(a)** Sample with 1% N_2 gas at 127 rpm. **(b)** Sample with 0.3% N_2 gas at 520 rpm.

6.2.7 Air Shot to Finalize the Quality of Single Phase Solution

There are many different approaches to begin the microcellular injection molding process. With the regular molding habit, many people use the traditional way to begin the molding cycle before being sure to make single-phase solution in the first stage [6]. It will waste time and material if the single-phase solution is not made well in the first stage. Therefore, the quick inspection of single-phase solution is a necessary procedure before moving to the next stage. It is a simple way to make an air shot to check if the job of the first stage is well done for making the single-phase solution. The methods to inspect the quality of the single-phase solution are introduced in the following.

6.2.7.1 Observation and Hearing. The observation for the purging of gas mixture under the pressure for good single-phase solution shows only a quick and quiet seeping out of the nozzle once it is opened. It results in a good-quality foaming part as well when this single-phase solution is used for the foam molding. It must be quiet air shooting from beginning to the end of the whole stroke. If there is some air pocket without mixing well with molten plastic, it will release the energy with a loud popping sound during the air shot. It may cause some hot air spray if there is an air pocket in the melt during purging or during air shot test. Therefore, the bad quality of gas-rich material may have some safety concerns as well.

6.2.7.2 Pressure Recording in Nozzle. The melt pressure is measured and recorded in the shut-off nozzle during the injection (see Figure 6.14). The melt back pressure during screw recovery is 6.9 MPa (1000 psi), which the is minimum pressure to make a single-phase solution for N_2 gas in GPPS melt with a 60-mm diameter and 28:1 *L*/*D* screw (see the first pressure curve of screw recovery in Figure 6.14). The melt temperature is 238 °C, and the injection speed is 0.076 m/sec consistently. The good single-phase solution is made by 0.25 wt% of N_2 gas at this processing conditions. It shows a quiet and smooth injection into air from beginning to the end through whole stroke, and the related pressure curve also gives a smooth continual pressure curve (see the second pressure curve in Figure 6.14a) about 9.7 MPa (1400 psi). The pressure curve in Figure 6.14b shows several sudden pressure decreases abruptly in the middle of the injection pressure curve. It indicates some gas pocket in the gas–melt mixture since 0.5 wt% of N_2 gas, instead of 0.25 wt% of N_2 gas, added into the melt at the same conditions in Figure 6.14a. However, the higher gas dosage may cause some uncompleted solution. This gas pocket is

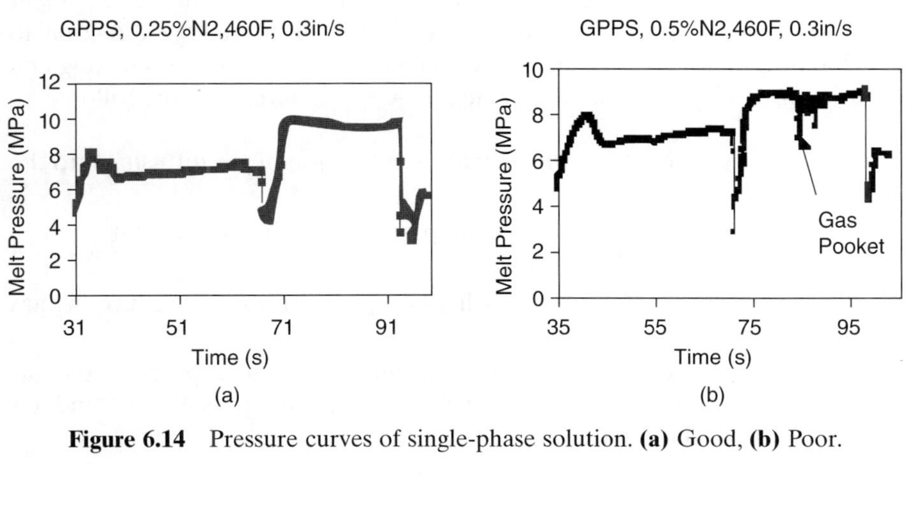

Figure 6.14 Pressure curves of single-phase solution. **(a)** Good, **(b)** Poor.

from the low back pressure since the gas dosage increased twofold when compared the gas dosage of the good single-phase solution. The unsafe operation with the gas pocket may also result from the loudly popping sound, and even hot material spraying or shooting out of the nozzle during injecting into air can easily detect the pressure decrease from the gas pocket. The pure gas under the same pressure will have much more energy stored than the gas–melt mixture so that it shows dangerous shooting with a loud pop sound. The molding test with the poor single-phase solution also makes some voids in the foamed part. However, if the melt pressure is increased to 13.8 MPa (2000 psi), the gas pocket is removed and single-phase solution is made well again. It verifies that the high percentage of gas dosage needs relative high pressure to increase the solubility and accelerate the gas diffusion.

6.2.7.3 Quenching the Sample for SEM Inspection. The most expensive way to inspect the quality in the first stage is to cause the air shot sample in the cooling container to freeze the microstructure of the air shot. Then, the scanning electron micrograph (SEM) picture can be made to check the gas dosing result in the screw. In Chapter 3, Figure 3.12 is the typical quenched sample of air shot from PC/ABS material. It shows a uniform cells distribution, and small cell sizes. With this excellent quality of air shot sample the processing conditions in the first stage can be recorded as the best setup document for this molding file.

In the practical operation, the first method of observation and hearing is the easiest way to check the quality of the first stage. The other two methods may use only the documentation of the mold trial to be kept either for customer requirement or to file a new material of patent application.

6.2.8 Conclusions for Gas Mixing and Diffusing

The analyses of the gas mixing and diffusing in the reciprocating screw are the combination of theory and practice. The theory is still in the development stage. However, the industry still uses some theory to get the direction to develop or to innovate new technology. The guidelines for the gas mixing and diffusing in RS injection molding machine are summarized as the follows:

- The mixing elements are the key for acceleration of gas diffusing into the plastics.
- The high shear rate is an important parameter to be preferred if the residence time has conflict with it.
- Generally, temperature rise may help for gas diffusion but reduce the gas solubility.
- Pressure is the easiest and more efficient parameter to promote the gas diffusing rate. The high pressure will increase both gas solubility and gas diffusivity in the plastics.

- A comprehensive mixing occurs in the screw mixing section. A modified mixing must have some upside-down mixing movements to exchange the top layer and bottom layer of material for uniform mixing.
- Back pressure is also a big help to promote the mixing, and it must be maintained at a certain level to keep the single-phase solution.
- A short screw provides acceptable quality of microcellular part for both amorphous and crystalline materials. Some material, such as GPPS, needs to run with either high back pressure or high gas dosage to avoid the voids and gas post pop in the part. The shortest residence time has been successfully tried as short as 6 sec. Generally, the single-phase solution can be made with higher pressure and more gas dosage if the mixing time is too short.
- It is strongly recommended to inspect the quality of the single-phase solution in the first stage before going to the next step for the second-stage process of molding.

6.3 NUCLEATION AND INITIAL CELL GROWTH

Once the first stage prepares the single-phase solution, the second stage needs to be focused on nucleation and cell growth. Additional mold-filling cooling analyses will be discussed in the molding stage. The nucleation quality of the foamed part is verified with the morphological pictures. The SEM picture is the most reliable proof for the quality of foams since the cell structure can be visibly checked by the morphology picture. In addition, it can clearly explain the property changes from the processing variation. Although the nucleation is the part of molding, the analyses of nucleation should separate the nucleation result from the mold filling result. It can be implemented to control the mold filling at only 80% of full cavity and can check the morphology picture in the runner system. Then, the microstructure in the runner system of the part represents the real result of nucleation through the restricted area, such as nozzle or valve gate.

6.3.1 Nucleation Theory in Injection Process

Previously, Shimbo [23] also found that the linear injection speed has a direct effect upon cell structure. The cell size and density of a foamed PS (polystyrene) sample by fast linear injection speed become small compared to that of the same PS sample produced by slow linear injection speed. In this paper, have a small effect on cell size distribution, and the overwhelming factor in controlling cell size is nucleation rate. Nucleation rate is controlled by pressure drop rate (*dp*/*dt*) that is directly related to the injection speed. It is easily shown that cell size can be controlled by shooting a single-phase solution into the air at varying injection rates. The air shot samples from the same nozzle tip with different injection volume rates will have different cell structures (cell size and cell density). Higher injection rates will produce finer cell structures. Shimbo [23] explained that the fast linear injection speed makes mold filling time short so that the distribution differences of temperature in the mold are small, cooling is uniform, and pressure in the mold is released evenly. Although temperature uniformity in the mold might be important, the proper nucleation of large number of cells has to dominate in the initial phase of bubble formation. Ideally, the homogeneous nucleation is favored when the activation energy to form and grow bubbles is uniform throughout the plastic matrix. However, the practical process is dominated by heterogeneous nucleation. There are many different additives in the polymer matrix, and most of them are excellent nucleates. Many nucleation sites are these foreign substances in the polymer matrix. These foreign substances can be organic or inorganic in nature. Some residues left in polymers during polymerization or during post-processing stages can be act as the nucleation sites as well. In some cases the additives can act as gas absorbers—that is, carbon black reduces the foaming agent effectiveness. Therefore, both homogeneous and heterogeneous nucleation happens simultaneously with sudden thermodynamic instability, and the heterogeneous nucleation is thermodynamically favored in microcellular injection molding. To help in understanding the nucleation phenomena, a general model is modified with injection processing parameters and is given based on the theory of homogeneous nucleation [24].

$$J_n = M_b \sqrt{\frac{2\sigma_b}{\pi m_g}} \exp\left(\frac{-16\pi\sigma_b^3}{3kT_{\text{poly}}(P_b - P_m)^2}\right) \tag{6.20}$$

where J_n is the rate of bubble nucleation, σ_b is the bubble surface tension, m_g is the mass of the gas molecule, T_{poly} is the melt absolute temperature, P_b is the internal gas pressure, P_m is the melt pressure, k is Boltzmann's constant, and M_b is the number of gas molecules per unit volume.

Although Equation (6.20) is for the homogeneous nucleation rate and assumes that steady-state bubble formation occurs in thermodynamic equilibrium, it can be used for qualitatively interpreting the heterogeneous nucleation results. There will be three key factors to be discussed for the nucleation with Equation (6.20): temperature, pressure, and gas content. The temperature and pressure are difficult to be separated completely in the process of microcellular injection molding. However, the gas content is easily controlled in the process.

Ideally, the pressure drop rate *dp*/*dt* needs to be controlled at the gate. Controlling the *dp*/*dt* at the gate allows for finer cell structure, more uniform cell size distribution, and less distortion of the bubbles. If the size of the nozzle orifice is comparable to that of the gate size, and both meet the minimum *dp*/*dt* requirement, the bubbles nucleated at the nozzle will be grossly distorted and ruptured in the gate. The distorted bubble may be recovered in the downstream of the mold filling. However, the ruptured bubble will either create a rougher surface with a swirl pattern of surface or be ironed to the smooth surface if the mold surface is hot, and mold venting is good.

6.3.2 Temperature Effect

It seems like the nucleation rate will be increased, with the melt temperature T_{poly} increasing based on Equation (6.20). It results in the cell density increasing and cell size decreasing. The classical nucleation theory has the similar conclusion that the high temperature will lead to the high nucleation rate. However, the temperature change will influence other parameters, such as surface tension, melt viscosity, and gas solubility. The fast injection process makes the effect of temperature for the nucleation more complex. In general, temperature rise will result in surface tension reduction, melt viscosity reduction, and gas solubility reduction as well. The final effects may cause the nucleation result change if the cell growth does not control well.

6.3.3 Injection Speed Effect (Pressure Drop Rate Effect)

It is well known that the pressure drop rate is the most important parameter contributed for the nucleation. Increasing of pressure drop rate in the nozzle will increase the nucleation rate as well. It is also the fact that is verified in the batch process nucleation strongly related to pressure drop rate. On the other hand, the nucleation rate is related to gas saturation pressure. The high gas saturation pressure results in high nucleation as well [12]. This result of batch process gives the direction of practical processing that requires not only a high rate of pressure drop but also a high initial pressure to be maintained in the accumulated material in front of screw tip to promote the nucleation.

When the pressure effect for the nucleation is the topic, the injection speed takes over since it is a direct parameter to be controlled and results in a direct effect of pressure drop rate. In fact, the injection speed change will definitely result in both temperature and pressure changes in the restricted nucleation area because the high volume rate of flow creates not only pressure drop but also the shearing heat. However, the pressure drop is an instant value change shown from measurement, while the temperature rise takes time to reach the balance. Therefore, the pressure drop rate still effectively sets up by different injection speeds. The pressure drop rate in the cylindrical channel of pressure-restricted area can be simplified as below:

$$\frac{dp}{dt} \propto f_{pt}\left(\eta_T, D^4, V^2, d^{-6}\right) \tag{6.21}$$

where dp/dt is the pressure drop rate, f_{pt} is the pressure drop rate function (see specific formula in the gate design of Chapter 5), D is the diameter of screw (or plunger for injection), V is the linear injection velocity of screw of plunger, d is the diameter of the orifice of nucleation element, and η_T is the non-Newtonian gas-laden polymer viscosity.

$$\eta_T = m_{p0} \exp(-k_c C) \exp\left(\frac{B_T}{T_{\text{poly}}}\right) \gamma^{n-1} \tag{6.22}$$

where m_{p0} is the power-law coefficient, k_c is the constant to be determined by experiment with gas concentrate C, C is the gas concentrate, B_T is the constant for Arrhenius temperature relation, γ is the apparent shear rate, and n is the power-law index.

The correct parameter of injection for the quality of the microcellular process is the injection volume rate. It is because the injection speed must be considered with the specific diameter of the injection elements, such as screw or plunger. The injection volume rate of high linear injection speed of a small-diameter screw may not be as big as the injection volume rate of low linear injection speed of a big-diameter screw. Therefore, if the injection linear speed is used for the injection performance, the screw diameter must be mentioned. Otherwise, the injection volume rate is only one precise definition without the necessity to specify the diameter of screw, or plunger, to evaluate the injection performance or nucleation.

With the increasing of injection speed, the viscosity will decrease. However, combining the effects in Equations (6.21) and (6.22) together, the overall pressure drop rate increases significantly with the rise of the linear velocity of injection. Nucleation occurs as the single-phase solution is injected into a mold through a restricted area of the nozzle or gate. The quantity analysis for the pressure drop rate is the tool here to find the quality of nucleation. To check the quality of the foamed part, an ISO test bar mold with two cavities was adapted. To avoid the possible second nucleation, a side gate has the same width of the bar (0.02 m) and same thickness (0.0026 m, with cross-section area 0.00005 m^2). Then, a 0.003-m-diameter orifice (with cross-section area 0.000007 m^2) in the nozzle tip becomes the true nucleation spot. All the samples were made with short shot (mold filling less than 80% of full volume of the mold). Therefore, there is no packing stage that may influence the final cell structure. The morphology of the sample is examined in the middle of a runner system that has a trapezoid section with an area that is three times larger than the nozzle orifice area. The injection speed varies from 0.0127 m/sec to 0.025 m/sec. The molding material is acrylonitrile butadiene styrene (ABS, GE, Cycolac AM). The blowing agent is nitrogen gas (N_2) at the supercritical state.

It is important to set up the processing conditions perfectly in the first stage and then focus on the study for processing parameters in the second stage. The ABS material is used and the processing conditions in the first stage are 127 rpm of screw speed, 0.5% of N_2 gas, 13.8 MPa of back pressure, and melt temperature at 480 °F. The results are shown in Figure 6.15 and Table 6.7.

(a)

(b)

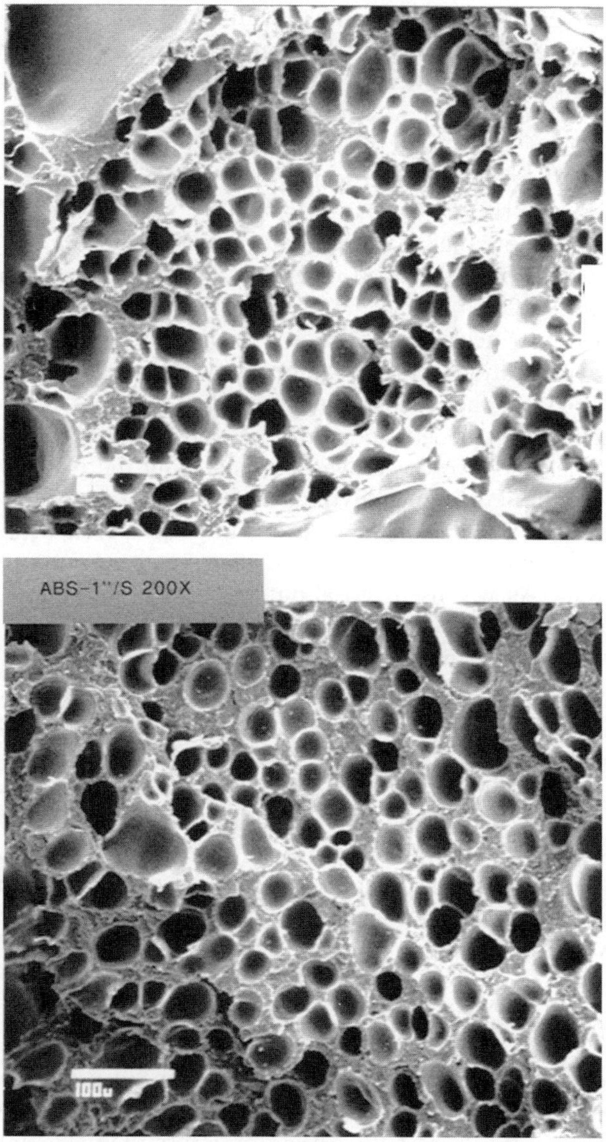

Figure 6.15 Cell structure for the ABS material (white bar indicates 100 μm). **(a)** Injection speed 0.0127 m/sec. **(b)** Injection speed 0.0254 m/sec.

TABLE 6.7 Results of Different Injection Speeds for Acceptable Cell Structure

Injection speed (m/sec)	0.0127	0.0254
dp/*dt* (Pa/sec), ABS	2.4×10^8	5.9×10^8
Shear rate (1/sec), ABS	3387	6773
Viscosity (Pa-sec), ABS	42.4	25.6

Park et al. [25] found minimum pressure drop rates required for the creation of microcellular foam, in fully saturated materials. For HIPS (high-impact polystyrene) with 10 wt% CO_2 gas, the cell density is increased from 10^8 cells/cm^3 to 10^9 cells/ cm^3 with the related *dp*/*dt* increased from 0.18×10^9 Pa/sec to 0.9×10^9 Pa/sec [25]. Theoretically, a cell density of 10^9 cells/cm^3 will produce cell sizes of about 10 μm [12]. Therefore, for HIPS saturated with SCF, the required pressure drop rate (*dp*/*dt*) is about 10^9 Pa/sec for 10 μm cells. It is important to note that at varying saturation pressures for the single-phase solution, different pressure drop rates may be required to achieve homogeneous nucleation [26].

Therefore, the critical pressure drop rate *dp*/*dt* may be about 1×10^9 Pa/sec [1, 7] at the nucleation spot for amorphous materials. Figure 6.15a shows the morphology of the cell structure of *dp*/*dt* of 2.4×10^8 Pa/sec at the 0.0127-m/sec injection speed. There are big voids around some local microcellular areas. The voids were removed by increasing the injection speed up to 0.0254 m/sec without changing anything else. The microstructure of an average 30- to 80-μm cell size shown in Figure 6.15b is injected with double speed of the result shown in Figure 6.15a, and that creates a pressure drop rate of 5.9×10^8 Pa/sec. As we discussed about the results in Equations (6.22) and (6.23), the shear rate for 0.0127 m/sec is 3387 1/sec, while the shear rate for 0.025 m/sec is 6773 1/sec without any slip. Therefore, the shear rate will increase with an increase in the injection speed. However, there is ABS shear thinning (see viscosity data in Table 6.7) mostly near the orifice wall of the nozzle where the shear is at maximum value. As a result, the core "plug flow" or slips along under the shear-thinned subsurface layer will reduce shear eventually. Therefore, the viscosity at the low injection speed is 42.4 Pa-sec, and the viscosity at the high injection speed reduces to 25.6 Pa-sec. Overall, the viscosity reduction at high injection speed does not change the tendency that the nucleation at high injection speed is increased almost twice as large as the nucleation at low injection speed.

6.3.4 Injection Speed Profile Effect

The injection speed profile may help for the uniform nucleation from the beginning to the end of the injection stroke. The injection flow rate determines the pressure drop rate across the nozzle or across the valve gate. However, a uniform nucleation is required from the beginning of mold filling to the end of mold filling. The general pattern should be slow–fast–slow. The first slow injection speed deals with the empty cavity of mold that is almost at the atmospheric pressure (or zero gage pressure). The slow injection speed will be enough to create a good pressure drop rate for nucleation. Then, fast speed needs to increase the pressure drop rate according to the cavity pressure increasing with the volume percentage of mold filling increasing. Finally, the slow speed of injection will help to relax the pressure near the gate to allow improved foaming. On the other hand, final slow injection speed leaves more foaming spaces in the front of mold filling.

Lee et al. [27] verified that this fast and slow injection speed profile helps to make uniform foam throughout in whole piece of the structural foam part. Similarly, as a result of injection speed profile from Lee et al. [27], microcellular foam shall follow the same principle of process as the structural foam. First of all, it is well known that the highest foaming area is always in the free-flow front of mold filling, or at the area of the beginning of the mold filling. However, the lowest foaming area is on the gate area. It is obvious that the pressure drop is the key factor for this kind of foaming distribution. At the free-flow front of mold filling, the pressure is zero or near zero, which determines the venting system of the mold. Regardless of the injection speeds, this phenomenon is always visible for every short shot of microcellular foam. The second interesting phenomenon is that the foam fraction in the free-flow end of mold filling is increased linearly when the injection speed is increased from low injection speed to some upper limit of high injection speed. This is most likely because the pressure drop rate increased when the injection speed is increased. Beyond the upper limit of high injection speed, the increase in foam fraction in the free-flow front will level off. In other words, once the upper limit of high injection speed is reached, there is no need to further increase the injection speed for more nucleation. Lee et al. [27] explained that the leveling off of the increasing foam fraction over the upper limit is the shearing force. Extremely highly injection speed may cause active cell coalescence and rupture because it may result in shearing hardening in extremely highly sheared material. On the other hand, even if the gate area always has a smaller foam fraction than does the free-flow front, it continues to have increasing foam fraction with the increasing of injection speed.

One more interesting phenomenon mentioned by Lee et al. [27] is the fast injection speed 200 mm/sec at the first 67% of stroke, and when a slow injection speed of 3 mm/sec is switched, the foam fraction can be found almost uniformly throughout the whole part. The slow speed actually is almost similar to the shrinkage compensation movement in regular molding. It is so slow that the gate area has enough time to relax, and foams occur in the gate area. This phenomenon has been found in microcellular molding as well. It was a PMMA microcellular molding. The solution for the un-uniformity of the foam fraction is not the only problem to be solved by the injection speed profile, but also we need to have a solution with a small nozzle orifice. Once the 9.525-mm nozzle orifice was changed to the 6.35-mm nozzle orifice, the whole part does have the nice cell throughout the part.

6.3.5 Nucleation Position Effect

The nucleation position needs to be checked to avoid the second nucleation. The second nucleation will cause the shearing on the existing initial cells after the first nucleation and will possibly distort the cells when the cells go through the small gate again.

There are three possible configurations among the nozzle, valve gates, and gates in the mold. One is that both a mold valve gate and a shut-off nozzle are used. Then, the nozzle orifice must be big enough to avoid any significant nucleation that occurs in the nozzle. Hence, the nucleation mostly occurs in the valve gate, which is the best spot of nucleation nearest to the mold. Another possible configuration of mold is the shut-off nozzle with the cold sprue and the small gate on the part. In this case the best way is still to use the big orifice of the shut-off nozzle to move most nucleation spots to the small cold gates near the part. The third configuration is the shut-off nozzle plus the cold sprue in the mold without small gates, and the nozzle orifice becomes the nucleation spot. Usually, the third configuration does not need to check the second nucleation because the nozzle orifice is truly the smallest diameter to control the nucleation rate there.

6.3.6 Material Effect

It is well known that with heterogeneous nucleation in filled material the same processing conditions can create better cell structure compared to the unfilled same material. Many experimental results of the batch process in Chapter 2 verify that it is true for all different materials. On the other hand, the practical injection molding process shows that the heterogeneous nucleation is favorable nucleation as well for both process stability and better cell structure.

Figure 6.16a shows the unfilled PA6 material that has cell size about 50 μm and low cell density. However, the 33 wt% of glass-fiber-reinforced PA6 material shows the cell size about 10–50 μm and much high cell density in Figure 6.16b. This result verifies the conclusion that glass fiber helps to reduce the gas dosage for the same quality of cell structure as the unfilled same material because it needs less gas to be nucleated.

There is another typical application example of talc-filled PP material. Figure 6.17a shows the unfilled PP cell structure that again has nonuniform cell size from 10 to 100 μm. However, the 20% talc-filled PP shown in Figure 6.17b indicates that the cell size varied from 10 to 50 μm.

Compared to the effect of improving nucleation results between glass fibers (result in Figure 6.16) and fillers like talc (result in Figure 6.17), glass fiber shows better improvement for nucleation than does filler.

Generally, the results above show again the filled material results in the better cell structure because the nucleation rates of the processing conditions are the same as those of unfilled material. On the other hand, the filled material can reduce the minimum injection speed significantly for the acceptable nucleation rate. It is obvious that the heterogeneous nucleation requires much less pressure drop rate than does the homogeneous nucleation. It may bring another benefit regarding energy saving since filled material requires less injection speed, which will save the peak energy consumption from highest

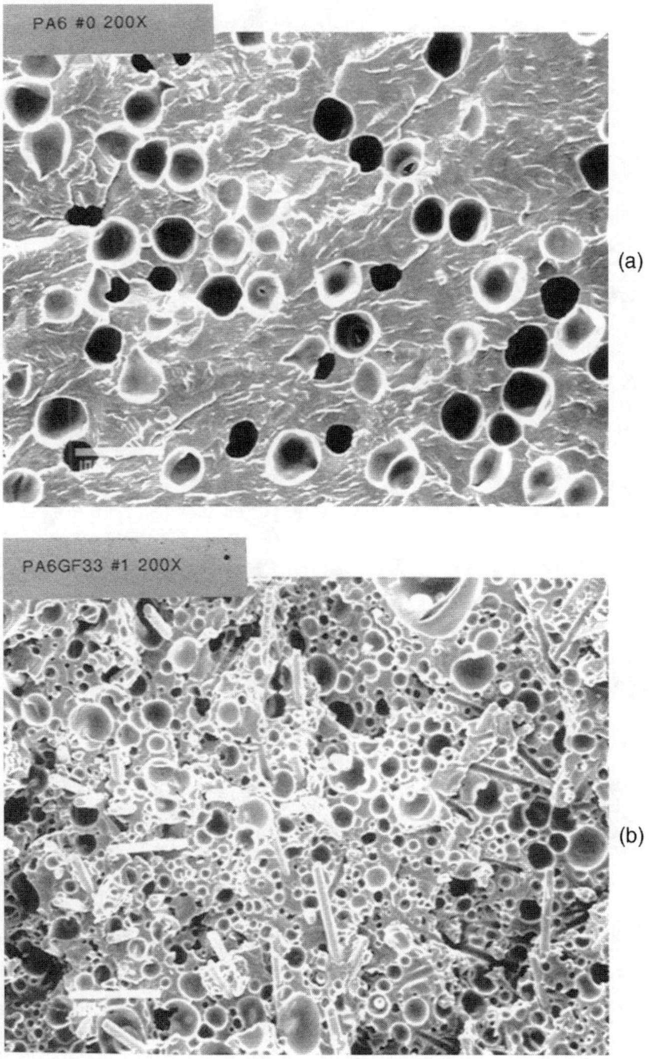

Figure 6.16 Cell structure for the PA6 material (white bar indicates 100 μm). **(a)** PA6. **(b)** PA6 + GF33.

injection speed. In addition, the lower gas weight percentage is required for filled material. It means stable processing because high weight percentage gas needs more precisely control for gas dosing and mixing.

6.3.7 Gas Concentration Effect

The high gas concentration will help for nucleation as well. It is predictable trend from the gas concentration terms in Equations (6.22), and (6.23). It is easily verified by the experiments with different gas weight percentage in the

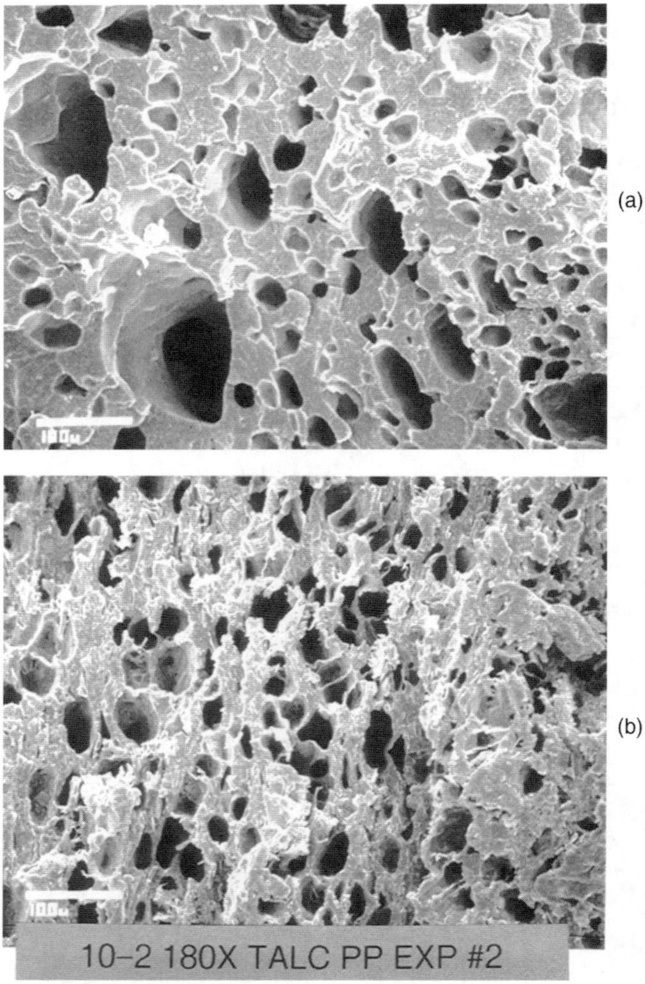

Figure 6.17 Cell structure for the PP material (white bar indicates 100 μm). **(a)** Unfilled PP. **(b)** 20% talc-filled PP.

same material at the same processing conditions. Generally, the higher gas concentration creates a greater number of small cells. It was verified by both CO_2 gas and N_2 gas in many different materials.

The results of gas concentration improving nucleation will be varied with different materials. The results discussed for PP material above are clearly showing the gas concentration effect for the nucleation. The 1 wt% of N_2 gas creates a huge number of nuclei, and the 0.6 wt% of N_2 gas produces a smaller number of nuclei.

For GPPS the gas concentration is also helpful for the nucleation. However, it is not obvious improvement for the nucleation compared to PP. One experiment has been carried out with GPPS material with different N_2 gas weight

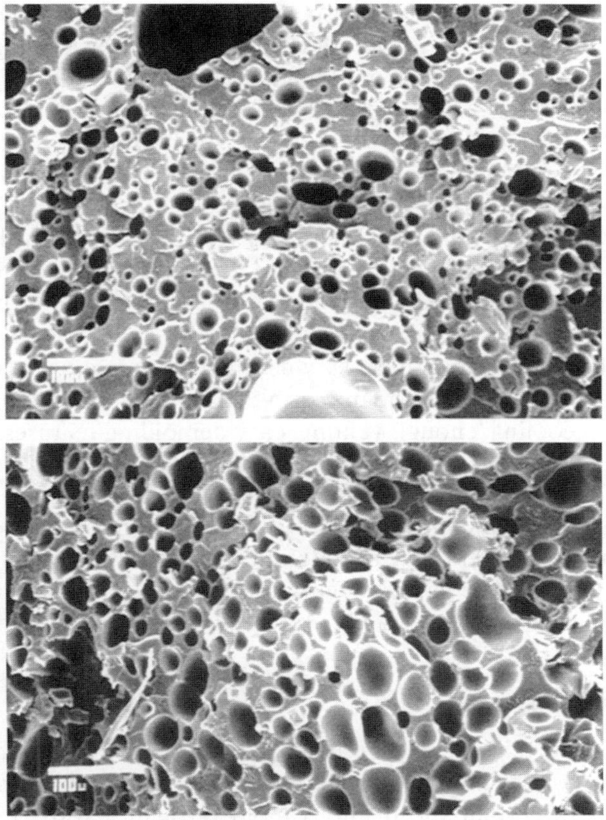

(a)

(b)

Figure 6.18 Cell structure for the GPPS material (white bar indicates 100 μm). **(a)** 0.34% of N_2 gas, **(b)** 1% of N_2 gas.

percentages in a 30-mm screw with a middle ring at the same processing conditions, screw rotation speed 520 rpm, and back pressure 13.8 MPa (2000 psi). Figure 6.18a is the cell structure of GPPS with 0.34 wt% of N_2 gas. It shows some big cells (or voids). However, it does create enough cells even if the gas percentage is low. Figure 6.18b displays the cell structure of high gas concentration that is as high as 1 wt% of N_2 gas. The cell structure with high gas percentage does show much uniform cell size and almost a packed pattern of cell structure. However, the number of cells does not have an obvious increase compared the cell number in Figure 6.18a.

6.4 MOLD FILLING ANALYSES

Mold filling is the last step of the microcellular injection molding process. It includes cell growth and shaping. The shaping process is similar to the regular injection molding process except shrinkage and warpage are definitely different from regular molding. Also, the cooling time and injection time will have some changes as well. The other parameters of processing may need to set up different from regular molding. According to DOE results in paper reference 3, injection speed and shot size are the important parameters. More experiments are carried out to verify these conclusions [7]. Since the shot size usually determines the weight reduction and the mechanical strength of the part, it may be the preselected parameter before molding. Therefore, the important variable molding parameter for foam quality control after the part is designed is injection speed. In addition, the injection speed tests show that the optimum injection speed exists for the best cell structure that determines the quality of foams [7].

We will focus on several necessary parameters to make good microcellular foams. It is well known that the microcellular mold and part must be designed with uniform cooling, enough venting, easy demolding features, and balanced gating system to avoid any possible interruption of cycling. These will be discussed in Chapter 5. Assume that all of the criteria regarding the mold and part design are qualified; the issues of nucleation and initial cell growth are the injection parameters that include injection speed, temperature, injection speed profile, nozzle or gate configurations, the single-phase solution quality from first stage, and so on.

6.4.1 Injection Speed

It is well known that the key performance during the injection is the nucleation in the nozzle or gate. Nucleation occurs as the single-phase solution is injected with certain speed into a mold through a restricted area at the nozzle or gate. The injection speed plays an important role not only for nucleation but also for molding. Homogeneous nucleation of the microcell occurs when conditions are provided such that it is easier for the new cells to nucleate than it is for existing cells to grow. During large-pressure drops, the single-phase solution will experience a decrease in solubility of the SCF in the polymer. The SCF, coming out of solution quickly, can either go into an existing nucleated cell or help to nucleate new cells. Microcellular foam requires that a largest number of cells must be created during this pressure drop. To create a large number of cells, conditions have to be such that the SCF coming out of solution prefers to form a new cell as opposed to migrating into an existing growing cell. The first condition for homogeneous nucleation to occur is that the time required to nucleate a nucleation site must be much, much faster than the time required for the SCF to diffuse into an existing cell. Secondly, the distance the SCF travels to go into a new cell must be much smaller than the spacing between stable growing nuclei [12]. In a practical sense, the only method to accomplish this is to ensure that a very high plastic pressure drop rate occurs during the injection phase of the cycle.

6.4.1.1 Injection Speed Effect on Cell Structure. The quantity analysis for the pressure drop rate is still the tool here to find the quality of nucleation in the mold, although it has already been discussed in the nucleation section above. To check the quality of the foamed part, the morphology of the part must be inspected. Therefore, an ISO test bar mold with two cavities was adapted, which is the same mold as the one used in the nucleation test. To avoid the possible second nucleation, a side gate has the same bar width (0.02 m), and thickness (0.0026 m; cross-section area 0.00005 m^2). Then, a 0.003-m-diameter orifice (cross-section area 0.000007 m^2) in the nozzle tip, not the gate area, becomes the true nucleation spot. All the samples were made with full shot after cell foaming with controlled weight reduction. There is no packing stage that may influence the final cell structure. The morphology of the sample is examined in the middle of the part. The injection speed varies from 0.013 m/sec, 0.025 m/sec, 0.051 m/sec, 0.076 m/sec, 0.102 m/sec, 0.127 m/sec, and up to 0.152 m/sec. The molding materials are general polystyrene (GPPS, Dow Chemicals, Styron 666D) and acrylonitrile butadiene styrene (ABS, GE, Cycolac AM). The blowing agent is nitrogen gas (N_2) at the supercritical state.

This the same conditions set up as the nucleation test above except the mold filling percentage. The ABS material is used and the processing conditions in the first stage are 127 rpm of screw speed, 0.5% of N_2 gas, 13.8 MPa of back pressure, and melt temperature at 480 °F. The results are shown in Figure 6.15 and Table 6.7.

The critical pressure drop rate dp/dt is 1×10^9 Pa/sec [1, 7] at the nucleation spot for amorphous materials. Figure 6.19a shows the morphology of the cell structure of dp/dt of 2.4×10^9 Pa/sec at the 0.076 m/sec of injection speed. The cell sizes varied from 25 μm to 50 μm and were uniformly distributed in the part. The interesting result is the optimum cell structure occurring with 0.102 m/sec of injection speed that creates 15 μm of uniform cell size in the sample in Figure 6.19b. Then, Figure 6.19c shows the result of higher injection speed of 0.127 m/sec with an acceptable 45 μm of cell size. Finally, the maximum injection speed at 0.152 m/sec as the machine limit has been tried and the result is almost the same as the one at 0.127 m/sec. These results show that the better cell structure is not proportional to the injection speed only. For the nucleation the optimum injection speed does exist. It may also be related to shear rate since the shear rate for 0.076 m/sec is 20,320 1/sec, while the shear rate for 0.127 m/sec is 33,867 1/sec without any slip. However, the optimum cell structure has the shear rate 27,093 1/sec. Excessive shear may create material degradation. On the other hand, there is ABS shear thinning (see viscosity data in Table 6.8 [7]) mostly near the orifice wall of nozzle where the shear rate reaches maximum. As a result, the core "plug flow" or slips along under the shear-thinned subsurface layer will reduce shear eventually. Therefore, once the critical pressure drop rate reaches the injection speed, it does not have to be raised too high for better cell structures. It is another challenge to

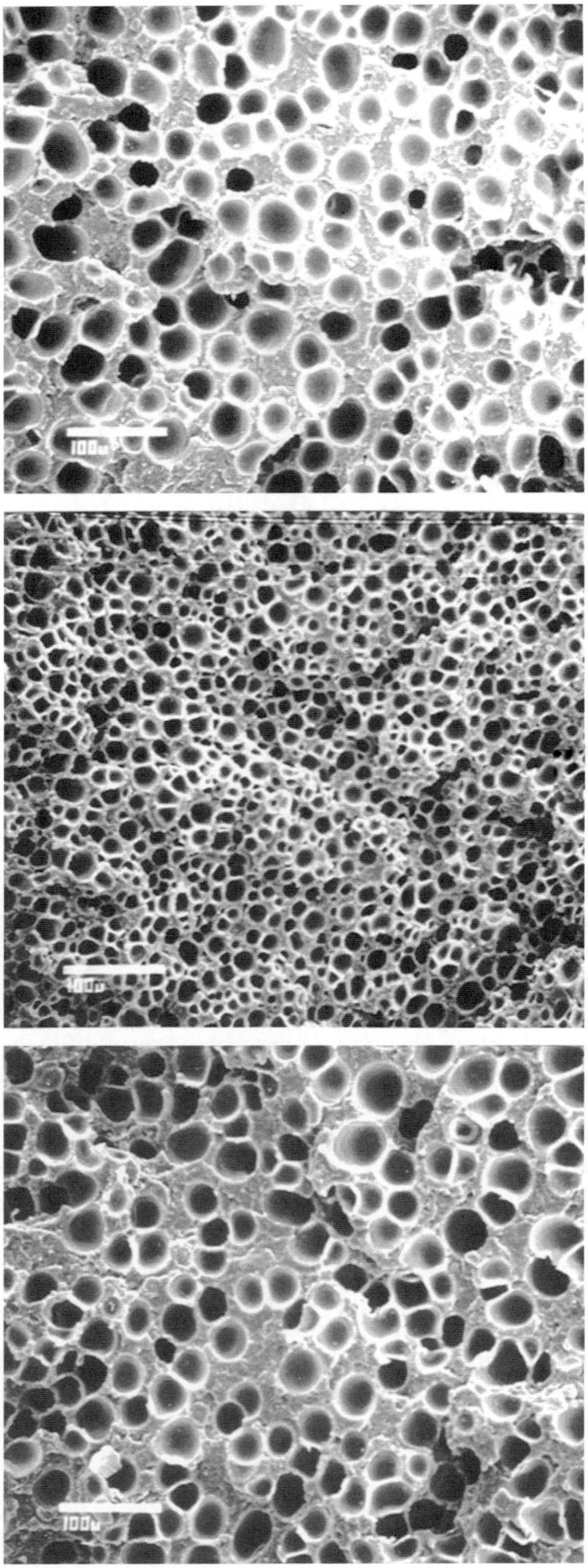

(a)

(b)

(c)

Figure 6.19 Cell structure for the ABS material. **(a)** Injection speed 0.076 m/sec. **(b)** Injection speed 0.101 m/sec. **(c)** Injection speed 0.127 m/sec.

TABLE 6.8 Results of Different Injection Speeds for Optimum Cell Structure

Injection speed (m/sec)	0.076	0.102	0.127
dp/*dt* (Pa/sec), ABS	2.4E9	3.4E9	4.5E9
Shear rate (1/sec), ABS	20,320	27,093	33,867
Viscosity (Pa-sec), ABS	11.4	9.3	7.8

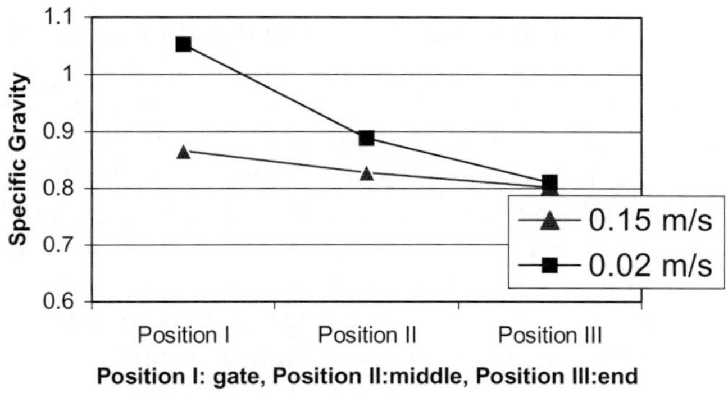

Figure 6.20 Density distribution over length of a molded part. Material: Polycarbonate.

find the optimum injection speed for the best cell structure in microcellular foams for all materials in the future.

6.4.1.2 Injection Speed Effect on Uniform Part Density. Linear injection speed also influences the density distribution. Figure 6.20 shows the specific gravity distribution of a standard tensile-bar sample for PC (polycarbonate). The sample was cut at three positions:

- Gate area (10 mm from gate)
- Middle area (70 mm from gate)
- End area (125 mm from gate)

In this particular instance and as the graph shows, the highest density is in the gate area, and the lowest density is at the end flow. Obviously, higher linear injection speeds result in a more uniform density distribution. To illustrate, at an injection speed of 0.02 m/sec, a 30% specific gravity difference between the gate and end of flow exists while only an 8% difference exists with a 0.15-m/sec linear injection speed at similar molding conditions. Pfannschmidt found the same results on his test for CO_2-preloaded PP [22]. Slow linear injection speeds fill the mold more slowly, which allows a thicker skin to form and thus a narrower flow channel results. This tends to create overpacking near the gate area and underpacking near the end of flow.

6.4.1.3 Injection Speed Effect on Weight Reductions. Higher injection speeds allow for a larger and instantaneous pressure drop across the part, due to a lower transmission loss of plastic pressure with a fast injection rate. If the injection rate is too slow, the material cools as it enters the mold and thus higher plastic pressures are required to fill the part. The higher plastic pressure means a lower pressure drop and thus lower weight reductions. The more important reason for higher weight reductions is that higher injection rates produce higher and more uniform nucleation sites.

6.4.1.4 Injection Speed Effect on Surface Finish. Linear injection speed is also an important factor for controlling the surface finish of the microcellular molded part. Linear injection speed needs to be large enough to keep the mold filling time at a minimum so that the gas in the flow front is not allowed to escape. If the gas bubble has enough time to become a big bubble, it will break through the front of the plastic melt flow. The hole left by the bubble is elongated by the shearing region, and it is moved to the mold surface by the differences in shear rate (largest near the mold surface and zero in the center layer of the melt flow) [28, 29]. Surface finish is a complicated phenomena, and it will be discussed as the special topic in this chapter.

6.4.1.5 Injection Speed Effect on Fiber Orientation. The injection speed also plays important role because it determines the nucleation rate at certain gate size or nozzle size. Figure 6.21 shows the results of different injection speeds for the PC 10% GF [30]. It shows that all fibers are orientated in the flow direction in the skin area regardless of the injection speed. For only the center area of the part, the high injection speed (0.101 m/sec) creates fine cell structure and improvement of fiber (Figure 6.21a). Figure 6.21b that the slow injection speed (0.0127 m/sec) creates cells that are not as good as those created by the fast injection speed. Then, the cells do not have enough expelling force to reorientate the glass fiber around the cells. On the other hand, PC has thick skin compared to other material so that usually disorientation only occurs in the center area and not in the skin layer. Therefore, for a thin-wall PC part, it may not be the effective way to use microcellular processing to improve the fiber orientation.

6.4.2 Injection Speed Profile

Injection speed profile also affects the some qualities of the microcellular part. It is specifically important to control the speed at the beginning and at the end. Some of the requests may conflict with nucleation. However, it may need to check which one is preferred if a part quality issue occurs.

6.4.2.1 Injection Speed Profile on Jetting. To avoid the jetting in the mold at the beginning of injection, the slow injection speed needs to run until the runner system of the mold filled. Then, the injection speed can be increased

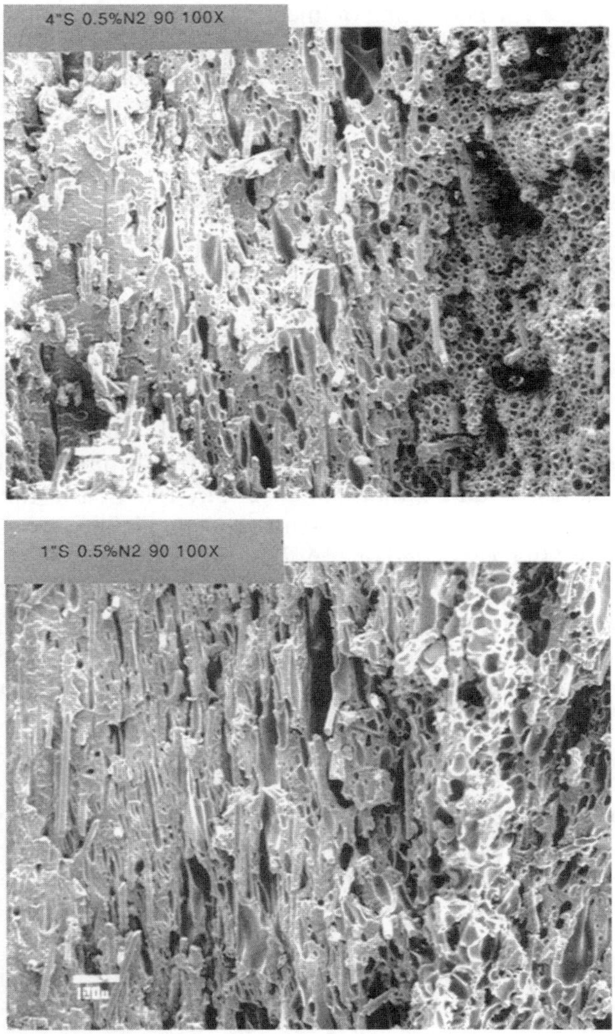

(a)

(b)

Figure 6.21 Morphology pictures of fiber distribution in thickness direction, 3-mm-thick part, foam PC GF10 with 15% weight reduction, section view parallel to the flow direction. **(a)** 0.102-m/sec injection speed [30]. **(b)** 0.025-m/sec injection speed (white bar indicates 100 μm) [30]. (Reproduced with copyright permission of Society of Plastics Engineers.)

to finish the mold filling as soon as possible with reasonable injection time and nucleation rate. This can be tested first with a short shot to determine the location of the switch point to change the injection speed from slow to fast.

6.4.2.2 Injection Speed Profile on Material Separation. The gas-rich material under the pressure has energy stored in it. This stored energy needs to be considered for correct setting of injection speed profile at the beginning of injection because stored energy creates initial injection as a result of energy release. Therefore, injection speed at the beginning may need to match possible gas surging at the beginning of injection. In other words, the injection speed should be set up at least as fast as the initial gas-laden material surging at the moment of nozzle or gate opening. With this initial injection speed matching between gas energy release and injection movement of the machine, there will be no material separation between the possible material surging by stored gas energy and the followed injection material. It may be estimated by the formula below:

$$U = \frac{1}{2} m_v V_{\text{free}}^2 \tag{6.23}$$

where U is the gas stored energy under pressure (using pressure to calculate the potential energy, assuming that it will be fully released to the air), m_v is the mass of the shot size under the pressure, and V_{free} is the free air shooting velocity driving by the gas stored energy U.

Then, based on the calculated value of V_{free} the first injection speed must match it or be a little higher than this value to be sure that the injection not only always follows possible gas surging at the beginning of the nozzle or at the open valve gate, but also pushes material continuously into the mold without any possible separation.

However, the easiest way to avoid doing this complicated estimation and trial is to delay the nozzle opening, or gate opening, and let injection occur first to further pressurized material and reach the necessary initial injection speed so that the gas pressure release will not be issue for the beginning of injection. This time delay usually is 0.5 sec or less to avoid too much pressure surging in the gate system and nozzle.

6.4.2.3 Injection Speed Profile on Surface Finish. Some crystalline material, such as nylon, will have a gas swirl spot obviously in the gate area if the injection speed is too fast at the beginning. This defect in gate area can be solved by the slow injection speed at the beginning. It may be explained by the slow injection speed filling the mold gradually without any gas separating from melt since the first injection into the mold is simply an air shot into the empty cavity of mold. On the other hand, the overall injection time cannot more than 3 or 4 sec so that the free-flow front will not have the overgrowing cells. The overgrowing cell will break to form the swirl mark on the surface of the part.

6.4.3 Mold Temperature

Mold temperature is an important parameter of molding. It determines the skin thickness, surface finish, and crystallization percentage. For the amor-

TABLE 6.9 Process Optimization of Microcellular Injection Molding

	Cycle Time	Strength	Weight Reduction	Surface Finish	Dimension Stability
Back Pressure	+	+	+	+	+
Gas Percentage	+	+	+	–	+
Screw Speed	+	+	+	+	+
Injection Speed	+	+	+	+	+
Mold Temperature	–	+ (for crystalline material) – (for amorphous material)	+	+	+
Melt Temperature	–	–	0	+	0
Cooling Time	–	+	0	+	+

+, increasing the setup of parameter.
–, decreasing the setup of parameter.
0, no effect of the parameter.

phous material the mold temperature can be as low as possible to cool the part soon unless the surface finish must be considered with high mold temperature. The crystalline material needs to consider both surface finish and crystallization in the skin during mold cooling. The regular injection molding offers the good reference range of mold temperature including the limit of mold temperature. The low limit of mold temperature can be even lower with microcellular injection molding as a result of the low viscosity of gas-laden melt. However, the high mold temperature is necessary just for the surface finish if it is required and longer cycle time is allowed. Table 6.9 shows the recommended mold temperature range for microcellular processing; these data have been taken from regular injection molding, and most of the low limits for making full use of microcellular processing benefits have been modified.

6.4.4 Shrinkage and Warpage

It is obvious that the microcellular injection molding will have much less shrinkage because it expands more than shrink during mold cooling stage. However, the warpage of the microcellular part may be determined by the comprehensive factors such as warpage direction and part stiffness reduction. The glass fiber orientation may influence the shrinkage and warpage as well. Usually, glass fiber significantly reduces the shrinkage.

Since shrink rates vary from part to part and from processing conditions, microcellular processing's affect on shrinkage will need to be taken into account as one more variable when designing a mold for the process. Generally, the shrinkage and warpage of any plastic can be reduced during the microcellular process [31]. Kramschuster and others did some quantitative study of shrinkage and warpage behavior for the microcellular process with fractional factorial design of experiments (DOE). It is found that the weight percentage of supercritical fluid (SCF) and injection speed affect the shrinkage and warpage of the microcellular part most significantly [23]. The hold pressure and hold time are usually the most significant effect on the shrink and warpage of the regular injection molding. However, the microcellular processing eliminates the holding stage. Therefore, the result from reference 31 is useful to focus on these two parameters above for reducing the shrink and warpage of the microcellular part. From the morphology point of view, these two parameters promote the cell architecture of small and uniform cells that can expand, instead of shrink, during cooling.

6.4.5 Demolding

Microcellular demolding needs to take into account the special issues of the microcellular foaming. It is because there are two characteristics of the microcellular foaming. Cells expanding will benefit the shrinkage. However, there are two disadvantages of the microcellular part. One is the soft surface of the microcellular part. It requires the slow ejecting movement at the beginning of ejection. In addition, the big contact ejecting area may be necessary that will be discussed in mold design in Chapter 5. Another disadvantage is that the cell expandsion will force the part to stick on the cavity, instead of on the core. Also, it is possible that the cold sprue breaks during demolding and stops the automatic cycling.

It is a unique benefit of the dimension stability from both gas assist and foam process. It is because of (a) the cell growth in foaming and (b) the gas pressure in gas-assist processing. However, the foam processing may result in a demoding problem since the gas pressure in the cells cannot be released as in the gas-assist process. The regular mold design makes use of the shrinkage to let the part stay on the core side, which is usually the moving half of the mold where there is an ejection system for demolding. If the expansion of cells is so strong that the foamed part sticks on the cavity of the mold, it causes a demolding difficulty. Therefore, the drafting angle shall be increased accordingly. The specific drafting angle must be determined by the material, depth of part, gas weight percent, and so on. Otherwise, some undercuts may be required in the surface of the core to keep the part in the core half of the mold if the part design is allowed to have these undercuts left on the part.

6.4.6 Surface Finish Improvement from Processing

The surface quality issue can be broken into three categories. One is the conventional foam surface that contains remains of burst or dragged cells leading to some streaking. Another is the elephant skin/swirling that contains previously frozen skin that has dislodged from its original position and buckled transverse to the direction of flow while it is displaced. Injection volume rate is a significant factor for this type of appearance. There may be a transition from fountain flow to plug flow at high injection speed. Finally, there is a spooge, or bold slug, which is the remains of aggressive/explosive pre-foam that develops before the flow path has been pressurized. The material created in the spooge is much lighter in color than the rest of the surface. The spooge may remain near the gate or be broken up and distribute throughout the part depending on injection rate. This can be solved by the delaying shut-off nozzle or valve gate opening action before injection truly occurs. It can remove any possible pre-foam material before injection movement reaches full speed.

There are clear solutions discussed above for spooge and elephant skin/swirling surface problems. However, the tough problem of surface quality for the microcellular foam is the same conventional foam surface problem as that for structural foam. There is no easy solution from processing methods for it yet. From a processing point of view, there are two major forming mechanisms of surface roughness: (a) broken bubble from free-flow front and (b) sheared bubble in the interface between mold wall and melt [32]. With the analyses for these mechanisms, most of the injection molding methods for the smooth surface of microcellular foam can be better understood.

The smooth surface is the restriction for the application of the injection molding microcellular foam. Similar to structural foam, most microcellular foam made by injection molding exhibits a splay-like appearance, or even a rougher swirl surface. The detailed descriptions of possible swirl forming mechanisms of structural foam are summarized and are well-defined by Semerdjiev [28] and others [29, 32, 33]. Stephen mentions two mechanisms for the bubbles formed on the surface adjacent to the mold wall. The first mechanism is that the surface bubble is deformed by shearing with a rapid flow and good adhesion to the surface of mold metal. The second mechanism is that the surface bubble moves along the interface if the adhesion is poor and the pressure is low.

Microcellular foam has special characteristics compared to structural foam, such as thin wall, small cell size, and huge cell density due to the gas in the melt at supercritical state before injecting into mold. The study here is to (a) understand the forming mechanism of poor surface to find the general guidelines to select the correct methods and (b) improve the current methods to smooth surface for microcellular foam. The assumption is that the single-phase solution has been well-prepared in the first stage of microcellular foaming so that the study focuses on the second stage that is the mold filling stage [7]. There are also two basic mechanisms in the formation of rough surface for microcellular foam [32]. One is that the gas escapes, or a bubble from the center layer of the free-flow front overgrows to break. Another is that the bubble away from center layer is sheared in the interface between mold wall and melt, which is similar to Stephen's first mechanism of roughness forming. The roughness is more or less related to the mold filling pattern, such as unidirectional filling or radial directional filling. The morphology picture approves the assumption that shearing bubble causes the roughness of microcellular foam. The flow patterns, such as radial directional filling and unidirectional filling, are analyzed with different thickness to show that injection processing conditions may also influence the surface quality significantly.

The analyses based on the mechanisms above are applied for three different methodologies to make smooth surface of microcellular foam. The first method is the solution for dealing with both forming mechanisms of roughness. The typical processes are co-injection and gas counterpressure moldings. The second method is for solving interface roughness that includes hot mold surface and coated mold surface. The third method is the surface improvement including processing, mold, material, and even part designing. The morphology results with methods of regular smooth surface mold and textured mold are discussed in this chapter. However, the details of co-injection and gas counterpressure are only discussed briefly in this chapter, and more details are discussed in Chapter 8 as the special processing methods.

6.4.6.1 Analyses for Forming Mechanisms of Surface Roughness. The broken bubble or roughness from the center layer of free-flow front may occur in the slow injection molding with a radial directional flow pattern. However, the interface roughness more or less always exists in the microcellular injection molding. In most cases, bubble shearing on the interface is the major source of roughness on the surface. Two different mechanisms are analyzed from the processing point of view as follows.

1. Interface Roughness. If the adhesion or friction force at the interface between the mold wall and the melt is high enough, it results in an assumption of the zero velocity for the melt in this interface. Then, the interface roughness mainly results from a fountain flow. Figure 6.22 is the SEM picture of GPPS sample with 3-mm thickness showing the bubble deformation distribution across the thickness direction in the part. It is clearly shown in Figure 6.22a that the center area is the zero shearing zone so that the cells keep the spherical shape. Then, the cell deformation increased with the distance close to the skin. In the very strong shear zone, where there is the interface between skin and melt, the cells almost are flattened in shearing field. This kind of bubble deformation trend across the thickness of the part is clearly displayed in Figure 6.22b, which is the local view of Figure 6.22a. It verifies the model of cell structure that has the spherical cells in the center layer. Any cell away from the center layer of the flow channel will be sheared more or less. As a result of this shearing, the shape of cell deforms to an oval shape and moves toward the skin. The strongest shearing layer is located near the interface between mold and melt. Compared to amorphous material, this strongest shearing layer is closer to the surface for the crystalline material since it has less skin than does amorphous material. The sheared cell in the surface either tears to break or stays in the oval shape below the skin. The broken cell in the surface becomes the major source of roughness for microcellular foam. Again, the

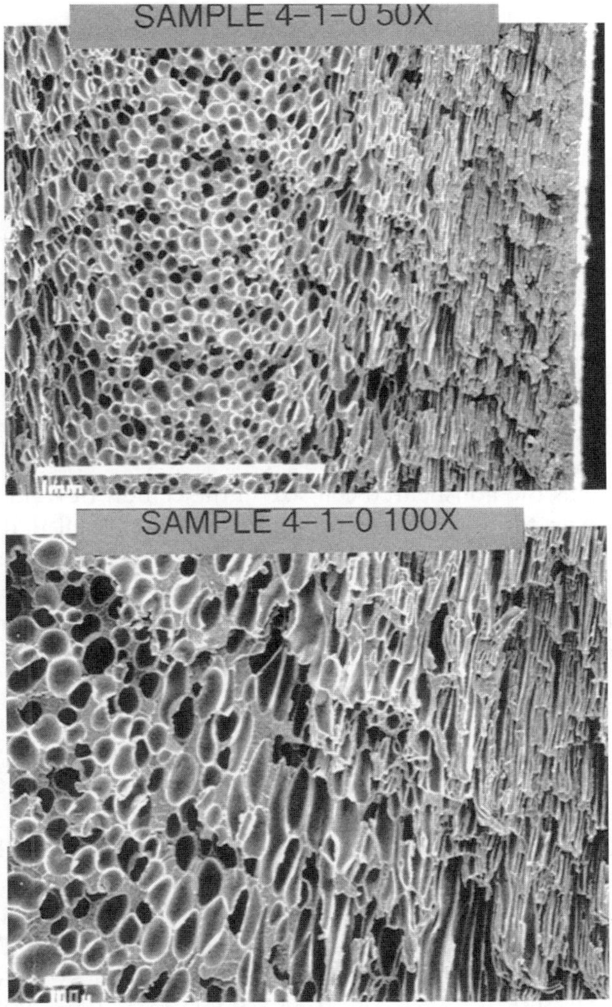

Figure 6.22 Cell deformation in the shearing field for the GPPS material (white bar indicates 1 mm).

result of GPPS in Figure 6.22 is similar to the result of ABS in Figure 3.11. Only reviewing the result showing the cells distribution in the section perpendicular to the flow direction may cause a misunderstanding that the bubble size is small near the skin while it is big in the center. The fact is that the cell near skin is elongated in flow direction, and the cell in the center has no deformation without shearing there (refer to Figure 3.11a). Figure 6.22b is the section local view parallel to the flow direction. It clearly shows the sheared cell along the flow direction in any layer away from center. The real overall size of sheared bubble near the skin looks similar to the overall size of the nonsheared bubble in the center. Therefore, the deformed cell size is almost

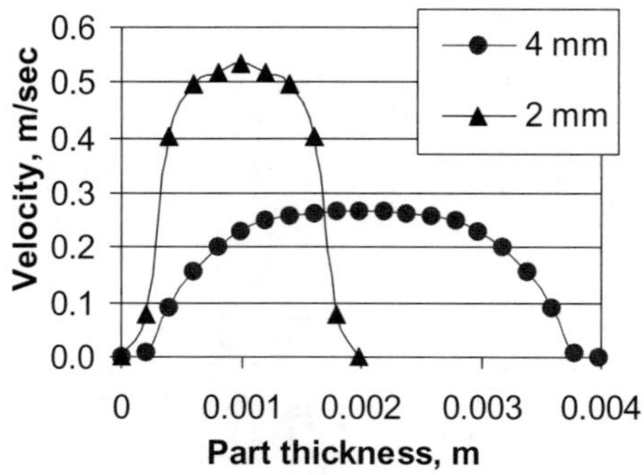

Figure 6.23 Velocity profile in depth direction of mold (mold width 150 mm, length 150 mm) [32]. (Reproduced with copyright permission of Society of Plastics Engineers.)

the same size as the nonsheared cell size in the center. Beyond the nonshearing center layer, there is a layer with an oval-shaped cell that is about half of the thickness center layer. Finally, it reaches the interface layer, which contains the zone with flattened cells. The cells in the interface are distorted so much that the cell does not have the cell shape at all. The strongest shearing is in the interface between mold and melt with an oval-shaped cell, and the least shearing or zero shearing zone is in the center layer with approximately spherical shape of cell. This is the fan-gated mold with the unidirectional flow so that it is truly the deformation from the shearing field caused by mold filling in one direction. It results in the interface roughness on the surface. The interface roughness dominates the surface quality problem. It is obvious that the sheared bubbles on the surface form the roughness and the swirl streak. Further investigation for the thickness affecting the bubble shearing also shows the difference between microcellular foam with a thin wall (2 mm typical) and regular foam with thick wall (4 mm typical). The simulation results for the velocity distribution in different thickness molds are shown in Figure 6.23. With the same injection volume rate, 2-mm thin mold has a sharp increase for the velocity near the interface. However, a 4-mm-thick mold has a slow velocity increase near the interface, which means weak shearing rate.

2. Free-Flow Front Roughness. In the model shown in reference 32, the cell in the center of the flow channel stays in the center position but continually grows with spherical shape without shearing in the center of flow channel. If the mold filling time is long, the center bubble grows continually until it breaks to form free-flow front roughness. Usually, microcellular foam has a short injection time so that the free-flow front roughness may occur only at the end of mold filling and the edge of the part. The free-flow front roughness may become the part of the surface roughness with the thick radial directional flow at slow injection speed. Figure 6.24 shows the schematic of radial directional

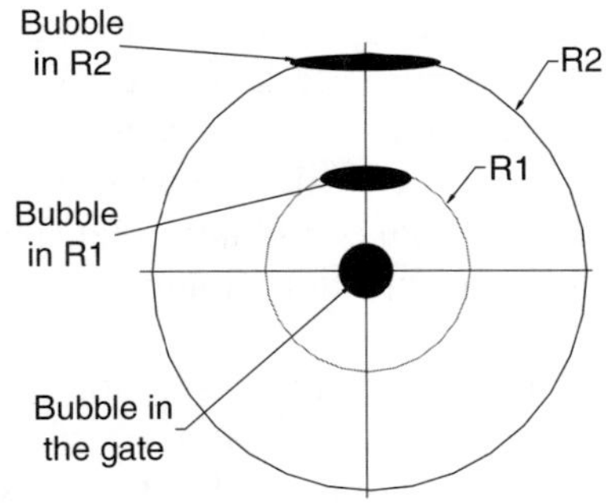

Figure 6.24 Schematic of the theoretical velocity profile in radial directional flow and related bubble distribution profile [32]. (Reproduced with copyright permission of Society of Plastics Engineers.)

flow with a center gate. A circumferential elongation of the bubble occurs on the surface of free-flow front. The bubble shown in Figure 6.24 is in the centerline of thickness direction so that there is no shearing in flow direction. Therefore, it has a spherical shape in the gate. In R1 position the bubble is deformed into an oval shape. Furthermore, in R2 position the bubble is stretched even more like a striation shape. It will be deformed further until breaking. Any layer away from the center will have the same circumferential deformation. Then, it may cancel some of the shearing deformation along the flow direction at the same position.

6.4.6.2 Processing Methods for a Smooth Surface. The methods for a smooth surface are summarized based on the mechanisms analyzed above. The first group includes co-injection and gas counterpressure, in which both roughness-forming mechanisms are well-controlled. The second group controls the interface roughness forming mechanism, such as hot surface mold and coated mold surface with unidirectional flow. Finally, there are many ways to improve the surface quality of microcellular foam, such as material modification with low viscosity, filled material, mold design, processing control, surface texture, extreme thin-wall molding, and so on.

1. Gas Counterpressure. This is a well-known method to improve the foam surface quality. The gas pressure in the mold of microcellular foam may be as low as 1.73 MPa (compared to 6.9 MPa gas pressure for traditional gas counterpressure for the structural foam). Then, the foaming occurs simultaneously with smooth skin forming. It is different from the method allowing foaming to occur after a smooth skin formed at high gas pressure in the mold. As long as the mold is sealed and venting channels in the mold is controlled with a venting valve, the gas counterpressure molding will achieve a smooth surface on parts with complicated geometry. Chapter 8 has a detailed introduction of this technology.

2. Co-injection. This method is used not only for a perfectly smooth surface but also for a better part property, such as (a) reinforced material as a core and (b) the pure material as a skin, or (a) recycled material in the core and (b) virgin material in the skin. It can also be used for different materials as skin and core as long as the material combination satisfies the rule of adhesion between them. The details of co-injection molding are given in Chapter 8.

3. Overlap Molding. The new overlapped smooth surface on the foamed bottom layer has been developed. It is similar to co-injection but only one side of smooth surface and another side of foam. The foam is controlled with better microcellular structure by opening the mold after the foam material filled the cavity so that the foam will be uniform in whole part. More details are given in Chapter 8.

4. Processing Optimization for Better Surface. Processing optimization is the direct way to improve the surface quality of microcellular foam. The free-flow front roughness can be used as a control factor for the maximum limit of injection time. If the edge surface roughness is obvious, the injection time must be decreased. As the parameter changes, the surface quality can be improved significantly. The parameters include: short injection time, 5% or less weight reduction, fast injection speed (slow at beginning if necessary), and high mold temperature. The general rule is to make cell size as small as possible. This helps to reduce the surface roughness. A typical example is a 0.5-mm-thick container made by PP filled with 20% talc. With a cell size of only about 5 μm, it has both smooth surface and small spherical cells.

There is one new developed technology for application of microcellular technology to make perfect surface quality with near 100% mold filling. It makes full use of gas-laden melt properties to fill the mold easily, and then the gas pressure takes care of the shrinkage and warpage issues. However, the cell growth does not occur since the mold is already filled without any space for cell growing. The real morphology may have some supermicrocellular structure in part; however, no such picture is available yet.

5. Mold, Part, and Material for Processing a Better Surface. Here are the analyses from the processing point of view for the better surface with special mold, part, and material selection. All of them are focused on the low friction and low adhesion on the mold surface so that the processing can have less bubble distortion and less rough surface.

The hot mold surface has less friction or adhesion between mold metal surface and melt, and most importantly it has no skin so that the surface of the part slides on such a surface acting as a plug flow (the theoretical velocity distribution is equal in the depth direction). Then, there is no shearing between flow layers in depth direction. In addition, the breaking bubble from the free-flow surface will be easily ironed or repaired by the hot mold surface under the melt pressure. On the other hand, the hot mold may not work for the smooth surface of crystalline material foam. This is because more crystallization on the hot mold surface will create high nonuniformity of cell structure and, then, a rough surface, which was also verified in the study by Saeed et al. [34]. A balance between percentage of crystallinity and small cell structure with a warm mold may result in better surface quality of crystal material microcellular foam.

The coated mold surface will have low friction so that it basically has no shearing for the surface bubble. Depending on the kind of coated layer, some of them act as thermo-insulator. In this way, the part surface will be soft enough so that even if some surface bubbles appear, the injection pressure will repair them. However, this method may leave the flow-free front as the possible problem for the rough surface in the end of filling unless the injection is fast enough.

Material is also a factor for the surface quality. A special developed material with low viscosity, or low adhesion to metal, tends to allow the sliding on the interface. Therefore, it is also an interface roughness improvement with sliding on the mold surface. However, the solution for free-flow front roughness is not necessary. The filled material is also helpful for a better surface finish because of small cell size with filled material.

The part may be redesigned for the improvement of surface quality, such as (a) using a texture on the cosmetic areas of the part or (b) switching to lighter colors. To verify the surface quality issue from the shearing bubble of interface roughness, three different mold surfaces have been tested. Figure 6.25 has shown three morphology pictures of ABS samples made by different mold surfaces. Figure 6.25a is the result of a smooth surface mold. It shows a broken bubble on the surface because of shearing. Then, stronger shearing of the bubble on the part surface occurs in Figure 6.25b for deep texture mold surface. Nonetheless, Figure 6.25c shows that a half-depth of texture mold surface still creates the same degree of the surface bubble shearing compared to the result of deep textured mold in Figure 6.25b. There may be some sliding action on the smooth surface mold since there is more surface bubble shearing for the textured surface mold. Overall, the textured surface foam may reduce the silver streaks by a diffuse backscatter [34], but it makes the bubble shearing worse from a processing point of view. On the other hand, all texture must have connected channels where the trapped air in the texture during mold filling is connected to the main venting channel.

6. Conclusions. There are two roughness-forming mechanisms for microcellular: interface roughness and free-flow front roughness. Interface roughness dominates most of the microcellular foaming process. Small and uniform cell structure is the key for improving the surface quality of microcellular foam. The possible methods to smooth surface of microcellular foam are summarized as follows:

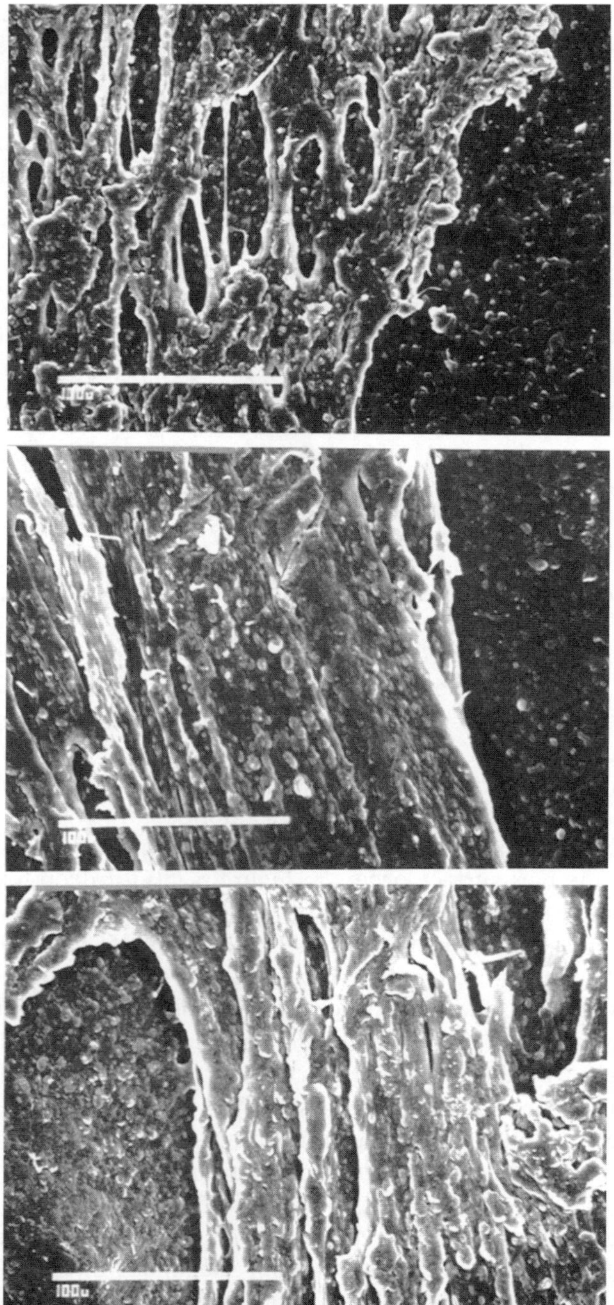

(a)

(b)

(c)

Figure 6.25 Morphology pictures of surface of ABS microcellular foam with three different mold surfaces. **(a)** Small surface [32]. **(b)** Deep textured surface [32]. **(c)** Shallow texture surface [32]. (Reproduced with copyright permission of Society of Plastics Engineers.)

- Hot and coated mold can eliminate the interface roughness for amorphous material. The free-flow front roughness may be considered at the end of mold filling for hot or coated molds if the injection time is too long. The free-flow front roughness can be used as the control factor for the maximum injection time limit.
- Texture surface mold may benefit from the optical effect of reducing the appearance of silver streak but making the severe bubble shearing.
- Both roughness mechanisms are controlled in co-injection and gas counterpressure molding. The smooth surface and spherical cells are made by both methods. However, both methods have less weight reduction and high cost of hardware. The best surface quality is achieved by co-injection molding. The gas counterpressure was successfully tested for smooth surface of microcellular foam with lowest gas pressure (1.73 MPa). In addition, a new overlapped smooth surface on the foamed bottom layer has been developed.
- The balance between percentage of crystallinity and small cell size is a way to improve the surface quality of crystal material.

6.4.7 Optimizing the Microcellular Injection Molding Process

The process optimization does exist for microcellular injection molding. However, it depends on the most desired part properties. Only a few part properties may be optimized without severely conflicting with other properties. Table 6.9 summarizes the processing parameter changing directions when one of the optimizing goals is emphasized.

6.4.7.1 Cycle Time. The cycle time is requested as short as possible to make the maximum benefit of microcellular processing. To reach this goal as the changes indicated in the Table 6.9, the parameters to be increased in the first stage of molding are screw speed, gas percentage, and back pressure of screw recovery. The high screw speed creates high shear rate, which is critical for the gas dosing and mixing well. The high back pressure during screw recovery also guarantees the gas dosing in the range of solubility pressure. Therefore, the gas is fully dissolved into the melt and forms single-phase solution. It results in a uniform cell structure that will help to cool down the part fast. Then, the second-stage parameter to be increased is the injection speed. The high injection speed suddenly releases the energy, which is the key to cool the part quickly. On the other hand, high injection speed generates the huge pressure drop rate to promote the nucleation that results in much better cell structure.

The other parameters, such as mold temperature, melt temperature, and cooling time, are obvious to be controlled as short as possible. They also contribute the cycle time reduction.

6.4.7.2 Strength. The processing conditions are optimized for the highest value of strength. The cell structure is the key for the high strength of the part. Therefore, the screw speed, gas percentage, back pressure of recovery, and injection speed are all to be increased. A longer cooling time helps to increase the skin and stabilize the cell structure near the skin.

Mold temperature is determined with the materials. If the crystalline material needs the completely crystallization in the skin to promote the strength, then the mold temperature shall be high. However, the amorphous material is easy to get thick skin so that the mold temperature may be set up low to have good skin for amorphous material. Similarly, melt temperature needs to be low to keep the higher gas solubility before nucleation. It is the advantage of microcellular processing because the high SCF level will lower the glass transition temperature.

6.4.7.3 Weight Reduction. The weight reduction optimization is obviously the goal of high percentage reduction. It may not be necessary to reduce maximum weight for the saving since cycle time is much more saving if the raw material is inexpensive. However, if the material is expensive engineering material, or biopolymer, then the weight reduction may become a valuable goal for the optimization. It is achievable by increasing all parameters in Table 6.9 except the cooling time. The key factor to reduce the weight of the part is the cell structure.

6.4.7.4 Surface Finish. Surface finish optimization is easy to be understood for the small cell size and lower gas percentage. Therefore, all parameters to be controlled are increasing except for the gas percentage, which is reduced. As long as the cell structure is acceptable, the gas percentage is as low as possible. For example, 0.1 weight percent of N_2 gas was successfully used for some part to use microcellular for mold filling purpose only. Therefore, only minimum gas percentage is used, and the mold filled almost 99% or more to have class A surface. Some test was carried out to verify that this gas percentage exists in very narrow range. If the gas percentage is too low, a hollow channel is created like a gas-assist part. Near this processing condition, either increase screw speed or increase back pressure, or add a little more gas percentage to get rid of the hollow channel but keep the perfect smooth skin [32].

6.4.7.5 Dimension Stability. It is also a complicated goal to optimize the processing conditions. Generally, it needs all parameters in Table 6.9 to increase except for the melt temperature for the final dimension stability. The principle is the cell structure uniformity. It is also important to have long cooling time for necessary dimension stability after the part is ejected from mold.

6.4.8 Influence of Mold Temperature for Microcellular Injection Process

Mold temperature will influence two major issues of microcellular processing. One is the surface finish of the microcellular part. The high mold temperature

TABLE 6.10 Recommendations for Resin Molding Temperatures Modified the Data with Microcellular Processing Conditions [35]

Resin	Mold Temperature (°C)	Resin	Mold Temperature (°C)
ABS	30–50	TPX	60
ABS, filled or reinforced	60–85	Polyethylene	20–60
Acrylic	35–75	Polyethersulfone	90–165
Cellulose acetate	40–70	Polyaryletherketone	145–195
Acetal	75–85	Polyimide	145–195
PC	70–105	Polypropylene	20–50
Ionomer	4–40	Polystyrene	35–75
SAN	30–70	Plastomer	15–25
Nylon 6	65–85	Polysulfone	90–140
Nylon 6/6	35–85	Polyurethane	15–50
Nylon 6/10	50–85	PPE	65–95
Nylon 4/6	80–140	PPS	90–140
Nylon 11	35–60	Styrene butadiene	15–45
Nylon 12	65–85	PETG	15–30
PBT filled	60–110	PET	80–110
PBT	54–80	PET (high heat)	140
ASA	40–80	PPO modified	80–95
LCP	29–90	PVC	15–60
LCP, filled	65–100	Polyetherimide	120–165
POM and copolymers	55–110	CA, CAB	40–50
PMMA	50–85	PCTFE	80–120
PFA, FEP	120–170		

Source: Reproduced with copyright permission from *Injection Molding Magazine*.

will promote the surface finish as discussed above. However, the crystalline material may increase the crystallization at the surface at high mold temperature at will expel the gas at the surface of the part and it may result in poor cell structure. Another issue is the skin thickness that is related to mold temperature setup. The cold mold temperature will increase the skin thickness that may be important for the strength of the microcellular part. As the general guidelines, the mold temperatures for microcellular materials are recommended in Table 6.10.

6.4.9 Influence of Moisture for Microcellular Injection Process

Adequate drying temperature and thorough drying are also basic requirements for making a good microcellular cell structure. It is because the moisture in plastic material may generate either (a) big cells and streaks, (b) spray marks on the surface, or (c) the blister on the surface. Moisture in the material will decrease the impact strength as well. Drying time and temperature for hygroscopic materials are the key parameters for the residual moisture before microcellular molding. The required drying time for microcellular material is usually the same as that for solid material. Generally, the residual moisture content of microcellular material should be less than 0.1%, or in the range of the material supplier's recommendation. The recommendations for the drying conditions of most popular materials are summarized in Table 6.11 [35].

Hygroscopic material requires drying because the pellet itself absorbs moisture from the ambient air. Some materials listed in Table 6.11 are not the hygroscopic materials; however, sometimes the nonhygroscopic material may still attract moisture that stays on the surface of pellets. For example, polyethylene and polypropylene still require drying, and even this surface moisture in polyethylene and polypropylene is easier to remove by the high-back-pressure setup in the screw recovery process.

The shape and size of resins are also the factors to be considered during the drying process. Larger pellets do not get as dry as small pellets at the same exposure. For example, when drying ABS, 5 hours of drying will take a batch of small pellets of ABS material down to 0.025% residual moisture, while the same period and temperature of drying will take larger pellets down to only 0.075% of residual moisture [35].

6.4.10 Influence of Melt Temperature for Microcellular Injection Process

The melt temperature for microcellular processing can be reduced to significantly low values because the supercritical fluid of blowing agent in the molten polymer can lower the viscosity up to 50% or more, as well as lower the glass transfer temperature of polymer. Then, there may be two choices to use this advantage: Either keep the same melt temperature of polymer without gas, or use the lowest melt temperature at which the normal flow ability (viscosity) maintains.

6.4.10.1 Lowest Melt Temperature with Normal Viscosity. To get the lowest melt temperature, the up limit of gas dosing must be applied in the gas dosing process. CO_2 gas is usually the best blowing agent to achieve the lowest melt temperature without changing the original polymer melt viscosity. For example, the maximum melt temperature reduction values for some materials are listed below:

PP: from 220 °C to 150 °C
PS: from 220 °C to 125 °C
PSU: from 370 °C to 295 °C

TABLE 6.11 Recommendations for Resin Drying Time and Conditions [35]

Resin	Drying Temperature (°C)	Time (hours)	Initial Moisture (%)	Maximum Residual Moisture (%)
ABS	82	2–3	0.45	0.02–0.1
Acetal	99	3	0.25	0.05–0.2
Acrylic	71–82	3–4	0.4	0.1
Cellulose acetate	71	2–3	0.7	0.1
Cellulose acetate butyrate	77	2–3	0.7	0.1
Cellulose propionate	77	2–3	1.0	0.1
Ionomer	77	8	0.32	0.05
LCP	149	8–24	NA[a]	0.01
Nylon 6	82	4–5	0.5	0.1–0.2
Nylon 6/6	82	4–5	NA	0.1–0.2
Nylon 6/10	85	4–5	NA	0.15–0.2
Nylon 4/6	121	2–4	NA	0.05
Nylon 11	77–93	4–5	0.7	0.02
Nylon 12	66–110	5	NA	0.1
Polyaryletherketone	149	3	NA	0.1
PBT	140	3	0.25	0.02
PC	121	3–4	0.16	0.02
PET	121	2–3	0.25	0.02
PETG	71	4–6	NA	0.08
Polyester elastomer	80–104	2–4	NA	0.03
Polyehtersulfone	150	3–6	0.43	0.02
Polyethylene	85	2	0.2	0.05
Polyethylene (40% black)	91	3	NA	0.05
Polyethylene (30% glass)	85	2.5–3.5	NA	0.05
Polyimide	182	5–10	0.32	0.05
Polypropylene	91	2	0.2	0.02–0.1
Polystyrene	82	2	0.1	0.02
Polysulfone	121	2–4	0.1	0.02
Polyurethane	82	3	0.9	0.01
PPE	88–110	2–4	NA	0.02
PPS	150	3–4	0.2	0.04
PVC	71	2	0.4	0.1
SAN	82	2	0.3	0.1
Styrene butadiene	60	2	0.6	0.02

[a]NA, not available.

Source: Reproduced with copyright permission from *Injection Molding Magazine*.

However, using the lowest melt temperatures above, all advantages from low viscosity or gas-laden molten polymer are not available and the regular molding will be carried out, and the only extra benefit will be low melt temperature to cool faster by the heat transfer in the mold.

6.4.10.2 Low Viscosity with Normal Melt Temperature. This is the current microcellular molding technology widely used in industry now. It seems like low-viscosity molding brings much more benefits than does low-melt-temperature molding. It is because it solves many inherent disadvantages in the regular molding, and all the advantages of microcellular materials discussed above rely on low viscosity and not on low melt temperature.

6.4.11 Troubleshooting the Microcellular Injection Molding Process

The troubleshooting of microcellular injection molding focuses on some typical defects of the parts related to the processing. Most of them are already discussed in reference [6]. However, all surface appearance issues addressed in reference 6 are truly processing issues. Table 6.12 gives the general directions to solve the problems. Each issue is discussed in the following.

6.4.11.1 Blister and Blowout. Blister and blowout are basically the large voids under the pressure. The primary cause is the gas out of solution. Therefore, the most effective solution is to make sure that the single-phase solution of the first stage of the process is well-prepared in the barrel. The back pressure needs to be set up high enough to force gas in the solution. Also, the screw speed is raised to have high shearing rate to mix gas well in the screw. To control the blowout, the melt temperature, mold temperature, and

TABLE 6.12 Troubleshooting for Microcellular Injection Molding Process

	Blister and Post Blow	Dimple	Warpage	Swirls	Inconsistent Part Weight
Back pressure	+	+	+	+	+
Gas percentage	– with lots of cells; + for hollow channel in the part	+	+	–	–
Screw speed	+	–	+	+	+
Injection speed	–	+	+	–	+
Mold temperature	–	0	–	+	0
Melt temperature	–	–	–	+	0
Cooling time	+	0	+	0	0
Drying time and temperature	+	0	+	+	+
Shot size	–	+	+	+	+

+, increasing the setup of parameter.
–, decreasing the setup of parameter.
0, no effect of the parameter.

injection speed shall be low for reducing the energy of gas blowout. Sometimes the shot size needs to be reduced to give more space for foaming.

The gas percentage will be considered in two different phenomena. One is the hollow channel in the part, which is the symptom of gas shortage. Therefore, the gas percentage needs to be increased for the hollow channel defect in the part. On the other hand, too much gas may cause the gas pocket as well. It can be found that the large void is surrounded by small cells. Then, the gas percentage needs to be decreased to remove the blowout and blister. The cooling time is adjusted to become longer if the post blow occurs, which is because the cooling time is too short.

The moisture is another possible reason for the surface blisters. It needs to follow the basic drying instructions from the material suppliers, and the ranges of drying conditions are listed in Table 6.11.

6.4.11.2 Dimple. Dimple usually indicates the shot size shortage. Therefore, first solution is to add more material or to increase the shot size. If only very little dimple exists, then some other methods may be tried. One is to increase the back pressure, which may help not only to speed up the gas dosing but also to make more dense material so that even the same volume of high-density material may already make up the dimple problem. It is the same reason to solve the dimple problem when the screw speed decreases to make denser shot size, or injection speed increases to shoot more material into the mold cavity. In addition, lowering the melt temperature can make denser material as well. However, the mold temperature and the cooling time make no significant contribution to the solution.

It is possible that the dimple is caused without enough gas pressure or cell expansion. If the sample of SEM shows few cells, the gas percentage needs to be increased to have nice cell structure.

6.4.11.3 Warpage. The warpage of microcellular material is usually less than that of the regular injection molding part. However, it may become worse if the cell structure is poor. To improve the cell structure of microcellular foam, the screw speed, back pressure of recovery, gas percentage, and injection speed need to increase. The melt and mold temperature prefer to be decreased for cooling more. Then, the cooling time is longer to freeze the cell structure to become more solid to have enough stiffness after the part is ejected into the mold. Sometime the weight reduction is too much to make the material too soft, so the shot size should be increased to have enough stiffness of material to resist the warpage.

6.4.11.4 Swirls. Swirls are the features of the microcellular foam. It is only controlled to minimize its depth and size. The quality in the first stage is the key to minimize the swirl in the mold. Therefore, the screw speed and the back pressure of recovery are usually high for the mixing and gas diffusing acceleration. However, the low percentage of gas is good for a small swirl as long as the cell structure is acceptable. The slow injection speed usually creates a small swirl on the surface. It is well known that the high mold temperature may help to iron the rough surface from foam and keep a relatively smooth surface compared to the result of a cold mold surface. It is also known that the cold melt temperature may create elephant surface.

The cooling time may not have an affect on the swirl on the surface. However, the shot size may need to be increased so that there are less cells on the surface of the part. It may create a good surface when 0.1% N_2 gas is used for 99% of mold filling. This is another application of microcellular injection molding that creates only low cell density while low residual stress and absence of sink marks are the major purpose of the microcellular processing.

6.4.11.5 Inconsistent Part Weight. The inconsistent part weight is a complicated issue that may be related to multifactors of processing. The major one is the gas dosing inconsistency. It is basically improved by the high screw speed, high back pressure during recovery, and increasing shot size. Sometimes it is because the gas percentage is too high so that the local dosing intensification may cause material slip on the screw so that the shot weight with gas percentage is not constant every cycle. Therefore, low gas percentage helps to stabilize the process. The temperature in mold and melt are not important for the weight to be consistent. It is a similar reason why the cooling time is not related to the inconsistent part weight, either.

The modern machine will provide injection pressure and speed profile that may help to keep the same shot size every cycle by controlling the slow–fast–slow injection speed profile. It is necessary to make sure that the material size is the same from cycle to cycle. The similar reason for the screw rotation profile, usually slow–fast–slow, will help for the screw stops at the same position every cycle. All of them are the key measures to keep the constant shot size.

6.5 COMPARISON WITH GAS-ASSIST MOLDING

As an entrained gas process, both the microcellular process and the gas-assist technology use gas in the melt to make cells and voids. The microcellular process may result in a big void in the part, which is a similar result of the hollow part from the gas-assist process. On the other hand, the gas assist may find some microcells in the layer of hollow channel made by gas. However, they are completely different processes. The differences are discussed in the following.

6.5.1 Gas and Melt Phases

The gas-assist technology uses gas and melt as separate phases. Therefore, gas is needed only when the melt is solidified at the skin layer, and then gas is injected with pressure to create a hollow channel inside of the part so that the solid skin should be fully developed and strong enough to hold the gas pressure in the part. Finally, the gas pressure and gas itself in the part will be fully released by a releasing valve [36]. However, the microcellular process needs gas and melt to become a good mixture like a single-phase solution. Once the single-phase solution changes to the separate phase, the microcellular foam becomes a gas-assist process.

6.5.2 Gas Pressure

Both gas-assist and foaming processes use pressurized gas that stores energy with potential hazard. Recently, the safety issue for an entrained gas process was discussed in the Entrained Gas Guideline Committee of the SPI (Society of the Plastics Industry) Machinery Division. The SPI Machinery Division has released the guideline for entrained gas processing in horizontal injection molding machines since May 2003 [37].

There is a unique benefit of the dimension stability from both gas-assist and foaming processes. The dimension stability benefit is from the cell growth in foaming and gas pressure in gas-assist processing. Therefore, one of the common technologies for both gas-assist and gas foaming processes is the application of highly pressurized gas. For the gas-assist process the pressure in the gas is about 2.5–30 MPa (363–4350 psi) for injecting gas and driving out the melt to a gas channel in the part. The melt pressure required for single-phase solution is in the range of 6.9–34.5 MPa (1000–5000 psi) for the foaming process. Therefore, the energy stored in the gas is approximately a similar level for both processes.

A typical low-pressure foaming process fills the mold partially, and the cell growth will finish the mold filling. The residual pressure in the cells will do the holding and overcome the shrinkage during cooling. It is well known that requirements for foaming mold include high injection rate, high pressure drop rate for nucleation, low injection pressure or low clamp force for mold, and no packing or holding stages. Similarly, for most popular gas-assist processing the cavity is partially filled with a defined amount of melt. After the skin is formed, the gas will be injected into the melt until the remainder of the mold cavity is completely filled. Then, the pressurized gas takes over the function of holding pressure.

6.5.3 Mold Design

Based on the unique performance of the molding with foaming and gas assist, the mold design shall be improved from a processing point of view. Both gas-assist and foaming molds require the gate to be located in the thinnest position of the mold and require a precisely balanced runner system and cavities. It is important not to locate the gate in the thickest area for foaming. The gate area is not only the latest cooling or hottest spot in the part but also the highest pressure spot; even foaming does not need a packing stage. Some post blow occurs in the gate area for the gate in the thick section of the mold. Without holding and packing stages for both processes, the mold runner system and cold sprue should be designed with less cooling time requirement compared to traditional molding. More details of mold designing for microcellular parts are discussed in Chapter 5.

6.5.4 Mold Cooling

Mold cooling is not the safety issue for the gas assist since the gas pressure will be released before demolding. However, it may be the issue of safety for microcellular foaming since the smaller the cells, the higher the residual pressure in the cell before demolding. The foamed part must be cooled enough not only reaching the ejection temperature but also forming the strong solid skin to hold the cells expanding. The post blow in the demolded foam part is dangerous for hot gas and hot plastic popping out. The post blow sometimes occurs after the part is demolded for a long time. Then, it causes some blister in the surface of the foamed part. A broken sprue is also a common problem occurring in a not properly designed mold. The sprue is always the hottest spot because it is the thick section in the mold and the last part of mold filling with pressure still maintained there. A broken sprue stays in the mold and immediately stops the automatic cycling. Unless a similar safety system is designed for this case [37], it is not safe for the operator to take the broken sprue from the mold because the safety guard is opened and single-phase solution is under the pressure.

Reducing the cooling time is also important for the gas-assist processing since the part contacts the cold mold wall by inside gas pressure before demolding. The foamed part has the same cooling benefit as the gas assist since the cell growth pushes the skin of the part touching the cold mold wall better than the solid part. The foamed part also benefits by quickly releasing the pressure and thereby generating a larger number of fine cells, which results in a dramatically cooling time reduction. But any thick foam part may have a longer cooling time than the solid because the heat transfer is not efficient for the foamed part when the insulation dominates the cooling stage. However, thickness is not the only restriction. The uniform thickness is a critical factor if the thick part is necessary. Longer cooling time may also be related to poor single-phase solution and wrong processing conditions or mold design.

However, the foam processing may result in a demoding problem since the gas pressure in the cells cannot be released as in the gas-assist process. The regular mold design makes use of the shrinkage to let the part stay on the core side, which is usually the moving half of the mold with an ejection system for demolding. If the expansion of cells is so strong that the foamed part sticks on the cavity of the mold, then it causes a demolding difficulty. In addition, a degassing requirement is necessary for the foamed part. It takes about 24 hours and up to 72 hours, depending on the thickness of the part and the permeability between gas and materials.

6.6 COMPARISON WITH STRUCTURAL FOAM MOLDING

Structural foam molding is process that is similar to microcellular injection molding. There are several obvious differences below:

6.6.1 Thickness and Cell Size

The structural foam is traditionally related to a thicker part and large cell sizes. The thickness of structural foam is usually larger than 4 mm. On the other hand, the cell size of structural foam is larger than 200 μm in the center because of the thicker part. The skin-interface-foamed center structure is the typical morphology for the structural foam.

6.6.2 Pressure in the First Stage

The low pressure in the first stage is another difference between structural foam and microcellular foam processes. Structural foam uses an extruder as the first-stage facility, and it uses the plunger as the second-stage tool for injection. However, microcellular processing uses the reciprocating screw for both first-stage and second-stage equipment. Therefore, the processing conditions are different from each other.

6.6.3 Property Change

The mechanical property changes varied in a big range for structural foam but varied in a small range for microcellular foam because the cell size and uniformity distribution are better in microcellular foam.

6.6.4 Equipment

Most microcellular injection molding machines are the reciprocating screw machines. However, the structural foam uses the extruder plus plunger machine.

6.7 COMPARISON WITH REGULAR INJECTION MOLDING

There is a concise comparison between microcellular and regular injection molding in Table 1.1, Chapter 1. However, more differences between regular and microcellular injection moldings from a processing point of view are summarized below.

6.7.1 Hold and Packing Stage

The most obvious benefit is the hold stage. Theoretically, the microcellular process does not need a hold stage at all. In some cases a short hold stage of about <1 sec is used to smoothly transfer the injection action to screw recovery in the hydraulic system.

6.7.2 Pressure

The pressure requirement for the microcellular injection molding is very low compared to regular injection molding. Usually only half of the regular injection pressure is necessary for the microcellular injection molding. It is because the mold filling of microcellular injection molding never reaches 100%. It is in the range from 85% to 95% of full shot size to be used for microcellular injection molding. Even if 100% mold filling is required for some surface quality and warp and sink mark issues, the peak pressure never reaches the same level of solid injection molding because packing is not necessary for this process.

6.7.3 Cooling

The cooling of microcellular foam is much faster than the regular injection molding. It may be a huge pressure drop rate, releasing energy so quickly so that the heat is also released quickly. However, if the part thickness is more than 4 mm, and the gate is still in the thickest spot of the part like in a regular injection mold, then it is possible to have longer cooling time for the microcellular process compared to regular molding. It is because the thick foam core may become a good insulator that increases cooling time significantly.

6.7.4 Demolding

Demolding is also different because the microcellular foam expands more than it shrinks during the mold cooling. However, the regular molding part shrinks during cooling. Therefore, the regular molding part usually stays with the half of the mold with core, and the ejecting system is also in the core side of the mold half. However, the microcellular part usually stays with cavity of the mold half. It will require a careful review of the effect between expansion of cell growth and the shrinkage of material itself.

Another significant difference is the strength and stiffness during the ejection. The strength of the part surface (or the hardness of the skin of microcellular foam) is weaker than the regular molding so that the eject pin needs to be designed with big a diameter and needs to eject slowly at the beginning of ejection.

6.7.5 Equipment

The equipment for microcellular injection molding requires less tonnage of clamp for the reason of mold short shot for microcellular injection molding. However, the microcellular process needs a high injection volume rate for nucleation if it is necessary. Except for the gas delivery system, the specially designed screw and screw tip are the key elements for the microcellular injection molding.

Then, the big difference is the plasticizing unit that requires special barrel, screw, and screw tip for the microcellular injection molding. A special high-performance shut-off nozzle, or valve gate, is also the key element for microcellular injection molding. The modified hydraulic system maintains the pressure in the barrel in the entire cycle time.

6.7.6 Cycle Time

The cycle-time reduction of microcellular injection molding is a significant benefit of the microcellular process. Figure 6.26 shows the cycle-time comparison between regular injection molding and microcellular injection molding. It is clear that the cycle-time saving is from the injection system of the microcellular molding machine. The clamp system usually does not have many changes for cycle-time reduction. It is usually kept the same as the regular injection molding.

First of all, it is obvious that cycle-time reduction from the hold stage is necessary for regular injection molding. For microcellular injection molding the hold stage is eliminated from microcellular injection molding because the material expands during final mold filling due to the cell growth. However, the

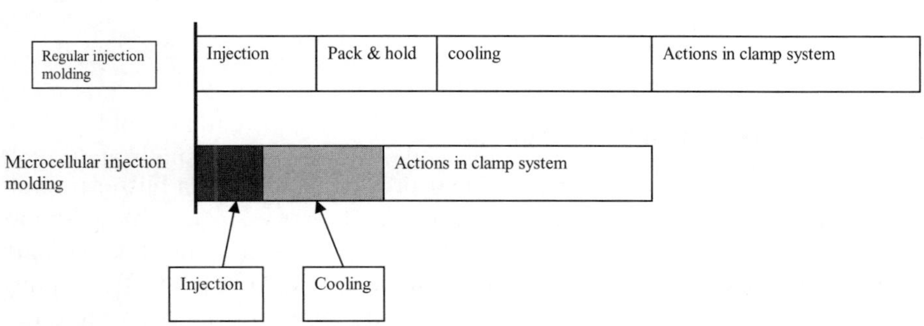

Figure 6.26 Cycle-time comparison between regular and microcellular injection moldings.

regular injection molding needs to hold about 70–80% of full injection pressure to pack the molding part for the compensation of material shrinkage during cooling. Then, for injection time it will be reduced for two factors. One factor is that the shot size of the microcellular part has about 5–15% less volume than the full shot volume of solid part. Another factor is that the microcellular injection molding needs quick injection for nucleation of the high cell density part and prevention of overgrow cells so that injection time is usually short as well. Finally, the energy release during injection helps to cool the microcellular part quickly so that the cooling time is short compared to the regular molding as long as the part thickness does not exceed 3 mm. All of three factors above make the microcellular cycle time shorter than the regular injection molding cycle time. It is the biggest benefit of the microcellular process.

6.8 COMPARISON WITH MICROCELLULAR EXTRUSION

Extrusion was the one of the early continue processes developed for microcellular foam in the world. It is a unique process compared to the injection molding from the major processing differences, as discussed in the following.

6.8.1 A Continue Process (Extrusion) Versus a Noncontinue Process (Injection Molding)

The extrusion is a continue process so that the first stage of the microcellular is easy. The gas dosing is at the fixed position for both screw and barrel, and it is constantly running without any stop during the normal process. Therefore, the gas delivery system for the extrusion process is simple compared to the injection molding gas system. However, the injection molding is a periodic process. The gas dosing has stop and suddenly start during every cycle. Therefore, the gas surging at every start of dosing is more difficult to handle for the screw design and process.

6.8.2 Screw Design and Performance

The design for the extrusion microcellular process needs only a blister ring or simple pressure restriction element similar to the two-stage venting screw design. This is necessary to build high pressure in the middle of screw and prevent any possible gas leaks back through the upstream of screw. The wiping and mixing sections are similar to the screw designed for injection molding. The biggest difference between extrusion and injection molding is that the middle check valve and screw tip in the front of screw are not required for extrusion screw. On the other hand, the reciprocating screw of injection molding change the dosing position during screw recovery so that it has more difficulties to keep the constant dosing process from beginning to the end each cycle. It is necessary to design injection molding screw to deal with the effective mixing length changes during screw recovery.

6.8.3 Nucleation

High nucleation in the extruding die is made by the narrow die section area just before the outlet of die flow path, along with high melt pressure in the die to generate enough nucleation rate [20]. Therefore, the pressure in the extruder could be very high. In other words, the high nucleation rate is not only from the output rate of extruder but also from the high pressure in the screw. It is necessary to have a gear pump to keep high output and high discharge pressure in the die as well. On the other hand, the die of extrusion must be modified to eliminate any local pressure drop that is enough to lead to premature foam inside of the die. However, with the injection mold it is not necessary to worry about the local pressure drop during injection since the injection time is so short to cause any premature foam in the runner system. In addition, the injection volume rate in the small nozzle orifice or in small valve gates can reach a huge pressure drop rate to create a great number of nuclei, which is difficult for the matching performance between extrusion die pressure drop rate and screw output rate.

6.8.4 Shaping Stage

The shaping stage is much different between extrusion and injection. Injection molding has the restricted mold cavity to shape the microcellular foam. However, extrusion has a free surface after the microcellular foam, leaving the die until the shaping guiders holds the extrude material. The cell growth in the extruding part is much more difficult to be controlled in extrusion process.

On the other hand, the melt temperature may be too high to be cooled down enough before it leaves die outlet. This high temperature of molten plastic at extrusion die will result in rapid cell growth and then will cause some cell collapse on the surface because the melt strength is low at high melt temperature. Therefore, extrusion of microcellular foam may need the second extruder just for cooling the mixture of gas and melt in the second screw. The screw in the second extruder is designed with a deep channel to reduce the shear rate and cool the melt temperature before it reaches the die. Air cooling heat bands are used for the second extruder that helps to cool the melt temperature. On the other hand, the second screw must still maintain the minimum pressure to prevent the prefoaming before the gas melt mixture reaches the die outlet. The combination of two extruder-connected serials is called a tandem extruding system. Injection molding can survive with high melt temperature because it is injected into the mold so fast that no time is left for cells to grow too big or for cells to collapse. With the same reason the microcellular injection molding barrel does not need the air-cooled heat bands.

6.8.5 Materials

The material for extrusion microcellular process is high-viscosity material that is actually good for controlling of extruding material free expansion. The high viscosity of extruding grade of material may help to hold the shape of extruded material as well. The most popular materials for microcellular extrusion are polystyrene (PS), polyvinyl chloride (PVC), polyolefin (such as PP and HDPE), and thermoplastic vulcanizate (TPV). Other materials have been processed with extrusion of microcellular, such as thermoplastic polyurethane (TPU), nylon, and polycaprolactone [6].

However, the material viscosity for injection molding grade is usually much lower than any viscosity of extruding material. On the other hand, microcellular injection molding can process almost any thermoplastic plastic.

6.8.6 Pressure

The pressure of microcellular processing for extrusion is lower than injection molding. The typical die pressure of microcellular extrusion is about 24 MPa (3480 psi, or 240 bar). Therefore the pressure rupture disc of the safety device for microcellular extrusion is about 41.4 MPa (6000 psi). However, the injection molding requires a high-pressure rupture disc that is about 82.8 MPa (12,000 psi).

6.9 COMPARISON WITH MICROCELLULAR BLOWING MOLDING

Microcellular blow molding can be divided into two categories. One is the injection blow molding, and another is extrusion blow molding. The injection blow molding is similar to the microcellular injection molding to make foamed preform first. Then, this preform can be reheated and then blown to the right-size bottles or containers. However, the extrusion blow molding is a unique technology compared to injection molding and injection blow molding. Therefore, the comparisons will be given only for extrusion blow molding below.

There are two different possible extrusion microcellular blow-molding technologies based on the methods of making parison. One is a continue parison extrusion, which is a process that is similar to extrusion for the first stage of the microcellular process. Another is the intermittent parison extrusion, which actually has a first stage of microcellular processing that is similar to the injection molding because a reciprocating screw is used for microcellular blow-molding machine without a continuous process.

6.9.1 Continuous Parison Extrusion

The continuous extrusion screw makes the parison continuously, and the gas dosing is the same method as extrusion of the microcellular process. The die design needs to be modified with the same requirement as extrusion and usually has a restrict area in the die for nucleation, and the die channel is carefully designed without any local pressure drop in the die until it is out of die. The restricted application is that the parison size must be small enough not to allow too much the cell growth during parison-making stage. The major differences between injection molding and continuous parison extrusion blow molding are the same as the differences of extrusion.

The continuous parison blow molding may need to use high-viscosity material that has the melt flow index in the range from 0.16 to 0.7 g/10 min [6]. This high viscosity can help to make better cell size and avoid sagging and thinning problems during the parison formation stage of blow molding.

6.9.2 Intermittent Parison Process

This is the better method to make uniform and better cell structure of blow-molded microcellular parts. It is because the reciprocating screw must be used to avoid any extra sag of parison and cell overgrowth during parison extrusion. It is usually for the application of a large-size parison. On the other hand, the die still needs to be modified for nucleation purposes, although it is not as critical as the nucleation in the blow-molding die of the continuous parison process above. However, the nucleation rate is usually enough since the injection speed can be reached high enough to create the necessary nucleation rate. Therefore, the first stage of microcellular blow molding is similar to microcellular injection molding.

This intermittent parison process will make the parison much faster than the parison made by continuous extrusion. Therefore, the melt index of material for the intermittent parison process may not need to be as low as that for the continuous parison extrusion since the low viscosity is more tolerated by the fast parison made by injection. However, the viscosity of the intermittent parison process is still much higher than the viscosity of microcellular injection molding. Important factors to consider are the high torque of screw rotation and the relatively high injection pressure compared to those of microcellular injection molding.

REFERENCES

1. Xu, J., and Pierick, D. *J. Injection Molding Technol.* **5**, 152–159 (2001).
2. Shimbo, M., Nishida, K., Heraku, T., Iijima, K., Sekino, T., and Terayama, T. *First Int. Conf. Thermoplast. Foam*, 132–137 (1999).

3. Chandra, A., Gong, S., Turng, L. S., and Bramann, P. *SPE ANTEC, Tech. Papers*, 540–544 (2004).
4. Xu, J. *SPE ANTEC, Tech. Papers*, 594–598 (2004).
5. Xu, J. *SPE ANTEC, Tech. Papers*, 2660–2664 (2004).
6. Okamoto, T. K. *Microcellular Processing*, Hanser Publications, Cincinnati, 2003, pp. 38–60.
7. Xu, J. *SPE ANTEC, Tech. Papers*, 2770–2774 (2006).
8. Tadmor, Z., and Klein, I. *Engineering Principles of Plasticating Extrusion*, Robert E. Krieger Publishing Company, New York, 1978.
9. Chung, C. I. *Extrusion of Polymers: Theory and Practice*, Hanser/Gardner Publications, Cincinnati, 2000, pp. 254–272.
10. Durril, P. L., and Griskey, R. G. *AIChE J.* **12**, 1147 (1966).
11. Durril, P. L., and Griskey, R. G. *AIChE J.* **15**, 106 (1969).
12. Suh, N. P. edited by James F. Stevenson, *Innovation in Polymer Processing Molding*, Hanser/Gardner Publications, New York, Chapter 3, 1996.
13. Rosato, V. D., and Rosato, V. D. *Injection Molding Handbook*, second edition, Chapman & Hall, New York, 1995, pp. 187–188.
14. Xu, J. U.S. Patent No. 6,579,910 B2 (2003).
15. Johannaber, F. *Injection Molding Machines A User's Guide*, second edition, Hanser/Gardner Publications, New York, 1985, pp. 72–78.
16. Johnson, D. E., and Macedon, N. Y. U.S. Patent No. 4,124,336 (1978).
17. Saxton, R. L. U.S. Patent No. 3,006,029 (1961).
18. Dulmage, F. E. U.S. Patent No. 2,753,595 (1956).
19. Rauwendaal C., et al. *Mixing in Polymer Processing*, Marcel Dekker, New York, 1991, pp. 168–177.
20. Park, C. B., and Suh, N. P. *Cell. Polym.* **38**, 69–91 (1992).
21. Karan, H., and Bellinger, J. C. *Ind. Eng. Chem. Fundam.* **7**, 576 (1968).
22. Pfannschmidt, O., Michaeli, W. *SPE ANTEC, Tech. Papers*, 2100–2103 (1999).
23. Shimbo, M. *FOAMS 2000*, 162–168 (2000).
24. Colton, J. S., and Suh, N. P. *Polym. Eng. Sci.* **27**, 500–503 (1987).
25. Park, C. B., Baldwin, D. F., and Suh, N. P. *Polym. Eng. Sci.* **35**, 432 (1995).
26. Chen, L., Sheth, H., and Wang, X. *FOAMS 2000*, 127–131 (2000).
27. Lee, J. W. S., Wang, J., Park, C. B., and Tao, G. *SPE ANTEC, Tech. Papers*, 743–747 (2007).
28. Semerdjiev, S. *Introduction to Structural Foam*, SPE Inc., Towanda, PA, 1982, p. 34.
29. Throne, J. L. *Thermoplastic Foams*, Sherwood Publishers, Hertford, UK, 1996, p. 338.
30. Xu, J. *SPE ANTEC Tech. Papers*, 2158–2162 (2008).
31. Kramschuster, A., Cavitt, R., Ermer, D., Chen, Z., and Turng, L. S. *Polym. Eng. Sci.* **45**, 1408–1418 (2005).
32. Xu, J. *SPE ANTEC, Tech. Papers*, 2089–2093 (2007).
33. Michaeli, W., and Cramer, A. *SPE ANTEC, Tech. Papers*, 1210–1214 (2006).

34. Saeed, D., Park, C. B., and Kortschot, M. K. *Polym. Eng. Sci.* **36**, 2645–2662 (1996).
35. Witzler, S. *Injection Molding Mag.* 19–22 (2000).
36. Eckardt, H. *Innovation in Polymer Processing Molding*, edited by James F. Stevenson, Hanser/Gardner Publications, New York, Chapter 1, 1996.
37. SPI Machinery Division. Recommended Guideline for Entrained Gas Processing in Horizontal Injection Molding Machines (EGPHIMM), May 2003.

7

EQUIPMENT AND MACHINES FOR MICROCELLULAR INJECTION MOLDING

The complete molding equipment consists of an injection molding machine (IMM), an injection mold (IMO), a unit of mold temperature control (heat exchanger, HE), and a supercritical fluid (SCF) system including the dosing equipment (see Figure 7.1). These components exercise a direct influence on the process of microcellular injection molding and determine the quality of the microcellular part. They also interact with one another as a system. All components shown in Figure 7.1 will be discussed thoroughly in this chapter except for injection mold (IMO), which is presented in Chapter 5. Currently, the main machine for the microcellular process for injection molding is the reciprocal screw (RS) injection molding machine with physical blowing agent dosing system and gas pump. The machine with extruder plus plunger of injection is briefly introduced as well. In addition, this chapter will discuss the designing principles of key elements in the gas system and in the machine. Some useful guidelines will be given for the structural designs for the equipments. The new safety guidelines of the Society of the Plastics Industry (SPI) are introduced to help for applying this new technology successfully and safely. There are also rapid developments for some different microcellular injection molding technologies. Several successful alternative processes will be addressed in this book. These alternative technologies give readers other

Microcellular Injection Molding, Edited by Jingyi Xu
Copyright © 2010 John Wiley & Sons, Inc.

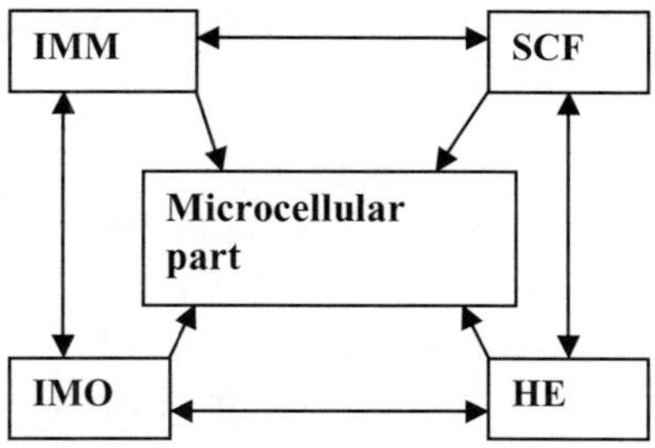

Figure 7.1 Microcellular injection molding system—interactions (IMM, injection molding machine; IMO, injection mold; HE, heat exchanger; SCF, supercritical fluid system).

choices of microcellular processes, or they bring more future challenges for researchers in this area.

7.1 TWO STAGES FOR MICROCELLULAR INJECTION MOLDING

The effects of injection process parameters for microcellular foams are clearly investigated in Chapter 6 by two stages. The equipments used in the first stage are plasticizing and gas dosing units. These units will make single-phase solution in the first stage. Specifically, the key parts in these units are the extruding screw, the screw tip, the gas injector, the gas dosing control system, and the hydraulic control system. The equipments used in the second stage are the molding system. It includes mold, shut-off nozzle or valve gates, injection unit, clamp unit, and mold heating-cooling unit.

The most popular method for making single-phase solution in the first stage is to add blowing agent in the plasticizing unit with the reciprocating screw (RS). Some papers provide information of real RS microcellular injection molding system [1–15]. Plasticizing in a reciprocating screw for creating single-phase solution is stressed as the main equipment in the first stage of microcellular injection molding. The other alternative methods, such as gas dosing in a nozzle sleeve and gas dosing in dynamic mixer, are addressed as the alternative equipments in the first stage of microcellular injection molding.

The important parameter of processing in the second stage is nucleation in the nozzle or gate during injection. Nucleation occurring as the single-phase solution is injected into a mold through a restricted channel. The quantity analysis for the pressure drop rate is a good tool to find the quality of nucleation in the nozzle, or the gate, at a certain injection volume rate. It is related to size of the nozzle (or the gate), number of the nozzles (or the gates), and the injection volume rate at certain pressure. All of these processing parameters related to equipments will be stressed in the equipment design in the second stage of microcellular processing.

7.2 RECIPROCATING SCREW INJECTION MOLDING MACHINE

The reciprocal screw injection molding machine with a physical blowing agent dosing MuCell® system* has been dominated the current microcellular injection molding market worldwide since 1998 [2, 5]. The MuCell® system is a total solution because it allows the molder to use either nitrogen (N_2) or carbon dioxide (CO_2) to create a single-phase solution of supercritical fluid (SCF) and molten polymer to make very small homogeneous cell structures in the part. It is implemented through a proven machine design that has been adopted and implemented by nearly all injection molding equipment suppliers worldwide.

The key function of this successful design is gas dissolution, where an SCF of an atmospheric gas (e.g., CO_2 or N_2) is injected into the polymer melt through the barrel to form a single-phase solution (a solution of SCF and polymer melt). To accomplish this, a specially designed plasticizing unit creates a uniform distribution of SCF, provides the conditions required for the gas diffusion, maintains the single-phase solution, and allows for a reasonable recovery rate. The software and hydraulic systems are modified to maintain the high melt pressure required to create and maintain the single-phase solution for the entire molding cycle. The injection unit of the machine requires a high volumetric injection rate to obtain better microcellular structure and to maximize the weight reduction. The other microcellular foaming methods to be discussed in this chapter failed to realize the importance of creating a uniform, single-phase solution in the first stage. Then, they failed to provide an environment where homogeneous nucleation can occur in the second stage. Typically, the microcellular injection molding system eliminates the packing and holding stages and relies solely on the nucleation rate to create and maintain uniform cell structure.

The major changes for the microcellular injection molding machine are plasticizing unit, injection unit, hydraulic system, and control system. There are some safety rules developed as the part of package of this technology to be offered by Trexel Inc. The clamp unit has no change necessary except for less tonnage requirement. Figure 7.2 shows the overall layout of these special units in the microcellular injection molding machine. A typical microcellular RS injection molding machine is MuCell® equipment, which is available on the market. The specific key elements in each unit will be discussed in this chapter.

The microcellular injection molding machine not only must be capable of creating the single-phase solution of polymer and gas, but also must maintain the required pressure to ensure a single-phase solution maintained throughout the entire cycle. Previously, the creation and maintenance of the single-phase solution was never considered important, but has been found to be critical in producing a consistent microcellular foam process [7, 10].

*MuCell® is a registered trademark of Trexel Inc., Woburn, Massachusetts.

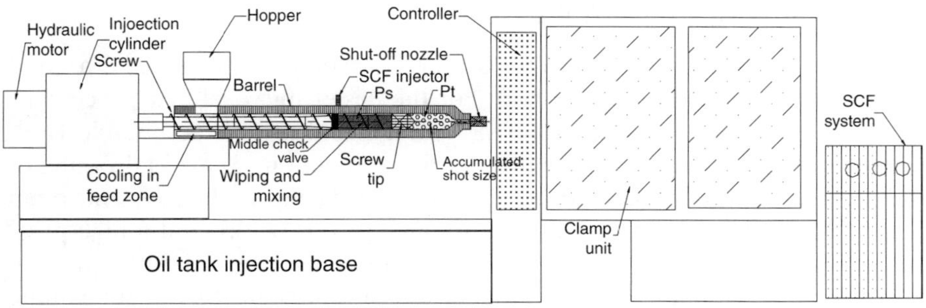

Figure 7.2 Layout of a microcellular (typical one: MuCell®) injection molding machine. (MuCell® is a registered trademark of Trexel Inc., Woburn, Massachusetts.)

The diffusion rate required for the creation of a single-phase solution for a particular gas–polymer system depends on three main factors, state of the blowing agent, temperature, and shear rate. The diffusion rate of gas in a polymer at higher temperature (i.e., at melt temperature) and when in a supercritical state is three to four orders of magnitude greater than the dissolution rate at room temperature [2]. The screw must provide the proper shear rate for the melt to speed up the gas diffusion process because shearing creates an appropriate bubble striation. The design of this screw–gas injector system is based on the well-known principles of plastic striation thickness versus diffusion time [2, 3].

The low-viscosity SCF acts as a plasticizing agent when it is injected into the melt of polymer in the screw. Once the SCF and molten polymer form a single-phase solution, the viscosity of the solution will drop significantly [10], allowing the screw torque to be decreased. Then, the requirements for torque are decreased by about 10% once enough CO_2 is added into the GPPS melt (typically 6%). This means that the microcellular screw speed can be increased without the need for increasing the horsepower of the hydraulic motor.

The shot size for the microcellular foam process is usually 5–10%, and perhaps 20%, smaller than a solid molded shot size. Therefore, if the microcellular screw has the same recovery rate as the standard OEM screw, the recovery time for microcellular process is shorter than the one for the solid molding process.

The end result of the above discoveries means that a standard RS injection molding machine typically only requires a new microcellular screw and barrel (with gas injectors) and does not typically require an upgrade of the hydraulic motor or the whole plasticizing unit.

The plasticizing unit is the most important hardware for a safe operation and a high-quality foaming process in the first stage of microcellular processing. Actually, the common safety device is a shut-off nozzle (or valve gate) and a pressure release device. The safety rules and design for these devices are discussed in the following.

7.2.1 Shut-off Nozzle and Valve Gate

A shut-off nozzle, or a valve gate, is a critical component to make a good microcellular injection molding part. For the easier microcellular processing, the valve gate is the best selection if mold design and cost are acceptable. Both nozzle and valve gate are used with traditional functions to avoid drooling of the melt and stringing and to be able to feed during mold cool and open period. However, the shut-off nozzle or valve gate is working at a high pressure environment for microcellular injection molding. The SCF in the molten polymer must maintain a minimum pressure that prevents SCF from coming out of the SCF/molten polymer solution in the barrel to cause the prefoam, or leaking during the whole first stage of the process. To satisfy this basic requirement specified above, the shut-off nozzle or the valve gate must be designed with the following features:

- Quick and reliable closing under high melt pressure up to a maximum of 34.5 MPa (5000 psi)
- Maintaining the minimum melt pressure required for the single-phase solution accumulated in front of the screw during plasticizing and screw idle period
- Minimum waste material left for the closing spot—in other words, closing the tip as close as possible to the gate or sprue bushing in the mold
- With normal close control as default setup to avoid dangerous hot spray once machine losing power or emergency stop executed
- A positive shut-off action required to be sure that nozzle is in close position at high pressure to prevent spraying of hot material with gas, overpacking for the part, and melt flowing back into the nozzle before gate freezing
- Permitting screw rotation at the end of injection, without delaying and preventing melt drool if the sprue break option is necessary for shut-off nozzle
- With small opening orifice to create necessary nucleation for the second-stage processing

7.2.1.1 Shut-off Nozzle. The typical shut-off nozzle used in current equipment is a needle shut-off nozzle (Figure 7.3). Every key element in this shut-off nozzle is indicated in Figure 7.3. Part 1 is the adapter between the end cap of the barrel and the shut-off nozzle. Part 2 is the nozzle tip that must match the radius and orifice diameter in the mold (usually 0.5 mm smaller for each of them). Part 3 is the needle, or sometimes called the pin, which is used for closing and opening the orifice of nozzle tip. Usually Parts 3 and 2 are a matched set to be changed at the same time. Part 4 is the needle guide body that has usually four channels inside as the inlets and outlets between adapter and nozzle tip. Part 5 is the lever that transfers the movement and force from

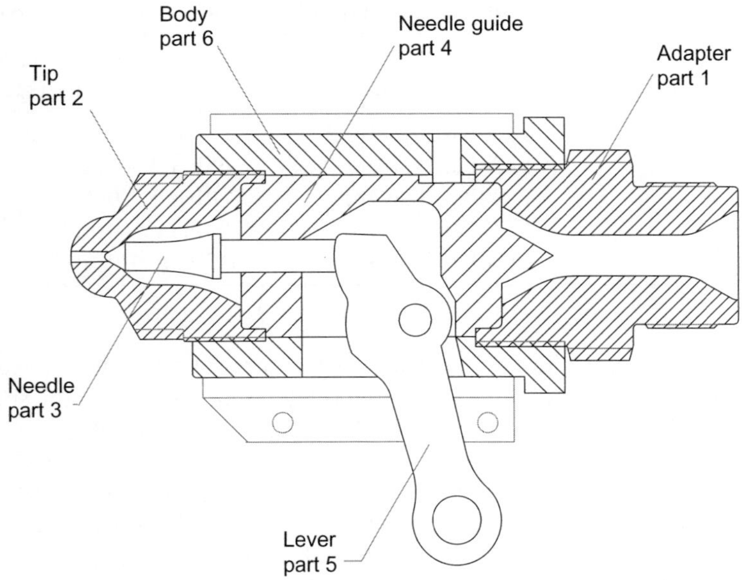

Figure 7.3 Needle shut-off nozzle. (Courtesy of Herzog AG, Degersheim, Switzerland.)

the actuator (either hydraulic or pneumatic cylinder that does not show in this layout for clarity) to the needle for closing. The lever must have contact with the end of the needle during a positive closing and must safely maintain the closing action in all phases of molding except for injection and holding.

When the injection signal is sent to the controller, a slight delay of nozzle opening will be necessary for microcellular operation. At least the injection and nozzle opening occur simultaneously if safety is an issue to avoid pressurizing the nozzle too much. The nozzle-opening-delaying requirement is necessary because the hydraulic oil in big injection cylinder needs time to be compressed to build pressure. In addition, moving the screw to overcome the resistance of viscous melt flow in the barrel plus other friction forces from mechanical parts of injection cylinder will take time, too. On the other hand, the gas-rich melt under the pressure will be immediately released if the nozzle opening occurs earlier than the beginning of real injection movement. Then, the material in the screw may spray out automatically before the injection action follows. It will create big cells in the flow front and result in bad quality of microcellular structure.

The special feature of the HPM series of shut-off nozzle designed for microcellular injection molding is high-pressure sealing. The gas injects into a plastic melt at a supercritical state with minimum pressure of about 6.9–13.8 MPa (1000–2000 psi). Sometimes this pressure in the single-phase solution in the barrel can be up to 34.5 MPa (5000 psi). The shut-off nozzle must be closed at this minimum pressure during the entire cycle time except for the injection period and possibly for a very short hold period (usually eliminated for

TABLE 7.1 HPM 100 Shut-Off Nozzle (Herzog Microcellular Shut-Off Nozzle)

Pressure at Hydraulic Actuator	Nozzle Maximum Back Pressure	Nozzle Closing Force Versus Hold Pressure (MPP)
110 bar/1,600 psi	1,220 bar/17,700 psi	305 bar/4,430 psi
120 bar/1,740 psi	1,330 bar/19,300 psi	335 bar/4,860 psi
130 bar/1,890 psi	1,440 bar/20,890 psi	360 bar/5,220 psi
140 bar/2,030 psi	1,550 bar/22,490 psi	390 bar/5,660 psi
150 bar/2,180 psi	1,660 bar/24,090 psi	415 bar/6,020 psi

microcellular molding). Table 7.1 shows the pressure relationship between the pressures in the hydraulic actuator and the related seal pressure in the shut-off pin chamber. The seal pressure of an HPM serial shut-off nozzle is very safe based on the data shown in Table 7.1.

There is a simple shut-off nozzle with a spring as the actuator, instead of a cylinder. The spring is installed in the spring chamber. The advantages of this spring-loaded shut-off nozzle are low cost and a compact design. More importantly, it will be closed always until the injection pressure builds up high enough to overcome the spring force. Therefore, the sequence of injection beginning first and then nozzle opening later is guaranteed in a spring-loaded shut-off nozzle. However, it is not good for high-temperature processing since the spring may lose the spring force significantly at a high processing temperature. One more restriction to use this spring-loaded shut-off nozzle is the leakage into the spring chamber for low-viscosity material. The melt leaking into the spring chamber may finally freeze the spring function to have a failed shut-off nozzle.

For microcellular injection molding, the spring-loaded shut-off nozzle can have modified the spring chamber as a melt filter. This filter actually is not only a melt filter but also an excellent additional static mixer for improving mixing quality of mixture of gas and melt. The result of the mixing performance with this kind of filter in a spring-loaded shut-off nozzle is discussed in Chapter 6.

There is also a special design to add a blender (or so-called static mixer) in the shut-off nozzle to make up mixing quality of gas and melt during injection through nozzle. This blender can be designed with whatever is necessary for the mixing degree, but the static mixer may significantly increase the length of the nozzle body.

For some tough material for microcellular processing, such as high percentage of rubber phase of ABS, the high back pressure up to 34.5 MPa (5000 psi) is necessary. Then, the big actuator may be needed. It does not have a compact design, and the cylinder needs to mount on the barrel track. A hydraulic or pneumatic control operates the lever mechanism with the maximum seal pressure in the outlet orifice up to 60 MPa (8700 psi). This closes the outlet orifice by means of a needle, independent of the pressure of the plastic.

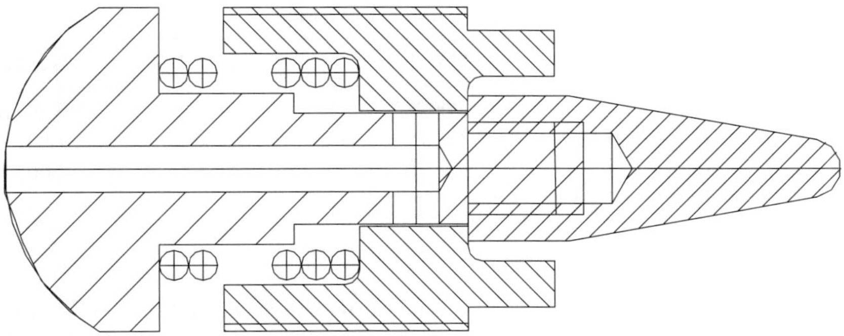

Figure 7.4 Compact design of shut-off nozzle. (Courtesy of Trexel Inc.).

The safety issue of gas-rich material processing generates a new requirement of the closing monitoring device. It is the special transducer that can detect the positions of the piston inside of shut-off cylinder. If the piston of the cylinder is not at the closed position, the high-pressure gas dosing will not allow the next cycle to begin.

There may be a situation that needs a compact design of shut-off nozzle working at high pressure and temperature. It has a new design shown in Figure 7.4. The closing force will be increased automatically with the pressure behind the cap pin. The spring just assists the closing. The spring is located in the open space so that the melt leaking will not freeze its action since the leakage can be released in the opening space. The nozzle opening position must be in the nozzle contact position with the mold so that the nozzle tip seals by the nozzle contact force on the mold sprue bushing. The only thing that needs to be set up is that the machine must run with sprue break because the closing position is only possible when the nozzle is free to contact the mold.

There is a key parameter of the shut-off nozzle to be discussed with the injection volume rate together. It is the nucleation rate, which is strongly related to the diameter of the nozzle orifice and the number of nozzles if the valve gates are not available. Calculation of *dp/dt* for injection nozzles and multi-gated parts are discussed as follows:

If the screw diameter is D, the nozzle orifice diameter is d, the linear injection speed is V, and the material viscosity is μ, then the formula to calculate the *dp/dt* through the nozzle orifice is (see Appendix A)

$$\frac{dp}{dt} = \frac{32\mu D^4 V^2}{d^6} \tag{7.1}$$

where μ is the Newtonian viscosity, D is the outside diameter of screw, d is the diameter of nozzle orifice, and V is the linear injection speed of screw.

Assume that the material melt viscosity μ is 350 Pa-sec, and that the screw diameter D is 0.06 m; then to keep the minimum *dp/dt* at 10^9 Pa/sec, the

TABLE 7.2 Injection Speeds Versus Nozzle Tip Orifice

	Nozzle 1	Nozzle 2	Nozzle 3
Orifice d (m)	0.00952	0.00635	0.00318
Min. injection rate (m^3/sec)	2.03E-4	6.02E-5	7.63E-6
Min. injection speed (m/sec)	0.0720	0.0213	0.0027

required linear injection speed V for different orifice diameters can be easily calculated (see Table 7.2). Compare nozzles 1 (d = 0.00952 m) and 3 (d = 0.00318 m). The required minimum injection speed for nozzle 1 is 26.6 times faster than the one for nozzle 3. Therefore, if the injection speed is already at maximum linear injection speed and the minimum dp/dt requirement has not been met, it can be easily reached by reducing the nozzle tip orifice.

7.2.1.2 Valve Gates. The shut-off nozzle is only good for the single cold sprue mold. Most of the applications for microcellular injection molding use the valve gates. For the multicavities mold, it is the best choice. The valve gates are also divided by driving apparatus with either cylinder actuated valve gate or spring-loaded valve gates. In addition, there is a mixing function designed in the valve gates as well. To avoid the double nucleation problem during injection, either valve gate only or shut-off nozzle only is recommended. Usually the best nucleation location is the valve gate since it is the closest spot to the mold. Then, the normal open nozzle shall be used, and the nozzle orifice in this case needs to be big enough without causing much nucleation before the melt flow travels into the valve gates. Otherwise, the cells are generated through the nucleation in the nozzle first and then go through high shearing again in the valve gates. The double nucleation in a nozzle and a valve gate will result in severely deformed bubbles across the thickness direction including the center layer of the part.

A typical spring-loaded valve gate used for microcellular injection molding shown in Figure 7.5 displays the closing position of the spring-loaded valve gate. When the injection action begins, the stem of valve gate slides back under higher injection pressure. In a good sequence, we can be sure that the valve gate will only open after the injection action is moved far enough to build pressure. This good sequence of spring-loaded valve gate prevents the premature foaming from the wrong sequence between injection and valve open. The valve gate is closed by a spring force after the injection pressure below a certain value that is usually the minimum pressure needed to maintain the melt pressure for single-phase solution in the first stage.

There is another popular valve gate design that has the actuator to close and open the gate positively. Figure 7.6 shows a typical design of valve gate with an actuator. It has a pneumatic cylinder, or hydraulic cylinder, actuate

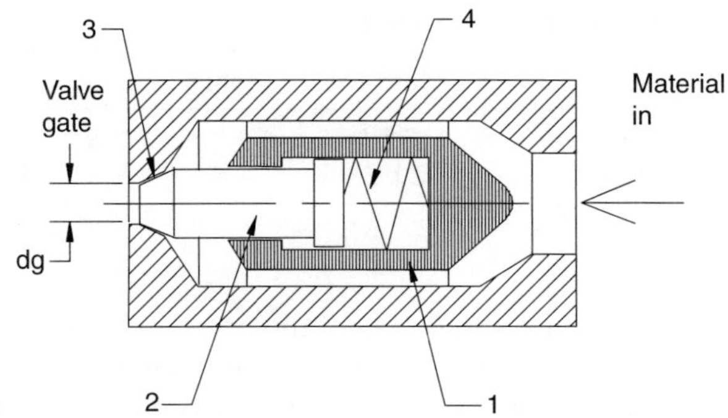

Figure 7.5 Valve gate with spring.

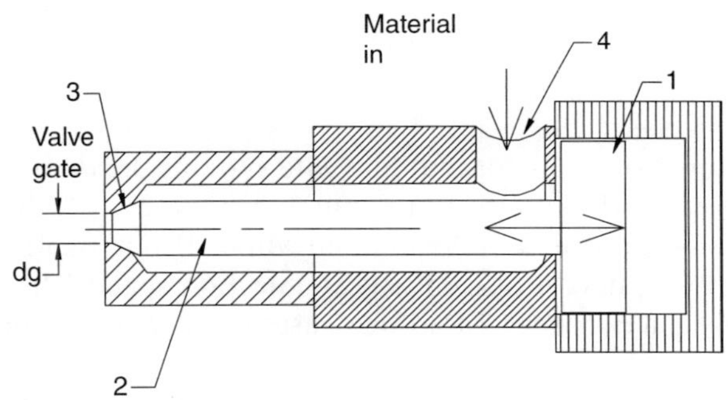

Figure 7.6 Valve gate with pneumatic cylinder.

the shut-off stem in both directions for positive positioning of both closing and opening. The pneumatic cylinder can be replaced by a hydraulic cylinder with much small space for a compact design and larger force. The critical design of valve gate is the tip of the stem shown in both Figures 7.5 and 7.6. The opening of the gate must satisfy the nucleation requirement calculated by Equation (7.2). For safety reasons, valve gate must follow the rule of default close in any case, such as power loss or emergency stop. Although the cost of the valve gate with actuators is more expensive, it is the only choice if the processing temperature is over 371 °C. The higher temperature of processing will significantly reduce the spring force, which will fail to close the valve gate positively under the high melt pressure.

The sequence of the valve gate with actuator and injection cylinder must be controlled to be sure that the injection builds enough pressure and that the injection piston has begun to move positively. Then, the valve gate will be opened without losing any pre-maintained pressure in the single-phase solution. On the other hand, the valve gate shall be open in time to avoid too much injection pressure buildup that would damage the valve gate system.

TABLE 7.3 *dp/dt* in Multi-gates

	Mold 1	Mold 2	Mold 3
Number of gates	1	4	8
dp/dt (Pa/sec)	2.2 E9	1.4 E8	3.5 E7

In a hot runner or a multi-gated system the pressure drop rate must occur at the entrance to the mold cavity. If a hot runner system with valve gates is used, the nozzle orifice will be selected as big as possible and the average gate size in the mold will be used to calculate *dp/dt*. If the diameter of the average size of gate is d_g and the number of the gates (gates opening simultaneously) of the mold is n_g, then the *dp/dt* will be (see Appendix A)

$$\frac{dp}{dt} = \frac{32\mu D^4 V^2}{d_g^6 n_g^2} \tag{7.2}$$

As shown in Equation (7.2), for multiple gate systems the *dp/dt* rate is related not only to gate size d_g but also to the number of gates n_g. Table 7.3 lists the results of a multi-gate structural foam mold with 0.0127 m (0.5-in.) gate orifice and 0.06-m screw diameter at 0.254-m/sec injection speed. The *dp/dt* for the mold with eight gates has a 63-fold decrease in *dp/dt* compared to the one-gated mold.

Ideally, the pressure drop rate *dp/dt* needs to be controlled at the gate. Controlling the *dp/dt* at the gate allows for finer cell structure, more uniform cell size distribution, and less distortion of the bubbles. If the size of the nozzle orifice is comparable to that of the gate size and both meet the minimum *dp/dt* requirement, the bubbles nucleated at the nozzle will be grossly distorted and ruptured.

It is an obvious fact that the smallest area in the valve gate determines the nucleation rate. Figure 7.7 shows a side view of a valve gate with the position of the stem and possible nucleation area in either (a) the end opening with a circular area or (b) the annular area between the tip of the stem and the chamber of the valve gate, whichever is the smaller spot that can be used to calculate the nucleation rate. If the smallest area is in the annular section (see Figure 7.7b), then the pressure drop rate estimation needs to use the formula below:

$$\frac{dp}{dt} = \frac{3\mu D^4 V^2}{d_{g1} n_g^2 (d_{g2} - d_{g1})^3 (d_{g2}^2 - d_{g1}^2)} \tag{7.3}$$

where d_{g1} is the diameter of stem and d_{g2} is the diameter of the valve chamber. It is similar to the shut-nozzle designed for microcellular injection molding. The valve gate must sustain the high seal pressure. The valve gate or shut-off

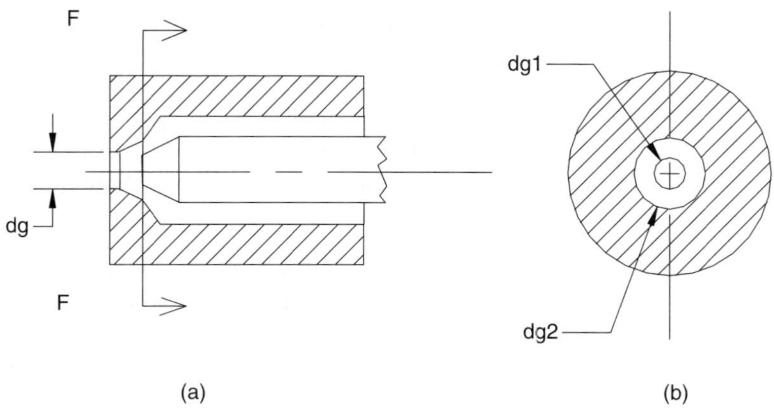

Figure 7.7 Valve gate open area. **(a)** Side view. **(b)** Section view showing annular flow channel.

nozzle requires a normal close feature because of a safety issue. The valve gates are located at the closest position to the mold cavity. Therefore, valve gate provides the best position for the nucleation during mold filling and has the least waste material for the runner system. On the other hand, it may be the most dangerous hot spray spots in the mold. The valve gates designed for the microcellular part must have better sealing performance to close the gates when the pressurized low-viscosity material is in the valve gate system.

7.2.2 Microcellular Screw and Barrel

The barrel and screw designed for microcellular injection molding have to carry out the same functions as the barrel and screw of regular injection molding machine. The solid plastic material is to be fed into the barrel, conveyed forward, heated and melted, compressed, mixed, and metered with a constant shot volume. However, the microcellular barrel and screw must be able to add gas into plastic melt (usually though the barrel in the metering zone has 90% or more melted plastics at least) at supercritical state with minimum pressure about 6.9–17.2 MPa (1000–2500 psi). The gas must be mixed and then diffused into the melt quickly to make a single-phase solution before the gas–melt mixture is accumulated in front of the screw tip [2, 5–8, 10]. The single-phase solution is also required to be maintained at minimum melt pressure to keep this single-phase solution in front of the screw tip before injection. Therefore, except for the molding issues, all the foaming equipments developed so far are focused on two key technologies: Create a single-phase solution during screw recovery and maintain the single-phase solution in the screw idle period. To make a single-phase solution is a challenge for all foaming processes. The single-phase solution is made by a necessary balance among time, gas weight percentage (wt%), temperature, and pressure. The profitable process for industry applies to an even higher pressure than the data recommended in reference 11 to reach the economic solution for making the single-phase solution at very short diffusion time. The microcellular foaming principles and safety operation rules are simply summarized as (a) higher pressure holding for single-phase solution and (b) fast pressure releasing for microcellular foam process. In this chapter, they are the technical details to be focused on.

7.2.2.1 Screw Design for Microcellular Injection Molding. An injection molding screw works periodically, not like extrusion with continuing processing. There are three main stages of injection screw operation: recovery, heat soak, and injection. In addition, a microcellular injection screw for injection molding needs to finish extra SCF during screw recovery. Therefore, the new definitions for each stage can be explained as follows:

- The screw begins to rotate for plasticizing, which is called recovery. In the middle of the screw, SCF is injected into the wiping section of the screw and is mixed with melt. The mixture of SCF and melt is called the single-phase solution; enough single-phase solution will be built up in front of the screw for the shot.
- Then, the screw stays the position where the shot size has been accumulated in front of the screw. It is a heat-soaking period. The special requirement is to maintain the minimum pressure during the soaking period, instead of releasing the melt pressure immediately at the beginning of the soaking period. The soaking period needs to be kept as short as possible to have a minimum pressure loss.
- Finally, the screw is moved forward for injection; the single-phase solution is injected into the mold with preset speed and pressure. The injection stage is to make molding with necessary nucleation and injection time in the limit of mold filling. It is usually faster than regular injection molding.

The general configuration of a microcellular screw is displayed in Figure 7.8. The screw for the microcellular process consists of five zones: feed zone, transition (or compression) zone, metering zone, wiping zone, and mixing zone. Then, a critical restriction element needs to be considered in the position between metering and wiping section [12–15]. All of these sections and pressure restriction elements are designed to satisfy both the regular and microcellular injection molding screw performances. Therefore, the details of screw design must follow the general rules as follows:

- Longer feeding section length for the microcellular screw is preferred because the screw recovery stroke decreases the feeding length. The change of feeding length will change the solid bed pressure and temperature distribution and feeding efficiency. Therefore, there is a long feeding zone, usually 40–50% of full length of the first three zones, for minimum feed length of about 2–3D at the full screw stroke back position.

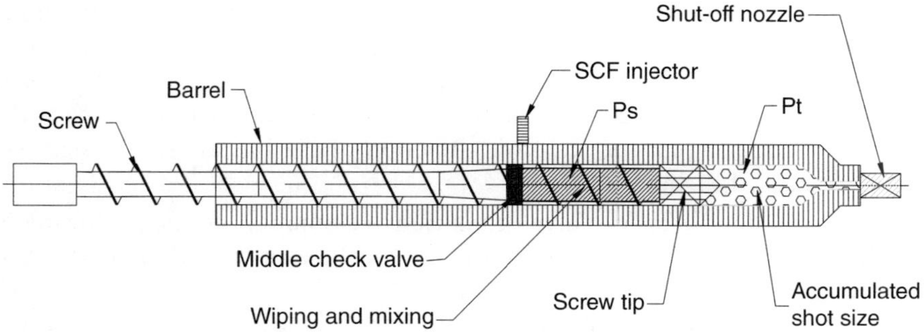

Figure 7.8 Microcellular screw configuration.

- Solid-fed (pumping capability determined by feeding section) is the preferred screw-feeding model for microcellular screw. It is because the back pressure for microcellular parts is about 6.9–17.2 MPa (1000–2500 psi), which is at about 10 times higher than normal back pressure of the injection molding screw so that melt-fed model (pumping capability determined by the metering section) is no good for such high back pressure. The solid-fed model will create high enough material pressure in the feeding zone so that the back pressure in front of the screw tip will not influence the output rate significantly. To reach the solid-fed model, more friction force must be created in the feeding zone and the melting should be delayed until enough pressure is built up. The specific details of the screw structure for this model could be small screw pitch, special coating on the screw root diameter, intensive cooling on feeding zone of barrel, feeding zone of screw, grooved barrel plus intensive cooling, and so on.
- Transition section length for the microcellular screw needs to be long enough because the screw position changes with the recovery stroke so that the beginning of the material melting position is varied with the screw stroke change. On the other hand, the different temperature profile of barrel will change the beginning of the material melting position as well. Therefore, the screw transition length in the microcellular screw is recommended to be about 25–30% of the full length of the first three zones.
- Shallow depth of the metering zone is also important for the microcellular injection screw. At high back pressure in front of the screw tip and in low-viscosity material with SCF, the pressure flow will have significant influence for the output rate of the screw in the deeper metering channel screw. Therefore, the shallow depth of metering in the microcellular screw is required.
- A blister or a similar structure is required for stabilizing downstream pressure and creating a pressure difference between metering and wiping sections in the screw. This restricted element is a key structure for

building up higher melt pressure in the metering section and maintaining lower melt pressure in the wiping section so that the gas can be added with more stability and consistency. If a ring valve is used, the blister becomes a seat of the ring. If a ball valve is used, the ball seat area is the restricted area that has the same function of the blister. This design will be discussed further below.

- A special middle check valve or restrictive element is necessary for maintaining the melt pressure inside the microcellular screw wherever the gas existed. In other words, once the gas is added into the screw, it must be kept at a certain level of pressure all the time until it is injected into the mold. The middle check valve must be chosen to match the screw tip performance to avoid either losing the melt pressure in the single-phase solution or building pressure surging during injection. The pressure surging during injection can easily blow off the rupture disc in the middle of barrel and can be avoided by the sequence of closing between screw tip and middle valve in the screw.
- A wiping section is designed with several continual flights and with enough length to cover the whole stroke of the screw. The depth of wiping section must be good for quick melt conveying, stable gas delivery (melt pressure constant), and controllable melt temperature raise. Therefore, the wiping depth is selected to be about twice the metering depth for less shearing heat and lower melt pressure for gas injection. Big helix angles for the wiping flights are used for quickly conveying melt forward to avoid any possible gas package. The length of wiping will be determined by full screw stroke and by the number of SCF injectors on the barrel. It at least needs to be long enough to cover full stroke of SCF dosing.
- A mixing section is important for the good quality of a microcellular injection molding screw. Usually the injection machine does not have a long enough mixing section. Therefore, the mixing strength for a MuCell injection molding screw must be strong enough to finish the mixing in the short mixing section. Usually a dispersion mixing is more important than a distribution mixing for a microcellular injection molding screw.

The shot size for the microcellular foam process is usually up to 20% smaller than that for a solid molded shot size. Therefore, if the microcellular screw has the same recovery rate as the standard OEM screw, the recovery time for microcellular process is shorter than the one for the solid molding process. In addition, three special requirements for the screw of microcellular injection molding are emphasized and discussed in detail as follows:

- The screw must generate adequate melt pressure profile.
- The screw needs to provide quick SCF dosing elements.
- The middle check valve must be designed to maintain the enough melt pressure to keep SCF in the solution.

1. Melt Pressure Profile in the Microcellular Screw. The first special requirement of a microcellular injection molding screw is to generate adequate melt pressure in the microcellular injection molding screw during recovery. There are two key issues for the pressures in the specific positions of the screw. The first issue is to keep a consistent pressure, or small variation of the pressure, at the position of the SCF injector in the barrel during screw recovery. This pressure variation will influence the SCF dosing rate variation from the beginning to the end of dosing. It is critical for SCF dosing constantly with the different screw stroke and screw speeds. The second issue is that the pressure on the upstream of the SCF dosing position must be kept higher in the entire dosing period. The high pressure at the upstream of the SCF dosing is the best seal to prevent SCF blowing agent leakage back through the screw to the hopper.

The typical pressure profile along the axis of screw is shown in Figure 7.9. It is a melt pressure profile measured in the whole microcellular screw at the axial position. The processing material is polystyrene. The screw is designed with the diameter $D = 0.105\,\text{m}$, the ratio of effective flight length to the diameter $L/D = 32{:}1$, and barrier section located in the transition zone. For both screw speeds, the same back pressure setup is used, which is 17.24 MPa (2500 psi). The pressures measured at positions P3, P4, and P5 are typical pressure values in three conventional zones of this screw. The slow screw rotation speed of 26 rpm generates higher pressure (13.1 MPa) at position P5, which is the end of transition zone. However, the fast screw rotation speed of 100 rpm results in low pressure (about 9.2 MPa) at position P5. It shows that with slow screw speed the melt pressure builds up early along the screw. With the blister as the pressure-restrictive part in the microcellular injection molding screw, the melt pressure at position P3 is kept almost the same at different

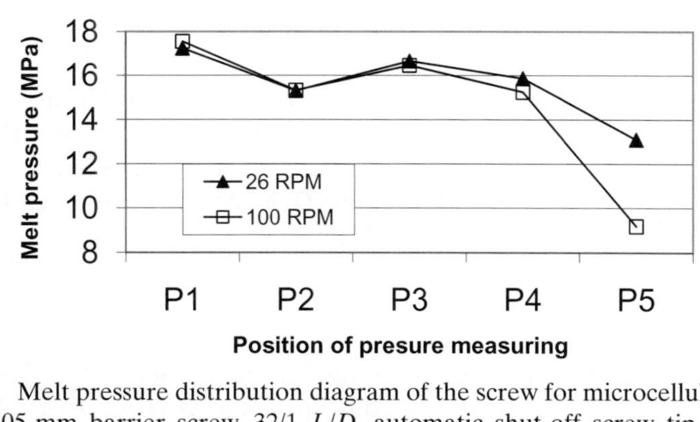

Figure 7.9 Melt pressure distribution diagram of the screw for microcellular injection molding, 105-mm barrier screw, 32/1 *L/D*, automatic shut-off screw tip, ring check middle valve. (Courtesy of Trexel Inc.) P5: 17D—end of transition zone (D is the outside diameter of screw). P4: 19D—middle of metering zone. P3: 20.25D—between end of metering and blister. P2: 22.75D—wiping zone. P1: 32D—in front of the screw tip.

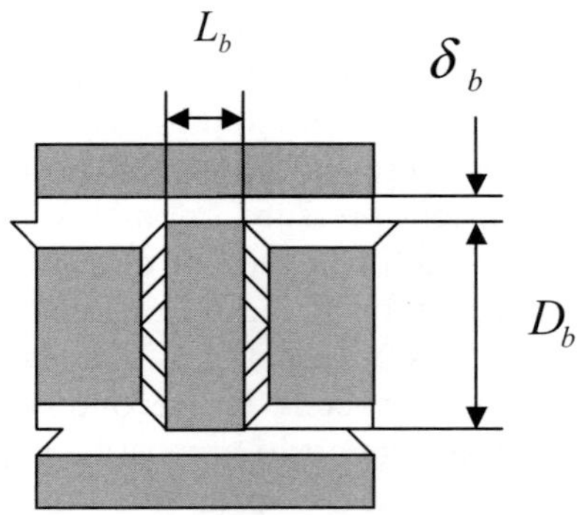

Figure 7.10 Typical blister ring geometry.

screw speeds. In addition, the melt pressure at position P2 with a deeper wiping section for SCF injection changes very little at different screw speeds. It verifies a balanced design between the blister ring geometry and wiping section geometry. It means that the SCF dosing position will have the constant melt pressure even if the screw speed varies from 26 rpm to 100 rpm.

A typical blister ring geometry is shown in Figure 7.10. It is located between the end of the metering zone and either a check valve or wiping section in the middle of screw. The geometry design of the blister ring can refer to the two-stage venting screw design for decompression, and it can also refer to the blister ring design for constant pressure drop [16]. The pressure drop over the blister ring needs to be calculated with a simplified formula without considering the effect of screw rotation.

$$\Delta P_b = \frac{2m_T \Delta L_b}{\delta_b}\left[\frac{2\dot{V}(s_n+2)}{\pi D_b \delta_b^2}\right]^n \tag{7.4}$$

where m_T is the consistency index, known as the viscosity at unit shear rate [refer to Equation (7.5)]; ΔL_b is the axial length of blister ring (m); δ_b is the clearance between inside diameter of barrel and outside diameter of blister ring (m); $s_n = 1/n$; n is the power-law index; D_b is the outside diameter of blister ring (m); and $\dot{V}$ = volumetric flow rate (m^3/sec).

$$m_T = m_{\text{ref}} e^{[-b_T(T_{\text{poly}}-T_{\text{ref}})]} \tag{7.5}$$

where m_{ref} is value of m_T at reference temperature T_{ref} (it is determined experimentally), b_T is the temperature coefficient of the viscosity (it is also determined experimentally), and T_{poly} is the absolute temperature of processing.

The estimation of pressure drop rate with Equation (7.3) may have up to 15% error because of not considering the circumferential linear speed of blister ring. It is good enough for engineering calculation. It can be simplified further to assume that the plastic melt is Newtonian fluid; then, the pressure drop is written as

$$\Delta P_b = \frac{12\mu \dot{V} \Delta L_b}{\pi D_b \delta_b^3} \tag{7.6}$$

The geometric parameters that need to be designed are axial length of blister ring ΔL_b and clearance between inside diameter of barrel and outside of blister δ_b. It is clear that the pressure drop ΔP_b has a linear relationship with the length of blister ring strongly related to ΔL_b. The empirical data are to have ΔL_b equal to 0.5–0.8 of the flight width, which is good enough for wear resistance and avoids material being sheared over the blister ring too long. On the other hand, the pressure drop ΔP_b over the length of the blister ring ΔL_b is inversely proportional to the δ_b^3 based on the relationships as shown in Equation (7.4). Therefore, the clearance δ_b is a sensitive parameter that can cause significant change for both pressure drop rate and shearing heat in the blister zone. The specific value recommended is to control the pressure drop ΔP_b of about 1.38 MPa (200 psi). It is enough to stabilize the pressure drop across the blister ring and overcome the SCF dosing pressure to prevent the SCF leaking backwards in the screw.

The pressure profile between positions P1 and P2 is important for the constant SCF dosing along the screw with the axial backup movement of the screw during screw recovery. It is well known that the reciprocating screw has two movements: One is rotation and another is backwards movement along the axial direction of barrel. However, the SCF dosing position is fixed on the barrel. Therefore, the SCF dosing position is changed in the screw during recovery because the screw moves backward relative to the barrel during recovery. The ideal pressure profile between P1 and P2 shall be horizontal, and then there will be no melt pressure change in the SCF dosing position along the screw recovery stroke. The real design will make it difficult to keep this requirement. The solution is to keep the pressure profile change gradually either higher (see Figure 7.9) or lower so that the pressure balance will be automatically adapted accordingly with slow change of melt pressure along the screw. This will be discussed with more designing details of wiping and mixing sections.

There is an alternative pressure-restrictive element design to create this kind of pressure profile similar to that of the blister ring. It is the reversal flights as the replacement of the blister ring [12]. The melt pressure is higher in the upstream of reversal flights and lower in the downstream of reversal flights. Although it is difficult to do the calculation, the design has been successfully used for both the single screw and the twin screw for microcellular foam processing. The advantage of this design is simple for manufacturing and design of screw strength. It also shows excellent mixing and shearing for speeding up the SCF diffusion in the melt. However, the processing becomes more complicated, and one design only works for a certain range of viscosity materials. The details will be discussed with the reversal channel section design for mixing and shearing below.

2. SCF Dosing in the Microcellular Screw. The definition of SCF dosing is work to add a certain percentage of SCF into the melt; the SCF constantly dissolves into the melt and finally mixes with the melt to form so-called single-phase solution. It is determined by the combination performances of the screw for the pressure profile, mixing, and shearing in both mixing and wiping sections. These basic SCF dosing requirements for screw design are discussed in detail below.

The key issue of pressure profile for SCF dosing is how to (a) create a stable melt pressure in the barrel where the SCF can be consistently dosed and (b) establish the correct SCF delivery pressure to initiate SCF flow into the melt through the barrel. Although the pressure drop across the blister ring can be estimated by Equation (7.3), or (7.5), the pressure transducer is still necessary in the SCF dosing spot, which is the wiping section or the downstream of the blister ring. Then, the SCF dosing pressure must be set up according the measured melt pressure at the same position of gas injector. This setup must have the gas dosing pressure lower than the pressure at the upstream of the blister ring, but higher than the real melt pressure at the downstream of the blister ring. The real production machine has only one pressure transducer installed in the SCF dosing position. Therefore, it is usually recommended to set up the pressure of SCF dosing in the range of 0.35 MPa (50 psi) to 0.69 MPa (100 psi) higher than the melt pressure reading from the pressure transducer at the same axial position of barrel.

To understand the pressure setup rules above, an observation for both gas pressure and melt pressure profiles is very helpful. Figure 6.5 [1] shows an important pressure profile that displays the relationship between SCF dosing pressure and related melt pressure in the same position during screw recovery and SCF dosing. The pressure difference between gas and melt at the gas injector position (same position as P2 in Figure 7.9) also plays an important role in gas droplet size. The traditional structural foam screw has a decompression zone for the gas injection spot. It creates huge gas surging at the beginning of gas dosing. Then, the gas mixing and diffusion in the melt become much more difficult because of the initial gas surging (big gas package). The microcellular screw does not have a decompression zone for a gas dosing spot, so the gas pressure cannot drop quickly at the beginning of the gas dosing. Figure 6.6 shows the typical pressure difference profiles in each cycle of the gas dosing process for both N_2 gas and GPPS melt in this experimental screw with 30-mm diameter, 26:1 *L*/*D*. The processing conditions in Figure 6.5 are listed below:

- 127 rpm screw speed (52 1/sec shear rate)
- 13.8-MPa back pressure with the melt pressure 16.2 MPa at the gas injector position
- 0.5% of N_2 gas at constant 17.7 MPa, 440 °F melt temperature

The pressure difference is about 1.45 MPa. It has three major features shown in Figure 6.5. One feature is that the initial pressure drop of gas gradually increases from the setup value when the gas injector opens. The melt pressure sharply rises about 0.67 MPa to balance the higher gas pressure, which is helpful to protect the gas surge for an initial gas injection. The screw will have pressure build-up elements to respond to the initial gas pressure surge when the gas injection opens. The actual melt pressure is pressurized to 16.9 MPa by gas pressure at the beginning of gas injection, and it stays in 16.9 MPa for only 3 sec. The second feature is that the gas pressure continually drops, and even melt pressure stays at 16.9 MPa. It means that gas flow rate continually increases slowly, which results in the melt pressure beginning to drop as well. It is because the extremely low viscosity gas layer lubricates the interface between melt and the surface of the inside diameter of barrel further, and the screw pumping capability is decreased slowly. Therefore, the screw will need enough metering length to stabilize the pumping capability, and even the viscosity is decreasing from gas lubrication. When the screw pumping and gas dosing generates new balance, the pressure stabilizes at the value of 16.6 MPa until the gas is shut off. The final stable pressure for both gas and melt takes almost half of the total 15 sec gas dosing time. The third feature is the constant 0.2 MPa of the pressure difference between gas and melt during gas dosing. This detailed pressure profile is valuable to understand the processing of gas dosing inside of the screw. Actually, the pressure profile in Figure 6.5 verifies that the screw design is successful to let melt pressure recover quickly to a stable pressure after gas dosing. The failure of screw design will have a continuous melt pressure drop during gas dosing. It can be corrected by the blister redesign. Some processing details of setting up the pressure properly are discussed in Chapter 6.

DESIGN OF WIPING SECTION. Burnham et al. [18] patented the design details with multiflights in the wiping section and multi-orifices in the injector tip in 2001. The real practical test shows that only multiflights in the wiping section are necessary, and multi-orifices in the injector may cause the nonconstant SCF dosing; consequently the quality of microcellular parts varies from cycle to cycle. It is because the degraded plastic may accumulate in some of the orifices and finally block them. The SCF is a kind of gas that always flows to the place that has the least resistance. Therefore, finally the number of truly opened orifices is not constant from cycle to cycle. The solution is to make only one orifice in one gas injector.

The wiping section shown in Figure 7.11 is designed with three functions: conveying melt away from SCF dosing, maintaining the pressure almost the same as downstream of blister ring, and breaking the SCF blowing agent into many small droplets. The first function is related to the flight pitch and the angle of flight with the axis of the screw. The second function needs to design the depth of flight channel that can either reduce depth in downstream or increase the depth in downstream that is determined by flight spiral angle.

These two functions can be designed using reference 17. The third function needs to have as many flights as possible to cut the SCF droplet to the smallest sizes. The necessary calculation formula can be derived as the follows:

$$W_t = \frac{N_f N_s}{60} \quad (1/\text{sec}) \tag{7.7}$$

where W_t is the frequency of wiping (1/sec), N_f is the number of flights in the wiping section, and N_s is the screw rotation speed (rpm). From Equation (7.7), it shows the frequency of wiping is proportional to the number of flights and screw rotation speed, respectively. Then, if we know the gas flow rate and orifice size of gas injector (here assume only one orifice in gas injector), the size of gas droplet is estimated as

$$L_j = \frac{240 Q_g}{\pi d_j^2 \rho_g N_j N_s} \tag{7.8}$$

where L_j is the size of droplet (assume the droplet with the orifice diameter), d_j is the diameter of the orifice in the gas injector, Q_g is the flow rate of gas in the gas injector, and ρ_g is the density of gas.

It is clear that the small gas droplet size needs high screw rotation speed N_s and more wiping flights N_f. An unwrapped circumferential layout of the screw wiping section (unwrapped cross-section view in Figure 7.11a) shows

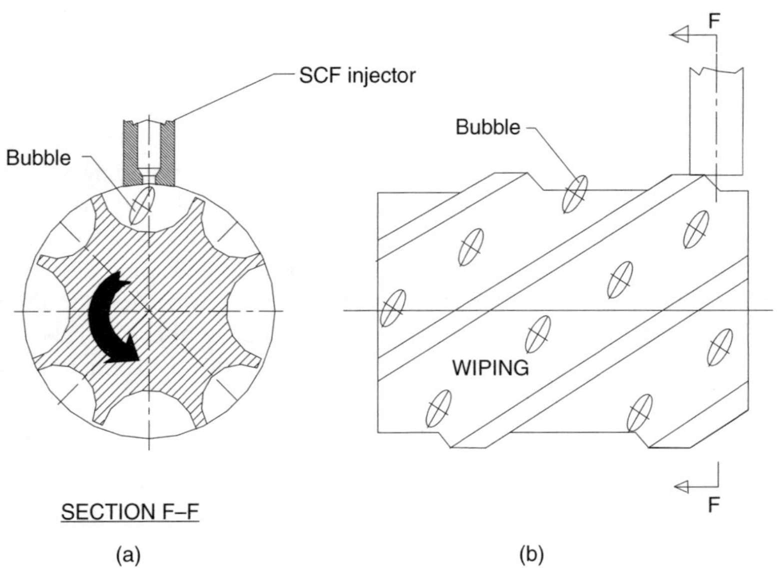

Figure 7.11 Schematic of wiping section related to SCF injector. **(a)** Section view. **(b)** Axial view.

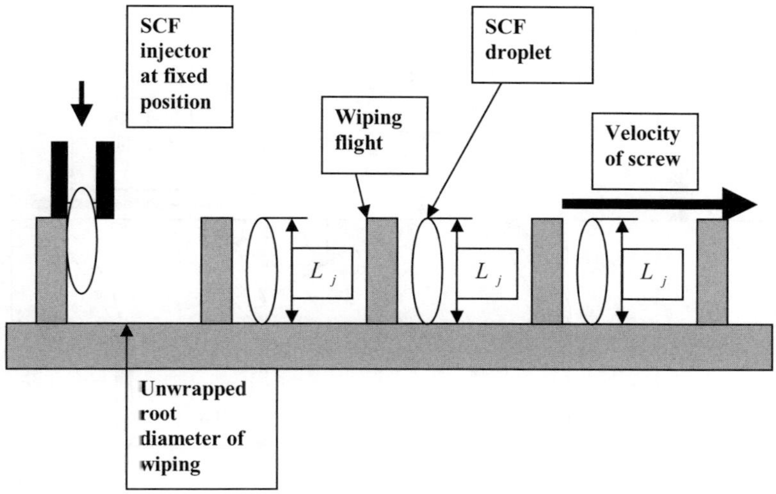

Figure 7.12 Unwrapped wiping section view for demonstration of SCF cutting by wiping flights with one SCF injector.

the droplet size and number in each channel of flight in Figure 7.12. It explains in one full turn of the screw that there is the same number of droplets to be cut as the number of wiping flights. It will be cut smaller if we add more gas injectors in the barrel in the same axial position but different location around the circumference of barrel.

To speed up the gas diffusion time with certain size of droplet, there is an idea to decrease the thickness of plastic l for a short SCF diffusion time t_d [11]:

$$t_d \sim l^2/\alpha \tag{7.9}$$

where t_d is the SCF diffusion time (sec), l is the thickness of plastic for gas diffusion (mm), and α is the gas (SCF) diffusivity [see Equation (7.10)].

$$\alpha = \alpha_0 \exp(-\Delta G/kT) \tag{7.10}$$

where ΔG is the activation energy, k is Boltzmann's constant, and T is the absolute temperature.

From Equations (7.9), and (7.10), the diffusion time t_d will increase at elevated temperature T. Therefore, for higher diffusivity of the SCF screw design needs to avoid overheat in the SCF dosing area. On the other hand, there are two ways for the screw design to reduce the l. One is to add more gas injectors at the same axial position of barrel. Another is to increase the screw rotation speed. The l between two adjacent gas droplets in axial direction in the same channel of wiping can be given as

$$l = U_m T_w \tag{7.11}$$

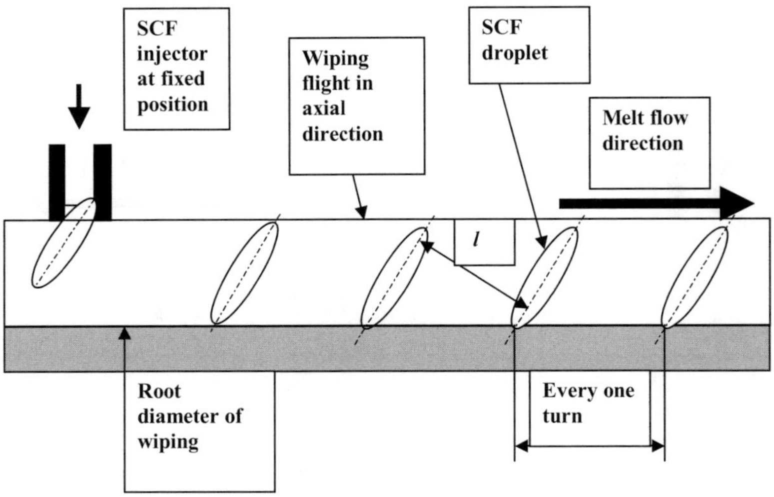

Figure 7.13 Axial section view for demonstration of SCF droplets in the same wiping channel at every full turn of screw with one SCF injector.

where U_m is the absolute velocity of melt in axial direction of screw (see Equations (6.5) and (6.7) [7]) and T_w is the period of wiping that is equal the reciprocal of W_t.

From Figure 7.13, it is obvious that the thickness of plastics l between two adjacent bubbles in the same channel of screw depends on screw rotating speed and melt axial velocity U_m. This critical thickness l is discussed well in Chapter 6 since it is strongly related to processing parameters in the screw.

In addition, the shearing field in the wiping section may result in further stretching of the bubbles to a critical length and break them into smaller ones when a critical Weber number *We* is reached in the shear field. The phenomenon of bubble stretching in the wiping section is demonstrated in Figure 7.13. A cross-section view of the screw channel is shown in Figure 7.14a. The depth of the wiping section is usually almost double the depth of metering so that there exists a pressure flow. The drag flow at the top of the channel and the pressure back flow at the bottom of the channel consist of a strong shearing field shown in Figure 7.14a. In Figure 7.14b, there are two initial non-sheared bubbles with the distance l_1 between them. They are stretched to thin oval shape, and the distance l_2 is significantly shortened as shown in Figure 7.14c. If this stretched bubble is stretched more to beyond the critical Weber number, the surface force is overcome by shear force, and one stretched bubble breaks into two small bubbles. Figure 7.14d shows the result that the droplet breaks into two small droplets. Therefore, the stretching in the shear field does two things: It reduces the distance l for short diffusion time, and it breaks the droplet into a small size for better distributive and dispersing mixes.

The Weber number is defined as [11]

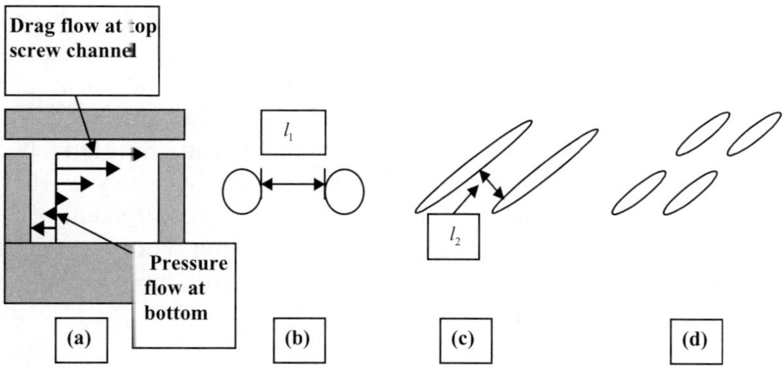

Figure 7.14 Bubble stretching in the shearing field in wiping section of screw. (a) melt flow profile. (b) initial nonsheared bubbles. (c) sheared bubbles. (d) broke bubbles.

$$W_e = \frac{\frac{d\gamma}{dt} d_p \eta_p f(\lambda)}{2\sigma} \tag{7.12}$$

where W_e is the Weber number (ratio of shear force to surface force), σ is the surface tension, η_p is the viscosity of polymer, γ is the shear strain, d_p is the bubble diameter, and λ is the viscosity ratio of gas to polymer.

$$\lambda = \frac{\eta_g}{\eta_p} \tag{7.13}$$

where η_g is the gas viscosity.

$$f(\lambda) = \frac{(19\lambda + 16)}{(16\lambda + 16)} \tag{7.14}$$

The critical Weber number W_e in the simple shear field is about 300. To reach the critical W_e, the strong shear field design in the screw geometry is necessary.

DESIGN OF THE MIXING SECTION. Even if the smallest drop size is cut in the wiping section, the mixing in the downstream is critical for the successful microcellular processing. It is well known that adequate mixing determines the final part properties, processing efficiency, and cost. The traditional mixing always pursues the turbulent flow as the effective mechanism of mixing. It is because turbulent flow is associated with the random fluid motion that is the most effective mechanism of mixing. The criterion of Reynolds number to reach turbulent flow is approximately the value of 2000 or larger for water or water-like liquid. However, the high viscosity of plastic melt eliminates the possibility of turbulent mixing. The laminar flow dominates the mixing analysis not only for the regular injection molding process but also for the microcellular injection molding. There is an extremely large difference between high viscosity of melt (in the range of 10–1,000,000 poise, g/cm sec [19]) and low viscosity of SCF (about 0.00005–0.05 poise, g/cm sec [19]). It is actually an advantage of such wide differences in viscosity for relatively easier mixing. However, it results in a more complicated mixing analysis. There are many actual mixing elements available for the mixing requirement of SCF in the molten polymers. The fundamentals for the mixing theory are discussed in the following, with particular emphasis on two basic mixing phenomena: dispersive and distributive mixings.

Dispersive Mixing. Dispersive mixing is also called intensive mixing. Dispersion is dominated by the history of fluid mechanical stresses in a mixture. A mechanical stress is applied to reduce the scale of the droplets of SCF wrapped by a plastic melt in the microcellular screw wiping and mixing sections. It involves changing the droplet size from big to small and then splitting it into a larger number of droplets. However, it does not involve changing the position of each droplet. The concept of this mixing is illustrated in Figure 6.8. It is a critical mixing to speed up the gas diffusion process in the screw or other elements with mixing requirement. The related mechanical design is the restricted barrier section that has a narrow gap left to force the gas–melt mixture to go through it with strong shearing and elongation. A dynamic mixer has a mixing function that is similar to that of dispersive mixing.

In addition, the restricted barrier, which is the narrow channel for all of the mixture to go through, helps to transport the molten material from the bottom of the screw channel in the wiping section to the upper layer because the initial gas droplet from gas injector is only in the upper layer of the wiping channel. Therefore, the melt at the bottom of screw channel will have a chance to mix with the gas as well through this dispersive mixing movement.

Distributive Mixing (Blending). The homogenization of the mixture is determined by the history of deformation imparted to the fluid. It is simply the strain. This process is also called extensive mixing, or blending. The principle is to distribute the component (here the gas droplet) uniformly in space without changing the size of the component. The concept layout can be referred from Figure 6.7. The typical distributive mixing element is the mixing pins, multichannels, and any mixing elements with the function to divide a main stream of flow into many small streams of flow. Most static mixers have the same function as distributive mixing.

Both mixings are required for microcellular processing in the injection screw. The real mixing in the screw is a comprehensive mixing including both of them, and it may be difficult to divide them clearly. One of the key factors is the shear rate. The screw must provide the proper shear rate for the melt to speed up the gas diffusion process because shearing creates the appropriate gas droplet striation. The design of this screw–gas injector system is based on the well-known principles of plastic striation thickness (see Figure 7.14) versus diffusion time. However, there is one important issue of shearing heat locally in the shearing area. If the shearing element is designed with regular rule used in conventional injection screw, it may be too aggressive for a microcellular injection screw. The extremely higher back pressure for the microcellular processing may cause some overheating problem with the regular shearing elements in the screw mixing zone. The solution is to design the same shearing strength that is mainly determined by the shearing clearance in the radial direction. To reduce the overheating in the shearing zone, the shearing time can be controlled as short as possible. It simply uses a short shearing land. The detailed processing analyses of mixing for gas droplets in the wiping and mixing sections are discussed in Chapter 6. Based on the comprehension of mixing principles, the mixer design in the microcellular screw is not the difficult task and can be referred from lots of published works [14–17]. However, the key features of the mixing design in the screw for microcellular injection molding can be summarized as follows:

- Low pressure drop across the overall length of mixing section
- Strong shearing and elongation deformations from both distributive and dispersive mixings without overheating issue
- The high number of striations divided by multichannel mixing sections
- The up–down movement to switch the material from the top layer to the bottom layer or vice versa in the mixing section
- No dead corner in the mixing section
- Partial pumping capability in the mixing section
- Necessary pressure flow backwards to promote the mixing efficiency

Another critical key parameter is the residence time of SCF in the screw. It is well known that the shearing in the screw speeds up the SCF dosing significantly. Hence, the residence time during the screw idle period is neglected since it is not important. Unless mentioned specifically, the residence time in this chapter is the residence time during screw rotation, known as shearing residence time. The method of calculation for the sharing residence time is to count the period from SCF injecting into the barrel to the SCF and melt mixture leaving the screw. Therefore, the governing equation to calculate the average residence time t_r in mixing and wiping sections is given by

$$t_r = V_m / \dot{V} \tag{7.15}$$

where V_m is the volume of mixing and wiping sections (m^3) and $\dot{V}$ is the volumetric output of the screw (m^3/sec).

Based on the results of an initial study in MIT, the residence time in the screw needs to be several minutes at least. However, it is almost impossible that the real shearing residence time in the screw will be greater than 1 min.

TABLE 7.4 Residence Time Versus Screw Speed

	Screw Speed 60 rpm	Screw Speed 118 rpm	Screw Speed 167 rpm
Residence time (sec) for 24:1 *L*/*D* screw	49.7	21.1	14.2
Residence time (sec) for 28:1 *L*/*D* screw rate (m^3/sec)	67.4	28.6	19.3

The general rule to design overall screw performance is to keep the relationship as

$$t_r \geq t_d \tag{7.16}$$

This means that the residence time of shearing the SCF-rich material must be longer than the diffusion time to make truly single-phase solution in the first stage of microcellular injection molding.

In the real practice of microcellular processing, the residence time requirement conflicts with the shearing strength requirement. It is because the shearing strength is proportional to the screw speed. Therefore, the strong shearing relies on the high screw speed while the residence time becomes short at high screw speed. It is obvious that the high screw speed results in short residence time are shown in Table 7.4 for both 24:1 *L*/*D* and 28:1 *L*/*D* screws. The residence time becomes reduced as the screw speed increases, showing a nonlinear relationship. It reduces sharply at the low to middle range of screw speed. The empirical data show that usually a fast SCF dosing is related to a high shearing, and longer residence time with small shear rate does not help much for SCF dosing. However, which one dominates the SCF dosing is still not scientifically clear. On the other hand, the modern microcellular screw design provides the wiping and mixing section as deep as possible to make the dynamic residence time of gas-rich material as long as possible at certain minimum screw speed. The deep channel in mixing and wiping also created more back flow at the bottom of the deep channels that actually increases the mixing quality by the increase in shearing on the top layer of the material flow field. However, the deep mixing and wiping channel may bring the risk of intensifying the dosage locally. Therefore, the depth of wiping shall be gradually increased from the upstream of the wiping section near the middle check valve to the downstream of the wiping section.

A modified mixing section has been designed for 30-mm diameter and 22:1 *L*/*D* of a short screw. This short screw has been run successfully with a special mixer. Compared to a long screw with 26:1 *L*/*D*, the short screw has the different depth of helix slot cut on the Saxton mixer [15]. The deep cut slot provides the distributive mixing. It simply splits and rejoins of the flow streams

TABLE 7.5 Output Rate and Screw Torque for 26:1 and 22:1, 30-mm Screws with 0.6% Gas

Material	*L/D* Ratio	Middle Check Valve	Output (g/turn)	Middle Pressure (MPa)	Ratio, Output, 22:1 to 26:1	Ratio, Torque, 22:1 to 26:1
PP, 20% talc	26	Middle ring	0.63	13.8	1	1
	22	Middle ring	0.89	13.8	1.412698	0.915493
	26	Middle ring	0.37	20.7	1	1
	22	Middle ring	0.66	20.7	1.783784	0.902778
PC	26	Middle ring	2.14	13.8	1	1
	22	Middle ring	2.88	13.8	1.345794	0.996732
	26	Middle ring	2.03	20.7	1	1
	22	Middle ring	2.56	20.7	1.261084	0.97561

through these deep cut slots in the mixing section. The shallow cut slots supply the dispersive mixing to break up the minor phase. The driving force of dispersive mixing mechanism is the shear stress. On the other hand, the shallow cut slot forces the material flow through the top slots locally and then the bottom layer material may move upwards. It increases the choice of material exchange from the upper layer to the bottom layer, and vice versa. Therefore, this upside-down movement in this special mixing section will keep the benefit of the deep mixing section (longer residence time overall) and makes up the nonuniformity residence time in the depth direction of the screw channel. In addition, the gas-laden material will be stretched or accelerated multiple times; this is the efficient way to break down the agglomerates, and it makes gas droplet smaller without creating shear heat significantly because the shearing time is controlled not very long over a narrow barrier with a deeper barrier depth.

This special short screw changes the screw performances significantly, as listed in Table 7.5. Only two materials are listed in Table 7.5: PP represents low viscosity, and PC represents high viscosity. Specific output is the output for every screw rotation, so it will not be related to screw rpm, and fare to be discussed the screw performance. It is obviously a trend that all output rates per turn of a 22:1 screw are higher than the ones of a 26:1 screw regardless of the pressure setup the materials that are processed. It verifies that the short mixing section of a 22:1 screw has less resistance to the pumping capability of the metering section in the screw because the conventional three zone sections are the same for both the 26:1 and the 22:1 screws.

In addition, the ratio of torque in the 22:1 *L/D* screw to the torque in the 26:1 *L/D* screw is always less than 1. It means that the driving torque of the 22:1 screw is lower than the torque of the 26:1 screw. It can be explained with the short mixing section. The processing results of this short *L/D* screw are discussed in Chapter 6.

DESIGN OF PRESSURE-RESTRICTIVE ELEMENT IN MICROCELLULAR SCREW. There are two different pressure-restrictive elements, with both being successful designs used in current industries. One is the reversal channel design, and another is the middle check valve.

Reversal Channel. A reversal channel designed in the wiping section is the simplest design for both the pressure-restrictive element and the wiping section in microcellular screws. It still needs the blister ring to create the decompression zone to add SCF in the reversal section easily. The wiping section for the 40-mm screw and smaller (as small as 25-mm-diameter screw) has a design that is similar to a Maddock mixing section but with both positive flights as inlet channels and negative flights as outlet reversal channels. This reversal channel creates a very different melt pressure profile at the point of the SCF injector as compared to a larger screw. The unique feature with this reversal channel design is with respect to the behavior of the melt pressure in the barrel at the SCF injector location. This in turn affects the delivery pressure and screw recovery settings. For larger screws, the melt pressure in the barrel at the SCF injector (SCF inlet pressure) is typically about 1.38–2.07 MPa (200–300 psi) greater than the back pressure setup. In the case of the reversal channel screw, that difference can be 5.5–6.9 MPa (800–1000 psi). This high pressure in the reversal channel screw creates a necessary measurement to keep the pressure from quick decaying to maintain the single-phase solution by this residual pressure in the reversal channel of screw. Since the delivery pressure needs to be set slightly above the SCF inlet pressure, the difference between the back pressure setting and the delivery pressure setting will be much larger with the smaller screws.

The other unique aspect of this reversal channel screw design is that at higher screw speeds, such as 50% of maximum screw speed, the SCF inlet melt pressure will continue to increase throughout the entire screw recovery stroke. Although the high pressure actually helps to speed up the gas diffusion and mixing, the long screw recovery stroke may have a processing issue. The continually increasing pressure at the gas dosing spot may become larger and larger if the screw recovery stroke is longer than the length of the positive channel of the reversal channel design. It is because the gas dosing spot may move to the position of the reversal channel that needs to be controlled only in the downstream of the gas dosing position. It will create a situation that the delivery pressure is always trying to adjust to match the increasing melt pressure in the barrel and ends up creating inconsistent SCF dosing. In this case, the adaptive SCF dosing control is necessary to automatically change the SCF pressure with the change of melt pressure accordingly. With adaptive gas pressure control, a constant pressure difference between SCF pressure and melt pressure will be kept the same no matter how many changes occur during screw recovery. However, if both screw stroke and cycle time are very short, then the screw idle period is short as well; reversal channel screw is a perfect economic solution for this application without any requirement for the adaptive pressure control.

On the other hand, the capability of mixing under high pressure is the important feature of reversal channel screw design. The reversal channels in the wiping section is not only increasing mixing instantly after SCF injected into melt but also create much high local pressure temporarily in wiping section. This reversal channel helps to slow down the decay of high melt pressure in screw idol period. It is a critical factor to maintain minimum melt pressure in single-phase solution between screw tip and reversal channel in the screw until injection begins. The best machine control method for reversal channel screw is the screw position control, not pressure control, after screw finishes the recovery.

In order to create a stable melt pressure region in which SCF can be dosed effectively, there should be a certain length in the reversal channel screw recovery stroke where screw speed is reduced to less than about 35% of the maximum speed for a minimum of 3 sec of screw recovery time. The purpose is to increase the dosing time as long as possible with acceptable shearing rate for reversal channel screw. Once SCF dosing is completed in this length of stroke, the screw speed can be increased at the last stroke of screw recovery to provide mixing of the SCF with the polymer melt. This high screw speed at last stroke also helps to move the existing SCF in the screw away from the SCF injector location. The screw speed after dosing can be any value required in order to achieve the necessary screw recovery time. In order to maintain the accuracy of shot, the screw speed can be slowed just before reaching final shot size. Typically, screw speed will be slowed to less than 20% of the maximum value for the last 1–3 mm of the shot size. This will provide a more consistent shot size. These processing parameters must be set up based on processing conditions for certain materials but will not add into the control software since it is processing adjustable parameters.

When working with small shot sizes, less than 20% of the barrel stroke, and low flow rates (<0.25 lb/hr), dosing starts to be controlled by pressure drop and not by flow rate. It is because this small dosage can be controlled with pressure drop through an orifice with almost the same precision as flow rate control but with a significant cost saving. The SCF gas dosing time is normally so short for the small gas dosage, such as <0.5 sec. Then, SCF gas is dosed solely through the initial pressure surge into the barrel. In this instance, SCF levels should be calculated based on a pressure surge formula and not a flow rate formula.

As with any application, SCF flow rate, SCF delivery pressure, SCF injector open position, and SCF injector open time need to be optimized for material and shot size in reversal screw processing. If the SCF delivery pressure is controlled by the back pressure in front of screw tip, the unique design of the smaller reversal channel screw will require a larger difference between the SCF delivery pressure and the melt pressure in front of screw tip because of the high pressure in the middle of screw. Precisely, the pressure difference between SCF delivery pressure and the middle pressure near the gas injector needs to be kept as small as possible if the SCF delivery pressure is controlled based on the pressure reading in the middle of screw. Also, profiling of the screw speed and back pressure during screw recovery will allow for better control of SCF dosing into the barrel as well as improved shot size control.

There are two most important applications for reversal screw design of microcellular processing. One is the small screw, whose strength does not allow the weak section of mixing with check valve. Another application is the very short dosage time, or short screw recovery time. Both of the cases above will not have long recovery stroke that is perfect for reversal screw to finish the gas dosing without the pressure change along the screw recovery stroke.

Middle Check Valve. For middle check valve design, both ball valve and ring valve are used for different applications successfully. Usually the ball valve performs better for low-viscosity material without fillers or glass fiber reinforcement. Generally, ball valve design needs to consider the issues of wearing problem, difficulty of high-viscosity material processing, and strength for small-diameter screws.

The ball valve is a very quick shut-off performance valve and is successfully used as a middle check valve for most low-viscosity and low-wearing-issue materials. Generally, the design principle of the ball valve in the middle of the screw is similar to the ball valve as a screw tip. The design of the ball valve with a cartridge that includes a ball inside of the cartridge as a kit will have features:

- There is no spring in the ball valve cartridge. It can be used for the high processing temperature and no dead corner in the ball valves.
- The ball valve includes everything inside of one cartridge, so that there is no wearing issue of the middle valve in the screw itself.
- The cartridge of the ball valve is simple for manufacturing for both screw and retainer itself, and it can be easily replaced with a new set of ball and cartridge.
- The flow area ratio of the inlet channel of the ball valve to the metering channel of the screw is 50–80%. The big screw requires a big ratio to have low resistance of material flow.
- The outlet of the ball valve has more opening area than does the ball valve inlet. It is necessary to create the pressure drop like blister ring in the check ring valve. Therefore, the ball valve itself is the pressure-restrictive element area to create the necessary pressure drop for SCF dosing.

Another successful middle valve design for the screw of microcellular injection molding is the ring style of the check valve shown in Figure 7.15 [8, 12–15]. There are two positions of the ring displayed in Figure 7.15; The top one shows

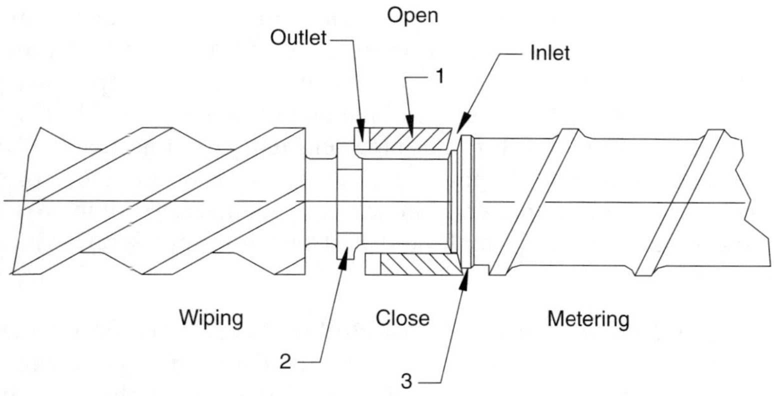

Figure 7.15 Middle ring check valve and blister ring [15]. (Courtesy of Trexel Inc.) 1, Check ring; 2, stop for open position; 3, seat of ring and a blister ring.

the opening position, and bottom one shows the closing position. The upstream (inlet side) of ring "1" is the metering section of the screw, where there is the molten material without gas. The downstream of ring "1" after the outlet is the wiping section for gas adding. Therefore, the top position of the valve opening is the normal screw recovery position. The pressure drop across the blister ahead of the ring forces the molten material without gas flowing freely through the ring in the top position shown in Figure 7.22. However, when the screw either moves forward for injection or stays idle, the ring slides back to block any possible backward flow of gas-rich material in the wiping section. To prevent the gas-rich material, flow back from this ring will solve two problems. One of the problems to be solved is that if the gas-rich material leaks back in the metering section of the screw, it will significantly reduce the pumping efficiency of the screw. This is because the extremely low viscosity of gas will build a lubrication layer in the ID of the barrel so that the drag flow as the positive output flow in the screw will be reduced significantly. Another problem to be solved is that if the gas-rich material loses the minimum melt pressure because of leaking, the single-phase solution in the whole wiping and mixing sections will be damaged, and it will cause more difficulties to rebuild a good single-phase solution in the next cycle. Therefore, the middle check valve is a true key element for microcellular processing of injection molding.

The most common design for the pressure-restrictive element includes the blister ring, which contains the rear seat of the check valve. The ring valve design is the durable design because it does not have many wearing problems like the ball valve Therefore, it can be used for heavily glass-fiber-reinforced material processing of microcellular application. The design principle of this two-closing-stage ring valve will be discussed in the screw tip section.

Using the same principle above, a split ring (the ring is cut into halves by the EDM method) has been designed for a small screw because the small screw such as a 30-mm screw must be made with one piece for the strength requirement. The EDM wire size is up to 0.1 mm (0.004 in.), which proved to be good size to easily cut the ring into halves without compromising the performance of sealing in the middle of the microcellular screw. The ring itself is cut with at least one tooth (any shape) engaging the ring together in the axial direction. This kind of split ring of check valve in one piece of screw almost covers all of the diameters of injection molding screws for microcellular processing because it is a simple and a strong non-split screw with good performance.

7.2.2.2 Barrel Design for the Microcellular Process. The barrel used for microcellular injection molding needs some modification. It is similar to the venting barrel that has holes for the venting valve in the middle of the barrel. The difference between the venting barrel and the microcellular foam barrel is the function of the hole: The hole is for the SCF injector for the microcellular foam barrel, instead of for the venting valve for venting moisture or chemical vapor from the barrel. Also, a special configuration for SCF injector, rupture disc, and pressure transducer is positioned in the same section as the gas injector as a standard configuration for most of the microcellular barrels. In addition, there are several special requirements for the SCF dosing. They are discussed in the following.

1. A pressure transducer must be installed at the same axial position of the SCF injector. It is critical to use this pressure reading to set up the correct SCF dosing pressure, which is the best way to decide the SCF deliver or dosing pressure regardless of the kind of screw designs. It is also the only reliable electronic signals used for SCF dosing pressure automatically follow up the changes of melt pressure in this position.
2. An air-cooled heat band is necessary whenever the processing temperature is lower than the normal processing temperature. The air-cooled heat band is located in the front of the barrel near the mixing screw section. In most cases it is not an economic solution and actually is not used widely.
3. More than one SCF injector maybe needed in both the axial and circumferential directions. The axial multi-SCF injector is necessary for long screw stroke. It is not only increasing the effective SCF dosing stroke, but also decreasing the length of the wiping section. A sequential analysis will be explained below. The circumferential multi-SCF injectors are used to add more gas to make heavy dosing. A rupture disc is also required in the barrel. The position of this rupture disc must be located between the screw tip and the middle check valve and should never allow the screw tip to pass the position of the rupture disc.
4. The injection pressure with the SCF-rich material in the barrel will be about at least 30% lower than normal injection pressure with solid material. However, enough barrel strength for normal injection pressure is

still recommended since the mold test may need to be for solids first sometimes, or in some cases the SCF-rich material is used for mold filling and dimension stability only with 100% filled molding that needs high injection pressure near the end of mold filling.

5. The SCF with extremely low viscosity will mix with plastic melt. It results in a low-viscosity SCF-rich melt that will significantly reduce the output rate of screw. To make up the loss of the output rate, sometimes it is necessary to have grooved barrel to promote the feed-dominated conveying system to increase the output rate of the screw without being influenced by the gas dosing process. Therefore, the output rate will be not only as high as the solid processing but also consistent for all different gas dosing percentage in different materials.

7.2.2.3 Safety Devices in Barrel Assembly. The traditional safety operation for the nonfoaming process requires the default nozzle to be open as a safety rule. It means that the pressure in the barrel will be released by the opening of nozzle whenever the unpredicted power loss or emergency stop (E-stop) is pressed. The gas–melt mixture (single-phase or partially single-phase solution) or pure gas (gas pocket) stores much more energy in the gas-rich melt than does the pure melt. Therefore, the safety operation for the foaming process changes the safety rule from default nozzle open to default nozzle closed. The safety issue for the entrained gas process was discussed in the Entrained Gas Guideline Committee of the Society of the Plastics Industry (SPI) Machinery Division. The SPI Machinery Division has released the guideline for entrained gas processing in horizontal injection molding machines since May 2003 [6, 20]. To better understand the guidelines, the basic principles behind the safety rules and some experiments related to safety are discussed below.

One of the common technologies for both gas-assist and gas-foaming processes is the application of highly pressurized gas. For gas assist the pressure in the gas is about 2.5–30 MPa (363–4350 psi) for injecting gas and driving out the melt to give a way to a gas channel in the part. The melt pressure required for the single-phase solution is in the range of 6.9–34.5 MPa (1000–5000 psi) for the foaming process. Therefore, the energy stored in the gas is approximately the same level for both processes. There is a formula to calculate the stored energy (joules) of gas under pressure [21]:

$$U = \frac{P_g v_1}{k_g - 1}\left[1 - \left(\frac{P_a}{P_g}\right)^{\frac{k_g - 1}{k_g}}\right] \tag{7.17}$$

where U is the energy stored in gas (joules), P_g is the gas-charged pressure (MPa), P_a is the atmospheric pressure (MPa), k_g is the adiabatic exponent (1.41 for N_2 gas), and v_1 is the volume of the gas (cm^3).

The calculation with Equation (7.17) is based on reversible adiabatic (isentropic) expansion of the confined nitrogen gas. It is obvious that the higher the pressure P_g and the higher the volume v_1, the greater the amount of energy U stored in the gas. Similarly, the stored energy in the liquid is given as

$$U = \frac{1}{2}\left(\frac{P_{ld}^2 v_1}{B}\right) \tag{7.18}$$

where B is the liquid bulk modulus (MPa) and P_{ld} is the liquid charged pressure (MPa).

Usually the energy stored in liquid is considerably less that the energy stored in gas under the same conditions. For general-purpose polystyrene melt with the bulk modulus 857 MPa at 230 °C temperature and 34.48 MPa pressure in 1 cm^3 of volume, the stored energy is about 0.69 joules. Under the same conditions, the energy content stored in 1 cm^3 of volume of N_2 gas at 34.48 MPa pressure is about 68.68 joules. Therefore, the pure gas can store the energy about 100 times higher than the plastic melt can do under the same conditions. It is also true for the gas–melt mixture, which stores more energy than melt itself does with the same pressure and volume.

Then, the default open nozzle and valve gate are potential hazards if someone is still working in the mold area with a guard opening during unpredicted power loss or E-stop pressed without warning. The foam process uses nozzle default-close as the safety rule to design and set up the machine control system. For the injection molding with reciprocating screw, the pressure of the material accumulated in front of screw tip must be released by the screw free movements, not nozzle opening. To guarantee that the pressure will release in this way, the screw shall be free to move after power loss. It was verified on an experiment in the reciprocating screw of injection foaming machine with and without free stroke. The screw diameter is 60 mm and the material is GPPS with 0.5 wt% N_2 gas added into the melt at 20.7 MPa of pressure and 230 °C of melt temperature. Two different free strokes are selected for the test. One is the full shot size where the screw is in the full stroke with 0% free stroke. Another is the 90% of full shot size that the screw is in the position with 10% free stroke left. When the melt pressure is kept 20.7 MPa an emergency stop is pressed, that simulates the power loss unpredicted, and the barrel pressure is recorded with the time. One minute after power shuts down, the remaining pressure in the barrel decreases from 20.7 MPa to 13.4 MPa with the 0% free stroke, and it also quickly decays from 20.7 MPa to 6.3 MPa with 10% free stroke. In addition, the remaining pressure with 10% free stroke is only half of the one with 0% free stroke. It is recommended not to use full stroke of the reciprocating screw in the injection molding machine for microcellular molding.

The microcellular foam process utilizes the rupture disc or releasing valve in the pressure zone between the screw tip and the middle check valve or restrictive elements of the screw for unpredicted high pressure of more than 69–83 MPa (10,000–12,000 psi). However, this pressure-releasing valve is only used to protect the damage of the parts in the barrel assembly, not to protect persons working in the vicinity. It usually takes a long time to release the pressure in the gas-rich material completely in the barrel.

7.2.3 Microcellular Screw Tip

The screw-tip (also called a non-return valve) as the critical shut-off element of microcellular injection molding has three primary functions:

- Preventing a pressure surge in the middle of the screw during injection
- Maintaining a consistent shot size from cycle to cycle
- Maintaining the constant pressure during the screw idle period

Non-return valves are the key parts for the injection molding machine. It is mounted in the front end of screw as a tip with non-return function. There are many different designs for the non-return valves for regular injection molding process. Most of them are not functional parts for the microcellular process. Therefore, a quick closing, or pre-closing (also named automatic closing), screw tip is a mandatory requirement for the microcellular injection molding screw. Initially the pre-closing screw tips dominated the microcellular processing of injection molding until some failures occurred in special application cases. There is a new design of screw tip invented by Xu at Trexel in 2000, and it was released for the application in 2002. The unique feature of this screw tip is to have simple and reliable screw tip for a quick closing with a two-closing-stage principle. In this non-return valve there is a modified standard slide ring closing with two stages, pre-closing stage and final closing stage, which can be used either as a screw tip in the front of all injection molding screws or as a middle restrict element in all screws for conventional foam and microcellular foam processes. All screw tips successfully used for microcellular injection molding are discussed in detail below.

7.2.3.1 Pre-closing Screw Tip. Microcellular process requires either quick closing or pre-closing valves because the higher melt pressure must be maintained during the screw idle period and the entire injection stroke. In addition, a gas and plastic mixture with lower viscosity requires better performance of the non-return valve. The microcellular screw configuration with a middle restriction element makes the most non-return valves undependable for their performances. Here several typical pre-closing of the non-return valves used for microcellular processing are discussed in the following.

Figure 7.16 shows a typical non-return valve from Spirex with auto-shut function [22], When the screw rotates for recovery, the poppet “1” is forced open by melt pressure built in the screw to allow molten material to flow forward from the inlets of the valve body “2” to the outlet formed by the valve body “2” and the head of poppet “1”. Once screw stops turning, the poppet “1” retracts automatically by a spring “3”, the poppet head touches the seat of body “2”, and the outlet of valve is closed positively. It shuts off before

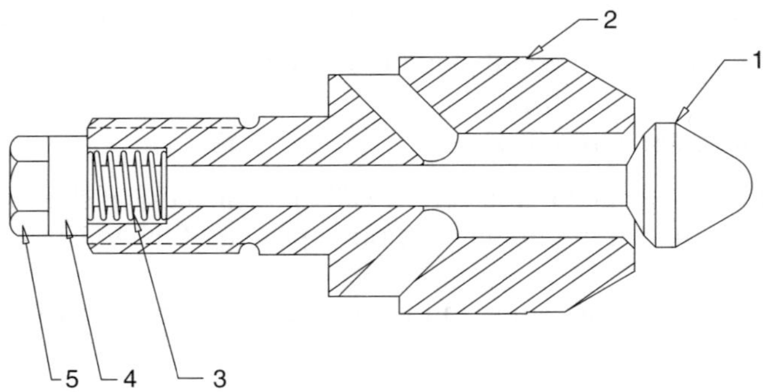

Figure 7.16 Auto-shut-off screw tip. (Courtesy of Spirex, U.S. Patent 5,164,207.) 1, Poppet; 2, screw tip body; 3, spring; 4, bearing; 5, lock nut.

injection and prevents back flow of molten material through this valve during both the screw idle period and the injection stroke. The processing drawback of this valve is a possible leakage through the gap between the shaft of poppet "1" and body "2" to the spring chamber that will cause the failure of the auto-shut off function. There are more issues during the application test, such as a possible breakage of body "2" from cold start, a wearing issue on the OD of valve body "2" because it turns with screw together, and the bending of shaft of poppet "1" during injection if the valve did not fully close. However, with careful design to take care the issues mentioned above, the auto-shut-off valve is still a useful screw tip for normal viscosity unfilled material processing.

There is a modified auto-shut-off screw tip (called a Posi-Trol valve) that is made from Md Plastics Inc. It has a ring that does not rotate with the screw so that the OD wearing problem is solved. The guiding tip with a spherical surface guides the ring centering with flexibility of rotation of the ring guided by the barrel. Two pins restrict the axial movement of the ring for the opening position, and it forces the shaft rotating with the screw together so that the nut to lock the spring position will not get lost. However, this design makes the shaft weak for torque, and either the pin or shaft in this section will break if the screw has a cold start. The strong disc spring closes the ring quickly once the screw stops turning. The main drawback of this valve is a possible leakage into the spring chamber (same as the problem of Spirex auto-shut-off valve above). This screw tip has excellent quick shut-off performance and the best sealing during the dead head test. However, it is a complicated design and is expensive for manufacturing and maintenance.

There are more non-return valves as auto-shut-off valves available in the market. One of the popular valves is the ball valve with a wet spring. It can close quickly and fully. The drawbacks are (a) valve OD wearing because of valve turning with a screw, (b) a complicated inside structure, and (c) restriction flow inside of a small ball valve channel.

Zeiger Industries developed a wet wire spring-loaded ring valve [23]. It closes the ring by a wire spring positively, but the ring rotates with the screw together, and the spring itself may have some dead spots to stagnate molten materials there At the high temperature above 435 °C, the spring may not function well.

Repeater valve [24] from U.S. Valves Inc. provides an automatic closing function from the melt pressure balance on the different areas in a piston. However, it only works in narrow viscosity range for certain area ratio of piston and has too much pressure loss across the piston during screw recovery. The tip body rotates with the screw together, and the OD of the body wears fast (same problems as the Spirex auto-shut-off valve, ball valve, etc.).

7.2.3.2 Two-Closing-Stage Screw Tip. A simple non-return valve (refer to Figure 7.17) is mounted in the front end of the injection screw to control the molten materials to flow in forward direction only. The non-return valve has a tip body "1", which is rigidly connected to the screw; a ring "2", which can move relative to the tip body "1", so as to (a) allow the molten material to flow in the forward direction when the ring is in the open position and (b) block molten material flow in the backward direction when the ring is in the closing position; and a rear seat "3", which is fixed between the tip body "1" and the front end of screw, to support the ring "2" in the closed position and form a seal surface between ring "2" and seat "3". A two-closing-stage principle is used here to design this screw tip. The first closing stage is called the pre-closing stage in that a narrow flow channel as a throttle (gap or orifice, etc.) will be set up. Once the valve finishes the first stage closing, the second closing stage begins. With the time of lasting, the second closing process is to keep going until the valve is closed completely.

The upper half of the section view in Figure 7.17 shows a full open position of the two-closing-stage ring design, and lower half of the section view in Figure 7.17 shows the full opening position of the regular OEM ring design. The rear seat needs to be modified with an extra closing step. The ring is the regular design without changing. The extra closing step creates a very narrow gap as the pre-closing distance or first closing-stage stroke. Regular close

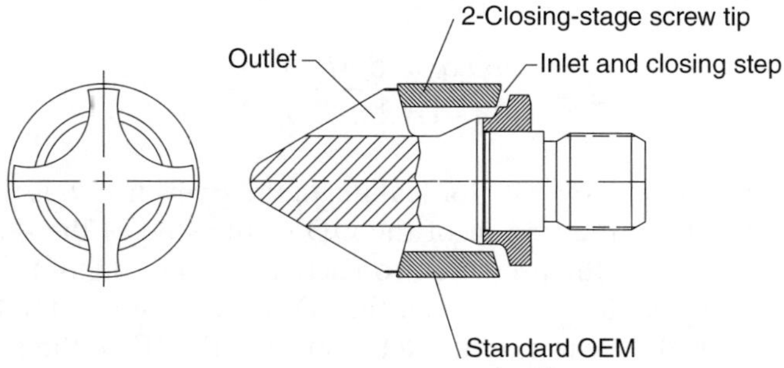

Figure 7.17 Two-closing-stage screw tip with modified rear seat. (Courtesy of Trexel Inc.)

TABLE 7.6 Output Rate of Plasticizing Unit with Different Screw Tips[a] and with a 60-mm Screw at 103 rpm, N_2 Gas in PP (MI = 35)

Melt Pressure (MPa)	V1 (kg/hr)	V2 (kg/hr)	V3 (kg/hr)	V4 (kg/hr)
10.345	34.24	36.10	33.27	38.68
13.793	30.00	36.80	30.10	35.62
20.690	21.60	24.32	21.78	27.60

[a]V1, auto-shut off valve of Spirex; V2, Trexel-modified two-closing stage valve; V3, Posi-Trol Md valve; V4, standard OEM 3 pieces ring valve.

stroke of all standard OEM designs is the full stroke of closing of the ring including both stages. It is a simple design except that either the ring or the rear seat needs to be modified.

The design theory is simple, too. First of all, the distance for pre-closing is very short, and it is about 20% of the regular full stroke of ring valve, which means that the pre-closing will be finished five times faster than the regular ring style valve fully closing. The opening area of the inlet of this valve is about 60–70% of metering across the channel area, which is good enough without the restriction problem for the free-flow area when the valve is open. In addition, the gap left during the opening of this screw tip is so short that the restriction effect is just like a hydraulic throttle valve with very little pressure loss of forward flow when the valve is open. Table 7.6 shows the output rate comparisons among all valves. The new two-closing-stage valve has the acceptable output rate compared to other non-return valves. The output rate of the two-closing-stage valve is higher than the output rates of all automatic shut-off valves, such as Spirex and Posi-Trol Md valves. However, the output rate of the two-closing-stage valve is about 7–13% less than the output rate of a standard ring valve (see Table 7.6).

This closing analysis can refer to the design in Figure 7.17. The leaking channel geometry can be simplified as narrow gap with the width πD_2 and height δ, which satisfies the condition $\pi D_2 \gg \delta$. The leaking volume rate Q_a can be calculated with the following simple formula:

$$Q_a = \frac{\pi n D_2 \delta_r^2}{2(2n+1)}\left(\frac{\delta_r \Delta P_r}{2m_T L_a}\right) \tag{7.19}$$

where Q_a is the volume flow rate for leaking in the screw tip, D_2 is the nominal diameter of the ID of the ring or of the OD of the step of the seat, n is the power law index, ΔP_r is the pressure drop across the the ring of the screw tip, L_a is the overlap axial length between the ID of the ring and the OD of the rear seat, and δ_r is the clearance in radius between the ID of the ring and the OD of the rear seat.

Compared to a standard ring valve, the opening of the ring before injection is at least 2 mm while the clearance for the two-closing-stage valve before injection is as low as 0.2 mm, which is about 10 times smaller. The output rate is proportional to the square of the clearance so that the leaking rate is about 1% of the regular leaking rate of standard ring valve. In this case this leaking may only cause initial drifting and then close fully after the ring moves more than the distance of the initial opening.

The pressure drop in the whole screw tip is simply the pressure difference on the downstream side of ring and the upstream side of ring. It will be increased, with the ring continually closing until it is fully closed. Also, the clearance is so small that even if it is worn out, the pressure drop is always big enough to push the ring continually closing, because the larger the pressure drop, the larger the force acts on the downstream of ring (left side of ring in Figure 7.17).

The disadvantage of the two-closing-stage screw tip is that it only works well for the machine with the position control method after screw finishes the recovery. The pressure control method will have an initial drafting issue of this kind screw tip. Therefore, the pre-closing screw tip will be the only solution for the pressure control machine.

If the machine has position control after screw finishing the recovery, there will be no problem at all because the pressure will be equalized quickly from the front of the screw tip to the middle of the check valve. Then, the screw just keeps the position and so does the ring, and the opening distance is kept until the pressure in the middle of screw becomes lower than the pressure in front of the screw tip. This pressure difference, then, becomes a driving force to close the rings in both the front tip and the middle valve automatically.

If the key parameters are designed properly, a good balance between free flow and quick closing can be satisfied. From the performance viewpoint, all other valves cannot be made to benefit both quick closing and free flow. The two-closing stage provides the following features:

- Quick pre-closing maintains the melt pressure in front of the screw and middle of the screw.
- The ring valve is guided by the barrel only and does not turn with the screw, so the wearing of OD is light.
- Final closing relies on the seat surface that is strong enough to support very high injection pressure so that there is no issue of breaking parts such as breaking shaft in the auto-shut-off valve of Spirex.
- There is no risk of breaking the part for the two-closing-stage ring valve when cold start occurs.
- The wearing parts such as ring or rear seat are standard parts and are highly exchangeable spare parts.
- There is no dead corner in the two-closing-stage valve, so there is no issue for color change or material degrading problems.
- It can hold any dead head test so that there is no danger of blowing off the rupture disc in the middle of the barrel even if the nozzle freezes.

- The two-closing-stage valve is either a non-spring-loaded valve or only a partially (only 20% of the fully stroke) spring-loaded valve. It will work anyway without a spring for a little initial drifting, so there is no risk of blowing off a rupture disc or of inconsistency processing if the spring fails to respond in time.
- This two-closing-stage ring valve is also the most reliable design for all materials processing with either leaking (low viscosity material) or wearing (highly reinforced material) issues.

Another important application for this new idea is the restriction element in the middle of the microcellular screw. Figure 7.15 shows the configuration of this two-closing stage of the middle ring valve in the microcellular screw. The middle two-closing-stage ring valve performs even better pre-closing and final closing for the middle valve in the microcellular screw because the rear side (upstream) of middle valve is opened to the hopper where the upstream pressure drops immediately after the screw stops rotating. Then, the downstream pressure (about the same as melt pressure in the front of the screw) of the middle ring valve quickly becomes a big driving force to close the middle ring valve. If the key geometric parameters are designed properly, a good balance between free flow and quick closing can be satisfied. From the performance viewpoint, all other valves cannot be made to benefit both quick closing and free flow.

7.2.3.3 Closing Sequence for Screw Tip and Middle Check Valve. There may be an issue of closing sequence between the screw tip and the middle check valve since there is always either a check valve or a restriction element [12–15] in the middle of a microcellular screw. It can be clearly explained by the pressure differences between P_s and P_t in Figure 7.8. Pressure P_s is the pressure in molten plastic near the gas injector. Pressure P_t is the pressure of accumulated material in front of the screw tip. It is well known that both pressures P_t and P_s must be maintained as the minimum requirement value to keep the gas in the solution without prefoaming. However, the pressure P_s must be lower than P_t after screw finishing recovery and before injection. It is the critical pressure difference to keep the closing actions in order that the screw tip closes first and the middle valve closes after. If the middle valve closes earlier than the screw tip does, the first pressure spike at the beginning of injection will be past through the screw tip and transfer the pressure surging to the middle valve. The result is that a rupture disc for safety will blow off immediately by this pressure surging, and the automation of production will have to stop. The only exception for this pressure difference rule is that there is a reverse channel restrictive element in the small injection screw, which is always open with small shearing clearance between the inlet channel and the outlet channel of the reverse channel section. The first pressure spike at the beginning of injection will pass through the gap of the reverse channel section and avoid the rupture disc blowing off.

A special experiment to measure the pressure P_s between the screw tip and the middle check valve (near the gas injector) is carried out in a 60-mm-diameter screw with both a ring-style screw tip and a middle ball check valve [25]. Two different rings are tested in the same screw tip. One is the regular OEM ring with 5-mm closing stroke. Another one is the two-closing-stage ring that reduces the closing stroke from 5 mm down to 1.5 mm. The processing material is PP (Montell, XMA6170P, MI = 35). The same 0.076-m/sec injection speed is used for both testes. The pressure drop rate across the ring of screw tip during the beginning of injection is compared to the pressure surge due to material trapped in the mixing section with the assumption of the middle valve closed first while the screw tip does not completely close at that time. The experiments show that the longer closing ring stroke (5 mm) of the regular OEM ring causes a pressure peak behind the screw tip during injection (see Figure 7.18). Then, this pressure surge is removed with the two-closing-stage screw tip reducing the closing stroke down to 1.5 mm. A pressure dip in Figure 7.19 indicates a good ring-closing performance. This experimental result verifies that the design of the two-closing-stage screw tip is better than that of the regular screw tip for the microcellular screw with the middle check valve.

Finally, the sequence of the front screw tip and the middle check valve can be summarized as follows:

- Closing in time order: screw tip–middle check valve. Then, the melt pressure will be higher in front of the screw tip than the one in the region between the screw tip and the middle check valve. In this way the front screw tip is truly at the closed position because the higher melt pressure in front of the screw tip can overcome the low pressure behind the screw tip and keep the closing tightly. However, the melt pressure in the region between the screw tip and the middle check valve must be maintained a

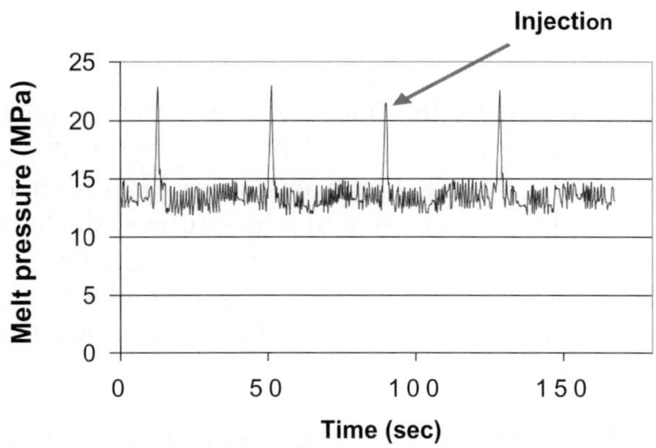

Figure 7.18 Pressure profile in ring-style screw tip with 5-mm closing stroke during injection [7].

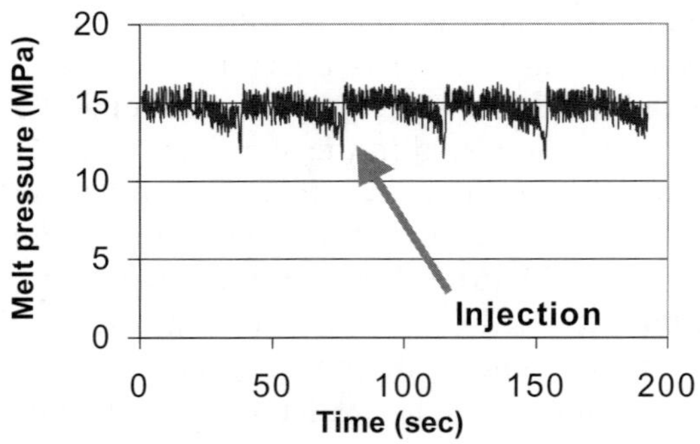

Figure 7.19 Pressure profile in ring-style screw tip with 1.5-mm closing stroke during injection [7].

TABLE 7.7 Configuration of Screw Tip and Middle Check Valve

Option Number	Screw Tip	Middle Restrictive Elements	Comments
1	Two-closing-stage ring screw tip	Two-closing-stage ring	It is the best configuration with position control in screw idle period
2	Two-closing-stage ring screw tip	Reversal channel	Recommended for small screw
3	Pre-closing non-return valve	Two-closing stage ring	Recommended with pressure control in screw idle period
4	Pre-closing non-return valve	Ball valve	Low-viscosity material
5	Pre-closing non-return valve	Reversal channel	Good for all applications

minimum pressure that is high enough to maintain the single-phase solution but lower than the pressure in front of the screw tip.

- Open in time order: middle check valve–screw tip. It is important to keep the front screw tip closed to avoid the pressurized single-phase solution in the region between the screw tip and the middle check valve leaking through the screw tip before the melt pressure builds up high enough. It also needs the machine designed with the melt pressure P_t and P_s always maintained even when the injection is finished.

Based on the discussion of the technical details, the practical configurations of the screw tip and the middle pressure-restrictive elements are recommended in Table 7.7.

7.2.3.4 System Analysis for the Injection Molding Screw and the Screw Tip. To analyze the performances of plasticizing plastics and gas diffusing into the melt in the plasticizing unit, the screw and screw tip (include the middle valve) need to be treated as a system. Reference 25 has briefly introduced the systematic analysis methodology and presented the example showing how to optimize the performance of the screw and the screw tip system for microcellular processing. It is the most complicated system with these components together in the same injection molding machine since there are screw, screw tip, and middle valve in the barrel as a typical microcellular processing tool. The modeling of these components together as a system can be found in reference 25.

7.2.4 Clamp Unit

The clamp unit of a microcellular injection molding machine has the same design as the regular injection molding machine except for low tonnage requirement. There are more special requirements in the clamp unit for the microcellular process. One is the mold crack opening for foaming (reversal coining, or reversal injection compression). Another is sequential injection compression for venting. The clamp unit not only accommodates the microcellular mold but also provides all motions needed for operation of the molding process. Since the microcellular foam process does not need to fill or pack the mold cavity to prevent sink marks and has a low melt viscosity, clamp tonnage can be reduced by up to 40–60% when compared to a solid molding [5]. Cavity pressure is the best parameter to explain why there is such a large reduction in clamp tonnage. As an example, a PBT part ran on a microcellular machine with and without gas. The maximum cavity pressure measured for the solid part is 105 MPa, and the maximum cavity pressure measured for microcellular foam part is only 45 MPa. The clamp tonnage of the microcellular foamed part was reduced 57%. In most applications, the actual clamp tonnage reduction depends on the mold design, flow ratio, viscosity of material, gas percentage, gas type, linear injection speed, cavity pressure, and mold and melt temperatures.

7.2.4.1 Platen Design. The reduction in required clamp tonnage of the microcellular process allows one to design an injection molding machine with either (a) larger spacing between tie bars or (b) thinner platens. Below are some calculations of required platen thickness and tie bar spacing for microcellular molding.

Assume that a platen #1 of a typical molding machine and a platen #2 of a microcellular machine have the following assumptions:

1. The mold load uniformly acts on the center part of the platen (see Figure 7.20).
2. The platen is a simple beam resting on supports at the ends (see Figure 7.20).

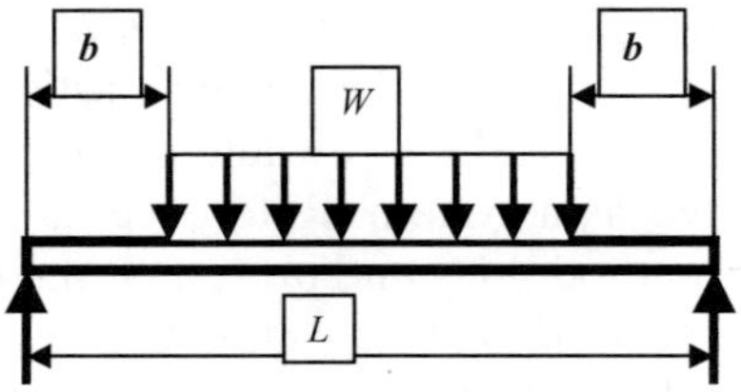

Figure 7.20 Simple beam for clamp platens. W, total load on the platen; L, length between supports of the platen; b, unloaded length of beam at each end.

3. The ratio of loaded length of mold to the total length of supports is fixed at about 0.65.
4. The deflection f of the platens is no more than 0.2 mm for 1-m distance.
5. The platens for both are the same material and have the same moment of inertia of the beam section.
6. The platens for a standard machine and a microcellular machine have different total length of supports L_1 and L_2 and different total load W_1 and W_2.

Then, the relationship between the ratio of loads and the ratio of total length between supports is (see Appendix B) [5]

$$\frac{W_1}{W_2} = \left(\frac{L_2}{L_1}\right)^3 \tag{7.20}$$

where W_1 is the load on the standard platen, W_2 is the load on the microcellular platen, L_1 is the length between supports of the standard platen, and L_2 is the length between supports of the microcellular platen.

For most microcellular foam molding, the clamp tonnage of the machine required is only half of the solid molding. Therefore, the distance between the tie bars for microcellular platens can be made 26% greater. It also means that the projected area for a microcellular mold can be 59% larger than the one for a standard mold. Another option is that the platen thickness can be reduced by 20% over the standard platen thickness.

7.2.4.2 Maximum and Minimum Mold Area. To control the platen flatness or deflection during mold filling under tonnage, a minimum mold area is specified for each clamp unit based on the distance of tie bar and maximum tonnage. It is just the defection calculation based on the work load W and the distance between support L. The microcellular injection process will not need maximum tonnage, so the minimum mold area can be reduced accordingly. However, in most cases a maximum mold area needs to be checked to make full use of the platen area for the big parts molding without buying a larger

TABLE 7.8 Maximum Mold Base Area Versus Clamping Force for Microcellular Machine and Standard Machine

Clamp Force (KN)	Standard Machine Maximum Mold Base Area (m^2)	Microcellular Machine Maximum Mold Base Area (m^2)
500	0.072	0.115
1,000	0.145	0.23
4,000	0.5	0.8
20,000	2	3.18

tonnage machine. It is the main advantage of a low tonnage microcellular clamp unit. Table 7.8 shows a recommended maximum mold base area relative to the maximum clamping force for both standard machines [26] and microcellular machines at the same unit deflection (0.2-mm deflection in 1-m span). From assumption 3 above, if the ratio of the loaded length of mold to the total length of supports is fixed at about 0.65, the maximum area of machine clamp platen for each maximum clamping force can be calculated by multiplying the mold base area data in Table 7.8 by 2.37. For the machine without a tie-bar, the benefit is even greater because the microcellular foam process allows the molder to make full use of the platen area.

7.2.4.3 Mold Opening Force. The mold opening force for microcellular injection molding usually does not change much. However, it may increase in some cases, instead of reducing. It occurs when the cells pressure does not release fully and the part expands in the mold to create friction between mold cavity and the part. This force usually is zero because plastic material shrinks a lot, so the cavity surface may not have contact with the part when the mold opens. Overall, the mold opening force for a microcellular part shall be less than that of a solid part because the most forces to be overcome by an opening force are mold deformation during high tonnage compression, vacuum in mold, local gate area deformation during high packing pressure, and so on. Therefore, the mold opening force for microcellular molding shall generally be reduced in most cases.

7.2.4.4 Special Functions in Clamp Unit for Microcellular Injection Molding. A newly developed technology known as Dolphin Skin has been an attractive recent topic. It uses reversal coining technology making a smooth surface on one side and soft-touch surface with foam on the another side. It is a dashboard panel in passenger car. The schematic layout of this molding is displayed in Figure 7.21. There are two sequential shots from two horizontally opposed injection units. The first stage of processing is for a glass-filled PBT/ASA blend (Ultradur S4090 IGX from BASF) that forms a solid smooth skin. After delaying for a while, the skin material is solidified enough to support the next shot overlaps on it. The second stage is the shot of gas-rich

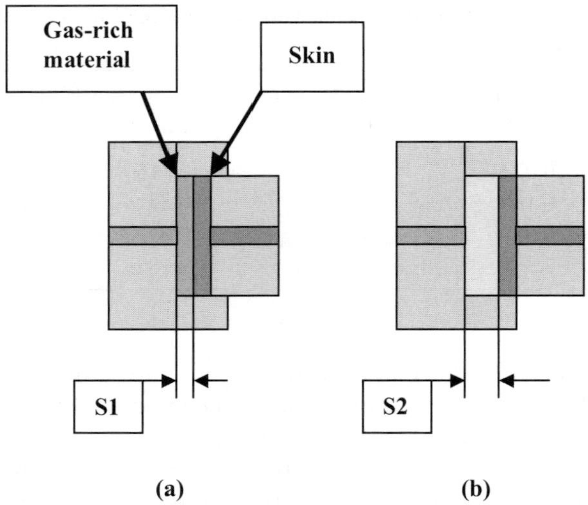

Figure 7.21 Schematic of the reversal coining mold layouts. **(a)** Mold closes with S1 thickness of gas-rich material without foaming. **(b)** Mold cracks open with S2 thickness of gas-rich material that expanded with foaming.

material that is overmolded on the first shot skin layer. The foam material is a special polyester (Pibiflex, a readily foamable thermoplastic polyester from P-Group). Both shots have 100% mold filling first, so there is no foaming at this cooling stage, and the gas-rich material keeps thickness S1. After the skin is solidified enough, the gas-rich material is still in the melt state since the delayed shot time. Then, the clamp unit provides the decompression movement to crack open from the S1 position to the S2 position, S2 > S1. Then, the gas-rich material behind solid skin expanded to fill the extra space of (S2–S1) by foaming uniformly. In this way, the foamed side will a have smooth surface as well since the skin is formed before foam. This will be discussed further as the one of the special processes in Chapter 8, and in Chapter 11 it will be studied for its advantages.

7.2.4.5 Energy Saving from Low Tonnage. Energy saving from the benefit of low tonnage for microcellular injection molding was an interesting topic regarding the advantage of the microcellular process. It definitely reduces the cost from thin platen, or large mold used in small tonnage clamp unit. In addition, the operation cost to run the low tonnage clamp unit is additional energy saving. From the part-wearing point of view, the low tonnage will increase the lifetime of the critical parts, such as tie-bar and platens, in the clamp unit. However, the energy saving is not enough from the clamp unit itself since the hydraulic system for the whole injection molding machine is designed with the most requirements from the injection unit. It is because injection pressure peak and high injection linear speed are the highest peak of energy to be used in the whole cycle, and it must be prepared by a machine power source. It may be a good idea to use an electric screw drive and an accumulator-bottle-modified hydraulic system to save more energy.

7.2.5 Microcellular Injection Unit

The layout of injection unit of microcellular is similar to the standard injection unit since it just needs high injection speed with lower injection pressure requirement. The related main components in the injection unit are an injection cylinder, a hydraulic motor, or an electric motor; thrust-bearing, normal specifications of the injection unit can be used for microcellular processing. However, because the viscosity of the melted polymer decreases once the single-phase solution is formed, and mold only fills 80–90% of full solid shot size, the average injection pressure of microcellular molding is up to 60% lower than the injection pressure for normal injection molding [5]. A more important feature of microcellular injection molding is that the injection peak, or packing pressure, is not necessary. This is critical for the packing stage to be eliminated, and the result is no residual stress in the microcellular part. However, for microcellular foam molding, a higher injection volume rate is preferred, but not always required, to improve weight reduction and good cell structure. Another basic microcellular processing requirement is that the back pressure acting on the screw tip is always maintained no matter when the screw is rotating or staying idle. In other words, the injection unit is always under pressure. In the following sections, we will discuss the theory and practice behind the preference for higher injection speeds and the design issues of injection unit, although some of the details have been discussed in Chapter 6 from a processing point of view.

7.2.5.1 Special Relationships with Screw and Nozzle (or Valve Gate). As we discussed in the screw design, the precision of injection action relies on a screw-tip shut-off element. In addition, the injection action must be controlled with the right sequence between injection and nozzle or valve gate opening.

1. Relationship with Screw Tip. From an injection point of view, the screw-tip shut-off element has two primary functions:

- Prevent a pressure surge in the middle of the screw at the beginning of injection by automatically separating, via a non-return valve, the screw tip from the middle section of the screw.
- Maintain a consistent shot size and minimum molten plastic pressure from cycle to cycle.

The quick shut-off, or pre-closing screw tip, is the key factor for the constant injection shot size and the same quality of microcells in the part. For the ring-style valve, the traditional method to quick close the screw tip through decompression before injection must be eliminated from consideration since it will immediately damage the single-phase solution with decompression to give extra space allowing pre-foaming. On the other hand, there is another way to quick close the screw tip through the fast injection at the beginning. For the low-viscosity material to be used for microcellular processing, the initial high injection speed does help to close the ring in the screw tip quickly.

2. Relationship with Screw. It is usually recommended that the diameter of the microcellular screw be selected as large as possible. It is because of the low injection pressure required for microcellular injection molding. The regular intensify ratio of the diameter of injection cylinder to the diameter of screw is about 8–12. The big diameter of the screw requires a short recovery stroke that makes microcellular processing a little easy. It is because the short recovery stroke of a big screw is not only good for SCF dosing that needs only one SCF injector in the barrel but also better for fast injection. The major benefits of a short injection stroke with a big screw are:

- High injection volume rate that is essential requirement for the microcellular injection molding
- Short injection time for better uniform cell growing history in the part

One more thing that we need to be aware of is the high back-pressure setup for screw recovery. It is the mandatory requirement for making a single-phase solution. The range of the back pressure is approximately from 6.9 MPa (1000 psi) of minimum to 34.5 MPa (5000 psi) of maximum as the melt pressure in front of the screw tip. This may generate extramechanical heat in the barrel. On the other hand, it increases the axial load on thrust bearing during recovery. Usually, the thrust bearing behind the end of the screw is designed with enough load capability and lift time.

3. Relationship with Shut-Off Nozzle and Valve Gate. The most important issue for the shut-off nozzle is to maintain the pressure in an entire period of cycle except for the injection and holding stages. Therefore, the sequence of the actions between injection and nozzle opening must be in the following order: injection first and nozzle opening second, with approximate delaying open action of about 0.5 sec. For a spring-loaded shut-off nozzle and a spring-loaded valve gate, this closing order is automatically maintained. However, for power actuated (hydraulic or pneumatic cylinders) the controller of the machine must be set up in the correct order above. Otherwise the pressurized gas-rich material in the nozzle or valve gate will be free to shoot into the mold at the beginning of nozzle or valve opening without any necessary pressure to follow in time.

7.2.5.2 High Injection Volume Rate with Low Injection Pressure. Injection speed controls homogeneous nucleation and smaller cells. Homogeneous nucleation of the microcells occurs when conditions are provided such that it is easier for the new cells to nucleate than for existing cells to grow. During large pressure drops, the single-phase solution will experience a decrease in solubility of the SCF in the polymer. The SCF, coming out of solution, can either go into existing nucleated cells or help to nucleate new cells. Microcellular foam requires that the largest number of cells be created during the pressure drop. To create a large number of cells, conditions have to be such that the SCF coming out of solution prefers to form a new cell as opposed to migrating into an existing growing cell. The first condition for homogeneous nucleation to occur is that the time required to nucleate a nucleation site must be much, much faster than the time required for the SCF to diffuse into an existing cell. Secondly, the distance the SCF travels to go into a new cell must be much smaller than the spacing between stable growing nuclei [11]. In a practical sense, the only method to accomplish this is to ensure that a very high plastic pressure drop rate occurs during the injection phase of the cycle. To go even further, it is possible to link the pressure drop rate to injection volume rate and the nozzle, runner, and gating design of the molding system. Using this theory, it can be quantitatively shown how the pressure drop rate concept effect the nucleation process, whereas previous models did not always have a full explanation of how to apply the theory to actual injection molding applications [2, 27–30].

Park et al. [30] found minimum pressure drop rates required for the creation of microcellular foam, in fully saturated materials. For HIPS (high-impact polystyrene) with 10 wt% CO_2 gas, the cell density is increased from 10^8 cells/cm^3 to 10^9 cells/cm^3 with the related *dp/dt* increased from 0.18×10^9 Pa/sec to 0.9×10^9 Pa/sec [30]. Theoretically, a cell density of 10^9 cells/cm^3 will produce cell sizes of about 10 μm [11]. Therefore, for HIPS saturated with SCF, the required pressure drop rate (*dp/dt*) is about 10^9 Pa/sec for 10-μm cells. It is important to note that at varying saturation pressures for the single-phase solution, different pressure drop rates may be required to achieve homogeneous nucleation [30].

In practice, the required injection rate to meet the minimum *dp/dt* is related to a number of parameters, including the mold gate size, material, gas type and percentage, and melt temperature. Based on actual molding results so far, the minimum required *dp/dt* is about 10^9 Pa/sec or higher. However, 10^9 Pa/sec of *dp/dt* is actually high enough to produce homogeneous nucleation for most thermoplastics. It is important to note that when this minimum pressure drop rate is achieved for homogeneous nucleation, the nucleation process is still considered an additive process of both homogeneous nucleation and heterogeneous nucleation. In other words, once homogeneous nucleation is achieved, you will have microcellular foam and the heterogeneous nucleation created by fillers, and the polymers multiphase nature will only add more to the total number of cell sites [31–33]. Therefore, once a pressure drop rate of 10^9 Pa/sec is achieved, microcellular foam is expected for most thermoplastic materials and applications (considering fully saturated polymer solutions).

Most practical cases show that any heterogeneous nucleation will help significantly for nucleation. Therefore, the real pressure drop rate of 10^9 Pa/sec may not be necessary for the filled material.

Linear injection speed also influences the density distribution that is discussed well in Chapter 6. The result in the injection molding shows that the highest density is in the gate area, and the lowest density is at the end flow. Obviously, higher linear injection speed results in a more uniform density distribution. On the other hand, slow linear injection speed fills the mold more slowly, which allows a thicker skin to form and thus a narrower flow channel results. This tends to create overpacking near the gate area and underpacking near the end of flow.

In addition, linear injection speed is an important factor for controlling the surface finish of the microcellular molded part. Linear injection speed needs to be large enough to keep the mold filling time at a minimum so that the gas in the flow front is not allowed to escape. If the gas bubble has enough time to become a big bubble, it will break through the front of the plastic melt flow. The hole left by the bubble is elongated by the shearing region, and it is moved to the mold surface by the differences in shear rate (largest near the mold surface and zero in the center layer of the melt flow) [34, 35].

For most RS injection molding machines the standard injection volume rate (without modification of injection unit and hydraulic system) is sufficient so that the available *dp/dt* in a single nozzle tip (regardless the mold gates) is higher than the minimum *dp/dt* of 10^9 Pa/sec (see Table 7.9). There are three nozzle tips with different orifice diameters listed in Table 7.9. They are the most popular sizes used for the injection molding process except for the combination of a 30-mm screw diameter plus a 9.525-mm nozzle tip orifice that does not match the *dp/dt* requirement. However, the fact is that the 9.525-mm nozzle tip orifice will never be used in a 30-mm screw, and this data listed in Table 7.9 is just showing the possible wrong combination to have the wrong *dp/dt* rate. Therefore, the combinations in Table 7.9 cover the real range of screws, nozzle tips, and reasonable linear injection speeds that are available in all OEM machine catalogs.

Figure 7.22 shows the comparison between standard OEM injection volume rate and empirical injection volume rate that is necessary for polyolefin. The polyolefin material is tough material to make microcellular part. Therefore, the empirical injection volume rate for microcellular polyolefin is a baseline to define the necessary injection volume rate for microcellular processing. At least the standard injection volume rate above an 80-mm-diameter screw is already larger than the empirical injection volume rate. However, even if the injection volume rate of all screw diameters is less than the empirical values, they are so close each other. The conclusion is that the standard injection volume rate is good enough to handle most of the materials to be processed with microcellular cell structure.

On the other hand, there are some cases when the regular machine cannot provide a high enough injection volume rate for a special material require-

TABLE 7.9 Screw Diameter Versus Nozzle Tip Orifice with Standard OEM Injection Volume Rate (Nozzle #1, 3.175-mm Orifice; Nozzle #2, 6.35-mm Orifice; Nozzle #3, 9.525-mm Orifice) [5]

Screw Diameter (mm)	*dp/dt* in Nozzle #1 (Pa/sec)	*dp/dt* in Nozzle #2 (Pa/sec)	*dp/dt* in Nozzle #3 (Pa/sec)	Linear Injection Speed (mm/sec)
30	4.4E11	6.9E9	6.0E8	157
40	1.4E12	2.2E10	1.9E9	157
50	1.8E12	2.8E10	2.4E9	114
60	2.2E12	3.5E10	3.1E9	89
70	4.1E12	6.5E10	5.7E9	89
80	7.1E12	1.1E11	9.7E9	89
90	1.1E13	1.8E11	1.6E10	89
105	2.1E13	3.3E11	2.9E10	89
120	3.6E13	5.6E11	4.9E10	89
135	5.7E13	9.0E11	7.9E10	89
150	8.7E13	1.4E12	1.2E11	89
160	1.1E14	1.8E12	1.6E11	89
170	1.4E14	2.3E12	2.0E11	89
180	1.8E14	2.8E12	2.5E11	89

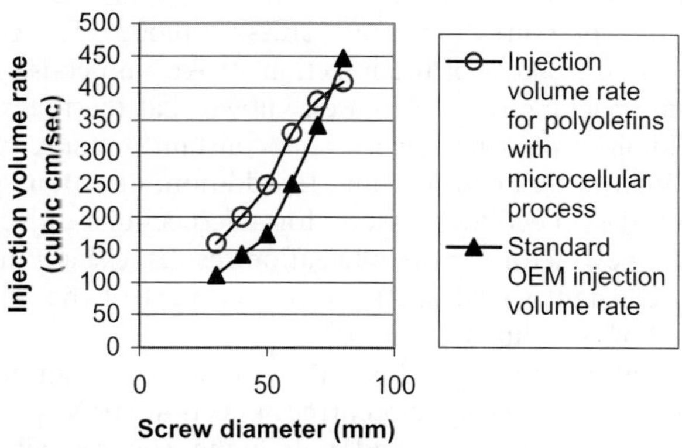

Figure 7.22 Comparison between standard OEM injection volume rate and required injection volume rate for microcellular polyolefin.

ment. Then, the hydraulic accumulator bottles are the best solution to add this higher injection volume rate that is beyond the standard OEM capacity. The technical details of a hydraulic accumulator bottle solution will be discussed in the section on hydraulic unit design.

7.2.5.3 Injection Linear Speed Profile. The gas-rich material under pressure will have significant stored energy that can be calculated with Equation

(7.17). Therefore, a special feature of microcellular injection molding is the initial injection shot just by this stored energy. It is a kind of surging of pressurized mixture of gas and molten plastic once the nozzle or valve gate is open. This is the evidence to eliminate the slow beginning of injection speed. With the known stored energy from Equation (7.17), the possible air shot speed can be estimated by Equation (6.23) in Chapter 6. On the other hand, if the injection speed is too fast at the beginning, the jetting is possible. The adequate initial injection speed is one of the key factors for the several possible defects:

- The wrong initial injection speed may cause the big swirl of gas on the gate area.
- Too fast initial injection speed results in jetting in the mold that may cause a weak spot in the part.
- The initial injection speed must catch up with the possible initial pressurized gas–melt surging to avoid the interruption of smooth continually injection.
- Delaying the nozzle or valve gate opening is the effective way to avoid trying the best initial injection speed, and a timer to control the sequence between the opening of a nozzle or a valve gate and the initial injection action is required.

The regular injection control is either constant speed (usually regenerative mode) or constant pressure (maximum pressure mode). The constant speed control will be used in most of the injection stroke, and constant pressure is only for the final packing stage. It is well known that the foam process does not need a holding or packing stage, so the constant pressure mode may not be needed in microcellular processing. In addition, injection speed is more important than the injection pressure for microcellular injection molding because it is the key factor for the nucleation results. Usually, it makes more sense to use a constant speed control mode in most of the injection stroke during a normal microcellular process.

Once the initial injection speed is set up to match gas surging, the rest of the injection stroke can set up the correct injection speed profile for some special requirement of nucleation and mold filling. Theoretically, the injection speed profile must satisfy the following criteria:

- Overall injection time must be in the limit of gas escaping time in the free-flow front.
- The pressure drop rate through the nozzle or valve gate keeps the same value, or varies in a certain range, in the whole injection stroke.

For the first criterion, the total injection time can be controlled in the limit based on the flow ratio and mold geometry. However, the second criterion is very complicated since it related the pressure drop rate needed to be kept the same from the beginning to the end of mold filling. The pressure increases with the increased mold filling volume in the mold. The most efficient way is to record the dynamic pressure profile from the controller. Then, find out the pressure at each position of injection stroke to determine the injection speed profile with the calculation from Equations (7.1)–(7.3). A successful injection speed profile of fast–slow (fast speed of 200 mm/sec in the first 67% of stroke, and slow speed of 3 mm/sec in the final 33% of stroke) makes a sample with nice foam in the whole part [36]. The other samples with all fast injection speed or with too short a stroke with slow injection speed could not make good uniform foaming in the whole part. There are many reasons to be able to explain this result. One of them is because the gate area is pressurized the most at high injection speed. If more extra space in the mold is left for the foaming expansion, it may not only give more free space for foaming but also reduce the pressure on the gate area. On the other hand, since the foaming in the final 33% of stroke will have enough time to expand in free space, at the same time very slow injection speed provides the chance to release more pressure in the gate area by cell growth and push materials flowing inside of the mold. In other words, the pressure release rate from foam processing may be quicker than the pressure buildup rate in the gate area at such small injection speed.

7.2.5.4 Holding Stage and Packing Pressure. The holding stage is not necessary for microcellular injection molding. Then, packing pressure is also not needed. However, sometimes the machine needs a very short time of holding stage to enable the hydraulic system to smoothly transfer from injection stage to holding a constant melt pressure in front of the screw tip before screw recovery stage begins. On the other hand, the holding time saving is one of the cycle time reductions that have been discussed in Chapter 6.

Packing pressure is not necessary either since the mold filling only needs about 80–90%, depending on the weight reduction target. There is a new idea to reduce the pressure in the gate area to get more cells in gate area. It is a kind of decompression from the screw moving back. Then, the shut of nozzle or valve gate must stop as soon as possible after a certain stroke of decompression to avoid further material flowing back from the mold gate area. This action must be finished quickly before there is too much of a pressure drop in the residence material in front of the screw tip. In other words, some minimum pressure will be kept as the low limit of the decompression pressure. The new pressure building up must be acted upon immediately to keep the pressure high again in front of the screw tip before the next recovery action begins. It may be referred to as reversal packing process. If the mold maker can make some releasing elements in the gate area to release the gate pressure, it will work as well to create more foam in the gate area. This is a special action required for microcellular injection molding as the decompression immediately after final injection. Therefore, It has not been applied in the machine control yet. However, the result of a slowdown of injection speed in reference 36 already indirectly verifies this new idea.

7.2.5.5 Energy Saving from Low Injection Pressure and Zero Packing and Holding Stages. The real energy saving is from the injection unit for the machine capable of performing microcellular injection molding. The data in Table 7.10 represent the energy saving analyses carried out in a typical modern full hydraulic clamp unit of a injection molding machine. For example, the injection pressure for microcellular processing is reduced as high as 50% with the injection speed unchanged. The power consumption in the injection unit (13.5% of total energy) is significantly larger than the clamping force buildup (3.1% of total energy), so the energy saving from injection unit will be significant as well. In addition, a hold stage is eliminated so that it has 6.9% of total energy to be eliminated. The detailed microcellular processing energy saving for this machine with the same mold is listed in Table 12.2.

In most toggle clamp units of the injection molding machine, the injection unit takes most of power in a very short time. The total power of the machine is determined by this power accordingly with an allowance of 50% overloading the electric motor in a short time. Now it may be not necessary to prepare the total power based on the injection peak load for the whole machine. There will be more energy saving from this kind of machine with microcellular processing. Therefore, the low injection pressure may involve an even higher

TABLE 7.10 Energy Consumption and Time for a Regular Injection Molding Machine, 1500 metric tons, 160-mm Diameter, 24:1 *L/D* Screw

Actions of Machine	Time Analysis (sec)	Energy (Wh)	Distribution Percentage in Total Energy Consumption (%)
Cores in	0.7	3.0	0.4
Closing	1.70	69.97	8.3
Locking	0.50	11.38	1.3
Clamping force buildup	0.60	26.19	3.1
Carriage unit forward	0.50	5.00	0.6
Injection unit	3.50	114.05	13.5
Post pressure unit	5.00	58.09	6.9
Feeding unit (hydraulic)	18.00	413.84	49.0
Cooling time	20.00	20.00	2.4
Clamping force reduction	1.40	9.72	1.2
Unlocking	0.80	5.56	0.7
Opening	1.60	59.10	7.0
Cores out	1.00	4.34	0.5
Ejector forward	1.00	2.89	0.3
Demolding	4.00	40.00	4.7
Ejector back	0.50	1.45	0.2
Total	60.8	844.58	100

percentage of saving compared to low clamp tonnage saving since the toggle itself is an energy-saving system. More details will be discussed in Chapter 12.

The cycle time will be reduced without the packing stage. In addition, the cooling time and injection time are usually reduced as well. The details of cooling time reduction are discussed in detail in Chapter 6. It is also a big cost saving that will be discussed in Chapter 12.

7.2.6 Microcellular Hydraulic System

The hydraulic and control system for the microcellular foam machine is designed with two mandatory requirements. One is a controllable back pressure during screw recovery and SCF dosing, along with a constant back pressure in the idle period. Another is a minimum injection speed, as previously discussed. On the other hand, the control system must match the SCF dosing requirements to open the gas injectors with a sequence according to the position of the wiping section inside of the screw. The hydraulic screw drive or electric screw drive must prepare the aggressive processing setup, which is usually high screw rotation speed to finish the screw recovery and SCF dosing by the end of the cooling period. It is critical to run the microcellular processing with the advantage of the cycle time reduction setting up every action under the best conditions. In addition, the hydraulic system of microcellular processing must be designed based on the safety guidelines recommended by SPI.

7.2.6.1 Hydraulic Back Pressure Control in Barrel. A foaming system will have special hydraulic design for the hydraulic pressure to be kept constant to maintain the minimum pressure required for single-phase solution after the screw stops rotation. It is usually above 6.9 MPa (1000 psi) and up to 34.5 MPa (5000 psi), depending the material and gas wt% in the plastic melt. On the other hand, a controllable screw back pressure during screw recovery is still required for microcellular injection molding.

1. Hydraulic Back-Pressure Profile. The hydraulic back-pressure profile for microcellular processing is available in a manner similar to that of a standard injection molding machine. We can be set up as many as 10 different backup pressure profiles based on the material processing and screw design to make homogenous single-phase solution and precise shot size. In most cases, an SCF dosing needs to have the same back pressure set up in the whole screw recovery stroke (see Figure 7.9) since the constant SCF dosing does not like the melt pressure varied during the whole SCF dosing stroke. However, if the screw geometry generates a pressure variation in the SCF doing position, such as reversal channel screw design, then the back-pressure profile needs to be set up accordingly. The pressure profile in the fixed SCF injector position for the reversal channel screw in the whole of screw recovery period shows a continually increasing pressure profile with this screw in the whole screw stroke. If the pre-set-up SCF dosing pressure is only 0.35 MPa (51 psi) over the melt pressure, then the melt pressure at this position varies from 20.7 MPa (3000 psi) up to 23 MPa (3335 psi) in the whole screw stroke. The 2.34-MPa (339-psi) pressure increase definitely not only decreases the SCF dosing rate continually but also stops the SCF from dosing somewhere too early since the melt pressure increasing from the screw itself is already over the set up maximum SCF pressure 21 MPa (3045 psi) difference between SCF and melt pressure. However, if the back pressure is set up with continually reduced back pressure in the controller according to this known pressure profile, then, the horizontal pressure profile can be achieved like in Figure 7.9. Therefore, this set up profile of back pressure in the regular hydraulic control system is still necessary for microcellular injection molding.

2. Constant Hydraulic Back Pressure During Screw Idle Period. In microcellular injection molding, the hydraulic system must be modified to maintain the constant back pressure after the screw has completely recovered and is sitting idle to wait for injection. The hydraulic system needs either a proportional directional control valve or a servo-directional control valve to keep the final shot size the same from cycle to cycle and maintain it during the screw idle period. There are several different hydraulic systems to implement this requirement, as discussed below.

A position control of the hydraulic system is the simplest way for a constant hydraulic back pressure during the screw idle period. It just maintains the screw position after the screw finishes the recovery. In this way, it will maintain both screw position and the back pressure in the acceptable range of variation in the screw idle period. In fact, this hydraulic control is not only economical but is also a reliable solution to maintain the pressure in single-phase solution. The working principle is that the servo valve controls the screw final recovery position very precisely. The initial position of the end of screw will not be maintained when the screw stops rotation. The screw position will be pushed backward by gas pressure in the single-phase solution accumulated in front of the screw tip. Then, the position deviation will be immediately corrected by the servo valve through the injection cylinder forward movement to push the screw back to preset position. This action will automatically recover the back pressure that needs to be maintained. The final balanced back pressure may not be exactly the same as the preset value but will be very close. Once the pressure loss occurs because of position control final balance, this pressure loss can be added into the preset pressure to finally maintain both back pressure and position without any compromise. This position control method is the best one used for most small tonnage machines of microcellular injection molding.

Another common approach is the direct pressure control to maintain the pre-set-up constant back pressure exactly. It works for the microcellular processing because the constant back pressure is required. The difficulty of this method to maintain the pressure after screw recovery is from instant sealing

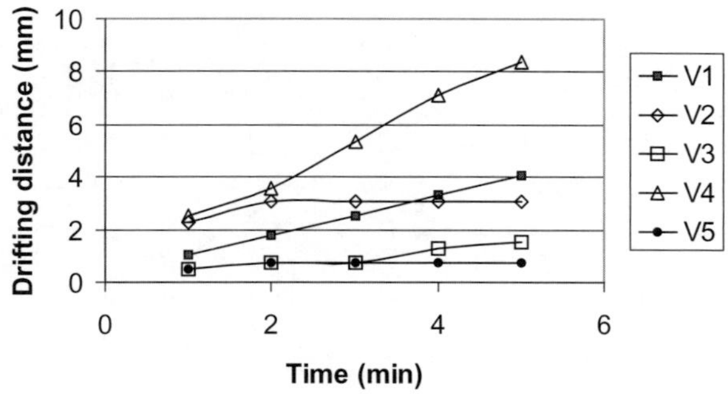

Figure 7.23 Comparison of drifting distances among the five screw tips at dead head test. (Courtesy of Trexel Inc.) V1: Auto-shut off valve. V2: Trexel-modified two-closing-stage valve V3: Posi-Trol Md valve. V4: Standard OEM three-piece ring valve. V5: Auto-shut off valve with piston seal ring in the valve body.

of the screw tip. Once the screw finishes the recovery, the screw tip must be closed immediately without any leaking in the screw tip under the set up pressure. Otherwise the screw position will drift forward because some leakage through the screw tip and the shot size is reduced. Then, the machine will be shut down once the deviation of the screw position changed beyond the limit. The drifting experimental results for typical microcellular screw tips in Figure 7.23 show that there is no perfect solution from the ring style of the screw tip for a strictly pressure control hydraulic system. The acceptable solution is to set up the wide range of screw position deviation allowance (about 2-mm maximum deviation allowance). It is still reasonable for the microcellular part as long as the drifting distance needed to pre-close the non-return valve is constant every cycle. With this special hydraulic system and control parameters re-setup, all screw tips in Figure 7.23 are qualified except for an OEM standard ring-style screw tip V4.

On the other hand, the position control of screw recovery process will not have drifting problems for all screw tip tested in Figure 7.23. It can tolerate more screw tip wearing and sealing issues than pressure control method.

Engel developed a safety system that was patented in 2002 [37]. It provides a process and an apparatus to avoid breaking down single-phase solution during a safety guard opening to remove a sprue or take out parts from the mold while maintaining high safety standards. Figure 7.24 shows the schematic of this special hydraulic circuit. It simply keeps the screw position after recovery with shutting off the oil inlet to the injection cylinder by valve 7. Then, the controller 10 (not shown in Figure 7.24) checks the signal of a position deviation from transducer 8 that is connected to injection piston and screw 9. If more oil and pressure are needed from the hydraulic system to make up the position change of screw 9, then valve 3 will open to allow the bypass oil supply to go through a pressure-reducing valve 4, a throat flow control valve 5, and a check valve 6 to correct the position of injection piston to maintain the screw

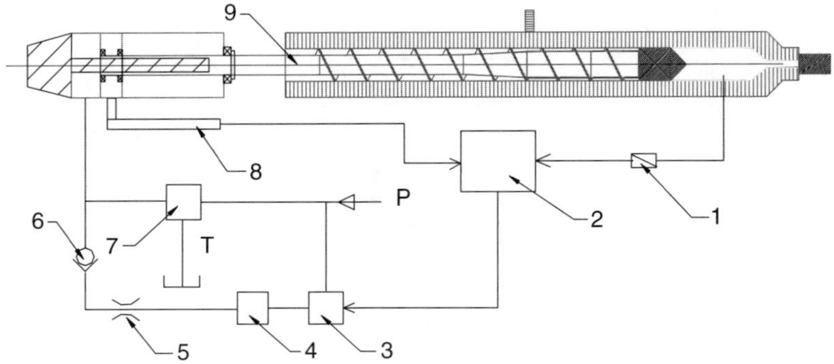

Figure 7.24 Hydraulic circuit modification for maintaining constant pressure and screw position with safety. 1, Pressure sensor; 2, controller; 3, direction valve; 4, pressure reducing valve; 5, flow control valve; 6, check valve; 7, shut-off valve; 8, position transducer for screw position control; 9, screw.

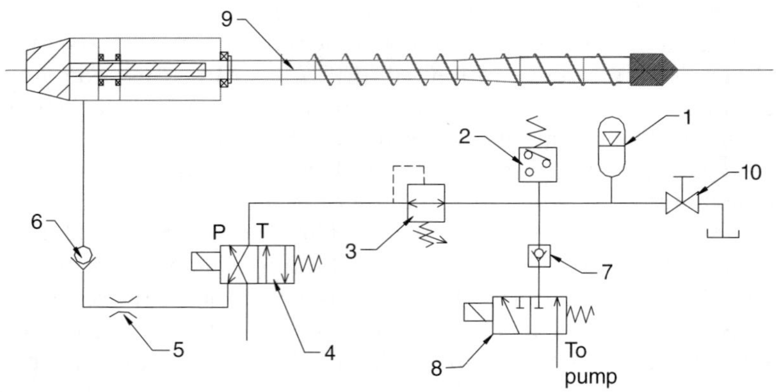

Figure 7.25 Hydraulic circuit modification for maintaining constant pressure with accumulator bottle. 1, Accumulator bottle; 2, pressure switch; 3, pressure reducing valve; 4, direction valve; 5, flow control valve; 6, check valve; 7, pilot-closed check valve; 8, direction valve; 9, screw; 10, emergency releasing or draining valve.

position. This hydraulic system provides a really low speed of screw movement (with low flow rate controlled by valve 5) that is only enough to maintain single-phase solution without hazards. The hydraulic pressure will be kept as low as possible as long as the single-phase solution is still maintained. The hydraulic pressure of the machine needs to be released immediately if unpredicted power loss or emergency stop button is pressed.

Another typical hydraulic design for maintaining the constant back pressure in front of the screw tip is the accumulator bottle circuit shown in Figure 7.25. Once the pressure in the injection cylinder is lower than the set pressure value, the solenoid of valve 4 actuates. The pressurized oil from accumulator bottle 1 goes through pressure reducing valve 3, directional control valve 4, throat valve 5, and check valve 6 to increase the pressure in the injection cylinder. The pressure switch 2 is actuated for the recharging of the accumulator bottle from the system pump. Valve 10 is an emergency releasing valve that can be either a manual releasing valve or an automatic one. For the safe operation of the microcellular process, it is always recommended to use an automatic releasing valve.

Finally, a simple hydraulic solution is introduced for the retrofit project to add the microcellular processing capability in the regular injection molding machine. Now it is widely used for the existing hydraulic system to be modified with this design. A pilot closed check valve is a two-way, solenoid-operated poppet valve that is "normally open." It is installed just between the inlet of the injection cylinder and the "A" port of the main valve (existed valve to connect to injection cylinder). Once the screw finishes the recovery, the solenoid of this check valve is quickly actuated to isolate the high-pressure oil in the hydraulic injection cylinder from the existing valve. This valve must be selected with a large size to handle the oil flow during normal injection, and it should not have any effect on the normal operation of the machine. Even the best seal valve has some leakage, so this solution only works for the screw idle time that is not very long, ideally less than 1 minute.

7.2.6.2 Hydraulic Design for Screw Driving. The low-viscosity SCF acts as a plasticizing agent when it is injected into the melt of polymer in the screw. Once the SCF and polymer form a single-phase solution, the viscosity of the solution will drop significantly [5], allowing the screw torque to be decreased. Table 7.11 shows that the screw torque requirements are decreased by about 10% once enough CO_2 gas (typically 6 wt%) is added into the GPPS melt. This means that the microcellular screw speed can be increased without the need for increasing the horsepower of the hydraulic motor. The end result of the above discoveries means that a standard RS injection molding machine typically only requires a new microcellular screw and barrel (with gas injectors) and does not typically require an upgrade of the hydraulic motor or plasticizing unit. It is necessary, then, to increase the hydraulic motor rotation speed to make up some output loss because of the gas lubrication effect and high back pressure for SCF dosing. Referring to the results from the hydraulic design for the screw driving system is also good when using an electric screw driving system where the electric motor and a gear box replace the hydraulic motor.

TABLE 7.11 Screw Torque Requirement for a 105-mm-Diameter Screw with and without Gas

Screw Speed (rpm)	Screw Torque with Gas (N-m)	Screw Torque without Gas (N-m)
26	4482	5172
51	4822	5518
100	5683	6206

There is a possibility to use a screw as the SCF dosing part. In this case the gas will go through the end of the screw to the middle of the screw for SCF dosing. Therefore, a hollow shaft of hydraulic motor is required for the screw dosing system.

7.2.6.3 Hydraulic Design for Injection. The hydraulic design for microcellular injection molding must provide a high injection volume rate. There are two hydraulic solutions, accumulator bottles, and high volume pumps. It is well known that the regular injection cylinder will have two different circuits available in the standard machine. One is the maximum pressure circuit, and another one is the regenerative circuit (also called the maximum speed circuit). Figure 7.26 shows schematics of two circuits. Then, the force and speed can be calculated as the following formula.

In maximum pressure circuit shown in Figure 7.26a the maximum injection pressure is

$$P_{\max} = P_{\text{oil}} A2 / A_s \tag{7.21}$$

$$A2 = \frac{\pi (d2)^2}{4} \tag{7.22}$$

$$A_s = \frac{\pi (D)^2}{4} \tag{7.23}$$

where $P_{\max}$ is the maximum injection pressure, A_s is the area of outside diameter of screw, $A2$ is the area of piston at blind side of injection cylinder, and $d2$ is the diameter of cylinder bore. The ratio of $A2/A_s$ is called the intensify

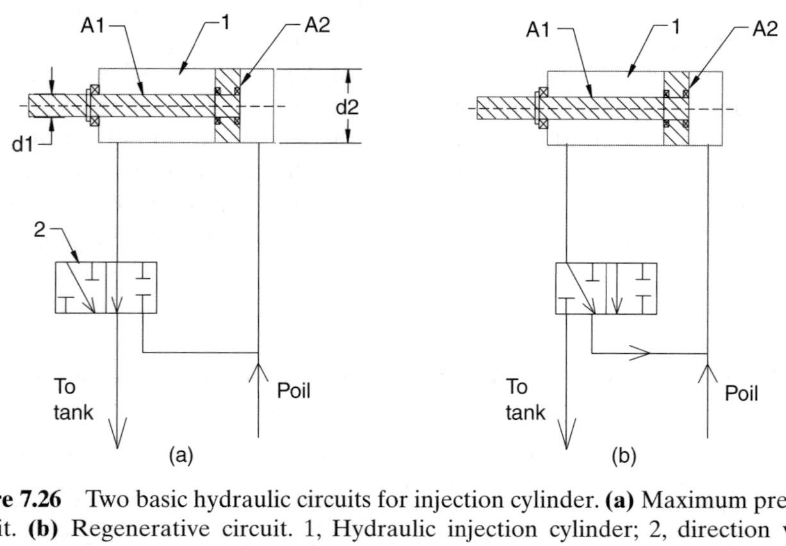

Figure 7.26 Two basic hydraulic circuits for injection cylinder. **(a)** Maximum pressure circuit. **(b)** Regenerative circuit. 1, Hydraulic injection cylinder; 2, direction valve; $A1$, area of piston rod; $A2$, area of piston at blind side; P_{oil}, pressure of hydraulic oil.

ratio of injection cylinder, which is about 8–12 as the industry standard. From Figure 7.26a, there is no resistance from the piston rod side if the oil goes to the tank directly. The back pressure for the return line is determined by the length of the return line, the diameter of return line, and possibly a check valve with very low cracking force. However, usually a zero pressure is used for the calculation in Equation (7.21) to simplify the calculation. However, this circuit always trades off between injection pressure and injection speed. If one increases, another must decrease in a proportional manner. Assuming that the oil flow rate is known, the injection speed at the maximum pressure circuit in Figure 7.26a will be

$$V_p = \frac{Q_p}{A2} \tag{7.24}$$

where V_p is the injection speed at operation set with maximum pressure and Q_p is the flow rate of hydraulic oil for the injection cylinder.

In the regenerative circuit in Figure 7.26b, the injection pressure at the regenerative operation setup is given as

$$P_{\text{reg}} = \frac{P_{\text{oil}}(A2 - A1)}{A_s} \tag{7.25}$$

$$A1 = \frac{\pi (d1)^2}{4} \tag{7.26}$$

where P_{reg} is the injection pressure at the regenerative operation setup, $A1$ is the area of the piston rod of the injection cylinder, and $d1$ is the diameter of the piston rod of the injection cylinder.

It is obvious that the injection pressure with the regenerative operation is lower than the injection pressure at the maximum pressure operation because the pressure in the piston rod side is not zero anymore and it is equal to the pressure on the blind side of the piston. Therefore, the effective area to be pushed by the same pressure P_{oil} is reduced from $A2$ to ($A2$–$A1$). However, the injection speed at the regenerative condition is increased to

$$V_{\text{reg}} = \frac{Q_p}{A2 - A1} \tag{7.27}$$

where V_{reg} is the injection speed at the regenerative operation setup.

If the piston has an area ratio $A2/(A2 - A1) = 2$, this will result in a doubling of the rate of speed. In general, microcellular injection molding needs only a regenerative operation since the high speed at low injection pressure dominates most microcellular processing applications. Then, even if accumulator bottles have to be used in the higher injection speed application, the design will focus on the regenerative circuit.

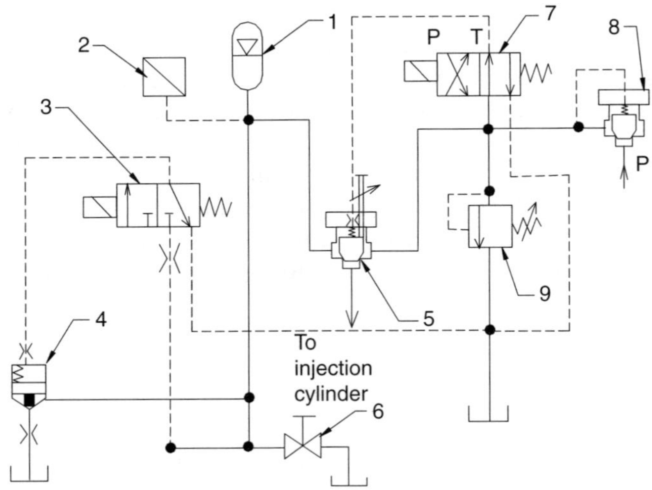

Figure 7.27 Typical accumulator hydraulic circuit for high-speed injection cylinder. 1, Hydraulic accumulator bottle; 2, pressure transducer; 3, valve control for accumulator automatic draining; 4, cartridge valve for automatic draining of accumulator; 5, cartridge valve for accumulator to discharge oil to injection cylinder with high-speed injection; 6, manual draining; 7, control valve for accumulator firing; 8, cartridge valve for quick accumulator recharging; 9, safety valve.

Figure 7.27 is a typical hydraulic schematic of accumulator bottles for high injection speed application. The control valve of the accumulator bottles for microcellular injection molding is recommended when using a cartridge valve that performs well for high flow rate with quick response. Valve 7 controls the opening of valve 5, which is the major cartridge valve for injection. Once the valve 7 is open, the oil in accumulator 1 will be released instantly to join the oil from the pump for fast injection. Valve 3 controls the automatic release valve 4 to empty the oil in the accumulator once the E-stop is actuated.

7.2.6.4 Hydraulic Design for Shut-Off Nozzle (or Valve Gates). The hydraulic design for the shut-off nozzle or for the valve gates is similar to the ejector or core pull circuit design. A typical hydraulic schematic is illustrated in Figure 7.28. The positive (or default) closing is necessary for the hydraulic design of the shut-off nozzle or the valve gates. Therefore, this special hydraulic circuit uses a little accumulator to keep the pressure in the shut-off nozzle cylinder when the system pressure *P* may be lower than the necessary setup closing pressure in the cylinder. On the other hand, this circuit has a pilot-to-close check valve 8. When the pump is running the system, pressure *P* will be on the pilot of check valve to keep it closed for the normal operation with the accumulator. However, when the pump is stopped and pressure is no longer available from the system, the check valve 8 will open to allow safe, automatic discharge of the accumulator. This is essential for safety when working on

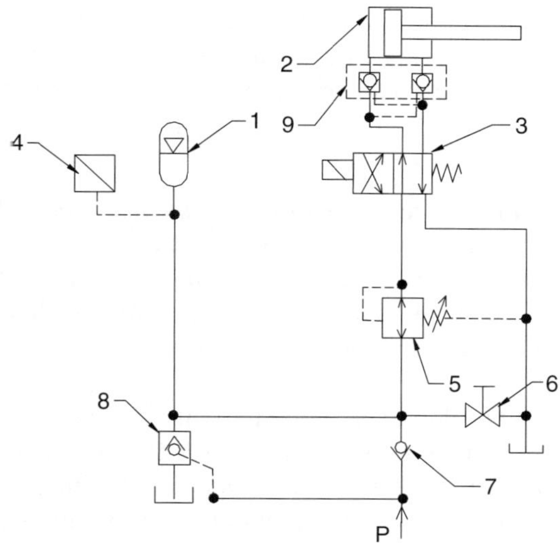

Figure 7.28 Typical accumulator hydraulic circuit for shut-off nozzle cylinder. 1, Hydraulic accumulator bottle; 2, cylinder of shut-off nozzle; 3, valve control for cylinder of shut-off nozzle; 4, pressure sensor and switch; 5, pressure reducing valve; 6, manual draining; 7, check valve to maintain pressure in the cylinder to keep shut-off function; 8, pilot-closed check valve; 9, pilot-operated check valve.

machines with an accumulator present. Note that a simple check valve 7 is virtually always used to isolate the accumulator from the pump. Another special design of this circuit is the pilot-operated check 9 valves for the shut-off nozzle cylinder. It can maintain the pressure in the shut-off nozzle cylinder if any emergency power loss occurs. This is the design for the microcellular process only. The regular injection molding will have the opposite way to operate the shut-off nozzle cylinder, which immediately releases the cylinder pressure; the shut-off nozzle will remain open after power loss.

To use this hydraulic circuit safely, it is necessary to have a manual pressure-releasing device in the barrel (or nozzle—Herzog standard shut-off nozzle releasing device) to finally release the pressure and the gas-rich material inside of the barrel if the machine operation stops for a long period.

7.2.7 Microcellular Injection Molding Control System

Compared with a regular injection molding machine, the control system for microcellular injection molding machine have several important changes. The sequence control between shut-off nozzle, or valve gates, and injection is one of the keys to guarantee the injection first and guarantee shut-off element opening second. Another one is the sequence opening–closing for the multi-SCF injectors. In addition, the SCF injector must be open with some delayed opening and ahead of closing during screw recovery. For some sensitive material it needs an adaptive control for the SCF pressure set up to automatically follow the melt pressure change in the barrel is necessary. Finally, a special SCF dosing named pulse dosing is introduced for a very short screw stroke.

7.2.7.1 Microcellular Injection Molding Sequence Control. Microcellular injection molding has a special control sequence for the actions of screw recovery, SCF dosing, shut-off nozzle open and close, mold open, and injection. If we check the special actions only for the microcellular process, there are several actions being discussed in the following:

1. Shut-off (S/O) nozzle delayed open about 0.5 sec relative to the start of injection time. This is a proven approach to prevent any possible free air shot of gas-rich material into the mold if the initial injection cannot match the automatic pressurized material discharging.
2. SCF dosing must be finished in the range of recovery time. It will be delayed to open at about 1 sec after recovery begins and close ahead of the end of recovery time. These are several critical issues for SCF dosing since gas under pressure will have surge releasing at the beginning of the gas injector opening. Therefore, it opens only after the screw buildup pressure and begins to move axially with setup recovery speed. Then, it guarantees that the SCF will be carried out by the flow of molten plastic in time away from the position of SCF dosing. For the same reason, the SCF dosing must be closed before the end of screw recovery.
3. When the hold stag ends, the shut-off nozzle must be closed to allow the screw to begin recovery.
4. When mold opens, the shut-off nozzle (or valve gates) must be closed for safe operation.
5. If there are multi-injectors for SCF dosing in the barrel, they must be acted upon with some overlap firing open time to match the length of the wiping section in the screw. If we assume that there are two SCF injectors on the barrel, the general recommendation for injector settings will be as follows:
    - First Injector opens at 0.2D (minimum opening position, D is the outside diameter of screw)
    - First Injector closes at 2.1D
    - Second Injector opens at 1.9D
    - Second Injector closes at 0.8 times the whole stroke

The injectors close on time or position, whichever occurs first. These settings are set up for a 2.2D maximum length of the wiping section. Without the second injector, the wiping section needs to be 4.2D longer at least to cover the whole SCF dosing stroke in 4.2D length. Then, this screw can be designed with only 2.5*D*-long wiping and save 2*D* for the extra mixing section that is needed for better SCF diffusion and mixing.

7.2.7.2 SCF Dosing Control. There may be two issues to be solved with SCF dosing control. One is the pressure adaptive control, and another is pulse SCF dosing for a short recovery time. Both are commercialized successfully and will be discussed below. On the other hand, there is a general rule for the SCF dosing control that needs to be set up for all kinds of microcellular screws.

1. Adaptive Pressure Control for SCF Dosing. This is specially designed control methodology for the screw with continuing pressure changes during screw recovery that needs this kind of adaptive control. The adaptive pressure control will keep the pressure difference the same all time of SCF dosing period. In other words, the SCF dosing pressure setting will dynamically vary with the change of the melt pressure measured by a pressure transducer and will then let the SCF setting pressure simply follow the measured melt pressure with the constant difference higher than the melt pressure. In this way the SCF flow rate will be kept the same and the dosing rate will be the same as well.

Therefore, an interface system needs to be developped to do the adaptive control. Basically this interface control is a system where the complete line or loop (out of the SCF unit and back to the SCF unit) controls a constant pressure (this is the delivery pressure). This pressure can be controlled from the SCF unit, by an electronic regulator. This regulator can take a feedback signal from the SCF inlet pressure regulator, and the controller can be programmed to maintain the differential pressure between the SCF flow line and the melt pressure. The injectors have to be redesigned to have an SCF inlet and outlet, as well as a bypass valve. It will be discussed in the SCF delivery system design in detail.

2. Pulse SCF Dosing. Another method is successfully used for a quick recovery time that is too short to stabilize the SCF dosing even if an adaptive methodology is applied. The easy way is actually to use a quick pulse of SCF injector opening time. It is approved to control the very short time of pulse to avoid too much SCF dosing in such a short time recovery.

3. General Rule of SCF Dosing Setup. The special software must be developed to control the sequence of SCF dosing time and screw recovery stroke (not the screw rotation time) for the microcellular process. Typically, the SCF injector should not be opened until a stable melt pressure has been reached in the barrel. There are several rules of writing this software for microcellular processing. A safe way to establish this open position correctly is to take the difference between the shot size and the screw position before screw rotating and add 5% of that difference to the screw position just before the start of screw rotation. The cushion setup in regular injection molding is never used for the microcellular processing since there is no packing and holding stages for microcellular process. It often occurs after the end of hold time, or it occurs after the end of injection time if there is no hold time; the screw moves back slightly in the barrel and decompresses. Then, this decompression moves the reversal channel of screw at the SCF injector open position before the start of screw rotation. When this happens in a reversal screw, there will be a significant pressure surge into the barrel because the reversal channel creates high pressure, and then barrel pressure will start to build faster than the SCF delivery pressure in the interface system. This situation is only for a reversal channel screw, and it creates the dosing process out of control. However, the other microcellular screw can tolerate the decompression as long as the minimum pressure is maintained for a good single-phase solution.

7.2.8 Guidelines of Gas Entrance Process of SPI

SPI released a document called "Recommended Guideline for Entrained Gas Processing in Horizontal Injection Molding Machine (EGPHIMM)" in 2003. This guideline provides recommendations for the design, installation, and application of an entrained gas process for horizontal injection molding machines.

This is a process where a pressurized gas is injected into the polymer melt in the barrel, accumulator, mixer, or nozzle of the injection molding machine. The gas is mixed into the polymer melt, and it is maintained under sufficient pressure to prevent foaming in any of the following components: barrel, accumulator, mixer, and nozzle of the injection molding machine. Similarly, a chemical blowing agent is added into the plastic raw material and is then heated and mixed. The mixture of plastic melt and chemical blowing agent is heated to degradation temperature of the chemical blowing agent and then releases a gas. The mixture of plastic melt and gas is, then, maintained under pressure to avoid prefoaming. Several key guidelines are discussed in the following:

7.2.8.1 Pressure Releasing Devices. A pressure-releasing device (rupture disk or equivalent) is recommended in the position wherever it is necessary to release the pressure or protect some weak parts in the machine.

A safety guard should be provided in the pressure-releasing spot. This guard is necessary to contain possible plastic melt spray in the event of released melt pressure without warning.

7.2.8.2 Shut off Nozzle. The entrained gas process always maintains pressure in the barrel. It is recommended that the nozzle shut-off action be positively actuated, and its close position and close force are monitored for processing reasons.

Possible redundant monitoring systems include, but are not limited to, monitoring the inlet pressure and monitoring the shut-off nozzle position.

A new rule for the shut-off nozzle is that the shut-off nozzle must be designed to have default of the closed position in case the emergency stop is pressed or the power is lost suddenly.

7.2.8.3 Pressure Release. The injection actuator should be free to move backward (10% of stroke) to facilitate pressure release in the event of E-stop actuation of HIMM safety gate opening.

The entrained gas process always maintains pressure in the barrel. Therefore the following recommendation is made: The purge guard (movable without tool) or non-operation side door is opened and the melt pressure in the nozzle should be dropped immediately.

7.2.8.4 Safety. Safety functions of EGPHIMM must meet all requirements of ANSI/SPI B151.1 with the exception of the shut-off nozzle.

7.2.8.5 Maintenance. Before shutdown, the injection molding machine for microcellular processing purges the barrel of all plastic melt and gas mixture. Then, refer to the OEM manual for further instructions on the maintenance schedules.

7.3 EXTRUDER WITH INJECTION PLUNGER MACHINE

This kind of machine uses an extruding screw for plasticizing and a plunger for injection. It is also called the two-stage screw–plunger injection molding machine [26]. Although it is an older design, it has been widely used for rubber, thermoset, structural foam, micro-molding, and packaging machines. It is an alternative solution to the microcellular injection molding machine. Trexel successfully developed the first microcellular injection molding with this screw–plunger machine in 1997 even if this screw–plunger never becomes the commercial machine for microcellular injection molding. Shimbo [29] reported results of microcellular foam process using a screw and plunger injection molding in 2000. Park and his students also developed this kind of screw–double-plunger foaming machine and called it the advanced structural foam molding machine [38, 39]. It would be a good idea to make not only a structural foam but also a microcellular foam.

In fact, this method of microcellular foam molding is somewhat less complicated than in-line foam molding because the requirement to maintain a single-phase solution is separated from the reciprocation step for both recovery and injection. Overall, there are a few modifications needed for the microcellular processing in the regular screw–plunger molding machine.

7.3.1 SCF Dosing System in the Extruder

There is a mandatory change for an SCF injector to be added on the extruder. The screw rotates to finish two functions, plasticizing the plastics and creating the single-phase solution with the SCF dosing. However, this screw will not have any reciprocating movement for either injection or recovery. The main advantage of this configuration is the plasticizing and SCF dosing either completely separating from the injection or reciprocating. It will make the extruder operation simpler and much more stable than the reciprocating screw extruder in the injection molding machine. The SCF dosing configuration is the same as the reciprocating screw except that it has no quick closing screw tip. All the wiping functions, blisters for the pressure profile suitable for SCF dosing, mixing for quick SCF diffusion, and uniform SCF–melt mixture are the same as for the reciprocating screw except that there is no axial relative movement. The extruder barrel can be designed similar to venting barrel. The pressure transducer is still required to set up SCF dosing correctly. It is always important to keep the safety device—that is, the rupture disc or the pressure releasing device—in the same location of SCF injector in the barrel. The design differences between extruding screw with SCF dosing and reciprocating screw with SCF dosing can be summarized as follows:

- The wiping section of the extruding screw is short and never needs multi-SCF injectors in the axial direction along the barrel because it has a fixed position for both the barrel and the screw.
- The sequence of SCF dosing is not necessary; it will continue the synchronous action as the screw rotation (except for slow rotation in the screw idle period waiting for injection of the plunger), with no axial movement anymore.
- There is the same history of SCF dosing, mixing for the whole shot size of the single-phase solution in the extruder. Therefore, the uniformity of the single-phase solution does not need to set up high back pressure, with screw speed profiles making up the difference because of the reciprocating movement.

7.3.2 Plunger Assembly for Injection

As the special configuration of this machine, there is a plunger with an injection system (double cylinders on both sides of the plunger, or a single cylinder at the same axis as the plunger) below the extruder. Then, the reciprocating movement will be from the plunger: (a) backwards to accumulate the shot size in front of the plunger tip and (b) forward for the injection to fill the mold. The injection movement will be provided by the injection system connected directly to the plunger. One key element to separate the extruder and the plunger is the ball valve, which allows the single-phase solution made by the extruder to go through it and accumulate in front of the plunger. It will be closed during two actions of the machine. One action is the injection of the plunger, which will create high pressure in the barrel of the plunger. This ball valve will protect the extruder without the risk of high injection pressure transferred into the extruder. It means that the extruder barrel can be designed with low pressure that is necessary for SCF dosing. Therefore, the maximum melt pressure inside of the extruder may be as high as 34.5 MPa (5000 psi) only. However, the plunger barrel will be as high as 138 MPa (20,000 psi) or larger. Even if the microcellular process needs low injection pressure, it is usually designed with the same injection pressure as for a reciprocating screw, which is 207 MPa (30,000 psi). It is very useful when the mold trial needs high pressure temporarily. On the other hand, once the screw finishes the shot-size recovery, the plunger may pressurize the existing single-phase solution in the barrel by using a little higher pressure, not only to maintain the pressure but also to force the ball valve to close before injection. Otherwise, this ball valve must have some assistance force such as a spring to actuate immediate closing after the screw recovery finishes and before beginning the injection.

The plunger diameter has a flexible design. It is not necessary to match the diameter of the extruder screw. For microcellular parts, it can be designed with a bigger diameter of plunger to make full use of microcellular benefits with low injection pressure. Then, with the same stroke, increased diameter of plunger can provide big oversize shot and high injection volume rate compared to reciprocating screw machine. With a bigger diameter plunger the injection time will be reduced as well since it reduces the injection stroke significantly.

7.3.3 Method to Maintain Pressure in the Extruder

The ball valve immediately separates the extruder and plunger chamber so that the pressure of the accumulated single-phase solution can be maintained at the necessary high value by a control system for the plunger–injection system. However, the pressure in the extruding screw may be decreased quickly once the screw stops rotation. The pressure dip too quick in extruder may cause the premature foaming in the already existing single-phase solution between SCF dosing position and the screw tip in the extruder of plunger machine. This prefoamed material is not acceptable for the next shot.

To solve the problems in this idle period of extruder, there are two methods to protect the existed single-phase solution already in the extruder between the SCF injector and the screw tip. One method is to keep screw rotate with necessary slow speed to maintain the pressure inside of the screw. Another option is to add a very reliable middle check valve in the extruding screw just on the upstream of the position of SCF dosing injector.

7.3.3.1 Continuous Screw Rotation During the Idle Period. This method has been tried in the screw–plunger injection molding machine, and it works with some special settings. First, there must be rotation at a certain screw speed to keep the necessary pressure in the zones with gas-rich material that is from the SCF dosing position in front of the screw tip. To avoid building too much pressure, the material in the hopper needs to be shut off temporarily so that the screw just pressurizes the existing material in the barrel. In this way, pressure is not only maintained but is also helpful for existing SCF shearing, mixing, and diffusion further into the melt. The disadvantage is the complex operation procedures that are required to modify the controller of the machine.

7.3.3.2 Add a Middle Check Valve. The same pressure-restrictive elements discussed for a reciprocating screw will work well for this requirement. Once the screw stops rotation, the pressure in the screw will quickly decay while the leaking flow through the middle valve will cause the valve to close automatically. In fact, this is the simple modification to solve the problem, and it is a proven method in many applications of different materials.

7.3.4 Other Developments for the Screw–Plunger-Style Injection Molding Machine

The difficulty of the injection molding machine is the periodic processing. It brings a noncontinuing process of SCF dosing. An advanced structural foam molding machine was developed in Park's group at the University of Toronto [38, 39]. Based on the configuration, this advanced structural foam machine can be called a screw–double-plunger molding machine. It may solve the SCF dosing problem of periodic processing from injection molding because it completely decouples the gas dissolution step from the injection operation. The new system consists of a positive displacement gear pump and an assist plunger. They are installed between the extrusion barrel and the first shut-off valve. The shut-off valve in this application should be positively actuated. The extruding screw will continue to rotate and will also continue to make single-phase solution without interruption by injection and mold cooling. The single-phase solution will be stored in the assist plunger chamber first if the injection and mold cooling are still going on. Then, the actuator discharges the accumulated material in the assist plunger into the main plunger with the valve opened and the shut-off nozzle to the mold remaining closed. It then prepares the next shot without interruption for the extruder to do the SCF dosing. When the main plunger injects the material into the mold, the valve remains closed and the shut-off nozzle stays open. The disadvantage is the extra action of material transferring from plunger to plunger, which takes extra time and may cause the second nucleation with big pressure difference between two plungers to create some elongated cells in the mold. On the other hand, one more plunger and actuator will increase the cost of the equipment.

7.4 SCF DELIVERY SYSTEM DESIGN

The gas at certain conditions will become supercritical fluid (SCF) that must be prepared before the gas is ready to be injected into the barrel. In this chapter, a SCF delivery system design will be explained with a specific gas type at the condition of supercritical fluid state unless otherwise specified with different state. The gas delivery system for microcellular foam machines must be controlled to consistently meter the gas by either a mass flow rate or a volume rate at a constant pressure. The gas used in the microcellular process must be pressurized to a certain high pressure at a certain temperature to become the supercritical fluid. This method allows the overall system of the SCF unit and injection molding machine to make a uniform single-phase solution and provide for consistent screw recovery. A generic schematic of the gas delivery system used for all foaming processes can be explained as follows: blowing agent source → pump to pressurize the blowing agent → blowing agent flow control → pressure regulator and injector. The blowing agent resource is usually either a single blowing agent container or a blowing agent station. Then, the pump is necessary to pressurize the blowing agent to the high pressure, which is about 34.5 MPa (5000 psi) to 41.4 MPa (6000 psi). Some systems use an even higher pressure up to 55.2 MPa (8000 psi). Most of gases at such high system pressure will become supercritical fluid that behaves liquid like fluid. The gas in the supercritical state is a necessary condition to do the next procedure, known as blowing agent control. It has two different control methods: One is a pressure control, and another is a mass flow control. Finally, there are an SCF pressure regulator and a special SCF injector on the barrel.

7.4.1 Physical Blowing Agents

To design the gas delivery system, the basic properties of physical blowing agents need to be reviewed first. More details on them are discussed in Chapter 4. As the physical blowing agent sources, different gases have been used as a blowing agent in the foaming industry by many professionals for a long time, such as nitrogen, carbon dioxide, and even air. Some exotic gases such as argon, helium, and hydrogen have been tested in the laboratory by some researchers, but none of them are used as commercialized application. Water is also a possible blowing agent source that was used for both practices and research. However, for all foaming industries, including microcellular injection molding, only carbon dioxide gas and nitrogen gas are by far the most widely used physical blowing agents.

Nitrogen gas is an inexpensive, nonflammable, nontoxic permanent gas. It can be easily made from the air and is chemically inert, which results in an environmentally safe blowing agent to replace some ozone depletion chemical blowing agents. The gas state of nitrogen is available at 13.8 MPa (2000 psi) to 20.7 MPa (3000 psi) as compressed gas in the steel cylinder. The liquid state of nitrogen is stored in the dewars as a cryogenic liquid at about −196 °C. For a heavily nitrogen flow rate application, the cryogenic nitrogen is preferred. However, the nitrogen vapor needs to be boiled off from the liquid state during the real usage, and the gas temperature will be warmed up to about room temperature prior to metering and injection into the machine. In other words, nitrogen as a blowing agent is used only in the gas state in the delivery equipment except in the storage equipment. Overall, nitrogen is preferred in many, if not most, technical applications because it results in a more consistent and uniform microcellular part.

Carbon dioxide can be useful for a number of special cases when gas diffusion, or viscosity, is the primary challenge. It is similar to nitrogen for the usage of an ideal foaming agent. It is also the inexpensive, chemical inert, environment acceptable, and intriguing physical blowing agent. However, carbon dioxide has some inherent handling problems, such as a relatively low critical point of 31 °C and 7.29 MPa (1027 psi). Carbon dioxide will be a vapor above the critical point. It may be used in the delivery system with either gas state or liquid state.

7.4.2 Pumps for Pressurization of Blowing Agent

In some cases the pressurized gas in a container is ready to be used as a blowing agent for the foaming process. However, for the long stable foaming process the pumps are necessary to store more volumes under higher pressure than the one in the vendor's standard container. There are two different pumps that can be used in the real application: liquid and gas. Both pumps will be discussed in the following.

7.4.2.1 Liquid Pump. The liquid pump is only possible for carbon dioxide. It is not new and was developed by different companies, including Trexel and Mitsui [8]. The key for the interest of using a liquid pump for carbon dioxide is the liquid state of carbon dioxide in which the mass flow can be metered easily as a liquid flow. This advantage of liquid metering will allow us to control a very low weight percentage of the supercritical fluid dosing. The basic equipment design includes a chiller to keep the liquid state of carbon dioxide in the whole system until it is out of the metering device, a liquid pump, a system pressure regulator, and an SCF dosing device. The pump itself may be good enough to be used for both (a) compression of the carbon dioxide liquid to the level of SCF and (b) metering the mass flow rate by the pump stroke adjustment.

Unlike the other permanent gases, the easy phase change of a semicryogenic liquid for carbon dioxide may cause dramatic processing condition changes unless a good coolant system is working consistently. The chiller increases the cost and complicates the process. Therefore, the saving of simplicity of liquid metering is overcome by the cooling system. It may be the major reason why the liquid pump does not become one of the commercialized SCF delivery systems on the market.

It is not possible to use liquid nitrogen in the liquid pump because its low cryogenic liquid temperature is as low as −196 °C. Another reason why the liquid pump is not acceptable as commercial equipment for microcellular injection molding is because nitrogen gas is a popular blowing agent in the foaming industry.

7.4.2.2 Gas Pump. A gas pump is just like an air pump that does not need any chiller. Different types of the pump are available. One of the popular gas pumps used for microcellular injection molding is the two-stage pump. This two-stage pump is an air-driven gas booster compressor. The concept is illustrated as a schematic of a single-stage gas pump in Figure 7.29. It shows a position of intensifying the gas. The principle of this gas booster compressor is the gas being intensified by the big diameter of air-driven cylinder 1 to pressurize the gas in a small-diameter piston chamber 2. The check valve 3 is closed by the intensify pressure, and check valve 4 is open to accumulate the pressurized gas into a gas tank at high pressure. When the air cylinder driven by air moves in the left direction, valve 4 closes and valve 3 opens to allow the low-pressure gas flow into the intensify chamber for the next intensify stroke. Then, more and more pressurized gas will be stored in the gas tank for the readily available blowing agent resource to be delivered to the SCF dosing system. The gas pump needs a high flow rate of air compressor, and the air pressure is required to be about 0.55 MPa (80 psi) to 1.035 MPa (150 psi).

With the same principle, the hydraulic cylinder can be used for this gas pump. Hence, it has the same schematic as in Figure 7.29 but has the hydraulic cylinder instead of the air cylinder, as the actuator. It may have a compact design since the size of the hydraulic cylinder can be much less than that of the air cylinder, and it may have low noise operation from the hydraulic system. However, it needs the oil tank of a hydraulic system, which may cost more compared to the gas pump.

The gas pump size must match the plasticizing unit output rate. The suggestion is to adjust the gas pump output rate percentage of the output rate in the plasticizing unit to be 1% for N_2 gas and 10% for CO_2 gas, respectively.

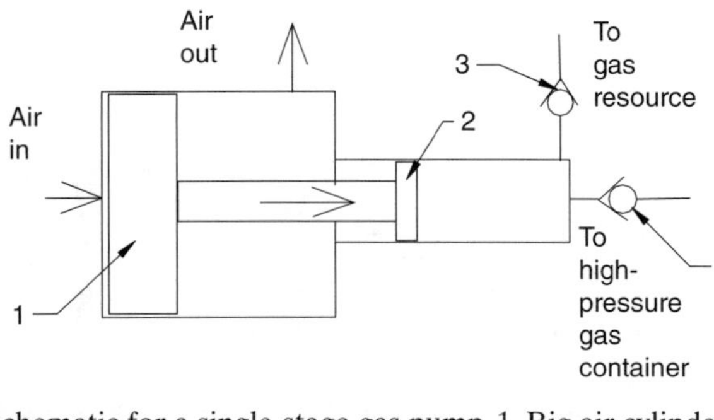

Figure 7.29 Schematic for a single-stage gas pump. 1, Big air cylinder with air-driving force; 2, small gas cylinder to intensify gas pressure; 3, check valve to allow low-pressure gas in; 4, check valve to allow pressurized gas to accumulate in high-pressure gas tank.

7.4.3 Gas Injector Design

The gas injector is the key element to safely inject gas at a supercritical fluid state into the barrel consistently from cycle to cycle. It often causes some special problems, such as a backflow of molten plastic in the gas injector. This material will stay in the injector or even in the line of pipe for the gas channel to the gas tank for a long time. Then, it will finally degrade and become solid in the injector. It will eventually clog the gas injector.

There is some help from process settings to clean up the clog of the injector. Typical molten plastics normally exhibit high viscosity in the range of 10–1,000,000 poise (g/cm sec). However, the gaseous blowing agent will normally exhibit viscosity in the very low range of 0.00005–0.05 poise (g/cm sec). Hence, the relative very low resistance for the blowing agent flows easily in the injector, and it helps to clean up the high-viscosity material in the flow lines of the injector. To protect the orifice of the SCF injector from clogging, the SCF injection pressure is always slightly higher (about 0.69 MPa) than the melt pressure at the position of SCF injector on the barrel. However, only processing settings are not enough to prevent this clogging in the gas injector. The gas injector can be designed to use the same advantage of viscosity difference to clean up the gas injection line efficiently.

7.4.3.1 Traditional Gas Injector for Structural Foam. There are many different designs of gas injectors used in the foaming industry. One of the reliable designs is the poppet-type check valve invented by David Johnson [19]. The blowing agent is introduced from the inside hole of the gas injector body (see Figure 7.30). It pushes the poppet and overcomes the force of the spring, and then it opens the poppet on the seal seat in the front end of the injector body. The blowing agent then commences to flow into the barrel, and it mixes with the molten plastic in the screw. The advantage of this design is evident when the periodic gas dosing is required for injection molding.

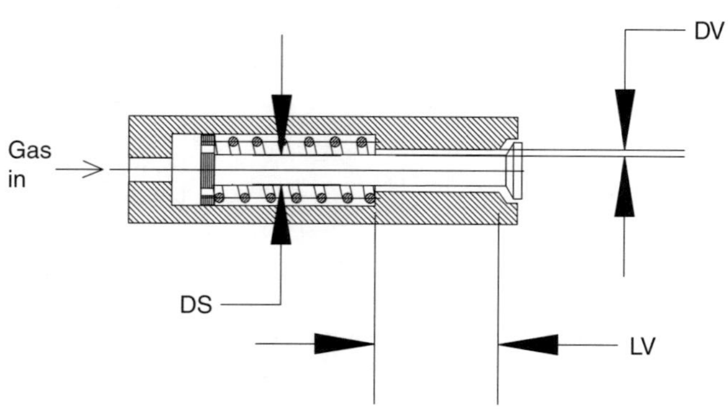

Figure 7.30 Parameters for the poppet type of gas injector in the open position.

As shown in Figure 7.30, the valve seat is so close to the end of the injector body. It will bring several advantages as below:

- The temperature of the seal seat is always as high as that of the barrel since this position is located deep inside of the barrel. It will guarantee that any residual material around the seal seat of the valve is in the molten state. In other words, there is no possible freezing material clogging the gas outlet.
- On the other hand, the poppet design will have the valve closed not only by spring but also by the plastic pressure inside of barrel. It is important that once the gas dosing is over, the pressure of the melt actually helps to close the poppet. A useful solution is to use the melt pressure to close the poppet valve when the processing temperature is so high that the spring force is reduced significantly.
- If any possible leaking occurs, the plastic melt stick on the poppet will be easily cleaned up by both high gas pressure and high temperature.
- There is almost no material left in the outlet of gas injector because the gas outlet is almost near the end of the injector body and the wiping spot of the screw. Therefore, it will be cleaned up more efficiently with every wiping of screw flights.

There is another unique feature of this design. An open position and some parameters necessary for a performance analysis of this poppet injector are displayed in Figure 7.30. The gas must go through a long and narrow annular channel from the bottom side of the spring chamber to the end of the poppet seat. The length of this narrow channel is *Lv*. The diameter of the valve is *ds*, and the clearance between the OD of the valve and the ID of the guide hole inside of the gas injector is *dv*. Then, the gas will have a significant pressure drop across the annular channel (simplifying as a wide narrow orifice). The gas pressure drop across the annular orifice is similar to Equation (7.4) and is given by

$$\Delta P_{\text{gas}} = \frac{12\mu_g \dot{V}_g L_v}{\pi d_s d_v^3} \tag{7.28}$$

where ΔP_{gas} is the gas pressure drop across the length of orifice of gas injector, μ_g is the gas viscosity in the gas injector, $\dot{V}_g$ is the gas volume flow rate in the gas injector, L_v is the length of the orifice in the gas injector, d_v is the clearance between the ID of the annular orifice and the OD of the poppet stem in the gas injector, and d_s is the diameter of the poppet stem in the orifice area of the gas injector. This pressure drop ΔP_{gas} will guarantee that the gas pressure in the annular orifice is higher than the pressure near the end of the annular orifice. It prevents any possible backflow of molten plastic into the deep position of the annular orifice of the gas injector. If by any chance the gas pressure at the outlet of gas injection is lower than the melt pressure inside of the barrel, the molten plastic pressure will force the poppet to move to the close direction. However, it is still possible that some molten material will leak into the gas injector before the poppet closes completely. With this pressure drop in the narrow orifice, the leakage may not go too far in this narrow annular channel since the pressure gradient across the narrow annular channel is still higher than the possible melt pressure leaking into the gas injector. On the other hand, this little residence material is still in the molten state because of the high temperature in this annular orifice of the gas injector. It can be easily cleaned up by the gas pressure during the gas dosing next cycle.

7.4.3.2 New Gas Injector for Microcellular Injection Molding. Trexel has developed a new gas injector specifically targeting as many gas splits as possible through the gas injector sleeve that has many mini-holes (orifices) for SCF dosing [18]. This new valve uses actuated injection valves. A pneumatic cylinder (or hydraulic, spring, and other actuators) is used to push a needle valve against the seat of the gas orifice every cycle to open and close the gas channel. It works fine with enough force from the pneumatic cylinder. On the other hand, the Trexel gas injector also uses a ball check valve on the downstream of the needle valve to double protect the material leaking back into the needle valve area. In addition, Trexel uses the multi-orifice sleeve to make the gas droplet small [18]. However, it may not have a consistent fixed number of opened orifices because of the stop-and-flow periodic cycle. The other features of new gas injectors on the market will follow the same rule of designing discussed above.

Some special experiments have been carried out to test the difference of the cell structure between one-orifice and multi-orifice gas injectors. There is no significant change in the samples tested with both methods. Most equipments on the market have used only the one-orifice gas injector; this orifice has been designed accordingly with good parameters, such as orifice diameter and orifice length.

7.4.3.3 Orifice Design. The orifice of the gas injector may determine the flow rate simple according to the pressure drop. The empirical data of the orifice diameter shall be 0.0005 m or less. It is an approved design that has been used in structural foam industry for many years. The simple gas delivery system in the structural foam machine only controls the pressure settings with a back-pressure regulator before it injects gas into the barrel.

7.4.4 Gas Dosing Control System

There are still two systems for gas dosing control used for microcellular injection molding machines. One is the mass metering system, and another is the volume flow control system. Theoretically, of course, the mass metering is the best way to control the gas dosing. However, it may be expensive for both maintenance and new installation.

The typical flow control valve is the Badger meter for the express purpose of controlling factions of the capacity of flow though 0.025 m (1 in.) and smaller line sizes. A flow metering apparatus from MicroMotion measures the SCF flow, and then the controller automatically adjusts the Badger valve opening to keep the constant SCF flow rate. The SCF controller also adjusts the differential between the system pressure and back pressure across a calibrated opening to maintain a consistent SCF flow.

A satellite gas system has been developed in Trexel. It can use one main pump system to supply the SCF flow to multi-injection molding machines. An economic solution for multi-equipments is to share the most expensive unit of the gas pump [8].

7.4.5 Gas Regulator

There are two different gas pressure regulators required for the gas delivery system. One is the pressure-reducing regulator (series 26–1000, Tescom), and another is the back-pressure regulator (series 26–1700, Tescom). The pressure-reducing regulator is used for setting the dosing pressure that is usually reduced from system pressure. The back-pressure regulator is used as a precision relief device that opens when the system pressure exceeds the set point and bleeds off gas at the rate required to maintain the system pressure at the set point. Both regulators are high-pressure models that can be used up to 20.7 MPa (3000 psi).

There is a special bypass valve to convert the continuous pump system for a periodic injection molding. The controller in the molding machine directs the incoming SCF flow either to the injector or to the bypass valve [40]. This is because only screw recovery needs the SCF entering the barrel through injector. When the injector shuts off, the SCF will bleed off through the bypass valve. This excess blowing agent either discharges into air (one may need to check if there is any safety issue) or recycles back to pump.

The SCF delivery pressure setting could be a function on the melt back pressure in front of the screw tip for a normal screw. However, for the reversal channel screw the SCF delivery pressure setting can be the function of screw speed as well. If enough right processing data are available, then some processing data can be added into the control software of microcellular processing; this can save lots of trial-and-error time for starting a process of a new microcellular mold.

It is possible to use back pressure to set up the starting point of gas delivery pressure if the machine only provides the back pressure in front of screw tip. For example, the normal melt pressure near the gas injector in a large screw will be 1.38 MPa (200 psi) above the back pressure in front of the screw tip. Since back pressure is always available from standard injection molding machine software, the gas delivery pressure can be set up accordingly at the beginning if there is no melt pressure reading in the gas injector position. However, the melt pressure in the gas injector in front of the screw tip for a small reversal channel screw could be 6.9 MPa (1000 psi) higher than the back pressure readable from the control screen. Hence, a starting point for SCF delivery pressure for a reversal channel screw should be about 6.9 MPa (1000 psi) above this value if no melt pressure reading in the barrel is available. SCF inlet melt pressure is usually displayed on the injection molding machine controller and also on the controller of gas pump. The SCF dosage time should be set to about at least 50% of the screw recovery time. The delivery pressure drop when the SCF injector opens must be monitored for the first 15 shots. As the SCF starts to dose into the barrel, the viscosity of the polymer will decrease, which may result in a reduction in SCF inlet pressure. If this happens, the delivery pressure drop will increase and result in screw recovery problems. This may require slight adjustments to the gas delivery pressure to maintain the delivery pressure drop to less than 1.38 MPa (200 psi). Normally, the process will be stable after the first 10–15 shots. It will be dependent on the flow rate and SCF dosage time.

7.4.6 Safety of Gas Delivery System

The gas delivery system is under the high pressure up to 55.2 MPa (8000 psi). The safety devices must be installed to protect any potential hazard. Except for the necessary guards, there are several safety devices in the gas delivery system:

- A rupture disc for the safety limit of high system pressure must be designed so that the rupture will release the system pressure quickly if the system pressure is over the maximum. The outlet of the rupture disc must be protected by a guard and a surrender so that it will not cause a dangerous situation when it releases high pressure either by a moving part or by noise.
- Any part may cause dangerous movement when there is a burst of gas hoses or gas pipes, then, it needs to be restrained by safety devices.
- There must be some venting device around the gas pump. Carbon dioxide and nitrogen gas may cause a local area to have insufficient oxygen for people working there if a large volume of idle gas is released so quickly and densely in a local area.
- The gas pump system will have both manual and automatic dump valves. It is always recommended to use a manual dump valve to slowly decrease the system pressure and dump the gas in the system without creating a huge noise. The automatic dump valve is only used for emergent system pressure release.
- All valves, SCF conduits, and fittings must use a high pressure grade to match the system pressure with a safety factor margin.

7.5 SINTER METAL SLEEVE OF GAS DOSING (OPTIFOAM®)

Dr. Walter Michaeli and co-workers at the Institut für Kunststoffverarbeitung (IKV) has developed another foam injection molding (FIM) process that utilizes a "fluid injection nozzle" between a regular plasticizing unit of injection molding and a shut-off nozzle to add gas into plastic melt during injection. The gas-rich material can be further mixed in the static mixer to make better microcells [41–43]. This is a special nozzle that has been patented in Germany [42] and commercialized by Sulzer Chemtech Inc. with a trade name Optifoam®. The principle is explained by the schematic layout in Figure 7.31. The gas-free material flows into the melt channel of the nozzle. Both the inside surface and the outside surface of a narrow annular channel provide a huge number of small gas droplets that enter into a thin molten polymer layer. Once the molten polymer flows out of the sintered rings, it becomes a uniform gas-rich material. The special key design is the sintered metal rings inside of the nozzle. The sintered upper ring and low ring form a narrow melt channel in the nozzle to allow molten polymer go through. There are several key features:

- The sintered rings are excellent gas dividers where the gas dosing is guaranteed to become small droplets entering the molten polymer. It is because the porous sintered ring has a huge number of small holes and they are distributed on whole contact surface between the ring and the molten polymer.
- The molten polymer must go through an annular channel formed by upper and lower sintered rings. The gas diffusion thickness of molten polymer becomes a uniform thin layer. This annular thin melt layer not only decreases the thickness of gas diffusion, but also maximizes the gas dosing surface-to-volume ratio. It dramatically speeds up the gas diffusion from both the top and bottom surfaces of sintered rings.

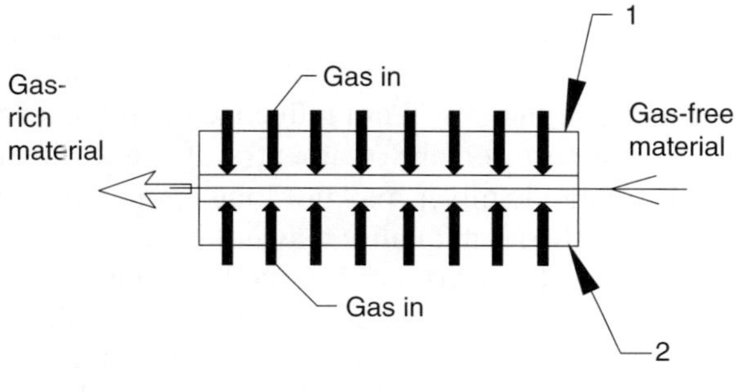

Figure 7.31 Schematic of the sintered rings in nozzle [43]. 1, Sintered upper ring; 2, sintered lower ring.

- This gas dosing occurs during injection. The injection speed guarantees that the gas will enter into the melt uniformly since there is no material stay in the nozzle during fast injection. Therefore, no big gas pocket is possible from the dosing period.

In this way, the single-phase solution idea may not be applicable here. It is because the gas droplet is so small once it comes out from the porous sintered metal ring. We may need a new theory to explain this excellent idea to be applied for real processing. Hence, we may need an additional static mixer [44] to make gas-rich material more uniform so that the gas residence time in the melt to be sheared becomes so short compared to the reciprocating screw or screw–plunger injection molding machine.

This technology can use both carbon dioxide and nitrogen gas at the pressure up to 40 MPa (5800 psi) with a special gas station. Another unique advantage of this design is the affordable retrofit solution to use standard screw and injection units. Just add the nozzle and gas station and then the machine is ready for making excellent foaming parts. The processing of Optifoam® technology will be discussed in detail in Chapter 6.

There is another application for Optifoam® technology. It is Optifoam® LSR (liquid silicone rubber). The raw liquid material A and B are supplied from LSR station. After the first check valve CO_2 gas is added into the A and B liquids, respectively. Then, gas-rich A and B liquids go through a special mixer and enter into the barrel.

7.6 DYNAMIC MIXER OF GAS DOSING (ERGOCELL®)

Demag Ergotech announced the Ergocell microcellular molding technology at K-Show 2001 in Germany. It makes a single-phase solution in a dynamic mixer where gas is added and mixed with the melt [45]. It uses the dynamic mixer to add carbon dioxide gas or liquid into a molten polymer that is from the regular plasticizing unit. The mixed gas-rich material is accumulated into a plunger. Once the shot size is reached in the plunger, the plunger will carry on the injection to force the material through the shut-off nozzle into the mold to make a foaming part. A check valve insulates the regular plasticizing unit from the dynamic mixer so that it will not influence the regular screw recovery. The disadvantage is the requirement of an extra dynamic mixer and plunger unit. In addition, the dynamic mixer may not long enough to mix the gas well with molten polymer, so the foam quality may not be as good as the one made in the reciprocating screw. The layout of Ergocell's initial development design is shown in reference 45. Recently, a new design of this technology is a horizontal plunger that acts as an extension of the screw. The dynamic mixer is still located between the shut-off nozzle and the screw [46]. Trexel has agreed that an Ergocell® user will obtain only patent rights, without any support from Trexel.

7.7 GAS FEED IN A SEALED HOPPER FOR GAS DOSING (PROFOAM®)

The newest developing foam technology from IKV is the ProFoam® process [47]. It is a new means of physically foaming injection molding technology. The goal is to commercialize a cheaper and easier foaming process than any existing microcellular foaming process. This cutting-edge technology can use either carbon dioxide or nitrogen gases. The gas as the blowing agent is directly added into the hopper and diffuses into the polymer during the normal plasticizing process. However, the different methodology of gas dosing, with the gas as the blowing agent, is not necessary at a supercritical state like the current microcellular process. The plasticizing unit of the molding machine is sealed off in the feeding section of the screw. Therefore, a pressurized chamber is installed between the material hopper and the plasticizing unit. The critical part is that the sluice in this pressurized chamber allows the blowing agent to set up the pressure, which is the sole additional parameter to be set up for this process. On the other hand, the sluice will be designed to keep the gas pressure but allow the pellets to feed through hopper under ambient conditions. With this ProFoam® process, the part can reduce up to 30% weight via the foaming.

7.8 RETROFIT MACHINE FOR MICROCELLULAR INJECTION MOLDING

Methods of retrofitting to satisfy conditions needed for processing microcellular materials are different among the technologies introduced above. All of these methods may also involve adding a blowing agent introduction system and modifying the polymer processing system including a control system more or less to be compatible therewith. The retrofitting methods advantageously enable production of a microcellular foam processing system at a considerably lower cost and in a shorter time than the production of a new microcellular foam processing system. A special polymer plasticizing system is provided for reciprocating the screw method of MuCell®.* For example, the methods generally involve replacing the conventional polymer processing screw with a new screw designed for microcellular processing. These new screws have been successfully designed with a short *L*/*D* ratio down to 22:1–24:1 [15]. This is to benefit the retrofit of the reciprocating screw machine with no change for the injection base, and even share the same barrel of the conventional screw. However, the plasticizing unit change is not necessary for both the sintered sleeve nozzle method Optifoam® and the dynamic mixer method Ergocell®. It is because both of them need only to add components in the nozzle area, not to the plasticizing unit. In addition, the MuCell® and Optifoam® customers

*MuCell® is a registered trademark of Trexel, Inc., Woburn, Massachusetts.

may switch between solid-parts molding and MuCell molding simply by turning the gas system on or off. An Ergocell® old customer must disassemble the machine and remove the mechanical system in order to do so. However, the newly designed Ergocell® configuration may allow a customer to use the same machine for both microcellular and solid processes.

7.9 GAS DOSING MIXER OF LIQUID SILICON

This is the special Sultz static mixer designed for mixing two liquid chemical components for liquid silicon. The A and B chemicals will be mixed in a static mixer where the CO_2 gas is also introduced.

7.10 ACCESSORIES FOR MICROCELLULAR INJECTION MOLDING

Most of the accessories for regular injection molding are acceptable for microcellular injection molding. Some of them need to be reconsidered for the estimation methods. For example, the capacity of the hopper, or hopper dryer, needs to match the production rate of microcellular molding. For the shot size, microcellular molding may only need 85% of regular molding if the weight reduction is 15%. However, if microcellular molding reaches 20% cycle time reduction, the production rate is actually increased. The final result must be the combination of these two factors together. The estimation of drying equipment may use some rules of thumb. Typical rules of thumb are listed as follows [48]:

- The volume of air required to dry resin is 1 ft^3/min/lb of material/hr.
- The regeneration air temperature in properly designed drying systems should be approximately 30 °C higher than that desired for the desiccant.
- The power consumption for cooling return air from the dryer before it goes to the hopper (a chiller heat exchanger is used) should be 1 kW to remove 4 kW of heat.

The cooling system for microcellular molding prefers high efficiency since microcellular processing can tolerate lower mold temperature. Some rules of thumb for regular molding are also good for microcellular molding. If 80% of materials are processed within a ±3 °C temperature range, a central chilling system may be considered. More than ±3 °C for a portable chilling system may be considered. Also, a chilling system requires 2% more capacity for every degree below the nominal ranting of 30 °C. Nominal design flow from a cooling tower is 3 gal/min/ton; from a chiller, it is 2.4 gal/min/ton. These are the flow rates necessary to achieve a 5.5 °C change in the process [48].

REFERENCES

1. Xu, J. *SPE ANTEC, Tech. Papers*, 2770–2774 (2006).
2. Shimbo, M., Nishida, K., Heraku, T., Iijima, K., Sekino, T., and Terayama, T. *First Int. Conf. Thermoplast. Foam*, 132–137 (1999).
3. Chandra, A., Gong, S., Turng, L-S., and Bramann, P. *SPE ANTEC, Tech. Papers*, 540–544 (2004).
4. Xu, J., and Kishbaugh, L. *J. Cell. Plastics* **39**(1), 29–47 (2003).
5. Xu, J., and Pierick, D. *J. Injection Molding Technol.* **5**, 152–159 (2001).
6. Xu, J. *SPE ANTEC, Tech. Papers*, 594–598 (2004).
7. Xu, J. *SPE ANTEC, Tech. Papers*, 2660–2664 (2004).
8. Okamoto, T. K. *Microcellular Processing*, Hanser/Gardner Publications, Cincinnati, 2003, pp. 30–37.
9. Xu, J. *SPE ANTEC, Tech. Papers*, 2089–2093 (2007).
10. Pierick, D., and Jacobsen, K. *Plastics Eng.* **57**(5), 46–51 (2001).
11. Suh, N. P. *Innovation in Polymer Processing*, edited by James F. Stevenson, Hanser/Gardner Publications, Cincinnati, 1996, Chapter 3, pp. 93–149.
12. Xu, J. U.S. Patent No. 6,322,347 (2001).
13. Xu, J. U.S. Patent No. 6,579,910 B2 (2003).
14. Xu, J. U.S. Patent No. 7,267,534 (2007).
15. Xu, J., et al. U.S. Patent No. 7,318,713 (2008).
16. Rauwendaal, C., et al. *Mixing in Polymer Processing*, Marcel Dekker, New York, 1991, pp. 182–183.
17. Chung, C. I. *Extrusion of Polymers: Theory and Practice*, Hanser/Gardner Publications, Cincinnati, 2000, pp. 254–272.
18. Burnham, T. B., et al. U.S. Patent No. 6,284,810 (2001).
19. Johnson, D. E. U.S. Patent No. 4,124,336 (1978).
20. SPI Machinery Division. Recommended Guideline for Entrained Gas Processing in Horizontal Injection Molding Machines (EGPHIMM), May 2003.
21. Baumeister, B. *Marks' Mechanics Engineers' Handbook*, McGraw-Hill, New York, 1966, pp. 4–25.
22. Durina, M. U.S. Patent No. 5,164,207 (1992).
23. Zeiger, D. J. U.S. Patent No. 5,441,400 (1995).
24. Dray, R. F. U.S. Patent No. 5,151,282 (1992) and 5,258,158 (1993).
25. Xu, J. *SPE ANTEC, Tech. Papers*, 2222–2226 (2005).
26. Johannaber, F. *Injection Molding Machines: A User's Guide*, Hanser/Gardner Publications, Cincinnati, 1985, p. 219.
27. Wang, C., Cox, K., and Campbell, G. A. *SPE ANTEC, Tech. Papers*, 406–410 (1995).
28. Park, C. B., and Suh, N. P. *Cell. Polym.* **38**, 69–91 (1992).
29. Shimbo, M. *FOAMS 2000*, 162–168 (2000).
30. Park, C. B., Baldwin, D. F., and Suh, N. P. *Polym. Eng. Sci.* **35**, 432 (1995).
31. Chen, L., Sheth, H., and Wang, X. *FOAMS 2000*, 127–131 (2000).

32. Chen, L., Sheth, H., and Kim, R. *SPE ANTEC, Tech. Papers*, 1950–1954 (2000).
33. Colton, J. S., and Suh, N. P. *Polym. Eng. Sci.* **27**, 485–492 (1987).
34. Semerdjiev, S. *Introduction to Structural Foam*, SPE., Towanda, PA, 1982.
35. Throne, L. J. *Thermoplastic Foams*, Sherwood Publishers, Hertford, UK, 1996, p. 338.
36. Lee, J. W., Wang, J., Park, C. B., and Tao, G. *SPE ANTEC, Tech. Papers*, 743–747 (2008).
37. Gruber, H., Voggeneder, J., and Kapfer, M. U.S. Patent, No. 0,056,935 A1 (2002).
38. Park, C. B., and Xu, X. Apparatus and Method for Advanced Structural Foam Molding. U.S. Patent application, Serial No. 11/219,309, filed on September 2, 2005; Canada Patent, Application No. CA 2,517,995, filed on September 2, 2005.
39. Park, C. B., Xu, X., Lee, J. W., and Zhu, X. Investigations on Advanced Structural Foam Molding Using a Continuous Polymer/Gas Melt Flow Stream. SPI, Plastic Parts Innovations Conference, Columbus, Ohio, April 2–4, 2006.
40. Cardona, J. C., et al. U.S. Patent No. 6,926,507 (2005).
41. Pfannschmidt, O., and Michaeli, W. *SPE ANTEC, Tech. Papers*, 2100–2103 (1999).
42. Michaeli, W., and Cramer, A. *SPE ANTEC Tech. Papers*, 1210–1214 (2006).
43. Michaeli, W., et al. German Patent DE 19 853 021 A1 (2000).
44. Habibi-Naini, S., and Schlummer, C. *SPE ANTEC, Tech. Papers*, 470–474 (2002).
45. Witzler, S. *Injection Molding Mag.* **December**, 80 (2001).
46. Mapleston, P. *Modern Plastics Worldwide* **November**, 31 (2002).
47. Defosse, M. *Modern Plastics Worldwide* **December**, 14–15 (2009).
48. Witzler, S. *Injection Molding Mag*. **April**, 19–20 (2000).

8

SPECIAL PROCESSES

This chapter discusses several successful special processes developed to improve the quality of a microcellular injection molding part. Typical special processes used for microcellular processing are co-injection, overlapping, gas counterpressure, reversal coining, hot and cold mold, super-microcellular foaming or low gas dosage foaming, local foaming, stress foaming, foam with chemical blowing agent, foam with water blowing agent, thin-wall foaming, and so on. All these special technologies attempt to make better surfaces on microcellular parts with uniform cell structure. The details of equipment and processing conditions are presented, and good reference data sheets are given.

8.1 CO-INJECTION (SANDWICH) MOLDING FOR MICROCELLULAR PART

The co-injection molding method (also called sandwich molding) has been in use since the early 1970s. It was the invention by Garner and Oxley of ICI [1]. Co-injection molding has been used not only for a perfectly smooth surface but also for a better part property or economic recycling parts, such as (a) reinforced material as the core and the pure material as the skin or (b) recycled material in the core and virgin material in the skin. It can also be used for different materials as skin and core as long as the material combination satisfies the rule of adhesion between them. The microcellular co-injection part usually has a nonfoamed skin material with a microcellular core. It follows

Microcellular Injection Molding, Edited by Jingyi Xu
Copyright © 2010 John Wiley & Sons, Inc.

most of the same rules of solid co-injection molding that have been studied in the literature [2–18]. The only difference that we must consider is that the foamed core made by low-viscosity gas-rich material is more difficult to make than a uniform core in the co-injection part.

8.1.1 Mold Filling Analysis for Microcellular Co-injection

The co-injection molding process is characterized as a process to completely encapsulate an inner core with an outer skin in the same molding cycle. The sequential process is skin–core–skin with three steps. The schematic of mold filling of co-injection is illustrated in Figure 8.1. The first step is to inject the skin material about 25–40% of full shot size of the part to form enough flow front and frozen skin on the wall of the mold. The free-flow front of the skin material during mold filling will have the fountain flow effect at the advancing melt front. This fountain flow helps to form more new skin before core material reaches the same position. The core material is injected secondly when the skin material in the center remains molten. There are two different ways during the second step. One is enough skin material already in the mold so that only core material will be injected into the mold at that time. The second way is to inject skin and core material simultaneously. However, the core material will be added more and more and skin material will be injected less and less with the injection speeds and pressure controls for skin and core separately. As the skin material is frozen around the core material, it prevents the core material from penetrating skin to become the surface of the part. However, the core material should never catch up to the skin material in the free-flow front since there is no solid skin on the wall of mold in the free-flow front. The ideal mold filling of co-injection is to keep the schematic shown in Figure 8.1 until the cavity is nearly filled. Finally, the core material stops being injected and the skin material is injected into the mold again at about 5–10% of full shot size to seal the gate area of co-injection part by skin; consequently

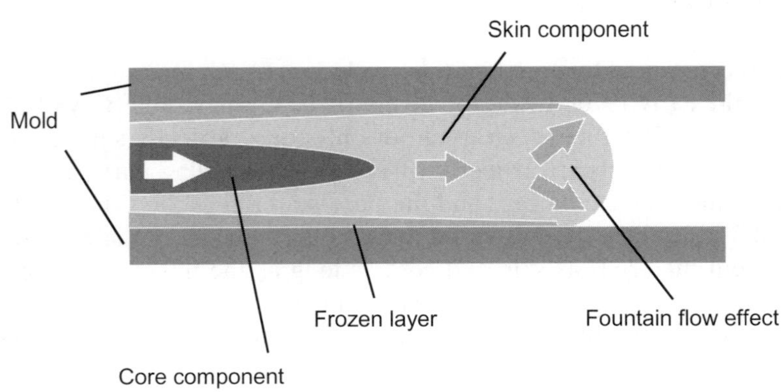

Figure 8.1 Mold filling schematic of co-injection. (Courtesy of Engel.)

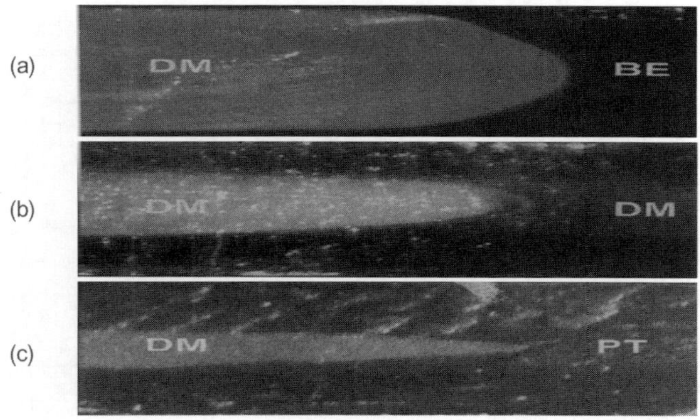

Figure 8.2 Results of experiments for the effects of different viscosity materials in co-injection process. (Courtesy of Engel) **(a)** Core viscosity lower than skin viscosity. **(b)** Core viscosity equal to skin viscosity, **(c)** Core viscosity higher than skin viscosity.

the skin material is cleaned in the runner and nozzle channels, with skin being the only material prepared for the next shot.

The mold filling analysis of co-injection molding is the good tool to determine the core–skin ratio and flow ratio. The part geometry and material viscosity are the key factors for a successful co-injected process. The experiments of skin and core materials co-injected in a visible mold with different viscosities were carried out in Engel's laboratory. The mold of the co-injection molding study has a glass to show the material filling in the mold. The dynamic pictures are taken during mold filling with both skin (transparent material) and core (dark colored material) in the mold. The different material combinations will result in the different thickness of core–skin ratio because of the viscosity differences, as shown in Figure 8.2. The thickness of the core becomes a maximum value among the tests if the core viscosity is lower than the skin viscosity in Figure 8.2a. The result in Figure 8.2a is the similar viscosity combination for the co-injection with foamed core in the unfoamed skin. Therefore, the low-viscosity gas-rich material can easily penetrate the skin if the materials of skin and core are the same. For the material combination of co-injection molding with equal viscosity of skin and core, the core thickness is close to the skin thickness, as shown in Figure 8.2b. Finally, Figure 8.2c shows the thinnest thickness of core in the experiment in which the core viscosity is higher than skin viscosity. The results in Figures 8.2b and 8.2c show that a high-viscosity core material reduces the risk of a core material penetrating the skin. Therefore, the combination of virgin material as a skin and reinforced same material as a foamed core, or low-viscosity material as a skin with a different material from skin as a high-viscosity foamed core, will be the best selection from a processing point of view.

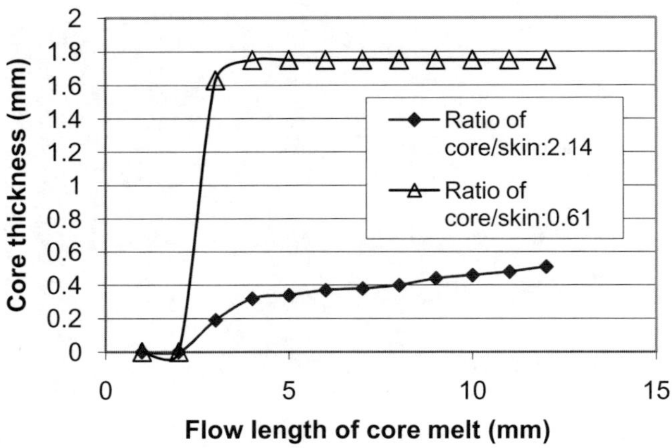

Figure 8.3 Core material thickness versus core material stroke.

The data in Figure 8.3 list the core thickness at each combination of different polypropylene (PP) materials, and they list the same materials measured at the position of 15 mm beyond the gate. The interesting results are the data of different materials with different viscosities. The top profile in Figure 8.3 represents the picture shown in Figure 8.2a, material combination is similar to the solid skin (high viscosity) and foamed core (low viscosity), with the viscosity ratio of core to skin 0.61. The low-viscosity core material occupies most thickness of the channel and the maximum core thickness is 1.75 mm in Figure 8.3. The low-viscosity core material only needs a 2-mm injection stroke of core material to stabilize the maximum core thickness 1.75 mm that is kept the same even if the injection stroke of core material is increasing. On the other hand, the thinnest core thickness (0.5 mm; see the bottom profile in Figure 8.3) is from the highest-core-viscosity material, whose viscosity ratio of core to skin is 2.14 (material combination). It matches the picture shown in Figure 8.3c. In addition, the thickness of high-viscosity core material will continually increase with the increasing of the stroke of the core material. The same material as skin and core will have the core thickness between the profiles of high ratio (larger than 1) and low ratio (less than 1). The important conclusion from Figure 8.3 is that the viscosity ratio of core to skin is effective for the core thickness in the co-injection molding. If the viscosity ratio of core to skin is less than one, the low-viscosity core material only needs a very short stroke to maintain the maximum thickness. However, the thickness of the high-viscosity core material will increase with the increasing of the core material stroke.

The skin thickness is also measured at the position of 15 mm beyond the gate. The skin thickness is formed instantly about 0.15 mm once material is injected into mold, as shown in Table 8.1. Then, the skin thickness is increased continually with the time of injection. Near the 1 sec of injection time, the skin thickness is almost twice the initial thickness up to 0.3 mm at the beginning of

TABLE 8.1 Skin Thickness Versus Filling Time Measured at 15 mm Away from the Gate

Fill Time (sec)	Skin Thickness (mm)
0.2	0.16
0.4	0.20
0.6	0.24
0.8	0.26
1.0	0.28

injection. The turning point occurs at about 0.6 sec. Before 0.6 sec, skin thickness increases 0.04 mm every 0.2 sec. However, after 0.6 sec, skin thickness increases 0.02 mm every 0.2 sec. It is because the skin thickness may build up more insulation for efficient heat transferring through the metal wall of mold so that the skin thickness forms slower at certain skin thickness build up.

The effects of molding parameters on the skin thickness are studied in reference 4. Also, Messaound et al. [11] studied the effects of injection speed, the skin and core strokes for the virgin PP, and reinforced PP with glass fibers, and they claimed that the skin injection speed has significant effect on the skin–core distribution. In addition, the skin stroke is twice as core stroke will get the most uniform core distribution.

Han [12] did some experiments with two different viscosity materials in the co-extruding die. One material is polystyrene (PS), and another one is the high-density polyethylene (HDPE). One test temperature is 200 °C, and the wall shear stress is lower than 9×10^6 dyne/cm^2. The viscosity of PS is higher than the viscosity of HDPE at this processing condition. Therefore, the high-viscosity material PS is wrapped by low-viscosity material HDPE stably at 200 °C. However, when the test temperature is changed to 240 °C, the viscosity of HDPE is higher than the viscosity of PS. Then, the HDPE material will be wrapped by PS material in the co-extrusion die. On the other hand, the elasticity of melts is almost unchanged for HDPE and PS at these two different test temperatures and other processing conditions. It is proved that the viscosity differences are the determining factor of the boundary shape changing between two materials. In addition, the important theory of co-injection is that the low-viscosity material is always trying to first wet the surface. In other words, the combination of low-viscosity core and high-viscosity skin will cause more core breakthrough defect in the co-injection part. The similar conclusions are given in the literature [2–4]. White and Dee found that the most uniform skin–core structure occurs with the following order: injection of low-viscosity material first and then high-viscosity material second.

The gas-laden melt has low viscosity compared to the same material without gas. Therefore, if the same material is used for the co-injection molding, the low-viscosity core tends to wrap the high-viscosity skin material. It means that the core material may eventually penetrate through the skin to show core material at surface if the flow ratio is large enough. This defect is called core surfacing, or core breakthrough [13, 15, 17–19]. If the viscosity of the core is very low compared to the viscosity of skin, the fingering effect of the core may easily occur [13, 15, 17–19]. To avoid the defects in the co-injection molding with microcellular foam as the core, the skin material must be sufficiently displaced to the melt front in the mold filling. Even if the skin material does have enough in the flow front so that the core material will not penetrate through the skin, the real core material distribution may not be symmetrical in the center of the part. One experiment makes a co-injection part with GPPS foamed core with pure GPPS skin in the 152.4-mm × 279.4-mm × 3.2-mm mold with center sprue gate. It is made with a skin–core method that accumulates skin and core in a specific order and then injects them together into the mold. It is obvious that the foamed core ends with the sharp tip that is similar to the end of the core shape in Figure 8.2c, except that the tip tends to near one side of the part. If the core material is displaced further, it may penetrate the skin front on one side only. On the other hand, the core material has relatively lower pressure drop while the skin material produces the higher pressure gradient in the skin domain and thus produces higher skin velocity. As the skin is displacing fast, the core penetration will be accelerated locally, with one side being penetrated more than another side. In other words, the foamed core will have an irregular penetration pattern in the co-injection molding. The flow front is usually the location where the lowest pressure, or even zero pressure, exists during mold filling. Therefore, the cell growth may be out of control, becoming a void.

Turng and Kharbas [17] analyzed the penetration behavior of (a) microcellular and solid cores during co-injection molding, (b) specimens from solid skin with microcellular core, and (c) solid skin to solid core. The mold used in this experiment was a two-cavity plate mold with tab gates. The dimensions of the plates were 120 mm × 40 mm × 3.5 mm. A general-purpose polystyrene (GPPS) resin (BASF 145 D) was used as both the skin and core materials [17]. The overall sample pictures are shown in Figure 8.4. Sample 3 is solid skin to solid core (45% of volume of core with dark color at top of Figure 8.4), and CM3 is solid skin with microcellular core (with milky microcellular core about 44% volume-wise sandwiched by the clear PS solid skin at the bottom of Figure 8.4). The maximum volume percent of core material encapsulated by skin completely is about 54% for solid to solid of sample and about 50% for solid to microcellular sample. Beyond the maximum volume percentage of core above core, breakthrough has occurred in both samples.

The samples of solid skin with microcellular core were sectioned along the flow direction through the center of the part (see Figure 8.5). The fraction of core thickness was then measured with the help of a light microscope. It can be seen that the solid–solid co-injection molded specimen exhibited a thinner skin layer near the gate and exhibited a thicker skin layer downstream. Recall that the solid core used in this study was essentially the same as the skin material except that it was blended with 1% coloring pigment by weight. Therefore,

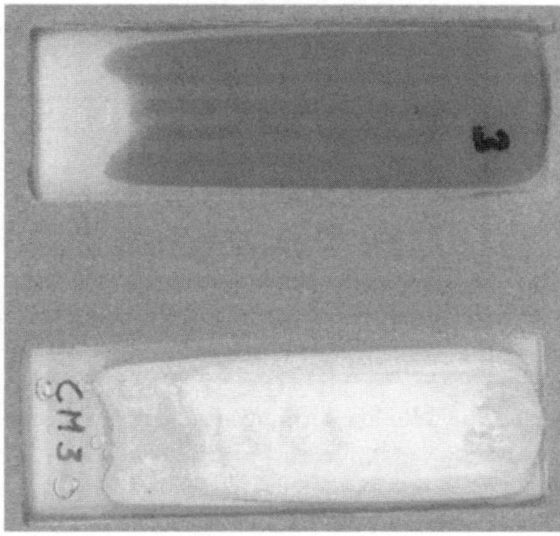

Figure 8.4 Microcellular co-injection molded parts [17]. (Reproduced with permission from Society of Plastics Engineers.)

Figure 8.5 Pictures of the cross sections of a co-injection molded part that contains a microcellular PS core sandwiched by clear, solid PS skin layers [17]. (Reproduced with permission from Society of Plastics Engineers.)

the penetration profile in the melt flow direction (which can be identified by the skin–core interface) is basically a stretched gap-wise velocity profile or a specific spatial contour of the residence-time distribution in the cavity. On the other hand, the microcellular core had a fairly even penetration with a uniform skin layer thickness along the flow length direction. It is speculated that the less viscous core material can only displace the skin material near the center of the part where the skin material remains hot and fluid, leaving a thicker

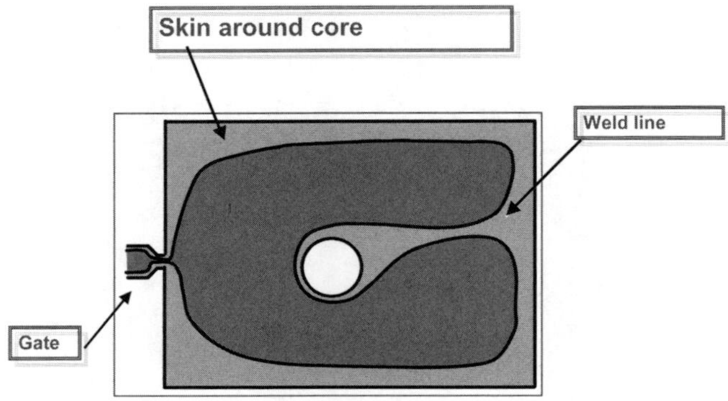

Figure 8.6 Schematic for the weld line forming around in co-injection mold. (Courtesy of Engel.)

skin layer, a thinner core, and a longer core penetration length. An analytical analysis has shown that such spatial core penetration profiles result from the different pressure gradients and velocity profiles in the skin and core regions.

The experimental results discussed above are simple mold without holes. The real mold has more details in the microcellular part. For example, the hole in the part will cause more mold filling issues, such as weld line position and strength. Figure 8.6 shows a schematic of the forming of a weld line because of the hole in the part. The weld line shown in Figure 8.6 is formed by skin material. It is the best to have a skin material joint for the weld line, instead of core material. It will be a weaker spot where the weld line is formed by foamed core material.

8.1.2 Material for Microcellular Co-injection

A modified material combination chart [solid to solid; see references 14 and 15] for microcellular processing in Table 8.2 shows the possible materials used together for the co-injection. It is a guideline chart for the material selection of microcellular injection molding. Although this chart is mostly from the data of structural foam, it actually performs better with microcellular foam because of the small cell size and large number of cells in microcellular foam. In addition, the pressure in small cells of microcellular foam is higher than the pressure in the cells of structural foam. The high pressure in the cell will help to push the core material against the skin and help for adhesion. The skin–core combination may have several possible functions as discussed the following.

Generally, addition of fillers or glass fiber reinforcements in material will result in reduction in bonding between the skin and core materials. However, the filled material is favored as the core material, with the unfilled same material as the skin for microcellular foam.

It is important that there be good adhesion between skin and core in co-injection molding. However, Table 8.2 shows that not all plastics have good

TABLE 8.2 Material Combinations for the Multiple Component Molding Method [1, 2]

RESIN	AES	ASA	CA	EVA	PA6	PA66	PC	HDPE	LDPE	PMMA	POM	PP	PPO	GPPS	HIPS	PBTP	RPVC	SPVC	SAN	TPU	PSU	Blend PC-PBTP	Blend PC-ABS	TPR
ABS	+	+	+				+	−	−	+		−		−	−	+	+	0	+	+		−	−	
ASA	+	+	+	+			+					−		0			+		+	+		−	−	
CA	+		+	0															+					
EVA		+	−	+				+	+			+		+			+	0						
PA6					+	+		−	−			−			−					+				
PA66					+	+	−	−	−			−			−					+				
PC	+					−	+							−	0				+	+	+	−	−	
HDPE	−			+	−	−		+	+	−	−	0												
LDPE	−			+	−	−		+	+	−	−	+		−				−						
PMMA	+							−	−	+		−		−	0		+	+	+					
POM								−	−		+		−											
PP	−	−		+	−	−		0	+	−	−	−	+	−	−	−		−	−					+
PPO													+	+	+	+			−					

TABLE 8.2 ***Continued***

RESIN	ABS	ASA	CA	EVA	PA6	PA66	PC	HDPE	LDPE	PMMA	POM	PP	PPO	GPPS	HIPS	PBTP	RPVC	SPVC	SAN	TPU	PSU	Blend PC-PBTP	Blend PC-ABS	TPR
GPPS	–			+			–	–	–	–		–	–	+	+	+		–	–					
HIPS	–	0			–	–	0			0		–		+	+	+		–	–					
PBTP	+															+	+		+					
RPVC	+			+						+		–				+	+						+	
SPVC	0			0				–		+		–		–	–		+	+	+	+				
SAN	+	+	+				+			+			–	–	–	+		+	+	+		–	–	
TPU																				+				
PSU							+														+			
PC-PBTP	–	–					–													–		+		
PC-ABS	–	–					–												–	–			+	
TPR												+												+

+, Good adhesion.
–, Moderate adhesion.
0, Poor adhesion.

adhesion, and they are divided into materials with good adhesion, bad adhesion, and no adhesion. The chemical structure of plastics also has an influence on the adhesion. Even if the adhesion between some plastics is no good, the addition of a special adhesive agent to one of the components may make them bondable. For example, soft PVC or thermoplastic polyurethane (TPU) has a good adhesion to polycarbonate, ABS, SAN or similar blends. Furthermore, thermoplastics rubber has a good adhesion to polypropylene.

8.1.3 Methods for Microcellular Co-injection

There are basically two major co-injection methods on the market. One is the skin–core injection simultaneously with two independent injection units: one for skin material and another for core material. The injection pattern for this method is skin–core and skin–skin. Another co-injection molding method is to accumulate skin and core materials in a specific order. The core material is accumulated first, which is usually a foam material. Then, the skin material is accumulated second. The result of accumulated material in the prepare shot volume is that the skin material is in the front position or upstream of the injection barrel, and the core material is behind the skin material. Then, only one injection unit is used to inject this material into the mold through one channel of nozzle. The details of both methods are discussed below.

8.1.3.1 Skin, Core Material Injected Simultaneously. This method is a traditional way to make solid co-injection part. The details of a co-injection head and a shut-off nozzle can be simplified as the layout in Figure 8.7. The outside channel is the skin channel that is connected to the skin injection unit. The center channel is the core channel that is from the core injection unit. The shut-off nozzle pin can close both skin and core channels and can control the core channel opening as well. For the two independent injection units used for co-head, there are two elements for independent shut-off control in the nozzle. One is the center pin and another is the sleeve outside of the pin, as

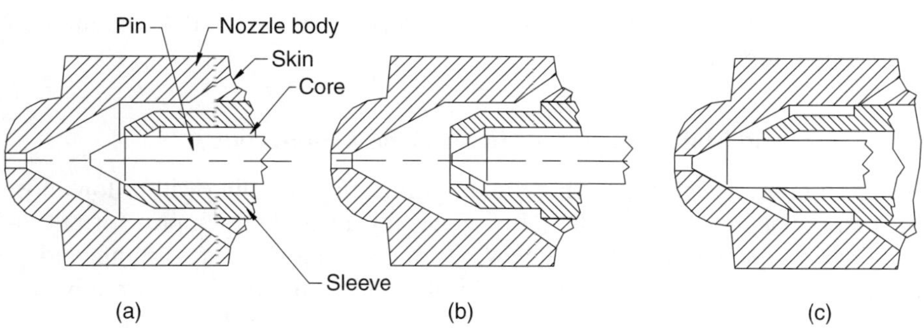

Figure 8.7 Schematic of co-injection head. **(a)** Skin injection. **(b)** Skin and core injection simultaneously. **(c)** Skin and core closed. (Courtesy of Engel.)

shown in Figure 8.7. The first step of co-injection is the skin injection that is only controlled by the sleeve retracted to open the skin channel shown in Figure 8.7a. On the other hand, the core channel is closed with the pin in Figure 8.7a. The second step of co-injection is shown in Figure 8.7b, where the pin is retracted further to an open core channel as the skin channel is still opened by the sleeve. The injection volume rates of skin and core are controlled with two injection units separately. The final step is to close the core channel again by the pin and keep the skin channel open as in Figure 8.7a again. Figure 8.7c shows the position of both the skin and the channel closed by the pin and the sleeve. It is necessary for screw recovery during mold cooling. This is the most popular design for microcellular co-injection molding since the opening and closing for skin and core channels are controlled separately.

Since the core material is a gas-laden melt that is difficult to be sealed, the tight tolerance is required for the fitting between the pin and the sleeve shown in Figure 8.7. Any contamination of gas on the surface will cause the quality issue of the smooth surface of the microcellular co-injection part.

8.1.3.2 Skin–Core Materials Accumulated in Order with Only One Injection Unit. An invention by Klaus and Stefan [20], which was later registered as the Mono-sandwich process by Ferromatik Milacron, provided the equipment. As shown in Figure 8.8, the skin and core materials are accumulated in a common barrel in a specific order. The core material with gas from gas injector “4” can be accumulated first from a reciprocating screw “3” and then accumulates the right shot size of the core material in the common barrel. Then, the skin material is plasticized from the extruder 1, and then it is accumulated in the common barrel ahead of the core material until the final total stacked shot size is reached. After the skin shot size is reached, the skin extruder stops. When injection cylinder “5” pushes the reciprocating screw forward, it guarantees that the skin material will be injected into the mold first and lays itself to the mold wall as skin. The following material constitutes the core of the part. This is a significant cost saving in relation to the sandwich co-injection process. It is also easy to change the materials since there is no such complicated co-injection head as the method mentioned above.

8.1.4 Part and Mold Design for Microcellular Co-injection

Part and mold design for microcellular co-injection molding will follow most of the regular injection molding. It is generally a simple thick part for microcellular co-injection molding. However, some special rules are still needed to design the microcellular part and mold of co-injection molding correctly.

8.1.4.1 Part Design. The part design for microcellular co-injection may have some basic targets as follows [15]:

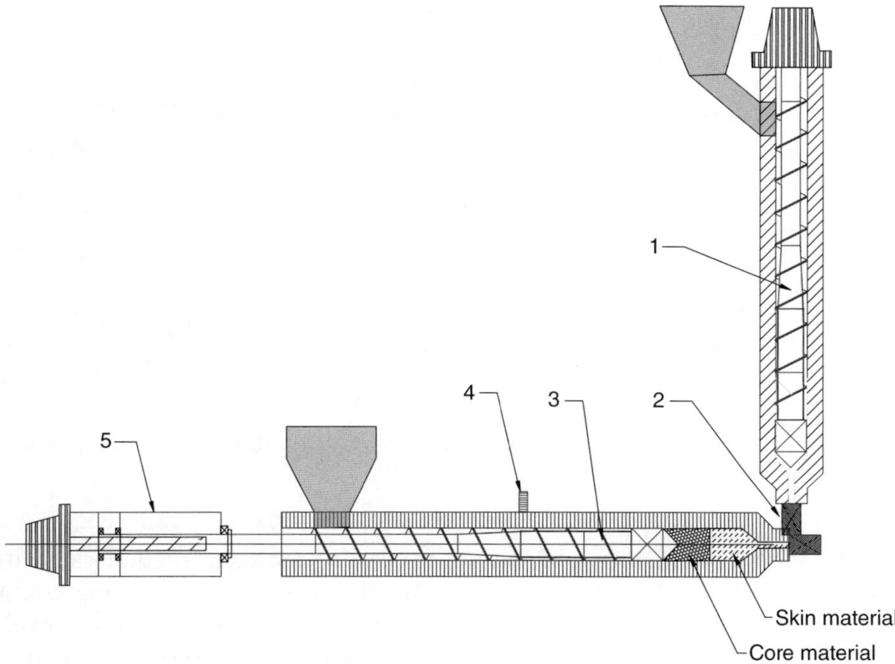

Figure 8.8 Layout of IKV co-injection machine. 1, Extruder for skin material; 2, check valve; 3, core material reciprocating screw; 4, gas injector; 5, injection cylinder.

1. Nonfoamed Skin Combined with Foam Core with the Same Material. This is the application for the part that needs perfect skin without a sink mark and warpage. The foamed core has the advantage to achieve the dimension stability and remove the sink marks because of the cell expansion during mold cooling. On the other hand, the solid material for skin creates the first-class finish of surface that is comparable to the surface quality of straight injection molding. This is one of the most popular applications with microcellular co-injection molding.

It has been observed that the internal pressure arising from foaming eliminates the typical sink marks and improves the surface flatness of the molded parts. A sink mark is a local surface depression that typically occurs when molding thicker sections, or at locations above ribs, bosses, or internal fillets. As an example, Figure 8.9 compares the surface depression profile at a selected thick section of a solid-microcellular co-injection molded sample to the same part molded conventionally. As can be seen, the sink mark in the solid-microcellular co-injection molded part was reduced by 90%. This is primarily because the internal gas pressure allows the material to expand as the part cools, thus holding the skin firmly against the mold walls. Although one can argue that sink marks in conventionally injection molded parts can be reduced with higher packing pressures, the microcellular process is more attractive as

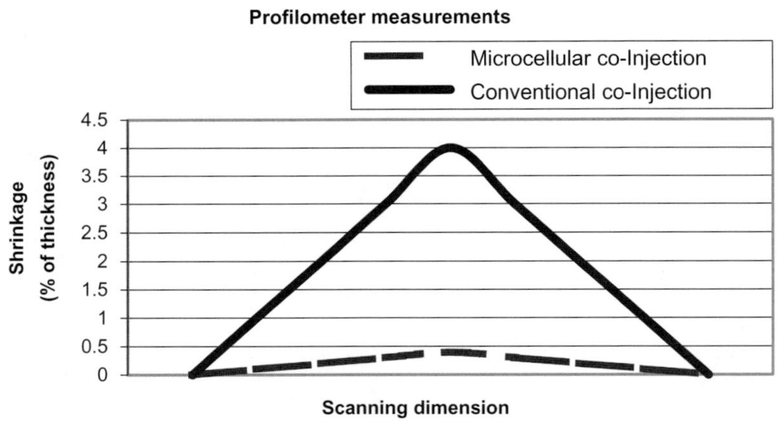

Figure 8.9 Measured surface depression (sink mark) [17]. (Reproduced with permission from Society of Plastics Engineers.)

far as energy consumption and ease of processing are concerned, because it does not require a packing phase during molding. Based on this finding, components of variable thickness could conceivably be consolidated into a single complex part (akin to gas-assisted injection molded parts) to cut production and assembly costs without concerns of sink marks or part warpage due to excessive or differential shrinkages. Additionally, since the viscosity of the MCP core is more comparable to that of the skin polymer than gas, it is reasonable to anticipate that solid-microcellular co-injection molding will be more stable and forgiving than gas-assisted injection molding.

As can be seen by injecting the solid resin and microcellular material sequentially into the mold with proper volume fractions, the microcellular core can be entirely encapsulated by the solid skin material. Such a structure eliminates the swirling surface patterns typical of microcellular parts and enables the production of components with class "A" surfaces. In addition, this hybrid process also retains the advantages of microcellular injection molding, such as lighter part weight, reduced cycle time, better surface flatness (i.e., reduced sink mark as discussed above), and improved dimensional stability. For example, a co-injection molded rectangular box part with a blue-colored polypropylene (PP) skin and a microcellular PP core is tested and warpage is measured. The experimental rectangular part mold has relatively thicker walls with a stepwise variation on the shorter sides of the part to intentionally induce race tracking and part warpage for research purposes. The shrinkage and warpage by employing the microcellular process can be significantly reduced when compared with a conventional injection molded part using the same PP resin and same mold-wall temperature.

However, the gas-laden material is sometimes difficult to clean up after every cycle. It needs to control the processing conditions tightly to make sure skin material will not be contaminated from the core material that has the pressurized gas in it. Otherwise, any gas swirl marks on the skin surface of co-injection molding part becomes the rejected part.

This application may also include the case of colored material saving. Since the colored skin is only determined by the skin material so that with the raw material used for the foamed core it is not necessary to use color concentrate, the cost of color concentration in the core material can be saved with the microcellular co-injection molding technology.

2. High-Quality Skin Combined with Low-Cost Foam Core. This is similar to the material combination above. However, it uses same the low-cost material as the microcellular foam core. It then, has the same advantages as the part above. In addition, the core has different material from the skin. Usually, core material is less expensive than skin material. This application includes more popular application with the same material but recycled waste for the foamed core.

3. Reinforced Skin Combined with Nonreinforced Foam Core. This special application just for the surface needs more hardness than the virgin unfilled material. Thus, the fillers are added to increase the surface hardness. Sometimes the wearing resistance is needed for the surface of the part as well. The solution is to add glass-fiber-reinforced material as the skin and to use foamed nonreinforced material in the core without contacting the wearing surface.

In addition, the reinforced material is much more expensive than the same material without reinforced fillers. If the bending load is the major concern of the part, the combination with reinforced material as skin is good enough to reach the rigidity requirement with the foamed nonreinforced material in the core. Then, this application can save the cost of using 100% of reinforced material.

4. Nonreinforced Skin Combined with Reinforced Core. This is another popular application to use the nonreinforced skin for the excellent surface appearance and to use reinforced core for the strength of the part. The co-injection processing prefers this combination because the viscosity of skin material is lower than the viscosity in the core material.

Furthermore, the reinforced material in the core increases the modulus of elasticity while the impact resistance of the part can be maintained. In addition, the reinforced material as the core will increase the thermal stability of the part because the reinforced material will reduce the thermal expansion.

5. Flexible Foam Skin Combined with Rigid Core without Foam. This usually requires the creation of a soft-touch surface that will need TPE or TPU microcellular foam skin. The rigid core without foam gives the strength and stiffness of the part and acts like a bone of the part. Very often, two moldings of different plastics are used, such as PVT foamed skin with ABS solid core. However, if the material itself is strong enough, the same material can be used for skin and core while the skin is made with microcellular foam, and the core is made without foam. Both of skin and core are made in a one-step operation of co-injection.

6. Rigid Skin Combined with Flexible Core with Foam. The rigid skin with flexible core with microcellular foam, which has excellent shock-absorbing properties, can be achieved. It has good sound and thermal insulation properties as well. This kind of part is the normal co-injection part with solid skin and foamed core. It may be the same material or different materials as skin and core.

7. Food- or Medical-Purpose Material of Skin Combined with Foam Core with Barrier Properties or Thermo Properties or with Low-Cost Materials. Anything for food and medical purposes must be nontoxic and may need some extra properties, such as barrier properties and impermeable properties. The medical material is usually expensive. Therefore, the foamed core may save the cost of medical material. On the other hand, other properties need to be added in the core material to make some special function of microcellular parts in the food and medical industries.

8. Morphology of Microcellular Co-injection Molding Part. The morphology of a co-injection part made by the skin–core method above shows a clear boundary between foamed core and skin, as shown in Figure 8.10. However, the cell structure of co-injection is not as good as the cell structure of regular microcellular injection molding. It is because the core pressure control may

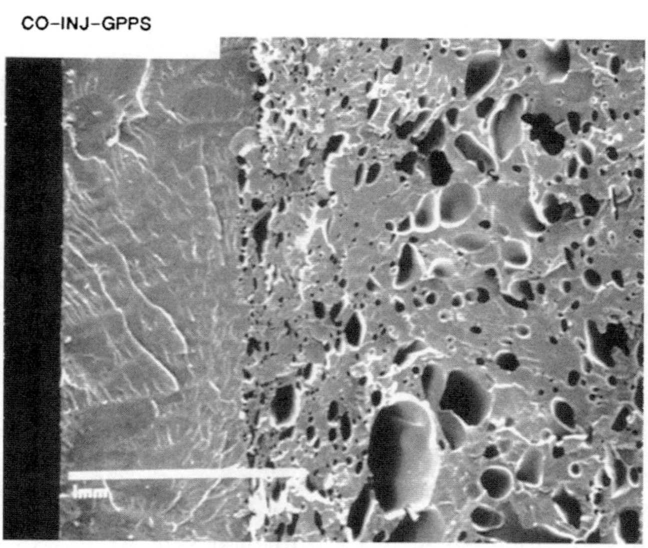

Figure 8.10 Clear skin–core boundary in GPPS co-injection part.

not be good enough to make the cell structure as good as the regular molding in a thick (3.2 mm) part. The core–skin viscosity ratio of this case is the restriction as well since the skin and core materials are the same. The core material is GPPS microcellular foam, which obviously has low viscosity compared to the skin of GPPS without any foam.

Turng and Kharbas [17] made a good microcellular co-injection molding part with two independent injection units. The principle is the same as the designs in Figures 8.7 and 8.8. However, the key element of the shut-off nozzle in Figure 8.7 is now in the platen as the switch valve to control different material opening channels. It is the skin–skin and core–skin method. The skin and core material can be controlled independently and more precisely. Therefore, the part morphology of microcellular injection molding with this method is much better than the morphology of the part made by the skin–core molding method above.

Analysis of the foamed core under a light microscope revealed a truly microcellular structure. Figure 8.11 shows a light micrographic image at a center layer of the microcellular core with 44% volume of full shot volume [17]. The average cell size observed in the light micrograph was around 8 μm. Different from the microcellular injection molding process, the foamed core in co-injection molded parts is encapsulated by a solid skin material. As a result, the growth of the microcells is constrained, thereby resulting in finer, more desirable cell sizes. The SEM image of a solid–microcellular co–injection molded sample with 50% volume of microcellular core shows results similar to those of the cell structure in Figure 8.11. The microstructure at the center

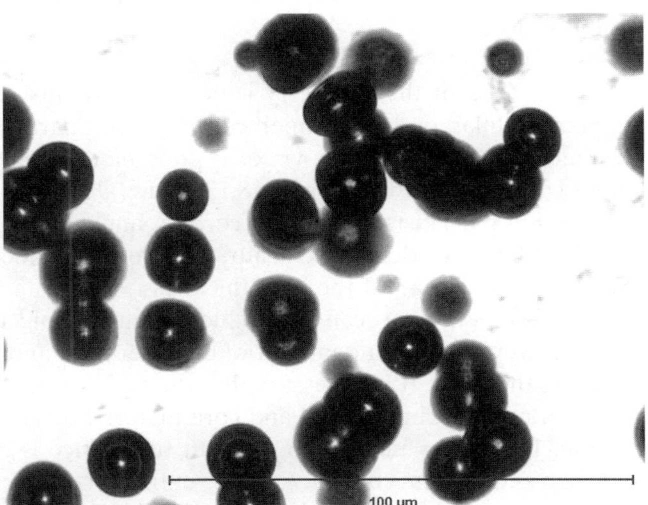

Figure 8.11 Light micrographic picture of the center layer of a microcellular co-injection molded part with 44% volume of core [17] (blue line indicates 100 μm). (Reproduced with permission from Society of Plastics Engineers.)

of the part's cross section exhibited fairly uniform and smaller cells as compared to those of regular microcellular injection molded parts. The average cell size observed in the SEM micrograph was around 10–12 μm, and the cell density varied from 2×10^8 to 3×10^9 cells/cm^3.

There is big improvement of cell structure if the core material uses the filled material, and then the skin material can use the same material without foam for class "A" surface. Figure 8.12 shows the morphology picture of a perfect example with PA 6 as the skin material and PA 6 + GF30 as the core material [18]. The reinforced PA 6 with 30% of glass fibers has slightly high viscosity even with gas dissolved into it compared to PA 6 skin. The part is a 2-mm-thick disk with a 350-mm diameter. It was made by the same skin–core method that was used for morphology in Figure 8.10. The volume ratio of co-injection is about 60% skin and 40% core. This is the best combination of two different viscosity materials. The skin and core still has a boundary between foamed core and nonfoam skin, as shown in Figure 8.12b. In addition, the local view in Figure 8.12b shows microstructure of the foam. The cell size in the core is about 5–20 μm, and cells are distributed uniformly across the thickness direction. The overall cell structure in the whole section view of a co-injection part shown in Figure 8.12a is excellent because the cell size is small and the distribution across the thickness of the sample is uniform. The end of mold filling is still fully covered by the PA solid skin material, as shown in Figure 8.12c. It is important to keep the end of mold filling with enough full solid skin so that the foamed core never penetrates the skin. The best material combination for processing is a high-viscosity material in the core and a low-viscosity material in the skin, which creates high core–skin ratio of co-injection. On the other hand, the cell structure of the filled material is very good and even the simple skin–core method is used.

9. Geometry of Part. One of the geometry issues is the radius of the inner corner of the microcellular part. The inner corner of the part needs to be designed with some radius since the low-viscosity core material tends to move close to the inner corner, as shown in Figure 8.13. Generally, fillets are necessary in the co-injection part, instead of sharp corners, and sudden changes in wall thickness should be avoided. Specifically, it is recommended that the inner radius be as big as the half of the near thickness.

The complicated geometric part can be produced using co-injection microcellular molding. However, a certain amount of consideration and know-how is involved in gating the parts. It is necessary that the laminar velocity profile be well-controlled for both skin material and core material during the whole stroke of injection. The stable mold filling without turbulence is determined by both mold design and process control. A good venting system is also important for mold design to avoid the air inclusion in the skin.

8.1.4.2 Mold Design. The mold design for co-injection microcellular molding is focused on the gate designing. It is basically to consider the gate

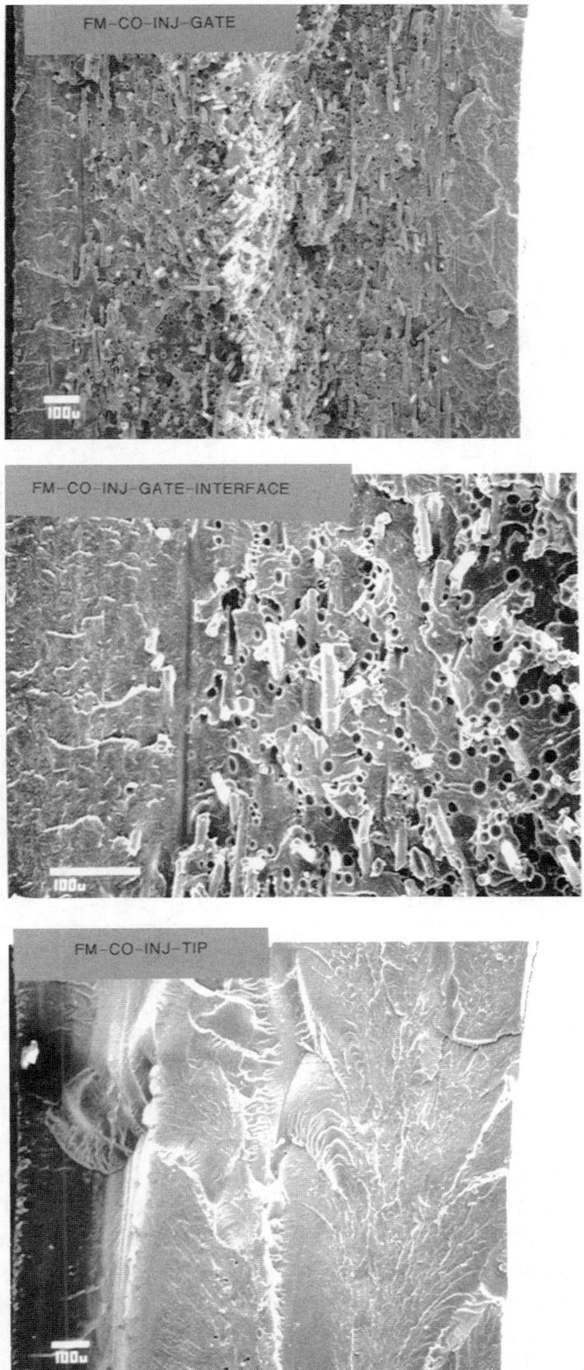

(a)

(b)

(c)

Figure 8.12 Morphology pictures of PA6 co-injection microcellular foam. **(a)** Whole section view of foamed core (PA6+GF30) and solid skin (PA6) [18]. **(b)** Local view of foamed core and solid skin [18]. **(c)** End of mold filling with full skin [18]. (Reproduced with permission from Society of Plastics Engineers.)

Figure 8.13 Core material is close to the inner corner.

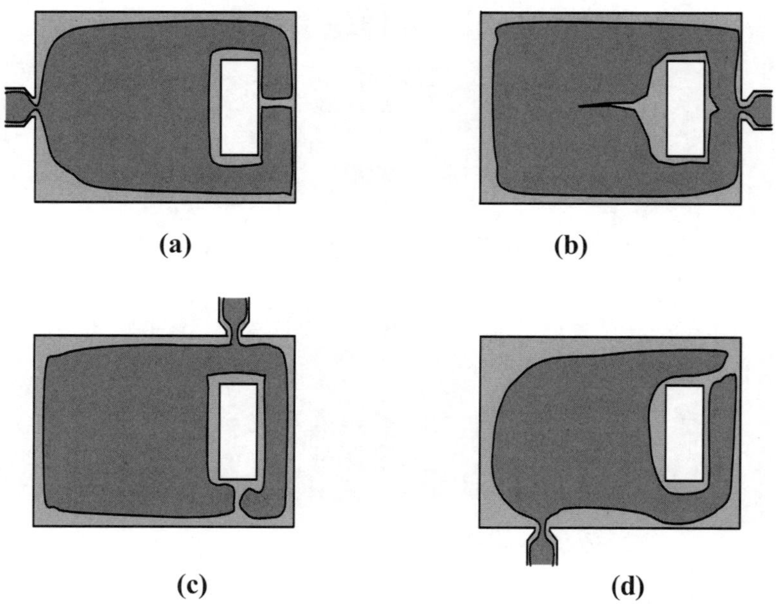

Figure 8.14 Weld line positions related to different gate locations.

locations for uniform distribution of core material and the weld line position in the part.

The weld line position is changed according to the gate position, as shown in Figure 8. 14. The weld line is usually the weak spot of the part, so it must be the position of the non-force support spot in the part. The Figure 8.14a is the normal gat position with the weld line position in the end of mold filling. If it is the position of force acting, then the gat may be arranged to the opposite position like Figure 8.14b. In this way, the weld line is located in the middle of the part. If the mold design allows the gate to be located in the side position as shown in Figures 8.14c and 8.14d, the weld line will be moved to the side or corner.

The possible gate location changes will have different core material distributions in the co-injection molding parts. For example, assume that the center gate is in the circular mold, square mold, and rectangular mold, respectively; then, if the mold temperature is uniform, the core material distribution is also uniform for all three center gate molds. It is obvious that the core material distribution is not symmetric if the gate is not in the center of the mold, as shown in Figure 8.14d.

All of the above are the result of one gate only in the mold. Sometimes, it is necessary to design (a) a multi-gate system to control the flow ratio not too long or (b) a good balanced mold filling. The multi-gate system in the mold can be designed to avoid some bad weld line positions. In addition, the multi-gate system may not be symmetric about the center of the part, such as when the left-hand gate-to-wall flow length is longer than the right-hand gate-to-flow length. Then the runner system for both gates must be balanced to get a uniform core distribution in the part. If an equal runner system is used in this nonsymmetric mold, the core material may break through on the side first. The balanced runner system for this mold with special nonsymmetrical gates can be reached by a special hot runner design from Dynisco Kona Hot Runner Systems [16]. This runner system keeps skin and core materials separate right up to the gates. Furthermore, the runner bore diameter for the left-hand side gate is increased to have high flow rate into the left-hand side. This balanced runner system can maximize the amount of core in the co-injection molding part. Finally, the gate for a core with gas-laden material must have an opening that can be controlled exactly by a valve gate.

8.1.5 Conclusions of Co-injection Molding of Microcellular Processing

The important guideline for the mold of co-injection molding is the flow ratio. The experimental result shows that the flow ratio will be determined by the factors below:

- Less than 100:1 if there is no hole
- Less than 50:1 for hole structure

As the summary of material combination in the co-injection molding, a slight high viscosity in the core is the best selection. The core material with the slightly higher viscosity compared to skin material will produce a plug-like core flow front inside of skin, which is more effectively pushing the skin material to the end of cavity without breakthrough. However, the low-viscosity core material can travel too fast and cause more breakthrough unless the core–skin ratio is reduced.

Generally, a hybrid solid–microcellular co-injection molding process has been developed and the processing benefits, skin–core spatial distributions, and resulting microstructures were well discussed in this chapter. It was found that with proper process conditions and skin–core volume fraction ratio, a microcellular core entirely encapsulated by the solid skin could be obtained, which would essentially retain the benefits of microcellular injection molding

TABLE 8.3 Effects of Various Processing Parameters on Co-injection Microcellular Part Quality

	Density	% of core Material	Surface Glossy
Injection speed fast	–	–	+
Mold temperature high	–	–	+
Melt viscosity higher for skin material	+	+	0
Wall thickness increasing	–	+	0

+, Increasing.
–, Decreasing.
0, No effect.

such as weight reduction, surface flatness, and dimensional stability while producing a perfect part surface [17, 18]. The foamed core exhibited a truly microcellular structure with average cell sizes of 8–12 μm and a cell density of up to 3×10^9 cells/cm^3. Visual and surface profile inspections of the molded samples showed a total absence of swirled patterns and a 90% reduction in the sink mark at the selected thick section, unless a breakthrough occurred. Comparison between solid–microcellular co-injection molding and solid–solid co-injection molding under similar process conditions showed that the former process generated a thinner, longer, and more uniform core in the sandwich structure.

A brief summary for the effect of various processing parameters on the microcellular co-injection part quality is listed in Table 8.3. It is the general trend and may be changed at different processing conditions.

8.2 GAS COUNTERPRESSURE

This is a well-known method to improve the surface quality of the microcellular foam. It was invented as a method to eliminate "swirl" on the surface. However, the physical properties can be improved significantly by this technology as well. The limitation of this technology is only from the mold that must be sealed to maintain the gas pressure in the mold before mold filling. It also involves special gas pressure and timing sequences controlling the molding machine. It is usually used for the heavy-walled parts of structural foam. However, the microcellular foam can widen the application to some thin-wall mold as long as the sealing mold is possible.

8.2.1 Gas Counterpressure Process

The gas counterpressure process may have many different techniques, but the underlying concept is the same. Several different technologies of the gas counterpressure processing are discussed in the following.

8.2.1.1 Gas Counterpressure with Short Shot. First of all, the common first stage must be prepared before mold filling. A pressure-tight mold is charged with gas counterpressure up to 6.9 MPa (1000 psi) prior to the injection of plastic melt. In order to maintain the necessary gas counterpressure, the mold parting line, ejectors, moving slides, and nozzle seat must be sealed well.

When the proper in-mold counterpressure is built up, the second stage can begin. A predetermined short shot containing a dispersed and compressed blowing agent is injected into the mold under the consistent resistance from gas counterpressure. During the injection sequence, controlled venting should occur to allow the counterpressure to be constant with increased shot volume into the mold. The constant counterpressure prevents the blowing agent from escaping in the flow front. Thus, it eliminates the swirl normally caused by the gas bubbles being trapped between the molded surface and the skin of the part.

The third stage occurs after the shut-off nozzle is closed, and the gas counterpressure in the mold is released through exhaust vents. In other words, the short shot finishes when shut-off nozzle is closed. Once the counterpressure is released, the pressurized blowing agent instantaneously expands to fill the voids and keep the part packed against the mold during the cooling stage. The gas counterpressure release may delay or decrease gradually to be sure that the blowing agent is encapsulated completely by solid skin.

The last stage is the same operation as the regular molding that includes normal cooling, part ejection, and mold actions. Then, the sequences of the cycle are repeated again.

8.2.1.2 Gas Counterpressure with Screw Decompression after Injection. This method requires the machine with a special option to allow screw moving back after injection finish. The differences from the method above are in stage 2 and stage 3. A full shot size is made in the second stage, and the gas pressure is released completely near the end of the full shot. Then, in the third stage, the screw retracts to suck part of the melt in the mold and gives the space for foaming inside of the part. After a predetermined decompression stroke of the screw, the nozzle shuts off to separate the mold and the screw, thereby allowing the blowing agent to expand and also allowing the part to become cold to finish the molding.

8.2.1.3 Gas Counterpressure with Mold Crack Open. This has the same first stage, second stage, and the fourth stage as in method 2 above in 8.2.1.2. However, stage 3 will be changed for foaming. It needs a machine designed with reversal coining (opposite to the injection compression molding) to be able to open the mold after injection is finished with full shot size without foaming. Therefore, this shot size is still short because the mold thickness is thinner than the thickness of the final part. It also needs a special mold designed for seal for the first stage and possible to move back to give space for foaming in the thickness direction. The mold crack opening time can be controlled to have enough cold skin and also to allow the foaming inside of the part because the center material of the part is still hot enough to foam once the pressure inside of the part is released. Therefore, this technology is only for the thick part, and it is not as economic as method 2 in 8.2.1.2.

The gas counterpressure molding part usually has 5–10% density reduction compared to a 15–20% reduction for a regular microcellular foaming part. This is due to the formation of thicker skins. However, the denser counterpressure microcellular part does display more uniform cell structure throughout the part, resulting in improved physical properties.

The traditional gas counterpressure required the gas pressure to be larger than the foaming pressure. Therefore, there is no foaming during mold filling while hard smooth skin is formed during mold filling. The gas used as the blowing agent is usually the N_2 gas. The pressurized N_2 gas filled the sealed mold before mold filling. The location of gas profiling is the same location of the mold venting outlet spot that must be considered in the mold gate design.

However, the microcellular molding may allow the cell growth with restriction from the much lower counterpressure compared to traditional foam gas counterpressure. The gas pressure in the mold of microcellular foam may be as low as 1.73 MPa for microcellular foam (compared to 6.9-MPa gas pressure for traditional structural foam). Then, the foaming occurs simultaneously with smooth skin forming as long as the gas is pressed in the flow front and there is no bubble breaking through the cold flow front by the gas. It is different from the method allowing foaming to occur after a smooth skin formed at high gas pressure in the mold.

The low gas counterpressure in microcellular injection molding still provides a controlled resistance to the flow front. This resistance pressures from gas already in the empty mold forces the flow front to move uniformly across the thickness direction, and it keeps the blowing agent in solution in the melt and behind the flow front. The schematic of this flow front restriction by counterpressure is illustrated in Figure 8.15. The gas pressure against the free flow front to force the melt in the center layer flows slowly and results in uniform plug flow, instead of fountain flow like the flow pattern shown in Figure 8.1. On the other hand, the uniform gas resistance across the thickness direction forces the melt to moves in every direction like the force perpendicular to the axial line of the mold (see Figure 8.15). Therefore, the melt will be forced to enter the thinner or harder-to-fill areas with good venting and gas collecting system during the mold filling. It is also forcing the skin against the cold wall surface of the mold. Then, the good contact with cold mold increases the efficiency of cooling, copies the mold surface well, and reduces the warpage and removes the sink marks.

The low gas counterpressure requires less clamp tonnage than the regular gas counterpressure molding. The regular gas counterpressure molding needs additional clamp tonnage compared to the structural foam of low pressure ranging from 25% to 100%, depending on the venting technique used [19]. Therefore, the reduced gas counterpressure for microcellular injection molding

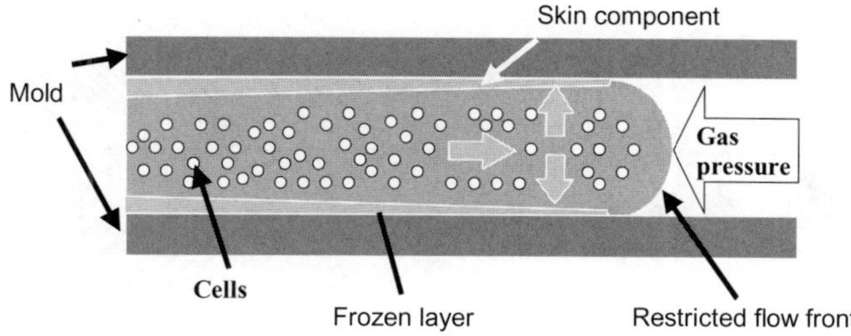

Figure 8.15 Mold filling schematic of gas counterpressure microcellular molding.

may maintain the advantage of the low clamp tonnage requirement of the microcellular molding process. In addition, the counterpressure molding method 1 in 8.2.1.1 with short shot counter is a true low-pressure process regardless of whether the gas counterpressure low or high. However, even the full-shot-size methods 2 in 8.2.1.2 and 3 in 8.2.1.3 need less clamp tonnages because both of them do not need the packing stage after injection.

Another advantage of low gas counterpressure is the low risk to blow out the mold cavity seal. Then, it is possible to start the process for the gas charging in the mold as the mold closing just touches each other, but before full mold tonnage is built. This is an extra cycle-time saver. On the other hand, low gas counterpressure uses low energy, so keep the gas counterpressure as low as necessary to save energy and cost of operation.

On the other hand, occasionally, the higher gas counterpressure may be needed for some microcellular process. This process is capable of increasing the maximum gas counterpressure as high as 3.45 MPa (500 psi), or more. This high gas counterpressure requirement is used for the application with a thicker and complicated geometry part. In addition, high gas pressure will slow down the mold filling and create thicker skin and smaller cell size. However, if the gas counterpressure is too high, the cell size can become bigger and the number of cells will decrease significantly if the gas counterpressure is higher than the gas pressure inside of the cells where cell growth is stopped. Therefore, the best approach is to first try the lower gas counterpressure that may match the gas-rich material in front of the screw tip before injection, and then gradually increase the gas counterpressure to improve the cell structure until the results become opposite.

8.2.2 Mold and Part Design

The mold and part design of gas counterpressure molding is basically the design for the mold sealing and gating systems. The material used for mold sealing depends on the gas pressure. The real gas pressure is determined from resin type, mold design, blowing agent, and sealing requirement. The necessary gas pressure can be determined by trial and error. Set the low gas pressure first, and check any splay that is an indication of too low gas counterpressure

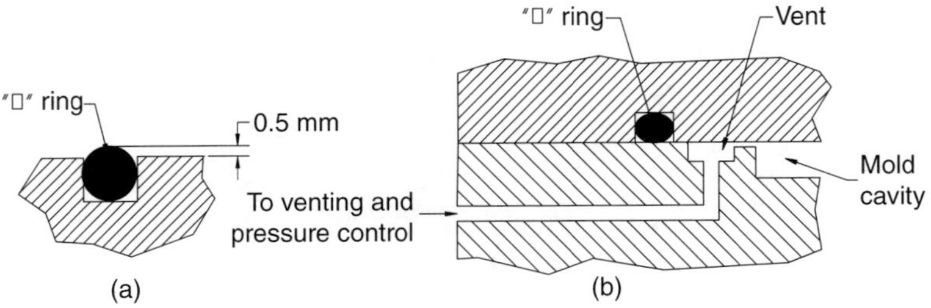

Figure 8.16 Gas counterpressure mold design. **(a)** O-ring groove. **(b)** Venting channel position.

in the mold. The set up gas pressure needs to be increased gradually until splay is removed and the part appearance and molding replication are acceptable. It is usually in the range of 0.55 to 2 MPa for microcellular injection molding. Then, microcellular structure may need to be checked if the material strength drops too much, which may be related to cell size in the part. The typical gas used in this technology is nitrogen gas or compressed air.

One of the biggest gas counterpressure hurdles to overcome is the sealing of complicated molds. If the seal fails, the flash, splay, and swirl marks on the surface will occur for the microcellular part. Generally, the normal O ring is good enough to be used for sealing the mold of counterpressure molding. The dimension of the O-ring groove is indicated in Figure 8.16a. The O ring on the parting line must consider the enough deformation for possible mold crack opening during injection. When the mold closes, the O ring is compressed into its groove. The groove should be big enough to hold the deformed O ring completely in the groove without overcompression. However, the injection pressure may force the mold crack to opening slightly. The O ring must have enough deformation to expand out of its groove to still make the airtight seal in the parting line of the mold. It is recommended that the normal deformation of the O ring in the mold be in the closed direction (0.5 mm), as shown in Figure 8.16a [21]. The different mold and processing condition may change this deformation value based on the mold crack opening gap and the gas pressure in the mold.

It is difficult to seal the stepped parting lines in the complicated geometry mold. Various-shaped seals (to compensate for the 90° bends) have been successful to some degree. To make sealing more stable, cut the O ring grooves on the stationary half of the mold, instead of on the moving half. However, if it must be done on the moving parts, it has always been a bigger challenge. The potential path of gas leaking must be identified first and sealed accordingly.

Gate placement and design along with the venting system are of special importance in the gas counterpressure mold. The gate and venting will be designed accordingly to place the venting channel at the end of mold filling.

The detail of the general gas counterpressure mold design is illustrated in Figure 8.16b. Only a local view is displayed with the necessary details of special requirements for the gas counterpressure mold. The O ring position must be located in the position encapsulating the whole cavity and all venting channels, as shown in Figure 8.16b. No matter how many venting channels are needed for the mold, all venting channels must be collected into one final groove that leads to the one outlet controlled by the pressure control circuit in Figure 8.17, which will be discussed next. However, if some independent venting area is to be controlled by a separate low gas pressure circuit, there may be several venting grooves separated by each other in the mold. Then, there are also several independent gas counterpressure control circuits needed in the gas pressure control system.

8.2.3 Gas Pressure Control System of Gas Counterpressure Mold

Managing the gas flow and pressure is a key for successful gas counterpressure molding. There are two means of accomplishing it: with pressure relief valve and with flow control valve. In the schematic displayed in Figure 8.17, the pressure relief valve "4" is located on the outlet side of the mold, near the outlet of venting of the mold "1". The function of valve "4" is to maintain a constant gas pressure in the cavity as the mold fills while releasing the compressed gas to atmosphere through a silencer (not showing in Figure 8.17). An alternative method to control the gas-releasing flow rate is to actuate the

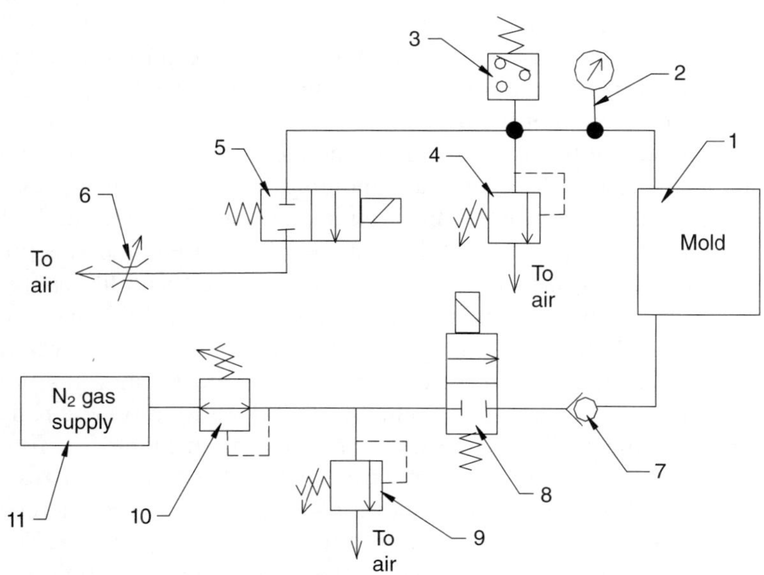

Figure 8.17 Gas pressure control circuit for gas counterpressure molding.

solenoid of valve “5” to release lead gas through flow valve “6”, which can be adjusted with different flow rate. The flow rate of the valve “6” is determined by the injection volume rate and gas pressure in the mold cavity. On the other hand, the pressure relief valve “4” is always functional even if the flow control takes over most of the time if valve “5” is actuated. Therefore, sometimes the flow control and pressure control may act together once the flow control valve does not act fast enough. The pressure switch is also used for actuating the valve “5” in case the pressure relief valve failed. The constant pressure in the mold will be maintained within 0.014–0.069 MPa (2–10 psi) as the range of variation of valve opening and closing sensitivity.

The N_2 gas supply tank “11” will have a pressure reducing valve “10” (called pressure regulator) to set up the pressure value. If the molding needs the 1.72 MPa gas counterpressure, the N_2 gas will be set up about 2 MPa as the initial pressurized gas in the mold cavity before injection. The limit switch on the clamp system of injection molding machine detects the mold position and control the actuating of solenoid of valve “8”. Once the mold is fully closed, the valve “8” is actuated by this clamp limit switch. Then, gas will fill the mold with the preset pressure. The check valve “7” allows the gas flow into the mold in one direction, and it will close to separate the gas supply with the mold inlet after the gas pressure reaches the set up value, also the valve “8” closes. The gas pressure in the mold cavity will be maintained constant. The pressure switch “3” measures the mold pressure. Once the gas pressure in the mold cavity reaches the set up value, the pressure switch “3” sends the signal to the solenoid of valve “8” to close the gas supply. At the same time a signal from pressure switch “3” is sent to the machine controller for injection. There is another limit switch on the injection cylinder to detect the position of injection. Once the injection is near the end, or where the position of the final injection stroke is set up the limit switch actuates the valve “5” to release the gas pressure completely in the mold.

There may be a difficult filling area in the mold. This area is more difficult to be filled with gas counterpressure molding because the gas pressure in the mold is equal everywhere and venting is closed during mold filling. Therefore, the mold filling will fill the least resistance area first. This problem can be solved with low pressure control locally in the difficult filling area. The separated circuit with low pressure will only act after mold filling is near the end. In other words, the mold filling will fill the mold almost everywhere that will separate the low gas pressure channel from the normal gas pressure channel. Then, the low gas pressure channel will open to lower the resistance locally in the difficult filling area only. The final mold filling will force the material to flow into a difficult filling area since it has the lowest resistance among all other flow directions. In addition, the pressure in the microcellular cells will expand to push material to fill this difficult filling corner. The final limit switch will release all the pressure channels at the same time to finish the gas counterpressure mold filling. The circuit analysis is discussed in references 21.

8.2.4 Morphology and Physical Properties of Gas Counterpressure for Microcellular Molding Part

Figure 8.18 shows the morphology pictures of clear PC material with the section view in the flow direction. This clear PC part made in the presence of 1.73 MPa gas pressure in Figure 8.18a has a smooth surface but fewer cells. The pressure drop rate with gas counterpressure molding is significantly reduced, so the weight reduction is only about 8%. The pressure drop rate reduction is the major reason why the number of cells is less than that in normal foam. On the other hand, the resistance of gas pressure in the mold

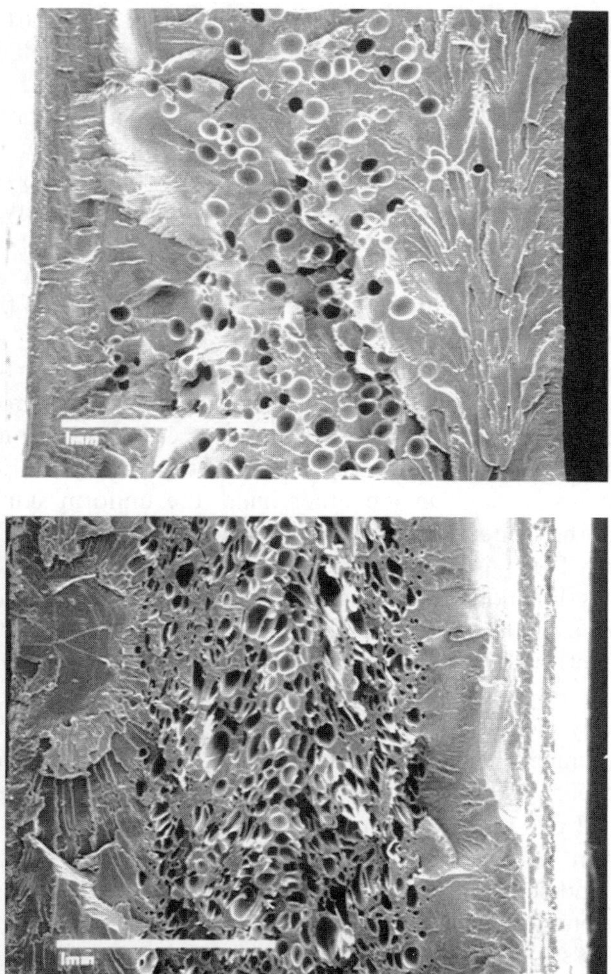

(a)

(b)

Figure 8.18 Morphology pictures of PC with gas counterpressure molding. **(a)** Gas pressure on [18]. **(b)** Gas pressure off [18]. (Reproduced with permission from Society of Plastics Engineers.)

keeps cells in the spherical shape. The morphology picture for the sample with gas pressure off is shown in Figure 8.18b. The weight reduction for the gas pressure off sample is increased up to 24% with rough surface but with many cells in the core. However, the cell shape in Figure 8.18b is deformed from the shearing in the fountain flow. The microcellular foam roughness is 1 μm or less with gas counterpressure molding, and it is over 20 μm without gas counterpressure molding. As long as the mold can be sealed and venting channels can be controlled by a venting valve gas, counterpressure molding can achieve smooth surface on parts with complicated geometry. The gas counterpressure pushes the flow front without any premature foam until the pressure is released completely. Therefore, the cell structure is uniform from the gate to the end of mold filling. The morphology picture in the same part at the end of mold flow position verifies this conclusion. The cell size and cell density are almost the same as the ones shown in Figure 8.18a.

Lee et al. [22] also presented the results of the gas counterpressure process with mold crack opening technology for structural foam for various PP materials. The morphology looks similar to the PC morphology in Figure 8.18. Since the mold cavity pressure remained higher than the solubility of the gas during mold filling, the cell nucleation is low because of a low pressure drop rate from gas counterpressure [22]. The cell density is obviously lower than the regular structural foam, although cell size is much better than the cell size in structural foam.

Except for a known best benefit of gas counterpressure molding for enhancing the aesthetics of microcellular foam, another benefit of this process is the physical property improvement. It is because the cell shape is spherical and the size is small enough with uniform distribution in the whole part, which results in good properties. On the other hand, the uniform skin thickness is denser than the general microcellular molding as a result of the gas counterpressure. There is also an increase in flexural modulus. Although the tensile strength is not affected significantly, the percentage elongation at break is enhanced significantly. In addition, the skin thickness and the bubble density will affect the impact strength of microcellular foam. Thin skin and larger bubbles result in lower impact strength, while the thick skin and small bubbles definitely yield higher impact strength. On the other hand, the gas counterpressure microcellular part has uniform cell structure and cell size throughout the part, yielding a more uniform impact strength. It may be the overall reason why the impact strength is improved for the gas counterpressure molding part.

The smooth and dense surface of the gas counterpressure molding part may have another benefit, which is to promote the fatigue load resistance. The roughness of the surface sometimes becomes the stress cracking spot, and it speeds up the failure under fatigue load.

The impact resistance of the microcellular part made by the gas counterpressure process has almost doubled compared to the impact resistance of the same solid part [23].

TABLE 8.4 Effects of Various Processing Parameters on Gas Counterpressure Microcellular Part Quality

	Density	Weld Line Visibility	Splay
Seal	+	–	None
Gas pressure high	+	–	None
Injection speed increasing	–	–	0
Mold temperature	–	–	0

+, Increasing the setup of parameter.
–, Decreasing the setup of parameter.
0, No effect of the parameter.

A special gas counterpressure unit from Engel has not only supplied gas pressure but also has supplied a gas recovery system. The pressure in the mold can be precisely controlled just high enough to keep the gas in the single-phase solution still in the solution on the flow front. This prevents the creation of surface swirls caused by dissolved gas between mold and melt [23].

A material used in the Engel unit with a gas counterpressure control system was an unreinforced grade of Bayer's Markrolon PC. The surface roughness of the microcellular samples with and without gas counterpressure have been measured and compared. Without using the gas counterpressure process, the microcellular part can be made reasonably smooth to the touch, but not to the eye and certainly not to the microscope. Adding the gas counterpressure process to the microcellular part appears to dramatically improve the part surface finish. It looks not only visually smooth and glossy, but also truly smooth under the microscope. The surface roughness for a gas counterpressure molded PC microcellular part is about 0.85 μm or less, while the surface roughness of a regular microcellular PC part without gas counterpressure is 23.11 μm (Rz) [23].

The gas counterpressure molding also saves the cycle time about 10–20% compared to the same solid part. A brief summary for the effect of various processing parameters on part quality made by the gas counterpressure microcellular part is listed in Table 8.4. It is the general trend and may be changed at different processing conditions.

8.3 OVERLAPPING (ALSO CALLED OVERMOLDING)

Overlapping technology is similar to the co-injection molding, except that the sandwich structure replaced by two layers overlapped each other. There is only one layer with solid and another layer with foam. The configuration can be either one on the top, depending on the application. If the soft material must be on the top surface, then foamed material will on the top of the solid material. In this case, the solid material acts as the solid bone to support the

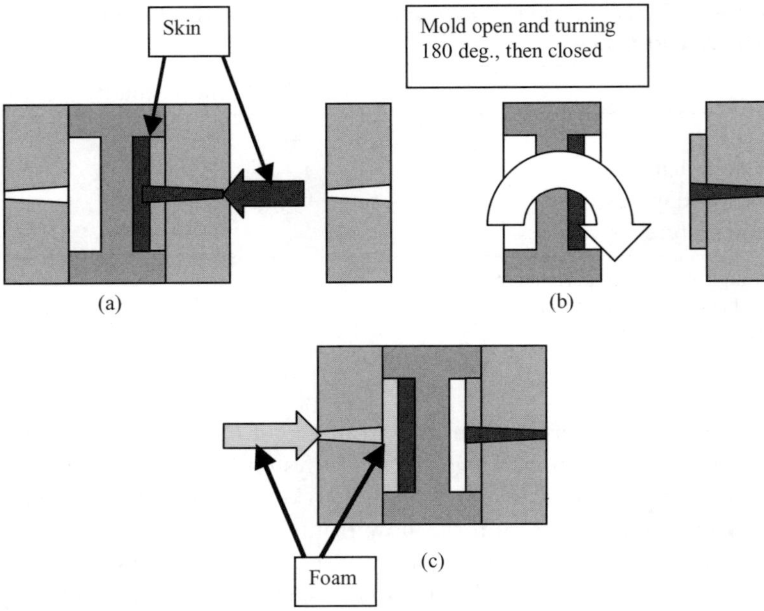

Figure 8.19 Overmolding layout. **(a)** Skin layer molded first. **(b)** Mold turning 180 degrees, and foam overmolded on the skin layer second.

foamed top surface. If perfect skin is needed, the solid skin will be on the top of the foam layer and the foam will take care of the other issues, such as sink mark, warpage, and so on.

Although it simplifies the process to just add one layer on another without considering the material penetration through the different layers the equipment must have a rotary table for either vertical rotation or horizontal rotation. Figure 8.19 shows a typical top-view layout of injection molding machine with a horizontal rotary table. This machine has two injection units. One is on the right side for skin material injection, while another one is on left side for foam material injection. Figure 8.19a is the molding position for skin. Usually the procedure for this overmolding technology is that the skin is injected first if the sink mark and warpage issues need to be addressed. The skin material needs to be cooled to be ready for the second overmolding. Then, the mold will open until the rotary table in the middle of the mold has enough space to horizontally rotate 180°, as shown in Figure 8.19b. After rotation, the mold closes and is ready for the foam material injection. The foam material will be injected as the substrate layer into the mold over the skin layer that is illustrated in Figure 8.19c. In this way the foam material will expand to remove possible sink marks on the first skin material since the skin material is still soft enough to be pushed by foaming force. Figure 8.19 shows just one molding at one position for the first shot and the second shot, respectively. It is used to clarify the concept of overmolding. The real case should begin with first shot of skin material when the other side of core cavity is empty. Once the skin molding was finished and the rotary table rotated 180° to the second position for foamed core material, the foam and skin materials injected simultaneously from both sides to save cycle time. In addition, this horizontal rotary table can rotate fast. A 600-mm-diameter table needs only less than 1 sec for 180° rotating. However, it takes approximately 2 sec for 180° rotation of a 1400-mm-diameter table.

Using this technology, the flow ratio of flow length to thickness can reach 500:1 in the newly developed Husky overmolding machine named Quadloc Tandem Index (QTI). It has the second injection unit along the side of a moving platen that reduces the overall length of the multi-material machines.

Another common way is to use the vertical rotary table that hangs on the moving platen. There are two or more injection units on the same side—usually on the stationary platen side, which is usually the normal single injection unit position. They may use several injection units in parallel arrangement to make even more than two layers of overmolding parts. The primary one is the largest capacity of injection unit. Then, several small ones are equipped with parallel injection units of varying sizes of screws.

Similar to the parallel style of a vertical rotary table machine, the so-called "piggyback" arrangement is also a popular configuration of multilayer injection molding. It is arranged to have a second injection unit positioned above the main injection unit with an angle. Both injection units move in parallel, ensuring a constantly equal distance between nozzles regardless of the respective depths of insertion. It is the same principle as the horizontal rotary table that is to mold one layer first and overmold another one on the top of the first layer.

If the overmolding is local, only the sliding split system may be useful. With this in-mold slide method, the cavity area of the second material is sealed by hydraulically actuated sliding inserts or locking slides, which are opened after injection and initial cooling of the first material. In contrast with the other methods, injection of the second material does not occur simultaneously, but instead occurs sequentially without opening and rotating the mold. With no need to transfer the parts and with a less complex mold construction, this method may be economically viable for lower volume production in spite of the longer cycle times generally associated with this method. Also the more compact mold construction allows—in some cases—the usage of smaller machines, and the feasibility of the in-mold slide method is heavily dependent on part design.

There are more applications for the overmolding ideas. One is to overmold a foam on the fabricate, such as a cloth, a fiber mat, and so on. This overlapping technology is very useful to apply the microcellular technology to a wide range of applications. There is no strict limit for the flow ratio anymore because it is the one material over another as long as the solid material can fill the mold first. There is no issue like sandwich molding a core material to penetrate a skin layer in a certain flow ratio in co-injection molding. Also, the viscosity matches do need to be considered for this overmolding technology. Some case studies that focus on the applications will be discussed in Chapter 11. Microcellular overmolding on the cloth is successful to reduce the valve gate from two to one only because of the low viscosity. In other words, the flow ratio of overmolding increases significantly. For substrates consisting of 2.5- to 3.5-mm-thick PP with 1.5-mm-thick overmolding TPE, the overmolding part can be created with flow-length–thickness ratio up to 500:1.

It is necessary to select materials to be bonded together by similar solubility so that the diffusion process will be favored to result in a higher adhesion resistance. In addition, the processing parameters that influence the bonding strength most are the melt and mold temperatures [24]. The high melt and mold temperature will result in high adhesion resistance between two overmolding materials.

8.4 REVERSAL COINING (ALSO CALLED EXPANDABLE MOLD OR BREATHING MOLD)

Reversal coining is the opposite of injection compression molding. The reversal coining technology is used for microcellular injection molding for perfect surface solution. Historically, it was the earliest technology used to improve the molded part surface with mold motion [25]. It is similar to the conventional injection molding with high injection pressure. The mold must be filled 100% and have packing pressure to ensure not only mold filling but also pressurization to prevent any foaming before mold cracking open. This will form the good-quality skin just like conventional injection molding. The mold will move back when enough skin is formed and the core is still hot and soft enough to making foam. The crack opening distance determines how much foamed core and cell structure are in the center of the part. The schematic in Figure 8.20 shows the principle of this operation. The mold is fully closed first. The mold filling is 100%, has the necessary packing stage to form nice skin, and has

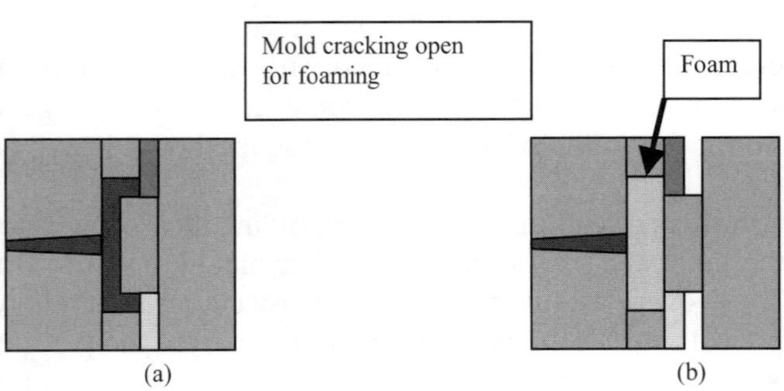

Figure 8.20 Expansion of mold with floating plate. **(a)** Mold closed for 100% mold filling and skin forming. **(b)** Mold cracks open and gives space for foaming.

pressurized the melt without any foaming, as shown in Figure 8.20a. Then, the mold opens to give the foaming space after the skin is formed and strong enough to hold the surface, which is shown in Figure 8.20b. This mold has a floating plate to keep the corner of the part formed nicely without any blemished edge because of mold moving. In this way the possible witness line is located on the backside of the mold.

There is an insert technology similar to the reversal coining. The insert is in the mold during full mold filling and retracts back to give space for foaming. It may be used for a complicated part to have some foaming for reducing the sink mark and warpage.

The more advanced development of the expandable mold allows the mold to expand in more than one plane so that a complex part can be molded with this method. Some molds even have moving elements (similar to insert technology) on the side of the mold to permit a radial expansion as well to give the foaming uniformly there [26].

The advantages of the reversal coining process for microcellular parts are as follows:

- The cooling time is shorter than the conventional low-pressure foam injection molding since it is full the mold first before foaming.
- The surface quality can be made as close as the surface quality of the regular injection molding part.
- Skin thickness is controllable with the time difference during the first stage to form the solid skin before reversal coining.
- There is no sink mark and less warpage because the foamed core will take care of the shrinkage caused defects.
- A local reversal coining is possible. That is just a simple way to remove an insert in the mold to give space for foam that will then have cell growth to overcome the sink mark defects.

The disadvantage of this method is the high injection pressure that is as high as the pressure for the regular injection molding. Generally, it is an economic solution for the nice surface microcellular part without sink mark and warpage. On the other hand, the reversal coining usually is the standard option for most injection molding machine manufacturers.

Mueller and others reported the good constancy of properties from breathing mold technology with the impact-modified PP that is a copolymer with a melt volume rate of $16\,cm^3/10\,min$. The breath mold can set up to 3-mm open stroke after injection. It is a good method to produce constancy of the final microcellular foam using special procedures [27]. By increasing the delay time for mold opening, the skin gets more chances to be solidified thicker, and an improvement of the mechanical properties is possible. The constancy of the foamed part properties is equally as high as that of a solid injection molding product when an appropriate foam molding approach is in service.

Michaeli and Cramer [28] also investigated the effect of breathing mold technology for the foam injects molding (FIM) that injects gas into the nozzle to make foam. The mold is completely filled with gas-laden material and held a short period to achieve compact border layers. Subsequently, the cavity is enlarged, initiating the foaming phase. The gas-swirled surface was repaired during the mold-filling phase. The surface quality of the foamed part can be improved significantly.

If the part is huge, the final injection action near the gate may create higher residual pressure in the gate area of the microcellular part. This pressure may reduce the nucleation and even compress the cells in the gate area, resulting in no visible cells or fewer cells. It may need to retract the screw when the nozzle is still opening. Then, certain molten polymer will be sucked back into the nozzle and the screw to give more space in the gate area for nice foaming in the gate area. It may be called reversing packing action. It is only used for big and long flow ratio parts to solve the problem of nonuniform cell structure in the whole part. If this action is performed quickly, then it may release enough energy in the gate area. It allows not only foaming but also quick cooling in the gate area because decompression may cause some nucleation and create more cells in the gate area.

8.5 A PROCESS WITH THE COMBINATION OF OVERMOLDING AND REVERSAL COINING

The new technology has been developed by Engel, Trexel, BASF, and P-Group and has been named Dolphin Skin technology. Processing-wise it is a combination of overmolding for one side perfect surface quality and reversal coining for microcellular foaming to create soft touch surface with thermal and sound insulation properties. The first sample was a passenger car dashboard panel with a soft-touch surface (foam with reversal coining) and strong glass-fiber-reinforced material underneath the soft-touch surface. The picture of the part morphology is shown in Figure 8.21.

The picture on the left side shows the glass-fiber-reinforced solid carrier underneath the foam layer. The glass-fiber-reinforced material is the PBT/ASA blend (Ultradur S4090 IGX from BASF), which is the material developed for this process; it is part of a pending patent. Beside this glass-fiber-reinforced solid carrier morphology, the interface morphology is illustrated in Figure 8.21. This interface is between foamed layer and solid carrier where nice skin forms from foamed side to contact and bond with the solid carrier layer. Then, the second layer is overmolded in the first layer in the second stage with the special polyester using microcellular (either MuCell® or chemical blowing agent) process. The second layer material is TPE-E, which is a readily foamable thermoplastic polyester provided by P-Group with trade name Pibiflex; this is again part of a pending patent to be a special material developed for this process. The close chemical affinity between these two

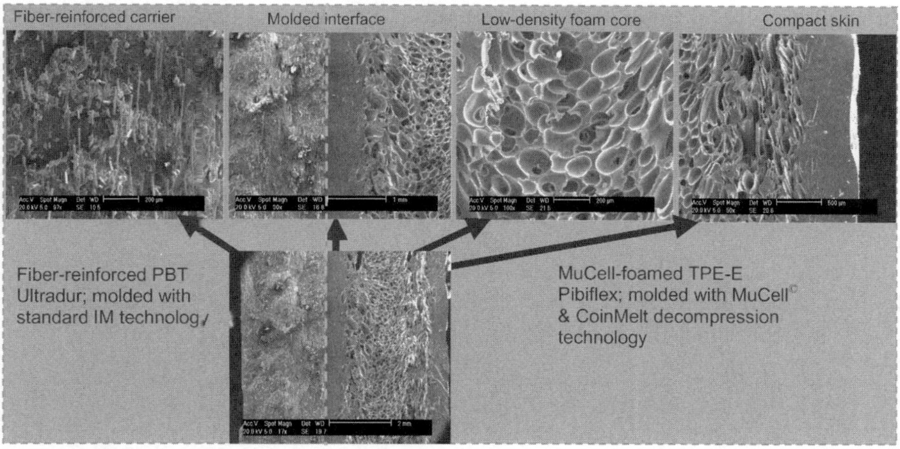

Figure 8.21 Morphology of overmolding part with soft skin and reinforced core made by Dolphin skin technology. (Courtesy of Engel.)

materials and pressurized regular injection molding process ensures an optimum bond between two layers and nice surface finish on both sides. And then, the foam will be made by the reversal coining process that Engel calls CoinMelt. It has perfect skin on the foam material, so the foam side can be used as surface as well. The foam side has a soft touch like rubber, and the other side is a smooth surface without sink marks but with a class "A" surface finish. The foamed core is made by reversal coining, which makes foam controllable to be either microcellular or even big cells, there by making soft surface even softer. In addition, the foam is uniformly distributed throughout the part since gas-laden material was injected into the mold, 100% filling the mold cavity, and then the mold cracks open to allow foam to occur.

The machine and mold features two horizontal opposed injection units, a double-daylight mold with a rotating central block, and a sensitive parallelism control integrated in the Engel machine control system (see Figure 8.19). Then, the cell structure clearly shows the foamed core, and the right most picture shows the interface between foam and solid skin where the soft touch surface named Dolphin Skin is created.

8.6 HOT AND COLD PROCESS (ALSO CALLED ALTERNATING MOLD TEMPERATURE, THERMAL CYCLING)

The hot and cold process is also called the alternating mold temperature process, or thermal cycling process. The hot mold surface obviously has less friction or adhesion between the mold metal surface and the melt, and most importantly it has no skin and thus the surface of the part slides on such a surface acting as a plug flow. This plug flow means that the theoretical velocity distributions are equal in the depth direction. Then, there is no shearing between flow layers in the depth direction. In addition, the breaking bubble from free-flow surface will be easily ironed or repaired by hot mold surface under the cavity pressure. Since the mold will be heated and cooled in the same cycle, it is also called thermal cycling (BASF method) [25, 26].

The concept of hot and cold process is well known for structural foam. It has been used since the early 1970s. This method heats the mold by steam, hot water, hot oil, and electrically heated plates in the mold. Then, cooling lines are installed behind the heating lines. Initially, the mold surface must be heated to within 50 °F of the polymer melt temperature. The gas-laden material is injected into the mold. There is still some gas escaped from the flow front and trapped between the plastic part surface and the mold surface. However, the hot mold surface will not freeze a bubble or trapped gas on the part surface. Instead, the gas on the hot mold surface will have two possible results. One is the gas diffusing back into melt towards the center of the part since it is the hottest place during mold filling. Another result is that the gas on the hot mold surface will be ironed flat. This traditional hot and cold process will keep the gas and bubble free on the foam surface. In this hot and cold molding, the injection speed can be very slow (as low as one-tenth of conventional injection molding speed) that there is no evidence of the boiling and churning of the wave front that is usually from high injection speed. After the mold filling is finished, the mold will be quenched by cold water as soon as possible. However, this method will have a primary drawback that has two to three times longer cycle time than the high-speed foam process [25].

However, a well-controlled thermal cycling processing can only increase the cycle time 60% longer than normal molding but can still improve the surface quality significantly. The heated mold surface temperature will be only a few degrees short of the injection melt temperature with steam as a heating resource. For example, for GPPS material the high mold temperature is 248–284 °F, and the cold mold temperature is about 60 °F (water as coolant). This method requires the mold designed for a uniform distribution of temperature throughout the mold surface.

With the same principle above, more companies developed rapid heat and cold molding technology to overcome the drawback of long cycle time of hot and cold process. For example, there is a new joint development between Trexel and a Japanese molding company Ono Sangyo, and this technology has a trade name of MuCell Gloss [8]. It finally opens the way to high-surface-quality microcellular part. The microcellular technology from Trexel is combined with the rapid heat cycle molding (RHCM) from Ono Sangyo. The RHCM molding process involves adding two fully independent temperature control circuits in the mold so that the surface can be maintained above the heat distortion temperature of the polymer during the mold cooling stage and then rapidly cools once mold filling is complete. Although there must be a small cycle time penalty for using this technology, perhaps 2 or 3 sec, an dimportant gain is the improvement in surface finish and a big improvement in removing the weld lines. While molding onto a warm mold surface, the microcellular part maintains the maximum surface gloss. This technology is also particularly suitable for the glass- or mineral-filled resins, where a hot mold surface prevents the reinforcement making the surface. It has an additional benefit with RHCM technology that provides slow cooling at mold surface allowing a high crystalline content layer to develop, giving significant improvements in flexural modulus and surface hardness of the microcellular part. The RHCM process also enables excellent replication of microscopic surface features to be achieved, prompting Sabic (previous GE) plastic to explore the microcellular plus RHCM for optical disc production. Key target applications are high-quality complex molding where weld lines currently make the painting necessary, or where a high-gloss surface must be achieved with low levels of molded-in stress. While the traditional microcellular process is used most often as a cost reduction, MuCell Gloss will be used predominantly where a higher-quality surface can displace a secondary process and reduce the overall system cost. It has been tried for a highly complex automotive center console fascia that has achieved class "A" surface with 6–10% weight reduction. Also, special mold material must be used to work on this highly thermo-stress environment.

On the other hand, hot mold may not always work for the smooth surface of crystalline material foam. This is because more crystallization on the hot mold surface will create high nonuniformity of cell structure and voids near the surface and, then, rough surface, which was also verified in the study by Selden, [6]. A balance between percentage of crystallinity and small cell structure with warm mold may result in better surface quality of crystal material microcellular foam.

8.7 SUPER-MICROCELLULAR (OR ULTRA-MICROCELLULAR) WITHOUT WEIGHT REDUCTION

This is developed by the initial trial for a good microcellular surface and a warpage-free part. The method fills the mold almost 100% with high injection speed. The result is a quality of surface similar to that of the glass-fiber-reinforced material, along with very nice cell architecture that shows almost 5 μm cell size with spherical shape. In addition, all cells are distributed uniformly throughout the part. If the SCF content continually reduced gradually, the cells number decreases accordingly, and the surface finish of the part becomes better. The final result is exactly like gas-assist molding, which is one hollow channel in the center of part with class "A" surface finish. The experiment is continued with screw speed change without changing the SCF level, and the cell size improved from hollow channel (or named a big void throughout the part) back to microcellular level while the surface finish of the part remained good as well. Another approach is to make nice foaming first and, then, continually reduce the gas dosage until the lowest gas dosage with good surface finish of the part is found. The usual procedure is to set up a gas dosage that is just above the gas dosage level, causing hollow channel inside of the part.

Another application of super-microcellular is the thin-wall packaging application. The think-wall part usually has the wall thickness of only 0.5 mm or less. The typical application is named Super Light Injection Molding (SLIM). It uses MuCell® microcellular injection molding technology to cut the finished part weight up to 4–5%, 30% less clamp tonnage, and 10% less injection pressure. This microcellular thin-wall technology runs the molding in a small tonnage machine that cannot be used for running regular thin-wall molding. The super cell architecture benefits the good insulation properties of microcellular for this thin-wall packaging part. A similar case study of thin wall also displayed in Okamoto's book is that the 0.5-mm thin-wall part has 10% weight reduction, with 27% of cycle time reduction without explaining the details [29]. This application was run with an initial one that could not fill the corner of the part in the regular molding machine. The trial successfully used microcellular technology that not only filled the mold perfectly with low tonnage but also reduced the weigh percentage with super-microcellular cell structure and nice smooth surface as well. The cell size in the thin-wall part is about 3–5 μm, but it is usually difficult to make such small cells in a normal microcellular part. All the thin-wall parts have a very smooth natural white color surface because the small cell size and uniform distribution does have a traditional swirl of the foamed part. However, the basic requirement is extremely high injection volume rate to fill the mold within 0.5 sec. If the injection cannot finish in time, the cells will be stretched severely and the cell size is only 20–100 μm which is still regular microcellular cell structure [30]. Most failures of thin-wall microcellular injection molding are from the slow acceleration of injection speed and not from the total injection time and the average injection speed. Ramping must be performed as quickly as 0.1 sec or less to avoid the freezing of flow channel because the thin-wall part usually has a high flow ratio over 200. Therefore, even with low-viscosity gas-laden melt for less resistance of mold filling, the most challenges for microcellular molding in thin-wall processing are to fill the mold before flow channel freezing. The interesting morphology of thin-wall molding at extremely high injection volume rate does not show much cell stretching but also does not show the perfect spherical shape [29]. It brings more research topics behind the thin-wall molding with super-microcellular structure.

If the SCF control can reach the precise level, then the super-microcellular structure with a nice surface part will be a new area for the microcellular technology application. It will be have a wide application market since it is the competitor of regular molding with good surface and no weight reduction. The big benefit is still the dimension stability. However, the full mold filling with gas-laden melt is much easier than regular molding without any gas in the melt. On the other hand, the clamp tonnage saving can make full use of the existing machine to do both, exceeding the capability of injection unit (less than 5–10% of injection volume used for microcellular part) and clamp unit (up to 50% lower tonnage requirement for microcellular process).

8.8 LOWEST GAS DOSAGE WITH MINIMUM FOAMING

The real processing conditions for the super-microcellular foam is very difficult to find out with minimum trials. Therefore, another method to have good surface finish and all other microcellular benefits is to find a lowest gas dosage with minimum foaming, instead no foaming at all.

Trexel did some studies for this methodology and successfully made some good parts [31]. The material is PA 6/6 with 25% mineral filler and 15% glass fiber. The blowing agent is N_2 gas. The microcellular screw is 105-mm diameter, with 32 *L/D*. The back pressure is about 13.8 MPa (2000 psi). The injection speed is normally set up as 89 mm/sec. A good surface finish sample was made with 7% weight reduction. The microcell size is 40 μm or less.

The different low gas dosage has been tried for other materials only filled materials, either filler-filled or glass-fiber-reinforced materials, are successful for the low dosage and low weight reduction with both acceptable surface finish and fine cell structure. Also, different gases have been used for this method as well. The recommendation is that only N_2 gas in filled material is good for this method without too much effort to make the final good microcellular parts. The unfilled crystalline material usually has big cells and a smaller number of cells with low dosage and minimum foaming method. The unfilled amorphous material will have both acceptable surface finish and cell size in the microcellular part if the processing conditions set up correctly. However, it is generally recommended to use filled materials for this method.

8.9 CHEMICAL BLOWING AGENT IN MICROCELLULAR FOAM

The chemical blowing agent (CBA) has been used in the plastic industry for a long time. However, most of the conventional chemical blowing agents are flammable, and they are regulated and require special handling techniques, long product-storage times, and regulatory approvals for their use and release. This needs to be changed because some chemical blowing agents can release some gases to damage the ozone layer. New chemical blowing agents have been developed and help the industry to change the chemical blowing agents from ozone-depleting chemical blowing agent to the environmentally friendly chemical blowing agent. Basically these environmentally friendly gases generated by a new chemical blowing agent are nitrogen gas and carbon dioxide gas. The new development of CBA is now widely used by some regular foaming process and even some microcellular process since it offers easier processing, better product quality, less or no byproduct during reaction, and the advantage of a drop-in additive without complicated hardware requirement.

8.9.1 Chemical Blowing Agent for Nitrogen Gas

One of the exothermic foaming agents, such as axodicarbonamide, generates N_2 when subject to temperature above their decomposition temperature. Although economical, N_2 gas is a highly insoluble gas that can create issues with surface roughness as the gas breaks through the skin. Therefore, the chemical blowing agent for generating nitrogen gas is not a popular CBA in industry.

8.9.2 Chemical Blowing Agent for Carbon Dioxide Gas

One of the good technologies using chemical blowing agents is chemical foam-assist (CFA) technology. It is developed from Reedy International Corporation (RIC) [32], Bergen International [33], Momentum International [34], and others. CFA technology enables foam more easily, and it improves the properties for injection molding processes. CFA uses carbon dioxide for the extensively used blowing agent in the production of thermoplastic foams. The studies have shown that CO_2 has a unique low-pressure solubility. When CO_2 is pressurized enough, it becomes a supercritical fluid and acts as a solvent, resulting in a lowering of the glass transition temperature and significantly improving the melt flow characteristics. The extra benefit from CO_2 is that the CO_2 will absorb heat energy during transition between a supercritical fluid and a gas that improves the melt cooling and consequently facilitates reduced cycle time and lower part stresses. Therefore, the thick part (4 mm or more) may use CFA technology to save the cost of hardware since the CFA may only need a good mixing screw without extra gas dosing equipment. The raw material can be mixed with a chemical blowing agent (CBA). The CBA using endothermic nucleating and blowing agents that decompose at a processing temperature in the range of 130–230 °C to release the CO_2 gas. This technology results in a microcellular structure with a smooth solid skin around a fine cellular core. Particle size, distribution purity, and a controlled gas release are tailored to provide a large number of very small nucleation sites that create this fine and uniform microcellular structure.

The permeation of the gas is also dependent on the polymer structure. Crystalline polymers exhibit much greater resistance to gas diffusion because the crystalline structure is tightly packed in lamellae crystals. The gaseous molecule is too large to penetrate these crystals. It is interesting to note that there is a significant research and development effort by many resin producers to develop a foamable HDPE, PP, nylons, and PET, which are best described as semicrystalline polymers. In amorphous polymer the interstitial space between the molecules is large enough to allow the gas molecules to form small bubbles.

8.9.3 Chemical Blowing Agent for Improving Mold Flow and Cooling Time

Chemical foam-assist (CFA) technology offers various advantages. This technology can use a chemical blowing agent (CBA) to improve melt flow characteristics, and ensure uniform cell formation. In many cases, the endothermic agent is not used to make foams, but will be used to improve the melt behavior of the polymer as a processing aid. However, whether if foam is made or not, the CBA does improve the mold flow. When most endothermic agents are added to the material blend, a solid pellet or powder is first converted to gaseous CO_2 though decomposition of the additive. Below 7.59 MPa (about 1100 psi), CO_2 acts as a lubricant improving melt flow. At high pressure, above 11.72 MPa (about 1700 psi), CO_2 is converted from gas to a supercritical fluid. In this state, the CO_2 gas is much more readily absorbed into the polymer melt and dramatically lowers the viscosity of most polymers. By lowering the viscosity of the polymer, one has the choice of reducing the processing temperature or utilizing the improved melt flow at the same temperature. Improved melt flow means fewer gates, thinner wall sections, less molded-in stress, and reduced burning through shear heat. In certain cases, the uses of endothermic agents make production of multicavity tools easier to balance.

Probably, the most important advantage of lowering viscosity is the possibility to achieve equivalent flow characteristics at lower melt temperatures. When high concentrations of gas molecules exist in the polymer matrix, there is significant plasticization of the polymer as evidenced by a decrease in the glass transition temperature (T_g). For specific microcellular PVC foams, the T_g drops from 78.2 °C (virgin PVC) to an estimated 14 °C upon saturation. Bubble nucleation and foam growth are possible below T_g of the original polymer through CO_2 saturation.

When CO_2 changes phase from a supercritical fluid to a gaseous bubble due to the drop in pressure, heat is an endothermic process of heat absorption. As the polymer enters the mold, it cools from the inside during foaming as well as from the physical contact to the mold surface. Here is interesting example of this CFA technology with a thick wall section: By adding 0.1–0.5% active level of an endothermic blowing agent, the cycle time was reduced by more than 10%.

The saturated CO_2 exerts an internal pressure in the product that results in improved material flow characteristics that enable the processor to feed polymeric material very consistently. This internal pressure is similar to the pressure inside of a soda bottle. Once the cap is removed, the pressure drops to atmospheric, and the CO_2 gas begins to escape the liquid. Since the product is still saturated with CO_2 and is significantly plasticized, lower molding stresses are evident in the product. The material has more time for the molecular chains to relax and for the stresses to dissipate.

8.9.4 Design Recommendation to Use Chemical Blowing Agent

In injection molding, we expect to see the major growth with endothermic blowing agents. This technology will enable designers to produce moldings with thin and thick sections, without built-in stresses or sink marks. To produce molded products that have wall sections between 0.04 in. and 0.4 in. or more in a single molding with ease, it is necessary to add only 0.1–0.5% active level of an endothermic blowing agent. Endothermic is often formulated with carrier resins that also improve the melt strength or properties of the diluting resin and that can be used with no additional equipment expense other than a high accuracy side blender/feeder at the hopper. However, simple tumble mixing will also give the desired result.

To produce Class "A" surface finish moldings with a good distribution of foam, it is necessary to observe both design and processing recommendations. In practice, most of these are similar to the conditions you would use for N_2 gas injection systems or gas counterpressure systems. To prevent splay or prefoaming inside of a mold, a gas pressure of 0.24 MPa on the melt front needs to be maintained. As the bubbles and skin form during the mold filling stage, the propensity for the bubble to appear on the surface is reduced, providing a better appearance on thick sections. Obtaining 0.24 MPa (35 psi) in the mold is easily achieved by reducing the open area of mold vents or by increasing injection speed. In most molds, what is normally regarded as inadequate venting is actually better for good surface finish of the part made by CFA molding. Sprue and runners should also be as short as possible to reduce the shear heat and allow for fast filling.

When designing products, gates should be as generous as possible but should still remain not more than two-thirds of the product thickness that is being fed. Always feed into the thinner wall sections of the mold, allowing material from there to flow into the thicker sections. Sharp transitions of the section thickness, especially close the point of injection will create more difficulty with regard to surface gassing. If possible, smooth transitions should be used. In considering the part design with CFA, just keep in mind that the flow of the material will be improved when using CFA, due to the reduced viscosity.

8.9.5 Processing Recommendation to Use Chemical Blowing Agent

From the processing aspect, most standard molding machines are capable of producing good-quality CFA moldings. Injection pressure and injection speed will need to be adjusted to reflect the changes in the viscosity of the polymer melt. When establishing molding conditions, it is important to change only one variable at a time. Best conditions are normally established by reducing the melt temperature by increment about 5 °C, reducing injection pressure, reducing shot weight, reducing cooling time, and increasing injection speed. For each shot weight/density reduction, a repeatable set of parameters can be quickly established. In all cases, machine operators will find that the processing window for each polymer will be considerably widened.

It has been found that the CFA technology produces similar weights of components and cycle times as gas-assist or gas injection systems. Mirror housing trials provided a real-world comparison of part finishes and foam behavior. The molding was produced in a natural manner to see how both the injected N_2 gas and CFA-generated gas behaved. First, a set of processing parameters was established that provided stable conditions and produced good appearance under identical conditions. The endothermic gas created a foam structure in those areas with the highest pressure drop. On close examination, there is no visible foam in the thin sections subject to high pressure, but the foam structure becomes more apparent in the thickest areas. The lowest density may be achieved in the thickest areas. This gas expansion is the driving force that prevents sink marks.

Another trial with an appliance producer demonstrates the comparison between the N_2 gas injection process and CFA. In this case, the transition from the thin-walled area to a thick-walled handle was even more dramatic. Even though a good appearance was achieved in both cases, the cycle time was shorter with the CFA endothermic agent than with the N_2 gas injection process.

In summary, CFA molding shall avoid too much flow constriction in the mold that may cause excessive shear heat and an undesired reaction of the chemical blowing agent. In addition, CFA molding needs to avoid suddenly uneven pressure drops during mold filling that will result in inadequate gas counterpressure on the melt face and cause prefoaming or spray inside the mold. Now with increased environmental pressure and a new emphasis on the use of CO_2, new opportunities are available to the processor. In some applications, CFA technology complements traditional direct gas injection applications. While high-pressure-injected gas acts as the accelerator for the polymer met, the CO_2 polymer solution improves melt flow characteristics that lead to improvements in overall performance.

8.9.6 Endo/Exothermic Blends

It is well known that there are two basic types of chemical blowing agents, endothermic and exothermic. However, there is a third type of better chemical blowing agent that is the blend of endothermic and exothermic. This blend of chemical blowing agent is called endo/exothermic. This blend of chemical blowing agent can be used in applications where the properties of both the endothermic and exothermic play the important roles for whole process. A typical example is the processing of rigid PVC [33]. The exothermic chemical blowing agent brings a larger volume of gas at a high pressure, allowing the gas to enter the PVC matrix that further makes the PVC material viscosity lower than the original PVC material. On the other hand, the endothermic chemical blowing agent provides gas and absorbs heat and thus increases the cooling rate for faster production.

8.9.7 Troubleshooting for Chemical Blowing Agent Processing

Every processing has a unique technology to improve the processing and yield the best part quality. The chemical blowing agent processing has long history of success processing and rich data files to be referred to for the processing. Reedy [32] summarized key foam processing troubleshooting actions with detailed explanations as follows

- Post blow in the foam is a possible problem for CFA molding. Because the part is ejected too early, parts removed from the mold will continue to expand and swell. The solution is to lower the processing temperature or to increase the cooling time. However, it may simply be the overdosing of blowing agent in the part, so reducing the blowing agent may solve the problem as well.
- Elephant skin is another common issue of surface roughness. It often appears near the end of the fill. This is usually an indication that the melt or mold temperature are too low.
- Sink mark is from the shrink on the thick ribs, or from bosses when the thermoplastic materials are in the cooling stage, which is more of a problem from semicrystalline materials. Add a little more blowing agent and make sure that the pack pressure is released so that the foam has the room to expand.
- Warpage is located in the position where the part side wall bows in or out. It may be from the nonuniform shrinkage or from high molding temperature and injection pressure. In this case, reducing the molding temperature and pressure may help. However, the general way is to cool the part longer.
- Voids, or big cells, are the areas of missing material, usually hollow areas formed from gas pockets. The solution is usually to increase the melt temperatures and lower the injection speed if jetting occurs. However, if it is from the nucleation problem, then increasing the injection speed is the solution. In addition, the nozzle open and injection action must be acted upon in the right sequence to avoid the premature foam before injection. It is also very important to match the foaming agent to the resin system.
- Plate out, or deposit on the screw and screen pack, is the problem with exothermic blowing agents. This can have a number of causes, including the resin and additive packages, but the quickest solution is to reduce the amount of blowing agent used. It is also important in this case to ensure that the blowing agent is matched to the resin emulsion system.

- Impact strength drop and brittleness is usually the result of big cells. Therefore, it may be the way to improve the cell structure. Typical fixes include lowering the temperature, decreasing the amount of blowing agent, or changing a different type of blowing agent. It may be from material degrading, so the material residence time and injection shearing rate through the gate needs to be checked as well. If it is the glass-fiber-reinforced material, the fiber orientation may need to be improved. In addition, the long glass fiber part needs to be processed with minimum damage to the length of fibers.
- Flash is the excess material usually at the mold knit lines seen after demolding. The solution is to lower the process temperatures, reduce shot size, or lower the blowing agent dosage. Avoid sharp transitions of section thickness, especially close to the point of injection because it will create more difficulty with regard to surface gassing; if possible, a smooth transition should be used.
- Burning shows that the part is discolored, and the surfaces display unevenness. It can be fixed by lowering temperature and by opening the gates and vents. In general, avoid the flow constrictions that may cause excessive shear heat and an undesired reaction of the blowing agent. Gate should be as generous as possible but should still remain not more than two-thirds of the product thickness that is being fed.

8.9.8 Future Work for Chemical Blowing Agent

The drawbacks of a chemical blowing agent are the possible byproducts after chemical reaction, so it may not be used for organoleptic applications. It also relies on the screw mixing and may have an inconsistent process due to the inherent difficulty controlling the levels and releasing of the blowing agent. It may result in a poor quality of cells if the screw mixing is not good enough. Sometimes the byproduct from a chemical reaction will plug up the mold vents, and it is possible that a chemical reaction with some engineering materials will degrade the polymer. The surface quality of the foam made by a chemical blowing agent may be the problem as well. Usually, a chemical blowing agent is used for a thick part, and sometimes we just want to add it to remove the sink mark and keep the stable dimensions of the part.

However, there is the new chemical blowing agent XO-230, which produces no moisture; it is the byproduct of the traditional chemical blowing agent that eliminates any brittleness due to polymer degradation from this moisture [34]. XO-230 was introduced by Bergen International, the manufacturer of Foamazol chemical foaming agent, and is specifically formulated for high-temperature polymer processing, such as PC foaming application.

There are other newly developed chemical foaming agents: Microcell 545 and 546 from Momentum International, a German manufacturer of plastics additives [34]. These new foaming agents provide endothermic reaction kinetics for applications in PP, HDPE, TPU, and nylon, and they produce very fine cell structures at low doses. With only 1% dosage of Microcell 545 foaming agent, the density of a foam product can be reduced significantly up to 25% for TPU, 10% for HDPE and PA 6, and 20% for PP. In addition, Mementum's foaming agent powders can act as nucleation when it is used as a blowing agent.

Reedy planned to study further how CFA technology utilizes unique surfactants and how high-surface-area particles can launch supercritical CO_2 into new areas of material science. These involve using supercritical CO_2 to improve the compatibility from polymer to polymer through interpenetrating networks. Also, supercritical CO_2 solutions are being evaluated as emulsifiers for organic and inorganic blowing agents and their contribution to the foamed materials.

8.10 WATER BLOWING AGENT

Research was carried out in the 1980s by Dow Chemical to produce PS foams with a water blowing agent [35–37]. The water as a blowing agent was combined with butane, and all of them are inexpensive and environmentally friendly. Except for some water-absorbing polymers, such as PC, PA, ABS, and so on, most polymers have big and nonuniform cells because the water has very low solubility in polymer. Adding inorganic particles may help to carry and disperse the water in the polymer [38–40].

Lee et al. [41] did try the combination of water and *n*-butane as the blowing agent to make some extruding foam. Lee reported that the bicellular polystyrene foam was successfully made in extruding foam with water, *n*-butane, and silica. Although the results are from an extruder, the first stage of making good water vapor–melt solution in extruder barrel is a good reference for the injection molding since the it is easier to make foam by injection molding than by extruding. The final result of the water blowing agent is the bicellular structure. The injection molding may improve the nucleation and control foam structure more than extruding process. This bicellular foam can be changed by varying the content ratio of *n*-butane, water, and silica, as well as by controlling the melt temperature. The optimum result for the bicellular structure was reported by Lee et al. [41] are 5–7% *n*-butane, 1–2% water, and 1–3% silica at 90–120 °C of temperature. It seems like the silica absorbs and disperses the water so that the content of silica increasing in a certain range will promote the cell density. It is a good reference researching result for the future water foaming injection molding.

VanHouten and Baird [42] tried a combination of water and supercritical CO_2 gas for the high-performance polymeric foams of poly(arylene either sulfone) (PAES). The PAES pellets are filled in the stainless steel mold with liquid water, and the mold is placed in the pressure vessel that is pressurized by 5.5 MPa of supercritical CO_2 gas. By controlling a vitrification of the poly(arylene either sulfone) through the diffusion of the plasticizers and foaming temperature, the cell size can be made between 1 μm and 200 μm, and the foam density varied between 15% and 85% of the un-foamed polymer. Although it was made in the batch process, it proves that the combination of supercritical carbon dioxide gas and water as the blowing agent is successful to make PAES microcellular foam. At low releasing temperature of about 160–165 °C, minimal cell growth occurred and the cell sizes ranged from 1 μm to 5 μm. As the release temperature increases, the cell size and the bulk density of the foam increase as well. In addition, VanHouten and Baird [42] found that the relative tensile modulus of the PAES foam was maximized when the cell size was minimized and the cell nucleation density was maximized simultaneously.

It is well known that a gas-saturated polymer will have low glass transition temperature T_g. Dr. Baird and his team found that when using only supercritical CO_2 gas as the blowing agent the T_g of PAES is reduced from 220 °C to about 205 °C. However, when using a dual plasticized system (both water and CO_2 gas are good plasticizers for PAES) the T_g of PAES is reduced further between 130 °C to 150 °C. Also, in this temperature range the onset of a large endotherm is observed. It is because of the rapid release of the blowing agents from the polymer matrix [42].

Water being be used as a blowing agent is not new, and it occurs when the moisture is not removed completely in material absorbing the water, such as nylon, PC, and so on. A special test was carried out with PC without drying, and the result is kind of microcellular if the back pressure still maintains the high pressure in the accumulated melt with moisture in. After injection, some cells are nucleated and shaped in the part. The size of the bubble from moisture is bigger than normal cells of microcellular foam, and the cell density is not as good as real microcellular foam. However, it just proves the possibility of using moisture of water as the blowing agent that may be developed in the future for microcellular injection molding.

8.11 STRESS FOAMING

This is a topic of theory and may need to be verified further for industry application. However, it seems like the existing phenomenon of molding for microcellular foam. It may explain some nice cell in the shearing area that is the interface between solid skin and foamed core. If the shear rate is in a certain range, there are some small cells existing in this shearing zone without stretching too much. However, if the cells are stretched severely, it may indicate that the shearing rate is over the range of cells resulting from shearing stress.

Lee and others did some experiments to simulate the shearing field and generate the shear stress on the spinning disc [43]. It results in some different cell densities at different shearing rate, as shown in Figure 8.22. It is obvious

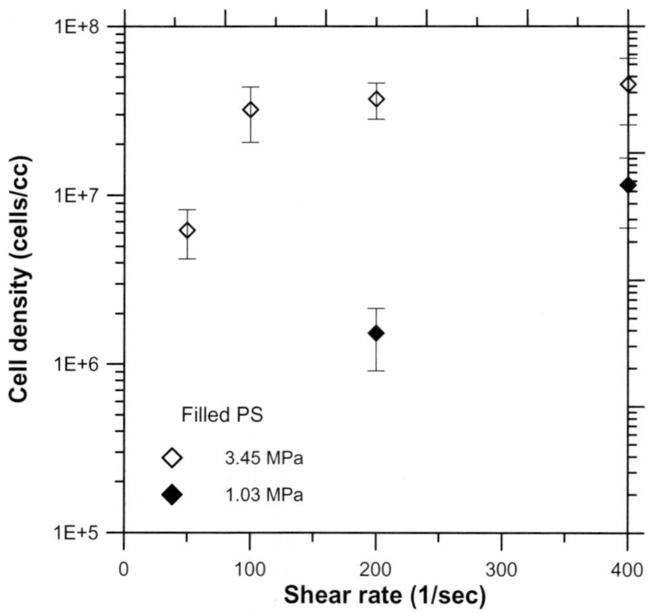

Figure 8.22 Cell density varies with the change of shear rate (stress) for filled PS material. (Courtesy of Trexel Inc.)

that the cell density will increase with a shear rate increase at low shearing rate range up to 100 1/sec under the melt pressure of 3.05 MPa. However, the cell density does not change with the continual increase of shear rate, which means that the shear stress is only effective at the shear rate 100 1/sec or lower at 3.45-MPa pressure. The interesting result is that when the low melt pressure is 1.05 MPa, the shear-stress-induced cell density is continually increasing, which is different from the data at high melt pressure. Chapter 2 discusses the physics behind these phenomena.

Other papers also introduce the stress foaming either in extrusion or in the batch process [44, 45]. The conclusion is similar to that in reference 43, which is helpful with stress to promote the foaming performance.

The concept of stress foaming may include the post-stress foaming process with 100% mold filling of a sample of a microcellular part. Then the part under the stress may cause local foam at the location of local stress. The effect of shear stress can be summarized below:

- At a low gas level, cell density can be one order of magnitude higher by applying shear stress.
- Higher cell density with higher shear stress.
- Mechanism: transformation of mechanical energy into surface energy, orientation of free volumes.

With sufficient gas pressure available, the shear stress is summarized as follows:

- Narrow gap is easiest to apply shear stress.
- Die forming or molding after shear.
- How long the shear effect lasts.
- Possibility to rearrange flow after high shear.
- Waiting time after high shear applied.
- Cell density as a function of waiting time.
- Relaxation of oriented free volumes.

A new stress foaming method has been developed with electromagnetic dynamic shearing by Zhang et al. [46]. A dynamic shear is induced from the rotor vibration that is vertically superposed on the static shear for the PP/HDPE blend with CO_2 gas as the blowing agent. The new methods show some interesting results [46]:

- The cell density increases with the increases of vibration frequency and amplitude. This may result from the melt shear rate and melt strength increases at higher dynamic shear.
- The vibration superposed vertically on the flow direction of material shear field enhanced the melt strength and improved the foamability and improved the cell structure.
- The shear without vibration is defined as static shear in this study. However, integrating the dynamic (vibration) shear with the static shear results in a multiple shear. The result of adding dynamic shear (axial vibration) to the existing static shear (circumferential shearing) shows that dynamic shear changes the cell shape from oval to round.

The ultrasonic vibration will help for nucleation and will then promote the foaming process. The ultrasonic vibration technology is similar to the vibration shearing technology. It is difficult to design a mold with this additional vibration shear with the direction perpendicular to the mold flow direction. It will be another challenge for the future microcellular injection molding.

8.12 MICROPOROUS METAL PART

The metal powder with porous structure made by injection molding is not new. However, the microporous process in metal powder injection molding (MIM or PIM) is the new idea [47, 48]. In the MIM process, fine metal powder is mixed with a binder phase to produce the feedstock for injection molding operation carried out at a later stage. The binder is a plastic material blend of polymers, waxes, and other additives. The binder phase essentially consists of a component that can hold the metal particles together after the molding process, and is early removed via chemical leaching or heat before sintering operation. This binder removal process is named debinding, and it is a slow process to avoid any distortion the molded part. Similar to the microcellular plastic processing, the gas at supercritical state, such as carbon dioxide or nitrogen, is injected into the molten feedstock, which then dissolves into the binder that is similar to plastic with flow ability. As the feedstock with dissolved gas travels through the nozzle, the gas bubbles nucleate, and brow to form pores. This pore growth halts when the melt freezes as it comes into contact with the cold mold surface. This will create a dense generally pore-free surface. The molded components are subsequently debindered and sintered with gas-generated pores preserved in the center of the part. The microcells left in the final part result in weight reduction without significantly changing the part property.

Two types of binders have been used in MIM feedstocks: thermoset and thermoplastic. The thermoplastic binders are by far the most popular. It is important for the chemistry of the binder to have high gas solubility when it is melted in the liquid state but have low gas solubility when it is in the solid state. In addition, it needs to be considered that the presence of metal powder will change the melting behavior of the plastic binder, which may influence the gas dosing process.

The metal powder injection molding uses metallic powder that is atomized with small size in the range of 0.5 to 25 μm. The best particle size for molding corresponds to the ranges from 2 to 8 μm. In addition to gas injection, the formation of microporous metal is similar to the microcellular in that both processes use the same type of molding machine. The barrel and screw, however, is upgraded to accommodate the materials used in MIM [49]. Overall, the powder particles are the excellent help for the heterogeneous nucleation.

The result is a metal part with a dense surface skin and core with small pores ranging from 5 to 200 μm in diameter. This dense surface is achieved upon sintering. It is because poor formation on the surface is minimized in the process. It results from a rapid surface cooling when the feedstock hits the mold walls and releases gas. During sintering, the small naturally occurring pores between metal particles are eliminated, leaving behind the larger gas-generated pores in the microstructure. This process saves significant material in feedstock consumption.

It is true that the sample shows that a dense surface skin is formed over the porous interior of a molded microporous metal part. The surface finish does not matter, and this technology has potential to be used for microporous metal, ceramic, and intermetallic components with dense surfaces.

It is suggested by the inventor that the process can be used for applications such as jewelry, sporting goods, lightweight structures, and heat insulating components.

The screw is designed with a structure similar to that in Chapter 7. The compression ratio is low for MIM. The microporous MIM screw still has five zones: feed, transition, metering, wiping, and mixing [49]. The special screw and mixer in the barrel for microporous metal powder processing were patented in 2007. Since the metal powder screw has a very low compression ratio, the shearing needs to be controlled for both mixing of gas and low shearing requirement by MIM material.

Mold design for MIM with microporous architecture will be similar to the regular MIM mold design [50]. The binders typically have a density of 1 g/cm^3, while the metal particles are much denser at about 8 g/cm^3. Parts filled in 0.2- to 0.6-sec range will produce high shear rate up to several hundred thousand 1/sec in gate and thin cross sections. As a result, the metal particles will stack up at every direction change in the flow pattern until pushed by the binder. Because the metal is heavier and has great inertia, metal particles require more energy to be set into motion. These dynamics can cause metal–binder separation, which leads to part distortion during sintering. Any area with too much binder shrinks more, while metal-rich areas shrink less. The gas pressure in the cells can help to push metal around, which may help to solve this separation between binder and metal during mold filling. However, the mold design may need to follow rules similar to those for regular MIM:

- Smooth transitions and minimal abrupt direction reversals. For corners, use diverter pins and smooth radiuses to guide the flow. Transitions to gates should be blended to avoid sponge effects, in which a metal-rich area squeezes the binder into the cavity.
- Gating that injects along cavity walls to avoid the metal–binder separation. Basically, do not shoot the melt directly against a wall or core pin. Instead, shoot the melt along the wall to form a good melt front and build plug flow.
- The vent depths of 0.013 mm or less is still needed because the low-viscosity binder with gas attains even lower viscosity, which is easily leaking into the gap of venting to cause flash on the parting line or other venting area. The solution between (a) more gas and tapped air venting for gas in a laden binder with high injection speed and (b) low-viscosity material leaking is to keep the same tight depth of venting but increase the width of venting. All mold inserts and ejector pins are built to an accuracy of 0.005 mm.
- A balanced gate size design for enough nucleation of binder, along with generous gating to avoid the feed prematurely cooling of the high-conductivity feedstock of metal. However, the rule of gate in the thickest section in the cavity must be changed since gas-pressurized cells in the binder does not like the thick position for gating. The full round runner system is still good for microporous metal powder injection molding.

- The metal powder fillers are abrasive, particularly in thin sections, such as gates. The hardened steel with 60–62 Rockwell C hardness range must be used for the mold. Gate blocks and cavities are typically made in D-2, critical slides are made in M-2, A-2, or CPM10V, and everything else is made in titanium-nitride-coated H-13 steel.

The low green strength of the part needs to be ejected from the mold without exposure to tensile, shear, or torsion stresses. It is because the MIM with a microporous part can withstand a high degree of compression stress but has low strength and limited elastic properties. Therefore, any tensile or shear stress can cause distortion, cracking, and residual stress problems. Usually a uniform ejection/core pull is required for entire part, such as a large number of ejector pins or blades. Similar to the microcellular plastic part design, the draft angle needs to be increased. The MIM feedstock has heavy metal powder at high filler content percentage up to 65–70% by volume. Metallic filler is incompressible and closely packed with very little shrinkage as they solidify. In addition, gas pressure in microporous will push the part, thereby expanding instead of shrinking. Therefore, the ejection of a MIM microcellular part may be more difficult than the ejection of a regular MIM part.

8.13 LOCAL MICROCELLULAR FOAM

There are several types of special processing to make microcellular parts. One of them is to make a potential microcellular part with injection molding. The part is the same as the regular solid part, except the gas is in the part without foaming. Then, before the de-gassing is completed, the part can be foamed locally by local heat, stress, or vibration. It is possible to make the partially foamed part for the sealing, insulation, or other purpose with one single piece of plastic part with different structure of solid and microcellular foam in a different position of the same part.

There was a special process used for local foaming concept. It was named Coralfoam technology (abbreviation CFT) [51, 52]. It was an innovative injection molding process that might have a radical effect on a foaming technology in injection molding. It improved the part rigidity and reduced the weight of the part at the same time. An endothermic blowing agent is used in the polymer. It selected the foam positions to do the local foam only at the initial part processing. The thick section was designed for local foaming where the rim is halfway up the cup's height. When the mold is opened for local foaming, the thin section of the cup has been solidified but the thick section is still hot enough to allow local foaming occurring in the thick sections [51]. This local foam part cut the cycle time in half and without enormous holding pressures and high clamp forces. It also removed any sink marks. Even if the local section was designed with thicker walls than most of the area, the thin-walled sections governed cooling time [51].

The newly developed CFT process can make a whole foam part with different foaming levels. The step of this technology is to make a squat preform in a regular injection molding with about 50% of the final height. Before the melt is solidified, the mold is stretched and the new sidewalls swing into the intervening gap. As the sidewall of the pre-form tried to expand in all directions, the preform is stretched and air pressure (applied within 0.7 sec after injection), which is balanced against the residual gas pressure in the cells, is provided. It had the local foam only in the rims and rings of a cup as the initial prototype product. In each process, gases are used in the polymer to create moldings that are stiffer and lighter. On the other hand, the process runs faster than the conventional injection molding. To allow gas–polymer mixtures to expand outside the confines of the mold, a controlled and repeatable foam area can be made [52]. Now this technology developed further to allow the full container wall to be foamed, followed by stretching of the walls of cup as the foam expands. The air pressure is the key to balance the gas pressure in the cells to control the wall thickness without crushing the foam.

Similar to the concept above, the photochemical foaming technology can be used to make a local microcellular foam in the part. A foamable plastic consisting of a polymer with *t*-butyl ester groups and a photoacid generator has been developed for the thin microcellular plastic sheet [53]. The cells are very fine in the range of 0.2–0.34 μm and are uniformly distributed in the specimen. Upon ultraviolet (UV) irradiation, the photoacid generator releases protons. The local foaming idea will use UV light to control the local area foaming. The UV light can be selected with special shape and area with a pattern mask, and then only the necessary foamed area will be exposed to UV light. The protons transform the *t*-butyl ester group attached to the polymer into isobutene gas upon heating. Then it results in a controlled microcellular formation [53]. With adjusting UV light exposure and heating temperature, it has been proved for the benefits of this technology [53]:

- High-resolution foaming pattern
- Smooth gradation of foam
- Controlled foaming in depth direction

If either local heat or local UV irradiation is provided for a controlled area to have a thermal-induced chemical reaction of a photoresist, only a local area will be foamed in a finished injection regular molded part. This may create the part with a sealing zone without making two components assembled together.

Similarly, the compression stress may be acted on the local area that needs foaming. The stress foaming technology can be used for the local microcellular foam cases with specially designed tools. However, it has not been applied in industry yet.

8.14 THIN-WALL MICROCELLULAR FOAM

Thin-wall foam is only possible for microcellular foam since the wall thickness is about 0.5 mm or less, and the big cell size, such as 500 μm, will not work for the thin-wall part. However, even if the microcellular foam process is used, the challenge still exists regarding the processing setup, which must match the criteria below:

- Mold must be filled before the freezing time of the thin channel.
- Cavity pressure needs to be high enough to overcome the resistance of the narrow channel for mold filling.
- Cell size must be small enough and uniform so that the thin-wall section will retain enough strength after foaming.

The key of the thin-wall microcellular process is that the mold filling volume in the mold cavity needs almost 100% with high injection speed. The result is smooth surface and very nice cell architecture that shows almost below a cell size of 5 μm with spherical shape. In addition, all cells are distributed uniformly throughout the part since the nucleation has no significant difference from beginning to the end in such a short time of mold filling (0.5 sec or less).

The typical application is named Super Light Injection Molding (SLIM) [54]. SLIM® technology uses the MuCell® process. The gas at supercritical state (nitrogen or carbon dioxide) with precisely metered quantities is injected into the barrel to make a single-phase solution, and then the single-phase solution is injected into the mold with extremely fast injection volume rate. The result is to create millions of almost invisible microcells in the thin-wall part. The microcells replace their equivalent volume of plastic in the microcellular part, resulting in a reduction of up to 5–10% in packaging weight without any perceptible difference in the final thin-wall part quality.

A similar thin-wall container manufacturer uses MuCell® microcellular injection molding technology, as reported in reference 29. This thin-wall part for the microcellular process is a 0.5-mm thin-wall container that has 10% weight reduction, with 27% of cycle time reduction, 36% clamp tonnage reduction, and 20% injection pressure reduction. This application was initially run with the issue of mold filling. The existing mold that is run in a regular process simply cannot fill the corner of the part. Therefore, the initial motivation of this thin-wall molding is to fill the corner without increasing the tonnage of the machine. The trial successfully used microcellular technology not only to fill the mold perfectly with low tonnage but also to reduce the weigh percentage with super-microcellular cell structure and nice smooth surface as well. The cell size is about 3–5 μm and is distributed uniformly across the thickness of the thin-wall part.

The entire thin-wall part has very smooth natural white color surface because of the small cell size and uniform distribution. It is possible that the contribution of extremely fast injection speed eliminates gas out of the solution on the skin, and the fast injection speed results in small cell sizes in the center as well. This critical injection time is preferred within 0.5 sec. If the injection cannot finish in time, the cells will be stretched severely and the cell size will vary from 20 to 100 μm [55]. Although the cell size in this experiment is still in regular microcellular cell structure, the cell size of 100 μm is almost 20% of the total wall thickness of 0.5 mm, so the part properties will be reduced significantly. Actually, most failures of thin-wall microcellular injection molding are from the slow acceleration of injection speed at the beginning of injection, not total injection time or average injection speed. Sometimes the successful thin-wall injection molding is determined by an important parameter of ramping time in addition to the total injection time. It must be ramping as quickly as 0.1 sec or less to avoid the freezing of flow channel because the thin-wall part has high flow ratio, usually 200 or more. Therefore, even with a low-viscosity gas-laden melt for less resistance of mold filling, the biggest challenges for microcellular molding in thin-wall processing are to fill the mold before flow channel freezing and to remove gas from the solution on the skin of the thin wall. The interesting morphology of thin-wall molding at extremely high injection volume rate shows no significant cell stretching in the foamed core. On the other hand, it does not show the perfect spherical shape cells either [29, 55].

Another key parameter of thin-wall molding is the gate size. It is usually thought that the shearing in such a small gate for thin-wall molding may be too high. However, there are several reasons to use small gate size for the thin-wall microcellular process:

- For high-viscosity material, such as PC and PC/ABS, the low viscosity of gas-laden material is not enough to lower the injection pressure during extremely fast injection. The shear thinning at the gate will be necessary to help successfully perform the thin-wall molding [55].
- Even the low-viscosity material (such as PP with melt index over 40) with enough gas in the melt needs the small gate to further reduce the viscosity during mold filling. The small valve gate was very helpful to finish the mold trial of microcellular that is reported in reference 29.
- A high gas percentage in the molten polymer is also necessary for thin-wall microcellular molding.
- A good venting system is also important for thin-wall molding since both air in the cavity and escaped gas from flow front must be vented in very short time during mold filling.

The supercell architecture in the thin-wall part benefits the good insulation properties for this thin-wall packaging part. It is because of the huge number of micro-close cells in the part with inherently low thermal conductivity.

REFERENCES

1. Garner, P. J., and Oxley, D. F. British Patent No. 1,156,217 (1971).
2. White, J. L., and Dee, H. B. *Polym. Eng. Sci.*, **14**, 233 (1974).
3. Young, S. S., White, J. L., Clark, E. S., and Oyanagi, Y. H. *Polym. Eng. Sci.*, **20**, 798 (1980).
4. Donavan, C., Rabe, K. S., Mammel, W. K., and Lord, H. A. *Polym. Eng. Sci.*, **15**, 774 (1975).
5. Hamada, H., Maekawa, Z., and Xia, M. *Int. Polym. Processing* **VIII**, 2 (1993).
6. Selden, R. *Polym. Eng. Sci.*, **40**, 5 (2000).
7. Derdouri, A., Carci-Rejon, A., Nauyen, K. T., Koppi, K. A., and Salamon, B. A. *SPE ANTEC Tech. Papers*, 481 (1999).
8. Nauyen, K. T., Turcot, E., Salamon, A., Ait Messaoud, D., Sanschagrin, B., Salamon, B. A., and Koppi, K. A. *SPE ANTEC Tech. Papers*, 533 (2000).
9. Turng, L. S., and Wang, V. W. *SPE ANTEC Tech. Papers*, 297 (1991).
10. Schlatter, G., Agassant, J. F., Davidoff, A., and Vincent, M. *Polym. Eng. Sci.*, **39**, 78 (1999).
11. Messaoud, D. Ait, Sanschagrin, B., and Derdouri, A. *SPE ANTEC Tech. Papers*, 671–675 (2004).
12. Han, C. D., *Rheology in Polymer Processing*, Academic Press, New York, 1976, pp. 264–270.
13. Turng, L. S. *J. Injection Molding Technol.*, **5**, 160–179 (2001).
14. Eckardt H., and Alex, K. *Advances in Plastics Technol.* April, 40–49 (1981).
15. Eckardt, H. *J. Cell. Plastics* **23**, 555–592 (1987).
16. Moss, M. D. *Mod. Plastics Mag.*, March, 81–83 (1997).
17. Turng, L. S., and Kharbas, H. *SPE ANTEC Tech. Papers*, 535–539 (2004).
18. Xu, J. *SPE ANTEC Tech. Papers*, 2089–2093 (2007).
19. Ham, S. *Plastics Eng.* June, 46–57 (2001).
20. Klaus, G., and Stefan, K. U.S. Patent, No. 6,419,869 (2002).
21. Caropreso, M. H. *SPE ANTEC Tech. Papers*, 79–81 (1987).
22. Lee, J. W. S., Wang, J., Kim, S. G., Yoon, J. D., and Park, C. B. Development of Structural Foams with a Class-A Surface Using Gas Counter Pressure and Mold Opening, *CSME Forum Paper No. 1569101377* (June 2008).
23. Neilley, R. *Injection Molding Mag.* November, 98–101 (2003).
24. Candal, M. V., Gordillo, A., Terife, G., and Santana, O. O. *SPE ANTEC Tech. Papers*, 620–624 (2007).
25. Throne, J. L., *Thermoplastic Foams*, Sherwood Publishers, Hertford, UK, 1996.
26. Semerdjiev, S. *Introduction to Structural Foam*, SPE. Towanda, PA, 1982.

27. Mueller, N., and Ehrenstein, G. W. *SPE ANTEC Tech. Papers*, 593–597 (2005).
28. Michaeli, W., and Cramer, A. *SPE ANTEC Tech. Papers*, 1210–1214 (2006).
29. Okamoto, T. K. *Microcellular Processing*, Hanser / Gardner Publications, Cincinnati, 2003, p. 120.
30. Wang, J., Lee, W. S., Yoon, J. D., and Park, C. B. *SPE ANTEC Tech. Papers*, 2168–2172 (2008).
31. Anderson J. R., et al. U.S. Patent, No. 6,593,384 (2003).
32. Reedy, M. E., Chemical Foaming Agents Improve Performance and Productivity, Reedy International Corporation, 2004.
33. LeMaster, N. *Modern Plastics Magazine—World Encyclopedia*, 97 (2009).
34. Stewart, R. *Plastics Engineering* December, 16–17 (2009).
35. Motani, S., Saito, T., and Itho, T. U.S. Patent No. 5,064,874 (1991).
36. Paquet, A. N., and Suh, K. U.S. Patent No. 5,240,968 (1993).
37. Suh, K., and Paquet, A. N. U.S. Patent No. 5,380,767 (1995).
38. Berghmans, H., Chorvath, I., Kelemen, P., Neijman, E., and Zijderveld, J. U.S. Patent No. 6,127,439 (2000).
39. Pallay, J., and Berghmans, H. *Cell. Polym.*, **21**, 1 (2002).
40. Pallay, J., and Berghmans, H. *Cellular Polymers* **21**, 19 (2002).
41. Lee, K. M, Lee, E. K., Park, C. B., and Naguib, H. E. *SPE ANTEC Tech. Papers*, 1920–1924 (2008).
42. VanHouten, D., and Baird, D. G. *SPE ANTEC Tech. Papers*, 1003–1007 (2008).
43. Chen, L., et al. *Polym. Eng. Sci.*, **42**, 1151–1158 (2002).
44. Lee, S. T. *Polym. Eng. Sci.*, **33**, 418 (1993).
45. Lee, S. T., and Kim, Y., Shear and Pressure Effects on Extruded Foam Nucleation *SPE ANTEC Tech. Papers*, 3527 (1998).
46. Zhang, P., Zhou, N. Q., Wen, S. P., Wang, M. Y., and Zhu W. L. *SPE ANTEC Tech. Papers*, 385–391 (2009).
47. Chitwood, A. *Injection Molding Mag.* April, 92 (2001).
48. Dwivedi, R. K. U.S. Patent No. 6,759,004 (2004).
49. Anderson, G., and Xu. J. U.S. Patent No. 7,172,333 (2007).
50. Maniscalco M. *Injection Molding Mag.*, 29–33 (1999).
51. Mapleston, P. *Mod. Plastics Mag.* October, 43–44 (1996).
52. Mapleston, P. *Mod. Plastics Mag.* June, 38 (1998).
53. Kojima, J., Takada, T., and Jinno, F. *J. Cell. Plastics* **43**, 103–109 (2007).
54. Dvorak, P. *Machine Design* January, 115 (2007).
55. Wang, J., Lee, W. S., Yoon, J. D., and Park, C. B. *SPE ANTEC Tech. Papers*, 2168–2172 (2008).

9

MODELING OF MICROCELLULAR INJECTION MOLDING

Essentially the microcellular foaming process involves the creation of a single-phase solution formed by adding supercritical atmospheric gas, such as carbon dioxide (CO_2) or nitrogen (N_2), to a molten polymer. A typical viscosity study shows that with 4–5 weight percent of CO_2 the viscosity of amorphous material will be decreased about 30%. This single-phase solution is then injected into the mold. The amount of material that is injected into the mold is less than the part volume. Nucleation of cells and the subsequent cell growth fill the rest of the cavity to create microcellular foam [1–4]. Such microcellular foams have cell sizes ranging from 5 to 100 μm and exhibit acceptable mechanical properties compared unfoamed part. There is no formal packing phase in this process. The benefits of using this process are reduction of weight and cycle time, excellent dimensional stability, reduction of warpage, and removing sink mark. In addition, due to the reduction in viscosity and the small cell size, the process may be used for thin-wall molding. In addition, this low-viscosity gas-rich material can be used in regular injection molding for mold-filling, residual stress-free, stable dimensions, and tonnage reduction purposes only. All of them may be predicted with proper models of simulation program with the support of a large database for gas-rich materials.

The rheology and PVT data for gas-rich molten plastics are the basic data files for the successful simulation. The rheology and PVT data for some of gas-rich materials have been investigated with a nozzle rheometer and special screw tip in the reciprocating screw of injection molding machine. Several

Microcellular Injection Molding, Edited by Jingyi Xu
Copyright © 2010 John Wiley & Sons, Inc.

typical rheology and PVT data of gas-rich GPPS are given for comparison with the rheology and PVT data of solid GPPS.

This simulation study presents a simulation model for the microcellular injection molding process in the second stage only. It is assumed that supercritical fluid such as carbon dioxide gas or nitrogen gas at supercritical state is well-mixed with molten polymer in the first stage. Then, the gas–melt mixture is injected into the mold as the second stage of this process. The model simulates the development of cells in the melt during injection molding. The effects of cell growth on material properties and mold flow have been investigated. Some simulation results such as melt pressure and final cell size distribution are compared with experimental results.

Finally, some simplified models of second stage for microcellular injection molding have been introduced for the quick estimation of the result of the microcellular part.

9.1 RHEOLOGY AND PVT DATA FOR GAS ENTRANCED MATERIAL

First of all, the database of materials with gas must be created for the modeling. The introduction of a physical blowing agent changes the thermodynamic and rheological properties of polymers. It is well known that the nitrogen or carbon dioxide gases at the supercritical fluid (SCF) state occupy the space between the polymer molecules that increase the mobility between these molecules. The effect of SCF dissolved in polymer is the viscosity reduction of material. Furthermore, this effect will allow for the reduction in processing temperature, both mold temperature and melt temperature, and increase in flow under equivalent processing conditions. It also results in a decrease in the grass transition temperature and a reduction of the crystallization temperature in semicrystalline polymers. These affects allow for polymer chain mobility at temperature below that of the same polymer without SCF. However, all these effects last only as long as the SCF is dissolved in the polymer.

The current study investigates the method of obtaining such data on an injection molding machine. A comparison between solid and gas-laden polymer melt is made. Various polymers are studied to investigate the properties of gas-laden melts and their effects on process abilities of those polymers.

9.1.1 Rheology for the Mixture of SCF and Plastic Melt

The database for gas-rich materials at supercritical fluid (SCF) state in the first stage is established with a special nozzle rheometer above. The rheology data are provided for the simulation program that will be discussed fully in the Moldflow simulation section.

1. Objectives

- To find the melt pressure effect on the viscosity for a single-phase solution at lower shear rate.
- To establish the relationship among melt pressure, melt temperature, gas weight percentage, and melt viscosity specifically for the single-phase solution well-prepared in the microcellular screw.
- To test an isothermal rheometer since it is running at a set up temperature in the nozzle rheometer for each set of data.

2. Material. Only general-purpose polystyrene (GPPS) is used in this analysis. Other material data can be obtained with this method, which is introduced below.

3. The Structure of Nozzle-Style Rheometer. An instrumented nozzle, as shown in Figure 9.1, consists of a capillary with three pressure transducers and one melt thermocouple. The inside diameter d is 12.7 mm (0.5 in.). The distance between pressure transducers is 114.3 mm (4.5 in.), so the length L_{ori} to diameter d (L_{ori}/d) ratio between pressure transducer in this nozzle is about 9. There are three pressure transducers total in this rheometer, and the distance between the first transducer and the third transducer is 228.6 mm (9 in.) and L_{ori}/d is 18. The first pressure transducer is located 52.3 mm (2.06 in.) away from the entrance of the nozzle, and the L_{ori}/d is 4.12. This pressure transducer is for the entrance correction of pressure profile during the calculation. The pressure and temperature at this position are called P_0 and T_0, respectively. The first measuring position is located position of the second melt thermocouple T_1 and the second pressure transducer P_1. Similarly, the second measuring position for temperature and pressure are located at the position of the third melt thermocouple T_2 and the third pressure transducer P_2, respectively.

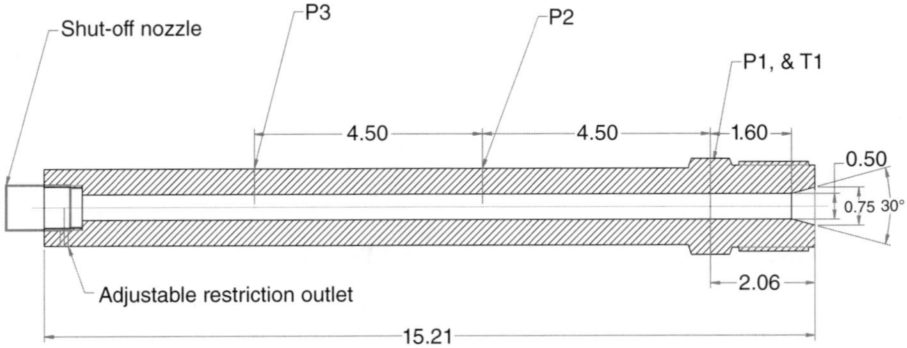

Figure 9.1 Layout of the nozzle viscometer (dimension in inches, converting to mm by multiply it by 25.4).

Then, an adjustable restrictive outlet of the nozzle tip will be used for creating the viscosity data with different pressure at the same shear rate and melt temperature. All three pressure transducers and melt thermocouples in this nozzle rheometer will be recorded for data analyses, and a viscosity data file will be created based on the result of data analyses.

Although many people studied this instrumented nozzle, which successfully generated quantitative rheological data in the reciprocating screw injection molding [5–9], the rheometer used for the gas-laden material is the first trial in the nozzle rheometer of injection molding. The difficulty of the nozzle rheometer for the microcellular process is the high pressure to be maintained in the shut-off nozzle during gas-laden melt accumulated in the nozzle. It is solved by a special nozzle tip designed to (a) seal the nozzle orifice during screw recovery and (b) open it with the nozzle contact a special bar on the machine. This special spring-loaded nozzle uses a spring to close the nozzle and to allow the nozzle to open by contacting the tip to the mold or a solid bar by purging the channel down (see the nozzle tip structure in Figure 7.4). There is special shut-off nozzle also used for this test. It is similar to the Herzog shut-off nozzle shown in Figure 7.3. However, it has a long needle shut-off pin to create an annular channel of nozzle rheometer. Therefore, the overall length of the annular channel nozzle rheometer is not as long as the cylindrical channel nozzle rheometer. Both cylindrical and annular channel nozzle rheometers have three pressure transducers in the nozzle body, and test results are discussed in this chapter. Then, similar to Malloy's et al.'s [5] instrumented nozzle, the nozzle rheometer is used for two different tests. One is the test with "air purge" mode only. It is essentially a large scale of laboratory capillary rheometer (LCR) with a much wider range of shearing rate. It has the advantages of (a) automation to run cycle of machine, (b) consistent injection volume rate by constant injection speed control of machine, (c) plasticizing material with different conditions, (d) mixing gas with melt at different weight percentages, and (e) temperatures and pressures. The pressure reading may be the only one in the middle of the nozzle, and the outlet pressure is zero for the air shot. The different nozzle orifice diameters are used to change the pressure drop at the same injection volume rate. This is an isothermal (nozzle) rheometer, and different temperatures need to be set up for the rheology data at different temperatures.

Another test is carried on with the mold. It is used as a nonisothermal rheometer that uses a modified Herzog shut-off nozzle. The modified nozzle channel has an annular channel for the nozzle rheometer. The mold is a plaque mold with the dimensions of 0.127-m width and 0.28-m length, with thickness varying from 3.18 mm to 6.35 mm. The data are collected only at the stable injection period.

4. The Viscosity Model of Cylindrical Channel of Rheometer. With the schematic of force balance shown in Figure 9.2, the wall shear stress can be derived as below [10]:

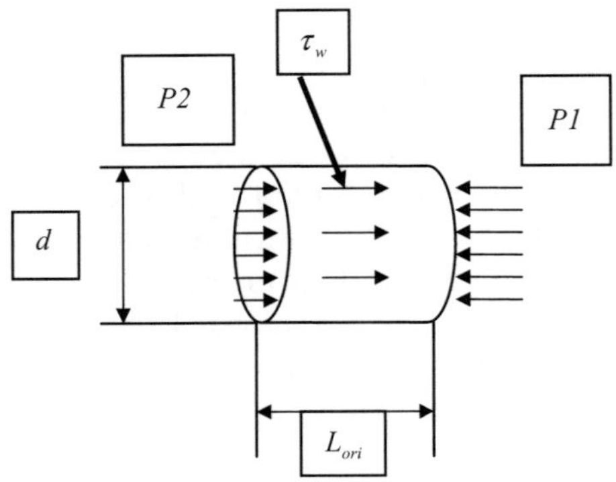

Figure 9.2 Schematic of force balance in the element of cylindrical channel of nozzle rheometer.

$$\tau_w = \frac{(P_1 - P_2)}{4L_{\text{ori}}} d \tag{9.1}$$

where d is the diameter of orifice in the nozzle tip, L_{ori} is the length of orifice, P_2 is the pressure at outlet of orifice, P_1 is the pressure at inlet of orifice, and τ_w is the wall shear stress in the nozzle viscometer.

A wall shear rate with Robinowitsch correction for non-Newtonian fluid is derived:

$$\gamma_w = \frac{8\dot{V}}{\pi(d^3)n}(3n+1) \tag{9.2}$$

where γ_w is the wall shear rate, $\dot{V}$ is the volume flow rate that is controlled by injection speed, and n is the power law index.

Finally, a viscosity can be calculated with the formula below:

$$\eta_T = \frac{\tau_w}{\gamma_w} \tag{9.3}$$

where, η_T is the true viscosity in the nozzle viscometer [see the detailed model for data processing in Equation (9.11)]. Since the orifice in the nozzle is not long enough, the geometric correction is necessary. The test data will be corrected based on the data from mold flow without gas first. Then the same correct factor is going to be used to correct all testing data of gas-laden materials.

5. The Viscosity Model of Annular Channel of Rheometer (see Appendix G). The annular channel of a rheometer will have a more complex model than that of the cylindrical channel above. The details of formulae developing can be found in Appendix G, and here is the direct result. The flow in the annular channel is assumed symmetric about the center of the annular channel; thus, the wall shear stress can be calculated as follows:

$$\tau_w = \frac{(R_{no} - R_{ni})(P_1 - P_2)}{2L_{\text{ori}}} \tag{9.4}$$

where R_{no} is the outside radius of annular channel in the nozzle rheometer, R_{ni} is the inside radius of annular channel in the nozzle rheometer, L_{ori} is the length of orifice (length of the effective annular), P_2 is the pressure at the outlet of the orifice, P_1 is the pressure at the inlet of the orifice, and τ_w is the wall shear stress in the nozzle rheometer.

A wall shear rate with Robinowitsch correction for non-Newtonian fluid is derived:

$$\gamma_w = \frac{(2n+1)\dot{V}}{\pi R_c n(R_{no} - R_{ni})^2} \tag{9.5}$$

where γ_w is the wall shear rate and R_c is the average radius of annular channel of nozzle rheometer:

$$R_c = \frac{R_{no} - R_{ni}}{2} \tag{9.6}$$

The final viscosity is given by Equation (9.3) above. However, one more geometric correction may be necessary if the relationship below is not satisfied [11]:

$$\frac{R_{no} + R_{ni}}{R_{no} - R_{ni}} \geq 37 \tag{9.7}$$

The correction factor F_p is determined by the ratio of thickness to the width. For example, for the annular channel it will be

$$\frac{H_{\text{ori}}}{W_{\text{ori}}} = \frac{R_{no} - R_{ni}}{\pi(R_{no} + R_{ni})} \tag{9.8}$$

where H_{ori} is the thickness of the annular channel and W_{ori} is the width of the annular channel that uses the average radius for the measurement.

Then, the shear rate is corrected as below:

$$\gamma_w = \frac{(2n+1)\dot{V}}{\pi R_c n F_p(R_{no} - R_{ni})^2} \tag{9.9}$$

If the annular channel is long enough to match the requirement in Equation (9.7), then, F_p equals approximately one, and Equation (9.9) becomes Equation (9.5).

Also, the shearing heat may be considered if the injection time is so short that the heat loss can be neglected. Then, the possible melt temperature increase from pressure drop can be calculated:

$$\Delta T_{\text{ori}} = \frac{\Delta P_{\text{ori}}}{\rho_{\text{melt}} C_{\text{melt}}} \tag{9.10}$$

where ΔP_{ori} is pressure drop across the nozzle orifice, ΔT_{ori} is the melt temperature increase in the nozzle orifice, ρ_{melt} is the melt density, and C_{melt} is the specific heat of the plastic melt. It can be verified quickly with the melt temperature coming out from the nozzle compared to the nozzle temperature setup.

6. Test Procedures. The Engel 200-ton tie-bar-less microcellular injection molding machine will be used for this test. Since the purge speed of the injection molding machine is not allowed for changes during air purge, the way to maintain the same shearing rate in air shot is to have two nozzles with different diameters of orifices.

- Use the air shot of GPPS under all conditions to be used for this test and measure the P1 (injection pressure), injection speed, and melt temperature. Assume that P_2 is equal to zero because the material is shot into the air; then, $P_1 - P_2$ can be calculated, which is the pressure loss in the orifice of nozzle during injection at certain shearing rate.
- One of the small-orifice nozzles was tested first to determine how much injection speed needs to create enough P_1. It is because the small-orifice nozzle needs low injection speed to reach the set up shearing rate. Then, the other nozzle with a bigger orifice will be set up with the same shearing rate as the nozzle with a small orifice. The big-orifice nozzle requires a high injection speed to match the same shearing rate in the small-orifice nozzle. Theoretically, if the shearing rate is the same for both nozzles, the pressure loss should be approximately same in both nozzle viscosity tests.
- These tests above need to be repeated several times with different shearing rate of air shot until a wide enough range of inlet pressure P_1 is covered.

7. Test Conditions. The test conditions will be set up based on the current database of unfoamed materials so that collected data are comparable each other. The testing conditions are listed as follows:

- Temperatures are set up as 440 °F (227 °C), 480 °F(249 °C), 520 °F(271 °C). The test will begin from the lowest temperature and increase it accordingly because raising the melt temperature is easier than lowering the melt temperature in the barrel.

- Melt pressures for screw recovery and pressure holding before injection are 6.9, 13.8, and 20.7 MPa.
- Injection speed is varied for three shearing rates that will create a high enough inlet pressure P_1 in the range of 0.006–0.154 m/sec.
- SCF (N_2 gas at supercritical fluid state) weight percentages are selected as 0%, 0.25%, 0.4%, and 0.75%.
- It is important to keep the testing injection molding machine in a continuous even cycling to avoid long residence time for the material in the barrel.
- The molding is controlled never to fill the mold 100% to avoid any pressure peak.

8. *Test Result.* The reciprocating screw used for rheometer tests provides a stable output that keeps a consistent melt quality for each test. In addition, the machine provides a constant injection volume rate with different gas percentage gas-laden melts. The nozzle rheometer in Figure 9.1 has been used for a series of injection velocities tests. The measured three pressures in the nozzle show that stable process occurs at 70 sec when the screw rotates at 101 rpm with 227 °C melt temperature, through the fixed nozzle orifice in the nozzle rheometer for GPPS solid material. The pressure profiles are kept constant at about 16.3, 12.1, and 3.3 MPa in the nozzle rheometer at 70 sec continual running time and after, respectively. Similarly, the measured pressures of three pressure transducers remained relatively constant over the majority of the injection stroke even with gas-laden materials.

The postprocessing viscosity data will be calculated from the logged temperature, orifice pressure, and injection ram displacement traces. A viscosity model is then fitted to these raw viscosity points. A second-order polynomial viscosity model has been used:

$$\ln \eta_T = A_1 + A_2 \ln(\dot{\gamma}) + A_3 T + A_4 (\ln(\dot{\gamma}))^2 + A_5 \ln(\dot{\gamma}) T + A_6 T^2 + A_7 C + A_8 C^2 + A_9 C \ln(\dot{\gamma}) + A_{10} T \ln(\dot{\gamma}) C \tag{9.11}$$

where A_i represents the constants for the polynomial model ($i = 1, 2, \ldots, 10$), $\dot{\gamma}$ is the apparent shear rate, and T is the melt temperature.

Ten reference points are taken from the second-order viscosity model. These ten points become the second-order standard points, and they sufficiently describe the second-order viscosity model. These points are then analyzed using melt flow index (MFI) in extrusion mode, and they are subsequently corrected to account for shear heating and non-Newtonian behavior. Thus, through an interactive process based on pressure modification, the viscosity values of the 10 standard points are modified to produce true temperature and shear rate corrected data. These 10 standard points are used to describe the viscosity data on the material database.

After the diagram of $\log(\tau_w)$ versus $\log(\dot{\gamma})$ is available, the correct power index n can be calculated with small increments:

$$n \approx \frac{\Delta \log(\tau_w)}{\Delta \log(\dot{\gamma})} \tag{9.12}$$

where $\Delta\log(\tau_w)$ is the small increment $\log(\tau_w)$ in the diagram of $\log(\tau_w) - \log(\dot{\gamma})$ and $\Delta\log(\dot{\gamma})$ is the small increment $\log(\dot{\gamma})$ in the diagram of $\log(\tau_w) - \log(\dot{\gamma})$. If the power index n is much different from different increments calculated with Equation (9.12), then it may be necessary to specify different n values in different shearing range. However, usually it is determined by the average value over a wide range of shear rate.

The result shown in Figure 9.3 is the viscosity data of GPPS at 520 °F (271 °C). It is obvious that a nonlinear relationship exists between viscosity and shear rate in these logarithmic plots. The viscosity of GPPS has a pronounced reduction at low shear rates in Figure 9.3. However, the viscosity does not have significant differences between solid and gas-laden melt after the shear rate is over 3000 1/sec at this melt temperature. It seems like the viscosity of the solid material is either almost equal to or even a little lower than that of the gas-laden material at extremely high shearing rate. It can be explained by the shear stress effect. The solid material has more significant shear thinning effect at high shearing rate than does the gas-laden material, which already has low viscosity with gas and will not generate more shearing heat at low viscosity. The other published papers and book also indicate a

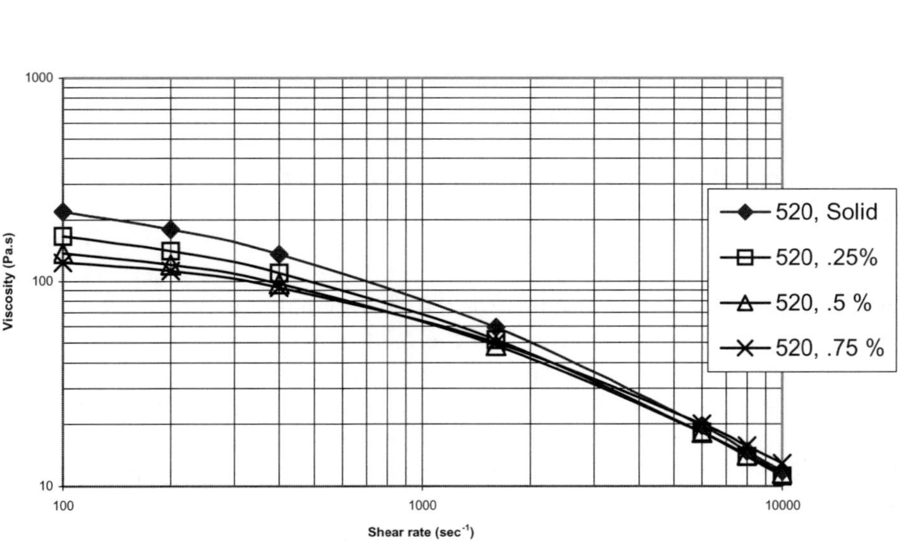

Figure 9.3 Viscosity data of GPPS at 520 °F (271 °C).

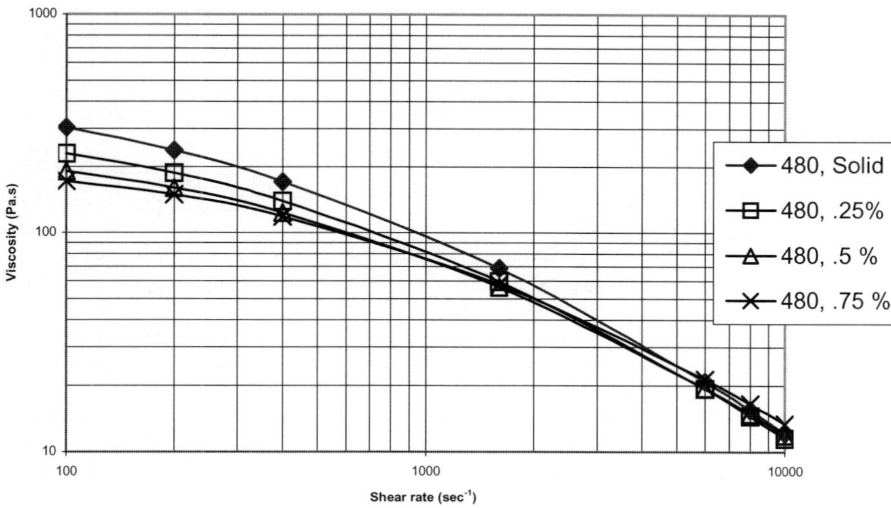

Figure 9.4 Viscosity data of GPPS at 480 °F (249 °C).

similar result of viscosity change trend at different shear rates [7]. In addition, the viscosity of gas-laden GPPS melt in Figure 9.3 shows a clear trend that the viscosity decreases with the increasing of SCF weight percentage at the same processing temperature and low shear rate until the shearing rate is over 4000 1/sec.

In the isothermal nozzle rheometer the viscosity data of gas-laden melt must be run at different melt temperatures to get the viscosity data related to the different melt temperatures. Figures 9.4 and 9.5 show the viscosity of gas-laden GPPS at 480 °F (249 °C), and 440 °F (227 °C), respectively. The results of these two different temperatures are similar to the result in Figure 9.3. The data shown in Figure 9.5 are the viscosity data of GPPS at 480 °F (249 °C). A nonlinear relationship is shown again between viscosity and shear rate in the logarithmic plots. The viscosity of GPPS is higher at low melt temperature. For example, the viscosities of gas-laden GPPS and solid GPPS at 480 °F (249 °C) are about 100 Pa-sec higher than the viscosity of gas-laden material at 520 °F (271 °C) with 100 1/sec shear rate as shown in Figure 9.4. Similarly, the viscosity of gas-laden and solid GPPS at 440 °F (227 °C) is about 100 Pa-sec higher than the viscosity data of gas-laden melt at 480 °F (249 °C) and at 100 1/sec shear rate (see Figure 9.5). It clearly illustrates that all the viscosities of solid and gas-laden materials decrease significantly at low shear rates in Figures 9.4 and 9.5. However, the viscosity does not have significantly differences between solid and gas-laden materials at low shearing rate. It is obvious that the viscosity difference between solid material and gas-laden material becomes less and less with the increasing of shear rate. On the other hand, at the same pressure, the viscosity of gas-laden GPPS melt in Figures 9.4 and 9.5 repeats the clear trend that the viscosity decreases with the increasing of SCF

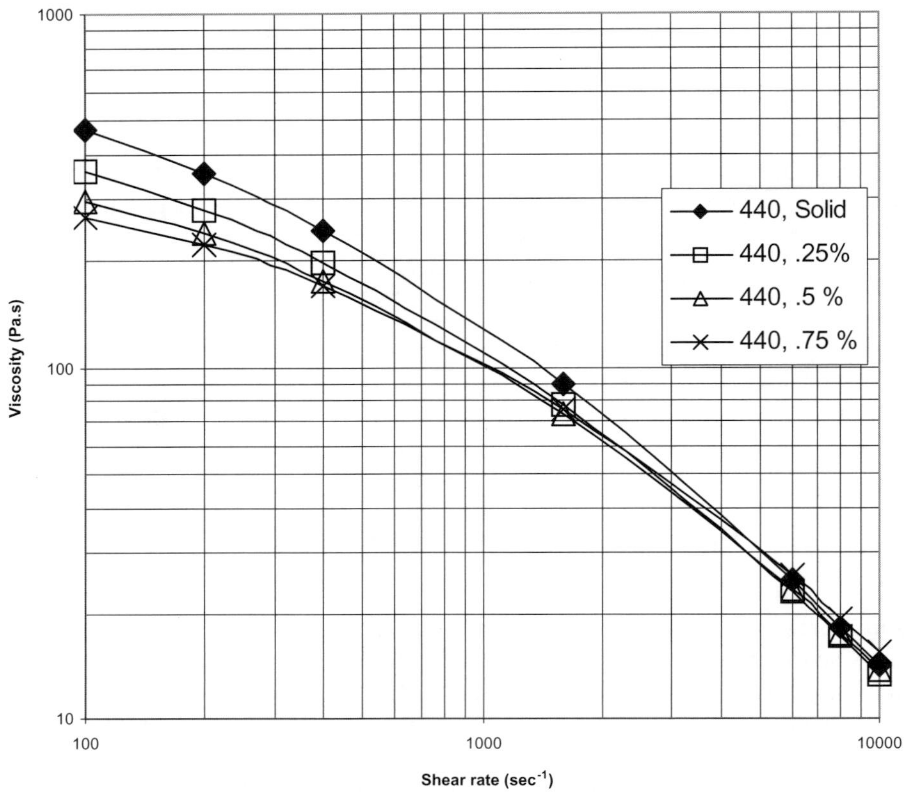

Figure 9.5 Viscosity data of GPPS at 440 °F (227 °C).

weight percentage at the same processing temperature and low shear rate until the shearing rate is over 5000 1/sec.

The results in Figures 9.3, 9.4, and 9.5 prove that the more weight percentage of gas, the lower the viscosity. However, this may not always true at high shear rate. The viscosities of solid and gas-laden materials keep a small difference that is hard to tell if this difference is significant for influencing the molding. The results in Figures 9.3, 9.4, and 9.5 prove that at high shearing rate the shear thinning effect dominates the viscosity change, so the gas weight percentage not the major factor necessary to reduce the viscosity.

There is a special test to check the relationships between viscosity and pressure drop rate at the same other processing conditions, such as the same shearing rate. Table 9.1 shows the interesting results of the tests with air shot tests in the cylindrical channel nozzle rheometer. At the 8000 1/sec fixed shearing rate test in Table 9.1 the viscosity of solid is always a little higher than the viscosity of the gas-laden melt with 0.5% weight percent N_2 gas. The first measuring point in Table 9.1 is about 1.5-mm diameter of the orifice that is the smallest orifice of the nozzle rheometer in this test. The solid material

TABLE 9.1 Viscosity Data Used for Moldflow Simulation

Shear Rate (1/sec)	Pressure (MPa)	Viscosity (Pa-sec)	Pressure (MPa)	Viscosity (Pa-sec)
3000 (0.5% N_2)	1.5	17	3.3	19.6
3000 (solid)	1.5	19.5	3.3	21.5
8000 (0.5% N_2)	3.6	7.3	6.2	7.8
8000 (solid)	3.6	8.7	6.2	9.3

requires a larger magnitude of pressure difference (about 3.5 MPa) to reach the same shearing rate than the gas-laden material does (about 2.8 MPa) in the same diameter of the nozzle rheometer. This fact is true in the microcellular injection process; thus the pressure difference required for solid material is always higher than that required for gas-laden material, which explains why the solid material needs higher injection pressure to fill the mold compared to gas-laden material.

There is a similar trend for the pressure drop required for solid and gas-laden melts in a big-diameter nozzle with 5-mm diameter of orifice in the nozzle tip. To reach the shearing rate 8000 1/sec, a big orifice diameter in the nozzle rheometer needs a higher pressure drop compared to a small orifice diameter of nozzle. It is because the big-diameter nozzle needs more flow rate to generate the same shear rate that will require more pressure to drive more mass flow. The overall viscosity in the big nozzle is a little higher than that in the small nozzle even if the shear rate is the same. It may be from two factors: One is the high pressure required to push more flow through the big nozzle to get the same shear rate. On the other hand, the high pressure in the nozzle will pressurize the melt to have high viscosity. The explaining for this trend is the free volume reduction in the melt changed by high pressure. In addition, the big nozzle will have more error from the calculation of viscosity because of the short *L*/*D* ratio of rheometer. One more interesting thing is that the data at 8000 1/sec shear rate in Table 9.1 has almost the same rate of viscosity increasing with the increasing of the pressure drop through the nozzle orifice for both solid material and gas-laden melt at this shear rate.

However, the data at a shear rate of 3000 1/sec in Table 9.1 has a different rate of viscosity increasing with the pressure drop increasing. It is because the viscosity measured at low shear rate of 3000 1/sec where the gas-laden material viscosity change is obviously different from solid viscosity as shown in Figures 9.3, 9.4, and 9.5. The conclusion is that the viscosity thinning effect at high shear rate for GPPS causes the gas dilution to have no significant effect on the viscosity reduction.

The absolute value of viscosity obtained using the method above varies with the actual mold being filled. Therefore, a final step was taken to normalize the data by dividing all of the resultant values of Pa-sec by its corresponding solid value. Then, it is much easier to use the normalized data because the resultant viscosity value is a percentage of the viscosity for the material without gas. It is similar to the regular viscosity versus shear rate that the final calculation provides normalized viscosity as a function of apparent shear rate.

As a general trend, amorphous materials exhibit a relatively flat viscosity response with the higher percentage nitrogen. The test for the typical amorphous material, such as GPPS, shows that the viscosity drops about 8% with the introduction of 0.2 weight percentage nitrogen gas in GPPS. Then, the viscosity of GPPS material only further falls to 13% on average when the gas content weight percentage of nitrogen is increased four times to 0.8%. However, the use of carbon dioxide in GPPS material produces a greater drop in viscosity due to the higher solubility limits of this gas. The minimum viscosity occurs at about 4.5 of weight percentage of carbon dioxide gas in amorphous material, with little benefit in further increases. The average viscosity reduction percentage with carbon dioxide in amorphous material is almost up to 30%.

On the other hand, semicrystalline engineering resins appear to have a more linear relationship between SCF concentration and viscosity of semicrystalline engineering resins. Unlike the amorphous material, these materials do not reach a viscosity plateau at gas concentrations below the gas saturation limit. It means that adding additional gas can always provide an avenue to better filling for the semicrystalline engineering materials. On the other hand, it is similar to the amorphous material; carbon dioxide always provides a larger viscosity reduction than nitrogen for semicrystalline engineering resins, too. However, this again is primarily a function of the higher solubility level.

There are different effects of the gas percentage, affecting the viscosity for polyolefin and engineering resins even if they are both semicrystalline materials. It is found that introducing nitrogen gas at supercritical state into HDPE and PP melts resulted in virtually no obvious viscosity reduction. The viscosity reduction of adding nitrogen gas in polyolefin is about 4%. The introduction of carbon dioxide creates more significant change of viscosity, as much as 15%. However, it is still well below those viscosity changes seen with other materials.

However, the real processing does not like too much gas percentage in the molten polymer because the low gas percentage in the molten polymer always provides more stable processing. Therefore, the viscosity reduction from adding SCF may not be necessary to reach the maximum in microcellular injection molding.

Furthermore, the effect of pressure on the apparent viscosity must be reviewed with the SCF solubility pressure [12]. The PVT test of the gas-laden material will be helpful to understand the relationships among the viscosity and parameters of pressure, volume, and temperature.

9.1.2 PVT Date Base for the Mixture of SCF and Plastic Melt

The viscosity and P (pressure) V (volume) T (temperature) relationship of gas-laden melt have been measured in a special screw tip and in a high-pressure shut-off nozzle. The constant injection pressure control of the reciprocating screw injection molding machine provides the perfect conditions to measure the PVT data at high pressure (simulate the real injection pressure) with not only the solid material but also the gas-laden material.

A leak-free screw tip is used to do this dead head test. It is just the simulation an injection at set up pressures. However, both the shut-off nozzle and the screw tip are closed completely, so the reciprocating screw is acting as a plunger to pressurize the melt between the screw tip and the shut-off nozzle. The screw tip layout in Figure 7.16 shows how a typical automatically shut-off screw works for this test with an extra piston ring on the OD of the tip body. The whole layout of this configuration of PVT test equipment can be seen in Figure 7.8. The principle of the instrument is simple. First of all, the gas-laden material (or 0% gas as solid material) is prepared by the screw rotation, and the correct shot size is accumulated in front of the screw tip precisely. Then, the reciprocating screw causes an injection action at a certain pressure and slow speed for maintaining an isothermal condition. The key element is the leak-free screw tip to be able to do the dead head test. The shut-off nozzle must be leak-free at high pressure as well, and the nozzle is closed. Then, the material accumulated in front of the screw tip is compressed with the exact injection pressure acting on it. Then, the displacement of the screw is measured during the compression. This test is carried out at certain melt temperature, so all three parameters—pressure, volume, and temperature—were recorded in the test. The final volume change is calculated by the screw displacement d_{scr}. The reduced volume of shot can be calculated:

$$d_{vol} = d_{scr}\left(\frac{\pi}{4}D^2\right) \tag{9.13}$$

where d_{scr} is the displacement of the screw during the PVT test and d_{vol} is the absolute volume reduction during the PVT test.

The GPPS with a melt index (MI) of 8 is used for this test in this special instrumental reciprocating screw machine. As the typical test data, the results of solid and the gas-laden melt with different gas weight percentages at 223 °C (433 °F) melt temperature are illustrated in Figure 9.6. The N_2 gas weight percentage is in the range of 0.25% up to 1.25%. The high gas compressibility gives the gas-laden melt significant volume reduction percentage (the absolute volume reduction divided by original volume before compression ×100) at high pressure. It is an obvious trend that the absolute volume reducing percentage is increased with the increasing of pressure. On the other hand, the high weight percentage of gas in the melt will have the relatively high volume reducing percentage as well. The other material, such as PP and HDPE, shows the similar trend. This is useful PVT data for the future shrinkage, warpage, and other simulation programs.

Park and Dealy [12] at McGill University tested different PP materials above with different gases. The supercritical carbon dioxide has been used for

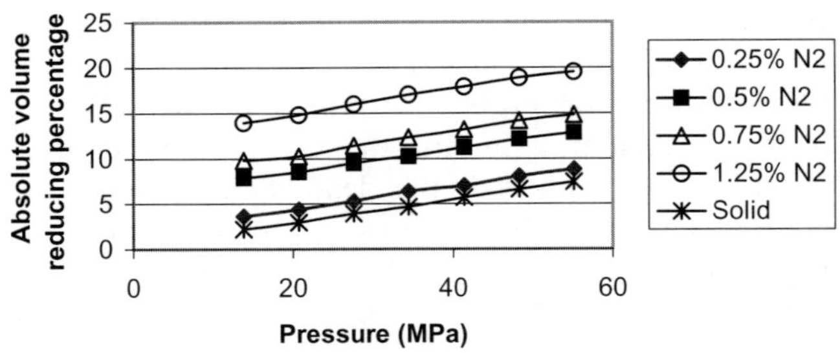

Figure 9.6 Volume reducing percentage at different pressure for GPPS at 433 °F.

the PP material PVT tests. A high-pressure sliding plate rheometer and a rotational rheometer were used to determine the combined effects of PP materials, gas contents percentage, pressure, and temperature. The effect of temperature could be described by the Arrhenius equation [13]. The pressure influence can be found by the Barus equation [14]. The effect of content percentage of CO_2 gas obeys the Fujita–Kishimoto equation [12, 15, 16].

9.2 MOLDFLOW MODELING OF MICROCELLULAR INJECTION MOLDING [7–9, 17–20, 25, 26]

Moldflow joined Trexel to develop the microcellular simulation program. As the pioneers for this modeling project of microcellular injection molding, Sejin Han, R. Zheng, Xiaoshi Jin, and Peter Kennedy of Moldflow Corp., along with Jingyi Xu, Himanshu R. Sheth, and Levi Kishbaugh of Trexel, have collaborated to develop a new simulation software of microcellular injection molding. Most research results are published in references 18 and 19. Okamoto also added the same contents from reference 18 into reference 17 later. The simulation successfully predicted the flow pattern and pressure, with acceptable cell structure prediction. Since then, Sejin Han continually worked on it and has modified the code of the program to make the result closer to the experimental results, although the same principal formulae are introduced below. The Moldflow models and more new results of simulation and verification after the year 2003 are discussed in this chapter. Unless specified, the most of formulae and figures in this session are the contents of reference 20 and results with permissions from Moldflow, Trexel, and Connector.

Attempts have been made to analyze this process numerically [18]. The current study here is an improvement of the earlier study [18]. The calculated results from the current study show good agreement with the experimental results in terms of cavity pressure and the final cell-size distribution in the filled material [19].

9.2.1 Theory

There is always an assumption that the short shot of injection in the mold will leave enough space for cell growth for the microcellular foam molding process. After injection, the nozzle is closed to separate the mold and screw without any packing stage. Then, the rest of the cavity is filled by the nucleation and expansion of cells. This is a new challenge for the mold filling analyses because the final mold filling is from cell growth and not from positive injection volume increasing. Therefore, the key issues of molding simulation are the models of nucleation and cell growth. It needs to calculate the diffusion of gas from the single-phase solution to the cell. The cell growth model is also coupled to the macrophase pressure distribution of the single-phase material [19]. Each aspect will be discussed in the following.

9.2.1.1 Cell Growth. There are several assumptions to simplify the modeling. One of the assumptions is a uniform nucleation throughout the melt and has a known value for the quantities of nuclei per unit mass. To determine the growth of cells, the cell model proposed by Amon and Denson [20] is used in the modeling. A unit cell model is defined and displayed in Figure 9.7. Another assumption is that foam is divided into spherical microscopic unit cells of equal and constant mass. There are a liquid shell and a concentric spherical gas bubble as the cell model. The schematic diagram of a unit cell shown in Figure 9.7 illustrates the bubble radius R, which is surrounded by the outer cell radius S. Furthermore, an assumption that the number of the cell equals the number of nuclei will neglect the initial cell growth during nucleation stage of injection. The real process will have both dynamic nuclei

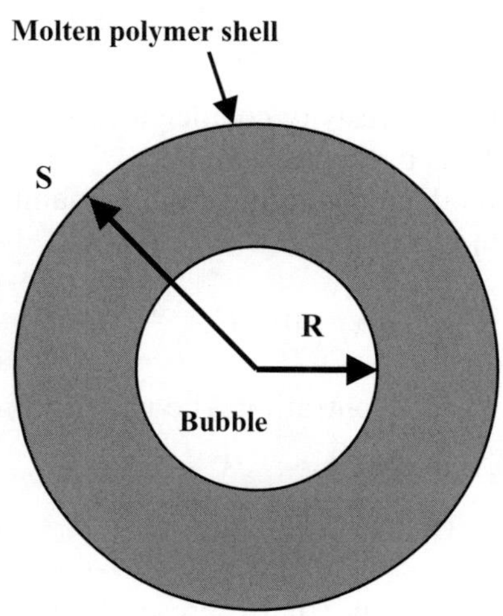

Figure 9.7 Schematic diagram of a unit cell [18]. (Reproduced with permission from Society of Plastics Engineers.)

and cell growth during injection. However, usually microcellular injection molding will need fast injection, so this short time may not be necessary to focus on the initial bubble growth. After injection (or nucleation), expansion of the bubble occurs due to two mechanisms:

- Hydrodynamically controlled growth
- Diffusion-controlled growth

9.2.1.2 Hydrodynamic Growth Equation. The rate of change of the radius of the gas–melt interface, $\dot{R}$, is given by [21]

$$4\eta_T \dot{R}/R = (P_g - P_m) - 2\sigma/R \tag{9.14}$$

where η_T is the non-Newtonian gas-laden polymer viscosity, P_g is the gas pressure inside the cell, σ_b is the surface tension at the interface of the melt and the gas, and P_m is the pressure of the melt at the outer boundary of the cell. The internal gas pressure P_g is determined by gas diffusion in the bubble. However, the macroscopic pressure equation of the mold filling process will determine the pressure at the outer boundary of the cell P_m.

9.2.1.3 Gas Diffusion Equation. The bubble radius R will determine the pressure inside the bubble P_g in Equation (9.14). The relationship between P_g and R can be derived from the mass balance in the diffusion process of the gas dissolved in the melt shell. The diffusion process is governed by Fick's law of diffusion:

$$\frac{\partial C}{\partial t} + v_r \frac{\partial C}{\partial r} = D_a \left[\frac{1}{r^2} \frac{\partial}{\partial r} \left(r^2 \frac{\partial C}{\partial r} \right) \right] \tag{9.15}$$

where D_a is the gas (SCF) diffusivity coefficient, and C is the weight concentration of the dissolved gas in the melt.

Assuming that thermodynamic equilibrium is maintained at all times, the initial condition is given by

$$C(r, t = 0) = C_0 \tag{9.16}$$

where C_0 is the initial gas concentration. The gas concentration at the bubble wall can be given based on

$$C(R, t) = C_w(t) = \kappa_h P_g(t) \tag{9.17}$$

where κ_h is Henry's law constant (that represents a gas solubility).

For a given gas, its solubility in different polymers is not much different. However, different gases in the same polymer have considerably different values of solubility. In the absence of experimental data, the following simple empirical equation has been used to estimate the solubility [22]:

$$\ln \kappa_h = -2.1 + 0.0074T_{\mathrm{cr}} \tag{9.18}$$

where T_{cr} is the critical temperature of SCF.

Assume that the gas concentrate at outer radius of cell is equal to the initial gas concentration,

$$C(r = S) = C_0 \tag{9.19}$$

and

$$\left.\frac{\partial C}{\partial r}\right|_{r=S} = 0 \tag{9.20}$$

Further assume that the gas follows the ideal gas law, then, the internal gas pressure is about

$$P_g(t) = (1000R_gT/M_w)\rho_g(t) = A\rho_g(t)T \tag{9.21}$$

where R_g is a universal gas constant (8.31451 J·mol^{-1}·K^{-1}), T is the temperature (in Kelvin), M_w is the molecular weight of the gas, ρ_g is the density of the gas in the bubble, and A is a constant for a given gas, $A = 1000R_g/M_w$.

At the bubble wall, $r = R$, the mass flux is given by [19]

$$\frac{d}{dt}(\rho_g R^3) = 3R^2\rho D_a\left(\frac{\partial C}{\partial r}\right)_{r=R} \tag{9.22}$$

To calculate the concentration gradient, solve Equation (9.15) first, and then, substitute this solution into the right-hand side of Equation (9.22). By integrating Equation (9.15) with respect to r [18]

$$\frac{d}{dt}\int_R^S (C_0 - C)r^2 dr = D_a R^2\left(\frac{\partial C}{\partial r}\right)_{r=R} \tag{9.23}$$

Combining Equations (9.22) and (9.23) and integrating the resulting equation with respect to t, that results [18]

$$3\rho\int_R^S \frac{C_0 - C}{C_0 - C_w} r^2 dr = \frac{\rho_g R^3 - \rho_{g0}R_0^3}{C_0 - C_w} \tag{9.24}$$

Refer to Rosner and Epstein [23], an assumption is made to use a second-order polynomial profile for the estimation of the gas concentration [18]:

$$\frac{C_0 - C}{C_0 - C_w} = \begin{cases} (1-\xi)^2, \ldots, R \le r < S \\ 0, \ldots, r \ge S \end{cases} \tag{9.25}$$

$$\xi = \frac{r - R}{S - R} \tag{9.26}$$

By substituting boundary conditions (9.25) into Equation (9.24), the expression for $S - R$ can be obtained. From the above equations, the result is given [18, 19]:

$$\frac{d}{dt}\left(\frac{p_g R^3}{R_g T}\right) = \frac{6\rho^2 D_a \kappa_h R_g T (p_{g0} - p_g)^2 R^4}{p_g R^3 - p_{g0} R_0^3} \tag{9.27}$$

Equations (9.14) and (9.27) can be solved together to obtain the results of bubble size and gas pressure inside of bubbles.

9.2.1.4 Rheology of the Mixture. In order to describe the viscosity of the gas–polymer mixture, a modified Cross–WLF equation is proposed [18]:

$$\eta(\dot{\gamma}, T, p, \phi) = \eta_0(T, p) f(\phi) \left[1 + \left(\frac{\eta_0(T, p)\dot{\gamma}}{\tau^*}\right)^{1-n}\right]^{(-1)} \tag{9.28}$$

where η_0 is the zero-shear viscosity, which is a function of temperature and pressure; n is the power-law index; τ^* is the model parameter; ϕ is the volume fraction of the gas; and f is a function of the volume fraction of the gas. The well-known zero-shear viscosity is given by

$$\eta_0(T, p) = D_1 \exp\left(-\frac{C_1(T - T_0)}{C_2 + (T - T_0)}\right) \tag{9.29}$$

where

$$T_0 = D_2 + D_3 p$$

D_1, D_2, and D_3 are model parameters and will be specified from the experiments for different materials.

The dissolved gas in the polymer melt acts as an internal plasticizer, and the plasticizing effect of the gas can significantly reduce the bulk viscosity. Then, the equation to define this effect is

$$f = (1 - \phi)^{\alpha} \tag{9.30}$$

where α is determined from experimental data. The volume fraction of the gas can be calculated by [18, 19]

$$\phi = \frac{4\pi R^3/3}{V_{\text{shell}} + 4\pi R^3/3} = \frac{4\pi R^3/3}{1/(\rho N_w) + 4\pi R^3/3} \tag{9.31}$$

9.2.1.5 Macroscopic Flow [18, 19]. To solve for the microcellular injection molding problems, the microscopic model of the cell is coupled with the macroscopic flow through the fluid pressure around the cell.

In the macroscopic simulation, is treated the microscopic multiphase fluid as a macroscopic single-phase fluid; thus all conventionally used governing equations, including the equations of continuity and momentum, as well as a constitutive equation, can be directly employed in the simulation [18]. The Hele–Shaw approximation can also be applied.

For mathematical convenience, assume that the foaming stage starts after the end of injection. Before the nozzle is closed, the mold filling analysis is therefore the same as for the conventional injection molding. After the nozzle closed, the bubble expansion is the only driving force behind the fluid flow. The pressure equation is given as follows:

$$\frac{\partial}{\partial t}\int_0^b \ln(\rho_{\text{cell}})dz - \nabla(S\nabla p) = 0 \tag{9.32}$$

with

$$S = \int_{-h/2}^{h/2} \frac{z^2}{\eta} dz \tag{9.33}$$

where $\nabla_{(x,y)}$ denotes the gradient operator with respect to the mid-plane, h is the thickness of the cavity, z the local thickness coordinate. Note that both the cell viscosity η and the cell density ρ_{cell} depend on the bubble size.

The cell viscosity is given by Equation (9.28), and the cell density is given by

$$\rho_{\text{cell}} = \frac{m_{\text{cell}}}{V_{\text{shell}} + 4\pi R^3/3} = \frac{3\rho}{3 + 4\pi R^3 N_m \rho} \tag{9.34}$$

The density of material ρ is calculated from the two-domain Tait equation. Also, the measurement of ρ is done for a solid sample without gas. The heat capacity and thermal conductivity are also from solid material data.

9.2.1.6 Numerical Algorithm. Most systems of governing equations are similar to the regular mold simulation [24]. The full system consists of the coupled differential equations [Equations (9.14), (9.31) and (9.32)] and the energy equation (which is not written in this paper for brevity and can be found in reference 24). The primary unknowns to be solved are fluid pressure P, the gas pressure inside the bubble P_g, the fluid temperature T, and the bubble radius R. Equations (9.28) and (9.34) are used to update the viscosity and density that are from experimental results with gas-laden materials.

9.2.1.7 Assumptions for Speed-Up Simulation. There are some assumptions to make this simulation reasonable long. They are summarized as the follows:

- Hele–Shaw method on mid-plane model
- Bubble growth neglected before foaming
- No flow from gate during foaming
- Heat capacity, thermal conductivity: using solid resin data
- Nucleation density: 2×10^{11} cells/m^3
- Initial bubble radius: 1.0×10^{-6} m
- Calculation time is longer because of the bubble growth calculation and slower convergence compared to regular injection molding

9.2.2 Experimental Work and Simulation

The experiments are planned accordingly for the different purposes. First mold is a simple rectangular mold without any boss or holes in the mold. The gate is just cold sprue, so the nucleation rate is controlled on the nozzle orifice only. This mold has Kisler pressure transducers in the mold cavity to check if the pressure simulation is right for gas-laden material. The second mold is the Trexel mold with holes and a big cut in the mold that will cause the detour of mold filling to join after the hole or big cut with a weld line. The third mold is from a customer connector, which is the complex mold with lots of details, and it uses glass-fiber-reinforced material. All of the experiments typically represent the theory verification and real parts simulation for this mold flow software.

9.2.2.1 Plaque Part [18]. This is the simplest geometry mold with special pressure transducers in the mold. The test was carried out to verify the relationship between the mold-filling position and pressure in the mold. This molding was made using a general-purpose polystyrene material (Dow Chemicals, Styron 666D) with 0.25% by weight of N_2. The mold is a rectangular plaque with the length of 279.4 mm, the width of 126.5 mm, and the thickness of 2.55 mm. A cold sprue is located in the center of this plaque. The sprue is of length 74.5 mm and has a tapered diameter of 4.11 mm at the nozzle and 7.08 mm joining with the part. The nozzle tip has the orifice of 3 mm, which is truly the nucleation position for this molding. Pressure transducers (Kistler 6157A) are located from the sprue centrally along the longest axis 25 mm (location 1), 72.5 mm (location 2), and 126.5 mm (location 3). Data from the transducers were dynamically stored using an Engel CC100 machine control system and compared with the simulation results from mold flow as first analysis to verify the model of mold flow for the microcellular process.

The molding parts were made on an Engel 150-ton machine in the extruder with a 40-mm screw and a 40-mm plunger for injection. A Trexel TR-10 gas pump system is used for (a) gas supplying at the supercritical state and (b) controlling for the dosing in the screw. Plaques were molded such that there were 5% and 15% weight reductions in the component, whereas a solid molding part was made when no gas was dissolved in the polymer. Pressures at the two locations in the mold were recorded in the whole period of injection. The pressure profile related to the injection time provides the traces of ram displacement and hydraulic pressure.

The Moldflow simulations were carried out with the models above and the processing conditions as follows. The gas–polymer solution was assumed to be a perfect single-phase solution during injection. Viscosity data were measured in a specially designed rheometer to allow measurement at varying levels of N_2. The measured viscosity data for 0.25% N_2 concentration are similar to the data shown in Figures 9.3–9.5. The data were fitted to the viscosity Equation (9.28), and the fitting constants are shown in Table 9.2. The dependence of viscosity on bubble growth is not available experimentally. Therefore, an empirical value of $\alpha = 2$ was used in this analysis. The thermal properties and PVT data were used with the same data as for solid resin. The nucleation theory is difficult to be used at this simulation since the data are not available at that time. Then, the nucleation density is assumed to be 2×10^{11} (cell/m^3) based on the SEM result of similar test. The other data were assumed as $c_0 = 0.0025$ and $\kappa_h = 8.4913 \times 10^{-10}$. In addition, the diffusion coefficient was assumed to be 1.5×10^{-9}, the initial bubble radius was at 1.0×10^{-6} m, and surface tension was assumed to be 0.05 (N/m). In actual case, the cell density varies with location, but it shall be uniform across the thickness direction and along the mold-filling path. In this study, the highest cell density values among the measured results were used, which correspond to the highest value at the farthest location from the sprue.

The weight reduction percentages were selected in this test based on the acceptable cell structures that must be in the range of microcellular foam. Although the weight percentage can be up to 25% in this test, only 5–15% of weight reductions were selected because they had the most stable processes to provide the repeatable experimental results. Therefore, injection of material was stopped at 85% and 95% of the part volume and the nozzle closed. The injection speed is selected as a constant from beginning to the end for

TABLE 9.2 Viscosity Data of GPPS Used for Moldflow Simulation [18]

D_1 (Pa·sec)	D_2 (K)	D_3 (K/Pa)	C_1	C_2(K)	τ^* (Pa)	n	α
3.102×10^{10}	373.15	0	24.03	51.6	2.839×10^4	0.1743	2

Source: Reproduced with permission from Society of Plastics Engineers.

easier simulation. Cells were assumed to nucleate when the pressure in the melt was reduced greatly such that the gas precipitates from the solution as well. Thereafter, a cell growth will fill the rest of mold cavity by the material expansion that may be difficult to be simulated because it needs a new model to simulate it.

The results of the 15% weight reduction test are discussed with both calculation and experimental data comparisons. It is an acceptable result of simulation overall. Pressure results from the experiment and the calculation are shown in Figure 9.8a,b. In this figure, a comparison is shown between the

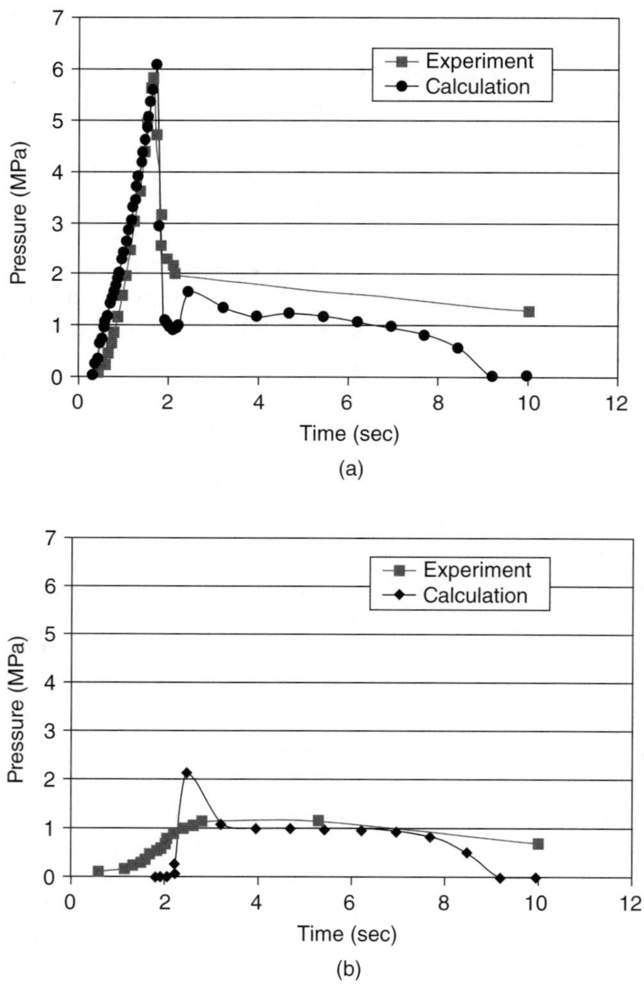

Figure 9.8 Pressure at different positions of mold during mold filling (15% weight reduction). (**a**) Pressure values at location 1. (**b**) Pressure values at location 3. (Courtesy of Moldflow.)

calculated results and the experimental results. Predicted pressure at the 25-mm location, which is location 1, is shown in Figure 9.8a. The pressure prediction from Moldflow is in good agreement with the measured value before the start of foaming at the gate area, location 1. After the injection, the nozzle closed and the pressure at gate area begins to drop sharply because there is no packing pressure to maintain the pressure in the mold. For the microcellular foaming process, it is the exact time for the start of foaming stage. During the foaming stage, the calculated pressure is generally lower than the measured pressure, which is the same conclusion as in reference 17. However, this model has been modified to have little pressure recovery after a sharp drop because the growing cells have a little pressure increase as shown in Figure 9.8a [25, 26]. On the other hand, Han modified the code of the Moldflow program for microcellular processing [19, 26]. Therefore, the calculation cavity pressure at location 1 in Figure 9.8a is closer to the experimental result than to the result in reference 17. At the end of the mold cooling, the experimental pressure still shows the residual pressure in the gate area but the calculation pressure reduces to zero.

The most interesting results in this study are in location 3, which is near the end of mold filling, and the pressure measured represents the cell-growth-driven pressure that pushes the final mold filling without any material injection action. The measured pressure at 126.5 mm from the sprue, which is location 3 shown in Figure 9.8b, rises almost instantly when the material reaches location 1. This may be due to increased pressure in the trapped air and gas ahead of the melt and, also, possibly due to the injection driven by compressed gas-laden material—some effects that are difficult to simulate and not taken into account in this simulation yet. The calculated pressure at this location is generally in good agreement with the measured pressure once the pressure reaches the peak value, except there is a pressure increase of calculated pressure that does not occur at experiment result. Similar to location 1, the residual pressure in the cells of location 3 in the mold is not zero when the cooling stage ends. This result confirms that the residual pressure in the cells in the whole microcellular part, whether in the gate area or at the end of mold filling, is not zero. It is good for removing sink marks or keeping the stable dimensions through the cell expansion, overwhelming the part shrink. However, it causes the post-blow and other surface defects such as blister. The calculated pressure at the end of cooling is zero, which does not match the real results. Therefore, the modification of the model for this simulation still needs to be studied, and some corrections must be made for better result of simulation.

Calculated cell size is shown in Figure 9.9. In this figure, the normalized half of the thickness is from 0 to 1 (0 at the center of core, 1 at the wall). The calculated size of cells increases with distance from the sprue, and the higher pressure encountered near the sprue reduces the rate of growth there. On the other hand, measured average cell radius decreases with distance from the sprue. The results of sample morphologies are illustrated in Figure 9.10. The cell structure at location 1 (25 mm away from sprue) is shown in Figure

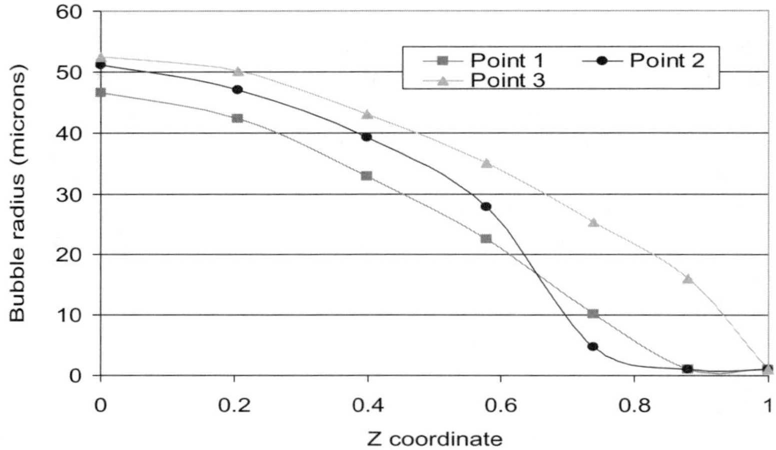

Figure 9.9 Calculated cell size (15% weight reduction) at locations 1, 2, and 3. (Courtesy of Moldflow.)

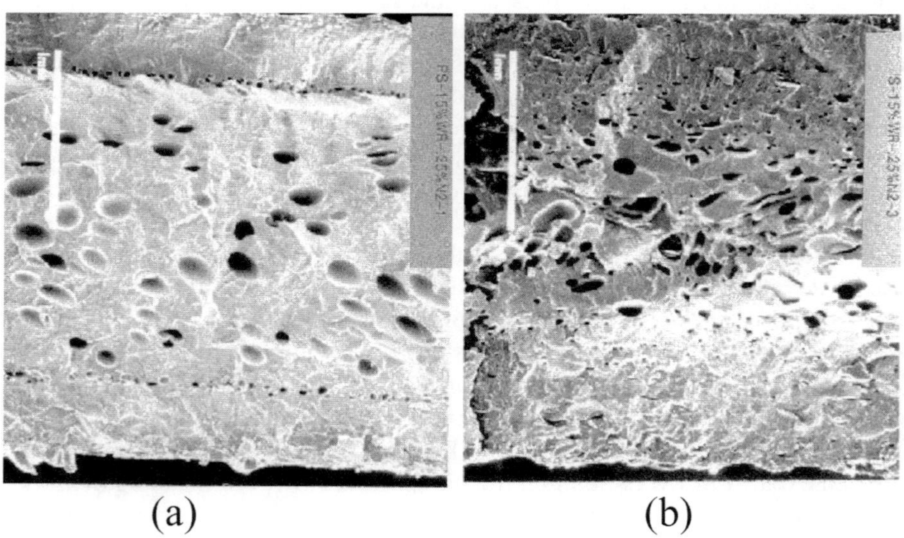

Figure 9.10 Cell sizes (15% weight reduction) at real samples for the verification of calculations. (**a**) Position 1. (**b**) Position 3, (Courtesy of Trexel.)

9.10a. It shows that the cell size varied from 20 to 120 μm. This is the pressurized zone where the cells are under higher pressure than the pressure in location 3, and there are fewer cells compared to the number of cells in location 3. On the other hand, the pressure difference in location 1 is the smallest compared to the pressure difference in location 3, so the nucleation in location

1 is not as good as the one in location 3. These two reasons are the main reasons for the difference of cell structure in location 1 (Figure 9.10a) and location 3 (Figure 9.10b). Although the calculated cell size in Figure 9.9 is about 48 μm in location 1, the predicted cell size distribution across the thickness of the part is right. In location 3 (126.5 mm away from sprue) the core region was found to be 35–200 μm. The trend of calculation of the cell size in Figure 9.9 also correctly indicates that is the cell size is small near the gate (location 1) but is big away from the gate (location 3). On the other hand, the prediction differences between the calculated and measured values are small near the sprue. However, the prediction difference is rather significant in regions far from the sprue. There are several reasons for this prediction error. One is the assumption of uniform nucleation in the simulation. Experimental results show that the nucleation is not uniform. The beginning of injection will have the highest pressure drop rate to create the largest nucleation. On the other hand, the flow front is almost zero pressure in the empty mold that does not have any resistance for the cell growth. The cell density far from the sprue is higher than that near the sprue. This can explain why the number of cells in location 3 is higher than the number in location 1. The second reason could be the assumption of atmospheric pressure at the melt front during the entire time of mold filling. It may be true at the beginning of injection. However, there will be some resistance from compressed air or gas in the mold when more and more mold cavity is filled by gas-laden material. Experiment verifies that the pressure at the melt front is higher than the atmospheric pressure because of the compressed air and gas at the unfilled region (Figure 9.8b). The third reason could be lack of material data, especially that the viscosity recovers to high because of losing the gas with the cell growth. The last factor could be the convection of cells during mold filling and continually foaming. The free-flow front will create both fine cells at the interface between center core and skin and coarse cells in the center layer. The small cells not in the center layer will move upwards to the skin layer because the fountain flow during mold filling makes the final bubble size at downstream smaller somewhere near the skin. In addition, the cell size generally grows big at the center core in location 3, which matches the prediction in Figure 9.9. It is obvious that the big cells are in the center core in the free-flow front because there is no resistance, or just low pressure from compressed air or gas; the most important factor is that is no shearing in the center core.

2. Five Percent Weight Reduction Result. Oth er results come from the 5% weight reduction test. The results are similar to the 15% weight reduction for the predicted pressure profile. Figure 9.11 shows the calculated pressure at location 1 and location 3 of the mold during mold filling. The injection peak pressure is higher than 15% weight reduction test because mold filling has 10% more volume to be filled positively by injection. Also, the injection time lasts about one more second than the 15% mold filling. Overall it is similar to the results of the 15% weight reduction test. However, it is obvious that the

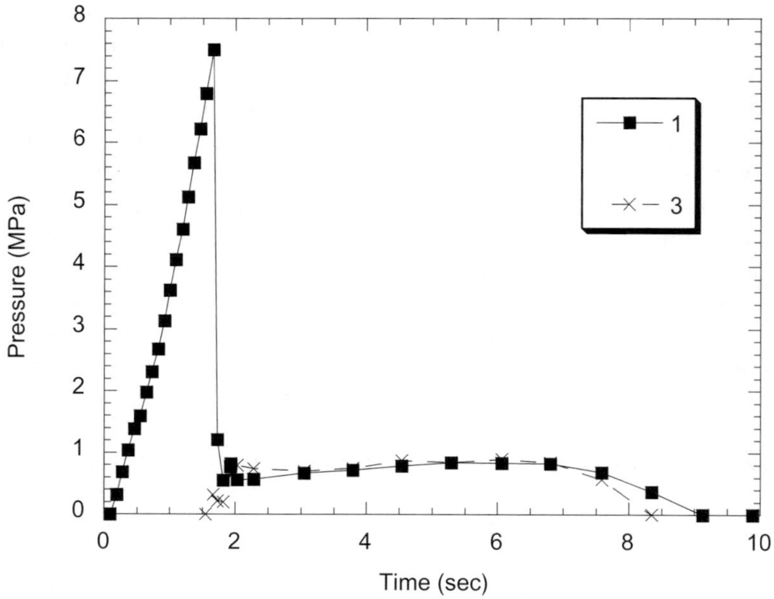

Figure 9.11 Pressure at different positions of mold during mold filling with 5% weight reduction. (Courtesy of Moldflow.)

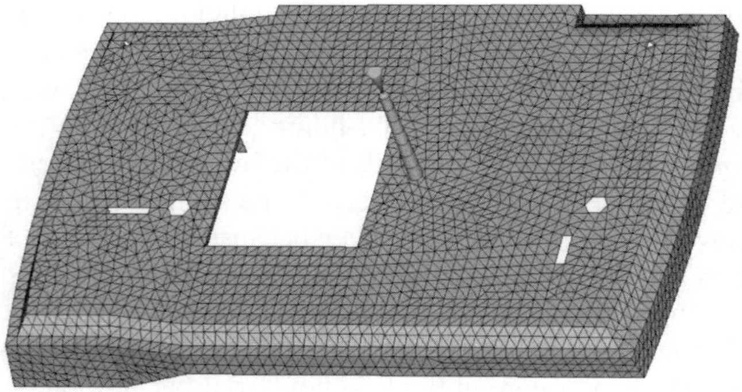

Figure 9.12 Geometry of the part used in the second case study. (Courtesy of Trexel and Moldflow.)

number of cells will be reduced for both locations 1 and 3 in 5% weight reduction sample because there is less space for the cell growth.

9.2.2.2 Clock Cover Mold. The clock cover mold from Trexel Incorporation is selected because there are several details, such as holes, a rectangular cut, some narrow slots, and 90 degrees of bending edge in the real part (see the part as shown in Figure 9.12). The objective is to provide the sample for verifying the MuCell® software simulation result with the real part. The focuses of the simulation are flow pattern of mold filling with and without gas. The material used in this test is the same material as the plaque mold, unfilled transparent GPPS (Dow666D). The Engel 200-ton tie-bar-less machine was used for this test. The test procedures are listed below:

- The solid part will first be run with filling volume 100%, 70%, 50% (this one may be adjusted to a higher percentage if the ejection system cannot eject such short shot size part).
- Then, 0.25% and 0.5% N_2 gas will be added, and three different weight reductions will be run at 5%, 10%, and 15%, respectively.
- Continual running with 15% weight reduction with 0.25% and 0.5% N_2 gas the cavity filling percentage will be decreased to the same shot size as the solid part, 70% and 50%, which will create some filled patterns comparable with solid mold filling.

The test conditions were set up based on the current database we already have so that it is comparable to them. The testing conditions are listed as follows:

- Temperature of melt: 227 °C for set up, and real melt temperature is about 229 °C
- Melt pressure during screw recovery: 20.7 MPa or 3000 psi
- Injection speed: 50.8 mm/sec
- Gas percentage: 0%, 0.25%, and 0.5%, respectively
- Mold temperature is set up 24 °C, (real measurement: 21 °C before starting the molding). After running the machine for a while, the core temperature will be 29 °C and cavity temperature will be 24.4 °C. Therefore, the 29 °C for core and 24.4 °C for cavity are used for the simulation conditions. The part temperature at ejection moment is about 32–34 °C for foamed part and 44 °C for the solid part. It is interesting that when measuring the final part temperature of the ejected part, the foamed part has a lower temperature than the solid.
- Cooling time is 50 sec for both solid and microcellular parts.
- Shot sizes are set based on the weight reduction percentages selected above. For 60-mm screw diameter, the shot sizes were set up with different screw strokes (see Table 9.3).

The measurements for all these tests will be the melt temperature, mold temperature, injection speed, and injection pressure. The injection pressure will be measured in the nozzle and will be collected with 5-Hz frequency sampling and recorded with Excel file. In addition, the samples will be collected, and the related SEM pictures will be prepared for cell structure verification.

The details of the part geometry and the location of sprue for injection are shown in Figure 9.12. The injection time to fill the whole mold is about 2.4 sec.

TABLE 9.3 Samples for MuCell Simulation Verification

Item #	N_2 Gas Weight percent	MPP (psi)	Injection Speed (in./sec)	Volume Percent of Mold Fill (screw stroke, in.)	Weight Reduction (%)	Comments
1	0	3000	2	50 (2.4)		
2	0	3000	2	70 (3.36)		
3	0	3000	2	100 (4.8)		
4	0.25	3000	2	100 (4.56)	5	
5	0.25	3000	2	100 (4.32)	10	
6	0.25	3000	2	100 (4.07)	15	
7	0.25	3000	2	50 (2.4)		The shot size will be the same as solid, 50% fill (see Figure 9.17)
8	0.25	3000	2	70 (3.36)		The shot size will be the same as solid, 70% fill (see Figure 9.18)
9	0.5	3000	2	100 (4.56)	5	
10	0.5	3000	2	100 (4.32)	10	
11	0.5	3000	2	100 (4.07)	15	
12	0.5	3000	2	70 (3.36)		The shot size will be the same as solid, 70% fill (see Figure 9.21)
13	0.5	3000	2	50 (2.4)		The shot size will be the same as solid, 50% fill (see Figure 9.20)

The nucleation density was determined to be 2×10^{10} (cells/m^3) from the SEM of the molded samples at the running conditions above. Actual nucleation density varies greatly with location, but a typical average value is used.

The short-shot comparisons between simulation and experiment are made in Figures 9.13 and 9.14. The result in Figure 9.13 is for 0.25% N_2 with 50% of mold filling that predicts the melt front position around the rectangular cut (a rectangular core in the mold) before the divided melt flows, joining to form a weld line. The results in Figure 9.13 show that simulation predicts a little smaller melt-front advancement than the experimental result. Generally, 0.25% N_2 in the GPPS melt creates extra force from cell expansion during foam.

The result in Figure 9.14 is for 0.25% N_2 with 70% of mold filling. The largest mold filling percentage is 85% with the same 0.25 weight percent of N_2 (not shown the results because the 70% is enough to represent the 85%

(a)

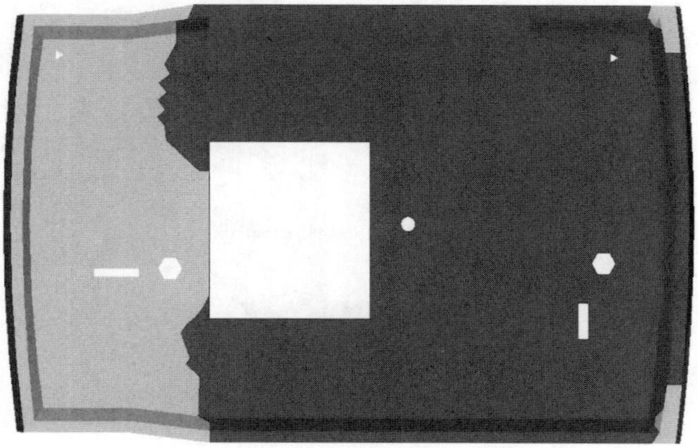

(b)

Figure 9.13 Short shot from experiment (**a**) and simulation (**b**) for N_2 gas concentration of 0.25% (weight) and shot size of 50%. (Courtesy of Trexel and Moldflow.)

result). It is interesting to see the simulation result for the melt front around a hole (a cylindrical pin in the mold) and a narrow slot (a narrow rectangular core in the mold). On the other hand, the 70% volume filling in the mold just shows the position of weld line formed by the two divided flows around the big cut. The simulation of Moldflow predicts the melt front and the welding line position very close to the real results, but they are a little shorter than the experimental melt front positions because of the foaming process. It indicated that the flow front is slowed from dividing the flow by the hole and slot, so the welding line is shifted toward the hole and slot positions, as shown in Figure 9.14b. The overall flow front from foamed material fills more volume

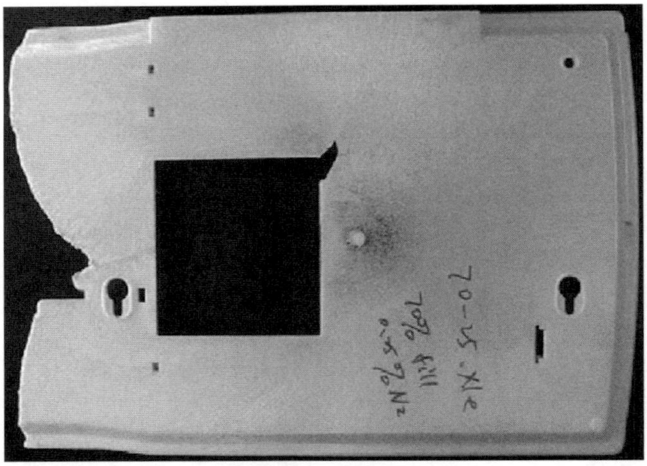

(a)

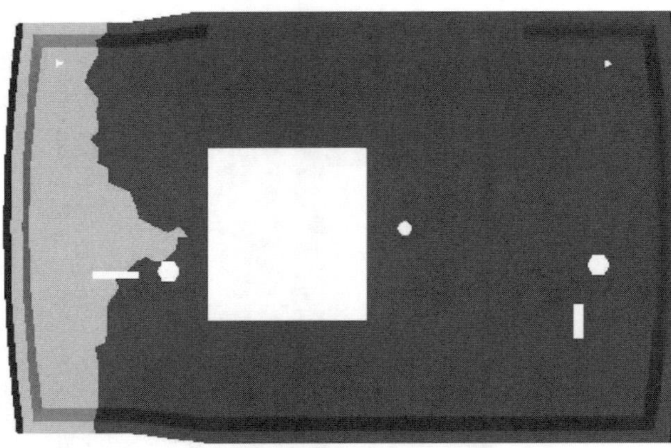

(b)

Figure 9.14 Short shot from experiment (**a**) and simulation (**b**) for N_2 gas concentration of 0.25% (weight) and shot size of 70%. (Courtesy of Trexel and Moldflow.)

than was predicted. This is because of the bubble expanding effects, which may not have a proper model for them in this simulation yet.

With the same GPPS material, but 0.5% weight of N_2 gas with 70% volume of shot size, the melt fronts from both experimental and prediction results indicate a more advanced melt front position (see Figure 9.15) than the result of 0.25% weight of N_2 gas with the same 70% volume material as the shot size (refer to Figure 9.14). It is because there are more driving forces from cell expansion as a result of double gas weight percentage in the cells. In addition, the agreement between experimental and prediction for the 0.5% N_2 gas sample is just a little better than for the 0.25 % N_2 gas sample. All three cases

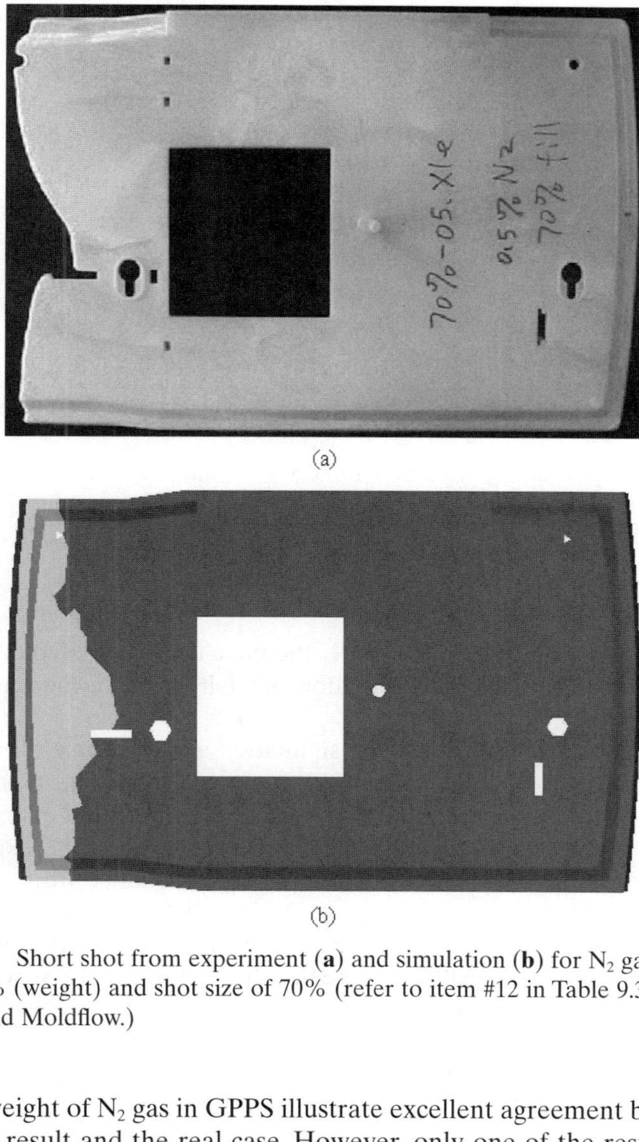

Figure 9.15 Short shot from experiment (**a**) and simulation (**b**) for N_2 gas concentration of 0.5% (weight) and shot size of 70% (refer to item #12 in Table 9.3). (Courtesy of Trexel and Moldflow.)

for 0.5% weight of N_2 gas in GPPS illustrate excellent agreement between the simulation result and the real case. However, only one of the results is illustrated in Figure 9.15. If the narrow slot area is focused on to examine the simulation result in the sample in Figure 9.15, the predicted flow front around the narrow slot in Figure 9.15b is so close to the flow front around the narrow slot of the real part shown in Figure 9.15a. There is one more factor to have these results become more predictable because the 0.5% weight of N_2 gas in the GPPS can create more uniform cell structure without voids so that the front of mold filling is more stable as well.

Overall, all the results of flow front of simulation are good enough to predict all three real mold filling results during the foam process, as shown in

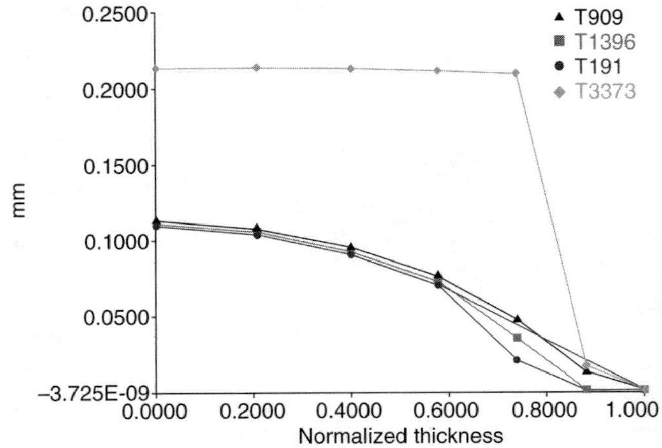

Figure 9.16 Bubble-size distribution for 0.5% N_2 and 70% shot case, simulation (T909, location A; T1396, location B; T191, location C; T3373, location D). (Courtesy of Trexel and Moldflow.)

Figures 9.13, 9.14, and 9.15. Therefore, the conclusion is that the simulation of Moldflow can be used for estimation of melt front during microcellular injection molding.

Bubble-size comparisons between simulation and experiment are made in Figure 9.16. It is for 0.5% N_2 gas in GPPS, and 70% shot size. Generally, the comparison between experiment and simulation is not good enough for the bubble-size distribution in current Moldflow software. As the experimental results show in Figure 9.16, the nucleation density obtained from experiment is not uniform. However, it is an acceptable result for cell size prediction in three locations (A, gate area; B and C, away from gate area) that have cell size of about 100 µm. The cell size in the flow front (location D) is complicated. The big cell size (>200 µm) in the center layer was predicted by the simulation. However, there are lots of small sizes in the free-flow front that cannot be predicted by Moldflow simulation software yet. In the future, a nucleation model that can describe different nucleation densities at different locations will be needed.

9.2.2.3 Connector Mold with Filled Material. This is the simulation for a complicated geometry with the glass-fiber-reinforced material. The part geometry and the gate system for this case are shown in Figure 9.17. The polymer used is Valox K4560, and the gas used as a blowing agent is CO_2. Three different percentages of full injection strokes (82%, 96%, and 100%) and one gas concentration (0.91% by weight) are used in this case. The melt temperature is 288 °C and mold temperature is 49 °C. The injection time is about 0.3 sec. The nucleation density was determined to be 2×10^{12} (cells/m^3) from

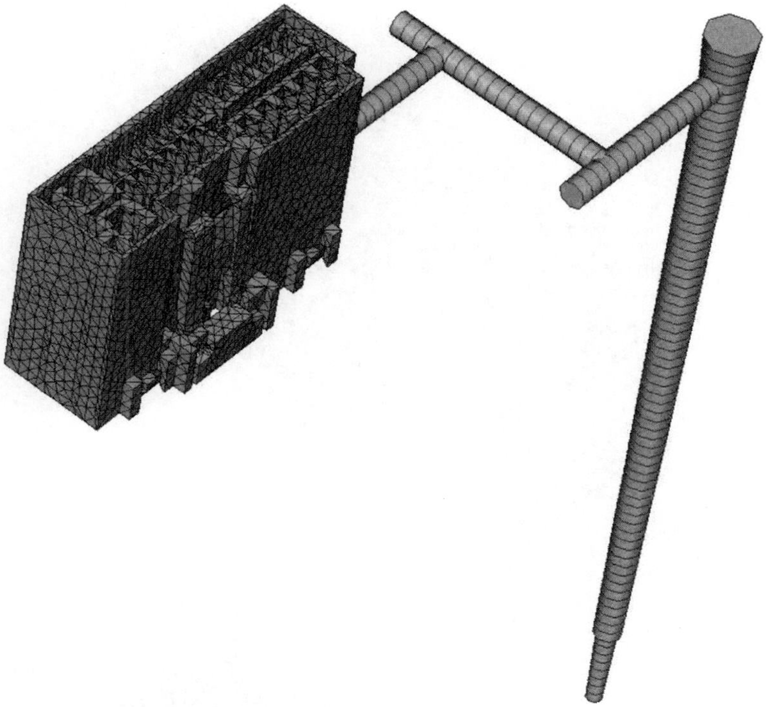

Figure 9.17 Geometry of the part used in a complex case made by glass fiber reinforced material. (Courtesy of Connector and Moldflow.)

the SEM of the molded samples. Actual nucleation density varies differently with location, but an average value is used for the simulation.

The 96% short-shot size comparisons between simulation and experiment are made in Figure 9.18. The final unfilled corners are predicted in the same locations as the experimental results shown. They are clearly the welding lines to be formed in the positions as the unfilled area. Therefore, even with the complicated geometric shape, the Moldflow software is good to do the prediction with foam process together. Generally, the short-shot agreement between prediction and experimental results is very good.

Bubble-size comparisons between experiment (see Figure 9.19) and simulation (see Figure 9.20) are made with this mold. It is also for 96% of full-shot size sample. Generally, the comparison between experiment and simulation is not good yet for the prediction of bubble-size distribution in unfilled material with current Moldflow software. However, the filled material has much better nucleation so that the prediction of cell sizes is better compared to unfilled GPPS material above. The cell size prediction at the end filling area, as shown in Figure 9.20, matches well with the experimental result, as shown in Figure 9.19. The cell size in the center of the sample as shown in the profile

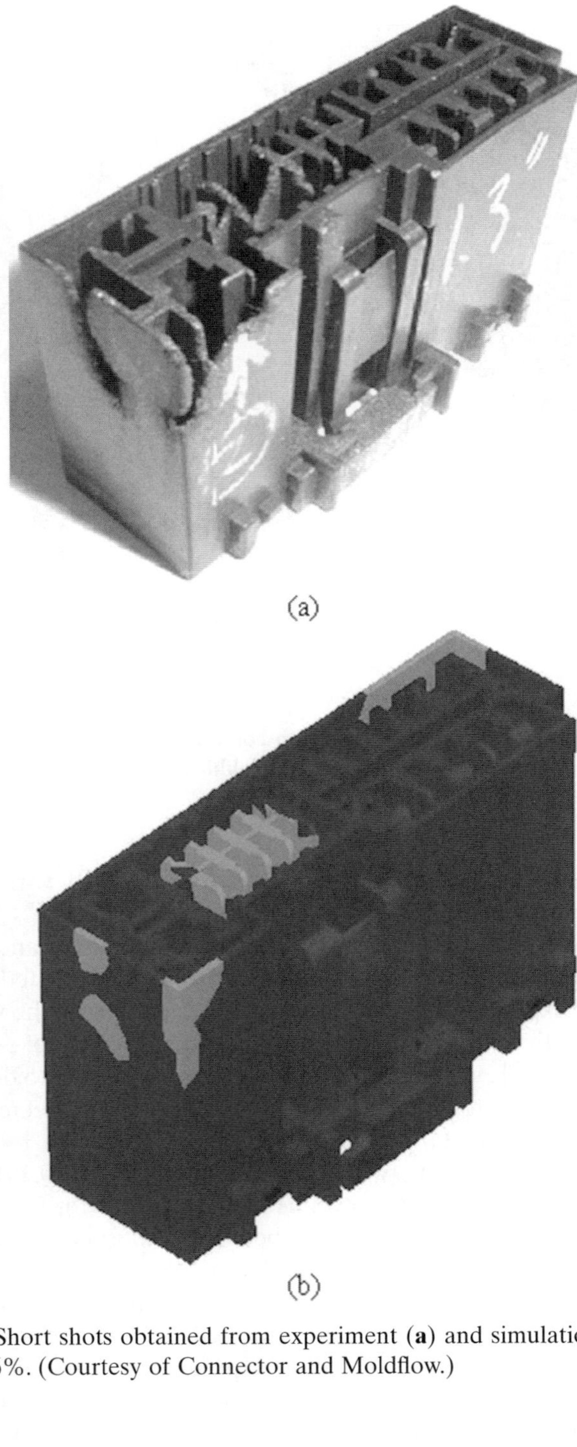

Figure 9.18 Short shots obtained from experiment (**a**) and simulation (**b**) for short-shot size of 96%. (Courtesy of Connector and Moldflow.)

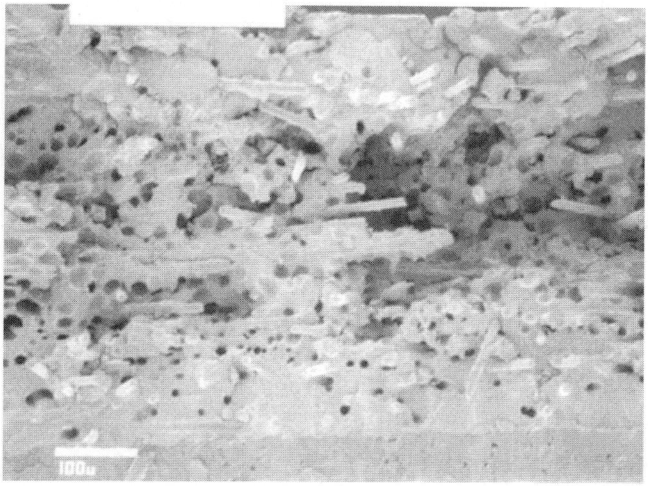

Figure 9.19 Bubble-size distribution obtained for 96% short-shot case by experiment. (Courtesy of Trexel, Connector, and Moldflow.)

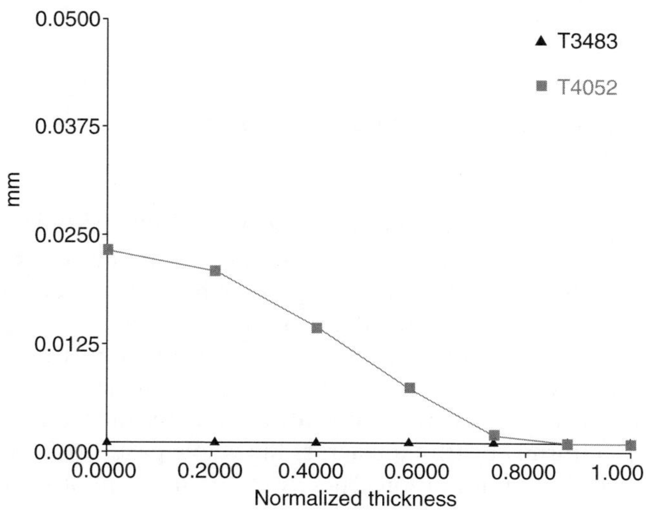

Figure 9.20 Bubble-size distribution obtained for 96% short-shot case by simulation. (Courtesy of Trexel, Connector, and Moldflow.)

T4052 of the cell size distribution in Figure 9.20 is predicted to be only 25 μm, and the cell sizes are continually changed to be even smaller near the skin. The cell distribution for the glass-fiber-reinforced material is much more uniform than the unfilled material, so the simulation result will be closer to the real result compared to the simulation result of unfilled materials. However, the cell size and cell size distribution across the whole part of the filled material is still a challenge to be modified for better prediction from this commercialized microcellular processing software from Moldflow.

9.2.2.4 Conclusion [25, 26]. The model developed by Moldflow can overall simulate the microcellular foam molding process reasonably well if the material properties are known, and the nucleation density is uniform and known. The short-shot comparisons show reasonable agreement between simulation and experiments in three cases studied here. However, in many cases the nucleation density is not uniform and is difficult to guess without actually examining the microstructures of the molded parts. Development of a better nucleation model may be needed in the future to resolve this problem.

Potential Future Improvement

- Speeding up the calculation
- More accurate material data
- More accurate process conditions
- Nonzero pressure at the melt front
- Bubble convection
- Nonuniform nucleation

9.3 SIMPLE MODELING OF MICROCELLULAR INJECTION MOLDING

In addition, some simplified models are introduced for the quick analysis of mold filling of gas-laden materials and for the estimation of cell structure. They are the quick tools to estimate the result of microcellular injection molding and the properties of the final microcellular part.

The most important estimation is the cell growth from the simulation. It determines the injection time limit for acceptable cell size in free melt flow front. The cell distribution is very difficult to simulate, and we usually assume that cells are uniformly distributed across the whole part. This can be reached from the processing optimization. Therefore, the assumptions are necessary to use those simplified models:

1. Nucleation is uniformly distributed in the whole injection size.
2. Pressure drop is constant in certain distances.
3. The injection control is only for constant speed, not for constant pressure.
4. There is no packing stage, which is replaced by cell growth stage.
5. There is neither significant shear heat generated nor heat loss during injection.

6. Cooling only affects the skin thickness not for foam and cell growth.
7. Surface tension uses the solid data unless data are available for the gas-laden material.
8. Viscosity is reduced 15% of original solid viscosity unless there are data available for gas-laden material.
9. Initial radius of cell begins from 10^{-4} (cm).
10. Minimum pressure drop rate *dp/dt* must be 10^8 Pa/sec or larger to determine the minimum injection speed and pressure with certain nozzle size or gate size (the smallest one will be the nucleation spot).
11. The cell growth model will determine injection time. If it is too long to have larger cells in the free-flow front, injection speed must increase.

The simulation program will need the schematic to do the logic iteration until the good results come. The major parameter for the good microcellular structure is the injection speed. It will be selected as the initial try input to calculate if the *dp/dt* is high enough to satisfy the nucleation rate requirement. Then, the injection speed may be adjusted again for controlling the total injection time to prevent too much cell growth in the free-flow front because the free-flow front has no pressure to restrict the cell overgrowth. This is the only easy way to control the cell overgrowth if the injection time is limited in the range of reasonable short time.

Then, the pressure–velocity distribution will be calculated without coupling the heat transfer equation but by solving the continuity and momentum equations. Zhu et al. [27] used the inlet pressure with observed experimental data, and they used the front pressure as the back pressure during mold filling. Through comparison with the solubility pressure, the foaming initiation line (cell growth beginning position) can be determined. However, the further simplified assumptions in this simple modeling are that the inlet pressure in the barrel is the injection pressure while the front pressure is the atmospheric pressure.

In the filling stage, the gas-laden molten polymer system generally consists of two regions: the unfoamed gas-laden molten polymer and the foamed polymer–bubble mixture. The difference between these two regions is the driving forces during injection. Zhu et al. [27] proposed reasonable assumptions in his simulation model. The unfoamed gas-laden molten polymer system is driven by the difference between the gate and solubility pressure, and it is a single-phase flow. However, the multiphase flow of the polymer–bubble mixture is driven by the difference between the solubility and back pressure. Even if he pointed out that the differences for these two difference regions existed in the microcellular mold-filling stage, he finally used the simple assumption that treated the microscopic multiphase fluid as a macroscopic, single-phase fluid [27]. This is the same assumption to be used in this simple modeling.

9.3.1 Viscosity Models

The simplified model of viscosity is necessary for this simulation as well. There are many different equations that can be used. Only several simple formulae are introduced for references as the simple models.

9.3.1.1 Power-Law Model. One simplified viscosity model is the power-law model, which is given for GPPS material with N_2 gas that was verified in microcellular injection molding machine:

$$\eta_T = 68631\dot{\gamma}^{(n-1)}\exp(-0.00371T)\exp(-0.22164C) \tag{9.35}$$

where η_T is the true viscosity of gas-laden melt (Pa-sec), C is the concentration (weight fraction) of the dissolved gas in the melt (for example, 0.75 wt% of gas, $C = 0.75$), $\dot{\gamma}$ is the apparent shear rate (1/sec), T is the melt temperature (use °F in this formula), and n is the power-law index (GPPS, $n = 0.236$).

9.3.1.2 Polynomial Model. Another simplified viscosity model is the polynomial model, which can be used for the case without knowing the value of power index n (or average n is not correct in a wide range of shear rates):

$$\begin{aligned}\eta_T = 6897\exp[&3.306157 - 1.479889\ln(\dot{\gamma}) + 0.006402\ln^2(\dot{\gamma}) + 0.004608T \\ &- 0.000019T^2 + 0.126237C + 0.396341C^2 + 0.001309T\ln(\dot{\gamma}) \\ &- 0.294978C\ln(\dot{\gamma}) + 0.000418TC\ln(\dot{\gamma})]\end{aligned} \tag{9.36}$$

The same units as in Equation (9.35) will be used in this formula.

9.3.1.3 Normalizing Viscosity with Weight Percentage of Gas as Variable. Furthermore, the viscosity can be simplified with normalizing viscosity. It is simply to calculate the viscosity ratio of gas-laden melt to the gas-zero melt. For example, the N_2 gas in material Questra WA212 will have the following relationship:

$$\eta_{N_2} = 0.3393W_x^2 - 0.5364W_x + 0.9951 \tag{9.37}$$

$$\eta_{CO_2} = 0.0184W_x^2 - 0.1527W_x + 0.9951 \tag{9.38}$$

where W_x is the weight percentage of gas in molten polymer.

9.3.1.4 Shift Factors of Pressure (P), Temperature (T), and Gas Concentration (C). Park and Dealy [12] used a high-pressure sliding-plate rheometer, in which the shear deformation, temperature (T), pressure (P), and gas concentration (C) are uniform. Then some shift factors for T, P, and C are used to see the effects of them for the viscosity change during processing [12].

TABLE 9.4 Activation Energy E_a for the Flow [12]

Material	Melt Temperature (°C DSC)	M_w (g/mol) (Absolute GPC)	M_w/M_n	E_a (kJ/mol)
HDPE	134	111,000	13.6	27.5
Linear PP	173	540,000	4.3	37.0
Branched PP	169	630,000	8.5	55.0

The effect of temperature for the viscosity change can be derived as below [12]:

$$\ln a_T = E_a\left(\frac{1}{T_{\text{poly}}} - \frac{1}{T_{\text{ref}}}\right)\Big/ R_g \tag{9.39}$$

where a_T is the horizontal shift factor for temperature coefficient of the viscosity, E_a is the activation energy for the flow (see Table 9.4), T_{poly} is the melt absolute temperature, and T_{ref} is the reference melt temperature at the shear zero condition.

Then, the zero-shear viscosity as a function of temperature is given [12]:

$$\eta_{T0}(T_{\text{poly}}) = \eta_{T0}(T_{\text{ref}})a_T(T_{\text{poly}}) \tag{9.40}$$

It is also well known that the pressure will increase the viscosity. The Barus equation [14] describes the P dependency of the pressure shift factor:

$$\ln a_p = \beta(P_{\text{poly}} - P_{\text{ref}}) \tag{9.41}$$

where β is the pressure coefficient (see Table 9.5), a_p is the pressure shift factor for viscosity change under the pressure, and P_{ref} is the reference pressure at certain temperature and 1 atm pressure.

Then, the zero-shear viscosity as a function of pressure is given [12]:

$$\eta_{P0}(P_{\text{poly}}) = \eta_{P0}(P_{\text{ref}})a_p(P_{\text{poly}}) \tag{9.42}$$

It is obvious that the temperature on the viscosity change is similar to the pressure but in the opposite direction. If the effects of temperature and pressure on the viscosity are modeled by the Arrhenius and Barus equations, and the effect from both parameters are separable, the relative influence of temperature and pressure can be calculated [12]:

$$\left(\frac{\partial T_{\text{poly}}}{\partial P_{\text{poly}}}\right)_{\eta T0,C} = \frac{R_g T_{\text{poly}}^2 \beta}{E_a} \tag{9.43}$$

Equation (9.43) can be used for quantity analysis for determining the optimum processing conditions by varying temperature and pressure.

Another important shift factor is the gas concentration C. The effect from CO_2 gas concentration is actually much larger than that of pressure alone for HDPE and PP materials [12]. To determine the effect of C alone at constant pressure P, it was assumed that the effects of P and C are separable [12]:

$$a_{P,C}(P,C) = a_p(P)a_C(C) \tag{9.44}$$

where $a_{P,C}$ is the shift factor of effect of gas concentration alone at constant P and a_C is the gas concentration as horizontal shift factor on the viscosity change.

The Fujita–Kishimoto equation [15] has the following relationship:

$$\ln(a_C b_C) = -g_c C(h_c + C) \tag{9.45}$$

where b_C is the gas concentration as vertical shift factor on the viscosity change and g_c and h_c are the fitting constants.

Therefore, the viscosity can be estimated at and P and C by

$$\eta_0(P_{\text{poly}}, C) = \eta_{P0}(P_{\text{ref}})a_p(P_{\text{poly}})a_C(C)b_C(C) \tag{9.46}$$

For the viscosity as a function of shear rate, the Cross model parameters obtained at 1 atm can be used along with Equation (9.46) [12].

Similarly, if the effects of P and C on the viscosity can be modeled using the Barus and Fujita–Kishimoto equations, and the effects of P and C are separable, we can formulate the relative influence of P and C on the viscosity as [12]

$$\left(\frac{\partial P}{\partial C}\right)_{\eta_0, T_{\text{poly}}} = \frac{g_c h_c}{\beta(h_c + C)^2} \tag{9.47}$$

Likewise, we can express the relative influence of C and T on the rheological properties as [12]

$$\left(\frac{\partial C}{\partial T}\right)_{\eta_0, C} = \frac{E_a(h_c + C)^2}{R_g h_c g_c T_{\text{poly}}^2} \tag{9.48}$$

Lee et al. [16] measured the effects of pressure and shear rate on the viscosity of GPPS with real microcellular extrusion system, and they used a least-squares method to determine the pressure factor and other parameters in the viscosity equation similar to Equation (9.28) (see Table 9.5).

TABLE 9.5 Pressure Coefficient β for the Pressure on the Viscosity Change [12, 16]

Material	Melt Temperature (°C DSC)	M_w (g/mol) (Absolute GPC)	M_w/M_n	β (1/GPa)
HDPE	134	111,000	13.6	15.5
Linear PP	173	540,000	4.3	22.0
Branched PP	169	630,000	8.5	42.0
GPPS				79.7

TABLE 9.6 Viscosity Reduction Percentage of Different Materials per Flow Number Differences[a]

Material	Filler	Maximum Viscosity Reduction with Nitrogen (Weight %)	Maximum Viscosity Reduction with Carbon Dioxide (Weight %)
PEEK		19	20
PET	15% Glass fiber	15	25
PBT	30% Glass fiber	10	25
PA/SPS	Filled	20	30
GPPS	Unfilled	12	20
PC	Unfilled	15	35
PA66	35% Glass fiber	8	23
PP	13% TF	5	11

[a]Data from Trexel, and it is the average viscosity difference between solid and gas-laden materials, only served as quick estimation for viscosity change with different gases.

Once η_0 is determined, then, Equation (9.28) will be used to fully determine the complex viscosity of the gas-laden molten polymer.

5. *Flow Number for Related Viscosity Changes.* Some OEM equipment like Engel offers a function that calculates the area under the pressure curve changing with the injection time during injection. It is basically that the injection pressure multiplies the injection time, which is defined as "flow number" in Engel's controller. Using this value to estimate the viscosity change is a direct comparison between gas-laden material and the same solid material. It can be used for a quick estimation of the viscosity reduction at certain processing conditions without any complicated nozzle rheometer and data calculation. The reference flow numbers of different materials are summarized in Table 9.6. As a practical comparison between solid and gas-laden materials, these viscosity data are good for quick estimation of the material viscosity change before knowing the exact material grade to be used for microcellular processing.

9.3.2 Gas Solubility and Concentration Calculation

Some general equations are introduced in Chapter 2. Sato et al. [28] recommend some correlation equations (from Equations 9.49 to 9.54) for nitrogen and carbon dioxide by using a least-squares method to determine a linear relationship between $\ln(H)$ and $(T_{cr}/T)^2$ for PS, HDPE, and PP, respectively:

For CO_2 + PS system

$$\ln(H) = 6.498 + 2.380(T_{cr}/T)^2 \qquad (\text{cm}^3(\text{STP})/(\text{kg}\cdot\text{MPa})) \tag{9.49}$$

For N_2 + PS system

$$\ln(H) = 6.970 - 9.587(T_{cr}/T)^2 \qquad (\text{cm}^3(\text{STP})/(\text{kg}\cdot\text{MPa})) \tag{9.50}$$

For CO_2 + HDPE system

$$\ln(H) = 6.571 + 2.764(T_{cr}/T)^2 \qquad (\text{cm}^3(\text{STP})/(\text{kg}\cdot\text{MPa})) \tag{9.51}$$

For CO_2 + PP system

$$\ln(H) = 6.255 + 3.706(T_{cr}/T)^2 \qquad (\text{cm}^3(\text{STP})/(\text{kg}\cdot\text{MPa})) \tag{9.52}$$

For N_2 + HDPE system

$$\ln(H) = 7.395 - 7.095(T_{cr}/T)^2 \qquad (\text{cm}^3(\text{STP})/(\text{kg}\cdot\text{MPa})) \tag{9.53}$$

For N_2 + PP system

$$\ln(H) = 8.407 - 20.39(T_{cr}/T)^2 \qquad (\text{cm}^3(\text{STP})/(\text{kg}\cdot\text{MPa})) \tag{9.54}$$

The critical temperature T_{cr} and processing temperature T are known for nitrogen and carbon dioxide, and the processing temperature is also known from set up. Therefore, the H value can be calculated from the equations above for HDPE, PP, and PS. Then, with the known molten polymer pressure, the gas concentration can be estimated by Equation (2.1).

9.3.3 Gas Diffusion Calculation

The pressure dependence of the diffusion coefficient can be neglected in the engineering calculation since thermodynamic and transport properties including the diffusion coefficient in liquid phase are not usually sensitive to pressure [27]. The diffusion coefficient shows weak concentration dependence, and it was fitted as follows [27]:

For N_2 + PP system

$$\log D_a = -8.298 + 25.56W_x \qquad (\mathrm{m^2/sec}) \tag{9.55}$$

For N_2 + HDPE system

$$\log D_a = -8.224 + 15.54W_x \qquad (\mathrm{m^2/sec}) \tag{9.56}$$

For CO_2 + HDPE system

$$\log D_a = -8.076 + 2.26W_x \qquad (\mathrm{m^2/sec}) \tag{9.57}$$

This will provide the estimation of gas dosing quality in the first stage of microcellular injection molding.

9.3.4 Cell Growth Calculation

The next calculation with simplified model is the cell growth. There will be two results to be checked with this calculation:

1. One is to estimate the final maximum cell size in the microcellular part. Then, it is necessary to know the injection time, or predetermined maximum injection time by processing because the skin thickness or gate freezing time restriction. It needs to calculate the cell growth without any restriction to see how big it will be in certain injection time. The maximum injection time allowed with a predetermined cell size will be decided by the final mechanical property as the target of the microcellular part. This approaching of the calculation may be the simplest way to just calculate the cell free growth to see how big will it be at certain injection time.
2. If there is a difference between calculation and set-up target in the preselected precision factor, the calculation is done. Otherwise change the set-up condition and do the calculation again until the difference between results of calculation and set-up target in the control factor range.
3. Also, the weight reduction needs to be selected first to know how much space is left for cell growth. Then, the result will be checked with the volume left in the mold with the weight reduction target.

Equilibrium concentration is obtained from simulator test results.

The equation of motion is similar to Equation (9.14), and it is rewritten for the pressure simply using P_t [pressure in the barrel, nozzle, or valve gate (if any valve gate is used, this is the position for the closest pressure as P_t] to represent the pressure in the cells P_g, as well as using P_a (atmospheric pressure) to be P_m as the pressure for the melt at the outer boundary of the cell:

$$\frac{4\eta_T(R_{i+1}-R_i)}{R_i\Delta t}=P_t-P_a-\frac{2\sigma}{R_i} \tag{9.58}$$

Therefore, the pressure in the barrel and atmospheric pressure are both known parameters now.

The initial conditions are

$$R(0)=R_{i=1}=R_0, \quad \text{and} \quad \dot{R}_{t=0}=0 \tag{9.59}$$

The equation of mass balance over the cell is

$$\frac{3}{4}\pi(R^3\rho_g-R_0^3\rho_{g0})=V_0C_0-\int_{R_0}^{R}4\pi r^2C\,dr \tag{9.60}$$

If we solve Equation (9.60), the cell growth can be estimated with this simple model.

9.4 MOLD-FILLING SIMULATION GUIDELINES FOR MUCELL® PROCESS

Trexel offers a real mold for microcellular MuCell® simulation developed by Moldflow [29]. The general guidelines are proposed below because it will help for anyone to use the standard mold-filling software to estimate the benefits of microcellular processing for their current molding conditions and molds.

1. Fundamental Differences from Regular Injection Molding. The microcellular process is different from regular injection molding in many ways. However, the fundamental differences may be summarized as follows:

- A significant melt viscosity reduction of gas-laden material is due to the single-phase solution as the mixture of molten polymer and supercritical foaming agents.
- The traditional pack/hold stage will not be necessary anymore for the microcellular process, and it will be replaced by the cell growth stage.

2. Simulation Approaches. Here are the summaries of the basic procedures to simulate the microcellular process:

- Identify the grade of material (same filler level and chemical composition) that may have the 12–15% lower viscosity than original grade of material to be used for microcellular processing. It is approximate viscosity drop with supercritical fluid of foaming agents.

- Pick an injection speed of microcellular process that should be faster than the solid molding (about 25–50% faster).
- Profiling of injection velocity will be necessary to avoid heavy surface splay around gate. At least the slow injection velocity is used first until the runner system is filled, and the full-flow front is formed near the gate; then injection velocity can be raised as high as possible.
- Decrease the hold pressure to 25% of the regular injection pressure for the microcellular process, and use a very short hold time. It is necessary to smooth transfer the hydraulic pressure from low to the set-up injection pressure. However, the electric screw driving system does not need any hold stage.
- Lower the packing and hold time to 2 sec, which is only for the simulation software request but is not really needed for molding.
- Make the gate of the microcellular process 30–50% larger than a normal solid part. It is necessary for minimizing the effect of gate shearing to induce the blister and imperfections around the gates (not for sub-gate design).
- Material and mold temperature should be set as expected for the process.

3. Evaluation the Results. The goal of this simulation of the microcellular process is to determine the types of processing conditions that can limit weight reduction in the part and remove them. The most common issues from the microcellular process are high flow length to thickness ratio, unbalanced fill, thin area at the end of fill, and a lack of appropriate venting. Therefore, the major issues to be checked when evaluating the result of mold-filling simulation from Trexel are as follows:

- Balance fill pattern. The areas of mold that are not filled until just before the end of simulation should be weld lines between gates, or the end of part, but not back-flow areas.
- It is necessary to run multiple simulations with adjustments to gates or through the use of flow leaders to achieve the best of results. The imbalanced fill pattern will limit the weight reduction percentages.
- Location of weld lines needs to match the vending positions. If the venting is not feasible, the simulation result must be rerun until the welding line area in the proper venting area as well.
- Look for race tracking that can close off the perimeter vents that will result in possible gas and air traps.
- Gating into thin sections is preferred for optimizing the microcellular processing benefits.

- Maximum pressure to fill the cavity is usually less than 20.7 MPa (8000 psi) at 8–10% weight reduction. When weight reduction percentage reduces to 3–6%, the pressure may be in the range of 69 MPa (about 10,005 psi) to 83 MPa (about 12,035 psi).

REFERENCES

1. Suh, N. P. *Innovation in Polymer Processing*, edited by James F. Stevenson, Hanser/Gardner Publications, Cincinnati, Chapter 3, 1996.
2. Martini-Vvedensky, J. E., Suh N. P., and Waldman, F. A. U.S. Patent No. 4,473,665 (1984).
3. Blizard, K., Okamoto, K. T., and Anderson, J. R. U.S. Patent No. 6,294,115 (2001).
4. Xu, J., and Pierick, D. J. *Injection Molding Technol.*, **5**, 152–159 (2001).
5. Malloy, R. A., Chen, S. J., and Orroth, S. A. *Melt Viscosity Measurements Using an Instrumented Injection Molding Nozzle*, University of Lowell, 1989.
6. Reynolds, A. R. *SPE ANTEC Tech. Papers*, 2523–2525 (1992).
7. Gottfert, A. (1990). Real time Rheometers: The second generation of on-line capillary rheometers, in *PPS Proceedings, 6th Annual Meeting, Nice, France.*
8. Tseng, H., Wang, K. K., Chiang, H. H., Grant, G. E., *SPE ANTEC Tech. Papers*, 716 (1985).
9. Gottfert, A. *SPE ANTEC Tech. Papers*, 2299 (1991).
10. Lenk, R. S. *Polymer Rheology*, Applied Science Publishers, London, 1978, pp. 70–84.
11. Rao, N., K. O'Brien, *Design Data for Plastic Engineers*, Hanser/Gardner Publications, Cincinnati, 1998, p. 127.
12. Park, H. E., and Dealy, J. M., *SPE ANTEC Tech. Papers*, 2534–2538 (2008).
13. Guzmán, J. de *Anales Soc. Espanola Fis. Quim.* **11**, 353 (1913).
14. Barus, C. *Am. J. Sci.* **45**, 87 (1883).
15. Fujita, H., and Kishimoto, A. *J. Polym. Sci.* **28**, 547 (1958).
16. Lee, M., Park, C. B., and Tzoganakis, C. *Polym. Eng. Sci.* **39**(1), 99–109 (1999).
17. Okamoto, T. K. *Microcellular Processing*, Hanser/Gardner Publications, Cincinnati, 2003, p. 42.
18. Zheng, R., Kennedy, P., Xu, J., and Kishbaugh, L. *SPE ANTEC Tech. Papers*, 498–502 (2002).
19. Han, S., Kennedy, P., Zheng, R., Xu, J., and Kishbaugh, L. *J. Cell. Plastics* **39**, 475–485 (2003).
20. Amon, M., and Denson, C. D. *Polym. Eng. Sci.* **24**, 1026–1034 (1984).
21. Eisenberg, P., and Tulin, M. P. Cavitation, in *Handbook of Fluid Dynamics*, edited by V. Streeter, McGraw-Hill, New York, 1961.
22. Van Kreevlan, D. W. *Properties of Polymers: Their Estimation and Correlation with Chemical Structure*, Elsevier Scientific Publishing, Armsterdam, 1976.
23. Rosner, D. E., Epstein, M. *Chem. Eng. Sci.* **27**, 69–88 (1972).
24. Kennedy, P. *Flow Analysis of Injection Molds*, Hanser, New York, 1995.

25. Han, S., Jin, X. S., and Xu, J. MuCell Validation Report 030703, Moldflow MPI 4.1, March 7, 2003.
26. Han, S., Jin, X. S., and Xu, J. Updated MuCell Validation Report 030703, Moldflow MPI 7.0, December 2, 2008.
27. Zhu, Z., Wang, J., Lee, J. W. S., and Park, C. B. *SPE ANTEC Tech. Papers*, 2178–2182 (2008).
28. Sato, Y., Fujiwara, K., Takikawa, T., Takishima, S. S., and Masuoka, H. *Fluid Phase Equilibria* **162**, 261–276 (1999).
29. Trexel Web Site, www. trexel.com.

10

POSTPROCESSING AND PROPERTY TEST OF MICROCELLULAR INJECTION MOLDING

This chapter provides the latest postprocessing technologies of microcellular injection molding. It includes the different welding methods, such as ultrasonic, vibration, hot plate, spin, laser welding, and so on. In addition, more postprocesses are necessary for some special microcellular applications. For example, for the perfect surface-quality postprocesses may include surface polishing and painting. To save cycle time, the thick microcellular part may need a postcooling process. Some microcellular part will have an index or label on the surface, and then a de-gas process must be finished before index or labeling on the surface of the microcellular part. Depending on the application requirements, the postprocess may be developed accordingly even more in the future. In addition, some test methods have been used for microcellular foam and will be briefly introduced in this chapter.

10.1 WELDING FOR MICROCELLULAR INJECTION MOLDING

Welding is defined as the process of uniting, fusing, or bringing (metal or plastic parts) into intimate contact [1]. Only thermoplastic materials can be welded with either thermal techniques or solvents. The principle is similar for both methods, which are used to soften material by heat or solvent and then joint together and hold under pressure until the joint is cooled or solvent evaporated. To join two parts together permanently, plastic thermal welding is still the most common method, and it is the only one that will be discussed

Microcellular Injection Molding, Edited by Jingyi Xu
Copyright © 2010 John Wiley & Sons, Inc.

TABLE 10.1 Welding Method Selection Guidelines [2]

Material	Ultrasonic Welding	Linear Vibration	Orbital Vibration	Hot Plate Welding	Electromagnetic Bonding
Amorphous thermoplastics	1	1	1	1	1
Semicrystalline thermoplastics	2	1	1	1	1
Olefins	2	1	1	1	1
TPRs	1	2	2	2	2
Composites	2	2	2	2	1
Part Details					
Thin walls	1	3	2	1	2
Complex geometry	2	1	1	1	1
Large parts	2	1	2	1	1
Small parts	1	1	1	1	1
Internal welds	1	2	2	1	1
Long, unsupported walls	1	3	2	1	1

Legend: 1, recommended; 2, limited; 3, not recommended.

Source: Reproduced with copyright permission of *Injection Molding Magazine* [2].

in this chapter. However, there are many different welding methods available. To select one logically to fit the requirement of a microcellular part joining needs, consider materials, the amount of welding area, the joint design, the desired strength of the weld, the aesthetic appearance, and the part geometric details. In Table 10.1 [2], some welding processing guidelines are listed with recommendations on how to select the specific welding methods based on different materials and part details. Since a microcellular part does not change the plastic welding principle, the welding process guidelines for a solid part in Table 10.1 are useful references for the reasonable selection of welding methods of microcellular parts as well.

10.1.1 Technology Fundamentals

The fundamentals of microcellular foam molding have been covered in many papers and articles including works done by M.I.T., Trexel, Sulzer Chemtech, and IKV Aachen [3]. The concept as it applies to the injection molding process involves introducing a physical foaming agent, typically nitrogen or carbon dioxide, in the form of a supercritical fluid into a molten polymer. This inert gas in the microcellular part may not be any issue during the welding since there is no chemical blowing agent residual left in the plastic part. That molten polymer is under the appropriate temperature and pressure conditions to cause the foaming agent to dissolve into the polymer to create a single-phase solution. Implied in the term of single-phase solution is not only that the foaming agent is completely dissolved into the polymer but also that it is uniformly dispersed throughout the polymer. When the single-phase solution of foaming agent and polymer are introduced into the mold, this mixture of gas and molten polymer undergoes a thermodynamic change, a pressure drop, which causes cell sites to form (cell nucleation) and then grow as the foaming agent comes out of solution. Cell density, the number of cells for a given volume of material, is a function of (a) the amount of foaming agent in its saturation limit and (b) the rate of pressure drop (change in pressure–time increment). This basic mechanism is applicable to any system using a physical foaming agent to create a microcellular injection molded part [3]. It is also important for welding because there is no chemical reaction to be considered during foaming with physical blowing agents.

Usually, only the same plastic materials are welded together, although rules similar to those of the co-injection process can be used to weld different plastics. A major exception is the welding of PVC to acrylic, which can be welded together. On the other hand, polytetrafluoroethylene (Teflon TFE) and other fluoroplastics may not be welded together [1].

It is necessary to know the differences between microcellular and solid parts before the welding methods can be selected. There are two key factors to be checked for microcellular parts:

- Skin thickness
- Cell structure underneath the skin

The skin thickness and cell structure in the weld area affect the welding strength and the welding method selections. The experimental results will be discussed in the following sessions.

10.1.2 Introduction for Welding Methods

There are a number of welding technologies used in the market today including vibration welding, ultrasonic welding, hot plate welding and spin welding, hot gas welding, and laser welding. A brief introduction for each welding method is discussed in the following.

10.1.2.1 Vibration Welding. The friction heat is employed in vibration welding. In vibration welding, the plastic parts are located in a fixture in such a manner that the weld surfaces are brought into contact under pressure. The weld surfaces are then rubbed together in either a linear or orbital manner at a given frequency, pressure, and time. As the parts move, friction builds, causing the surfaces in contact to melt. Once sufficient materials have been melted, the movement between parts stops and the molten plastic at the interface cools and solidifies to form a bond between the parts.

The friction heat in vibration welding is low frequency (90–120 Hz, up to possible 300 Hz) that is corresponding to normal AC line frequencies. It is simple equipment compared to ultrasonic because it does not need an energy director on the joint. It is applied for a big butt joint that is too large to be welded using ultrasonic welding. A clamping pressure of 1.4 MPa (200 psi) is commonly used in vibration welding.

Of these, vibration welding is the most common for large parts with complex parting lines such as air intake manifolds and is of more interest with microcellular foamed parts because these larger parts typically achieve greater benefits from the microcellular process.

10.1.2.2 Ultrasonic Welding. Ultrasonic welding is more common with small parts. In this technique, one part is held stationary while a second part is brought into contact. A high-frequency vibration is placed on the moving part, causing friction at the contact points that results in a melting of the material at the weld face. Parts being joined by ultrasonic welding typically have a design feature on the weld surface to focus the welding energy.

The same principle is used for both ultrasonic welding and vibration welding becuase frictional heat in both welding methods is generated by vibration. The basic difference between ultrasonic and vibration welding is the frequency. In ultrasonic operation, the power supply is used to change the low-frequency 60-Hz (60 cycles per second) line current to a high-frequency 20-kHz (20,000 cycles per second) line current. Such power supply is also called as a high-frequency generator or an oscillator. The transducer receives the high-frequency electrical oscillations and converts the electrical energy into mechanical vibration, by means of a piezoelectric ceramic, lead zirconium titanate, which converts 90% of the electrical energy into mechanical energy. The mechanical energy is then transferred from the transducer to the tool, a titanium horn that contacts the plastics to be welded. The horn serves two functions: (a) directing the ultrasonic vibration to the plastics and (b) applying pressure to the molten surfaces to form the weld and hold the pressure during the dwell time [1].

The harder, more rigid plastics, such as PC, PS, ABS, and acrylics, are generally good transmitters of acoustical energy, so they are bonded well ultrasonically. Therefore, the microcellular foam made by such rigid plastics are generally good for ultrasonic welding. However, any softer materials, such as HDPE, PP, soft PVC, and similar soft materials, effectively absorb the acoustical vibrations and do not form good welds from ultrasonic welding [1]. Therefore, in general, a microcellular part will be softer than a solid part made with the same material. It is why the skin thickness and cell structure may determine if ultrasonic welding can be used successfully for microcellular part welding. On the other hand, the filled PP, HDPE with fillers, or glass fiber will be a great improvement for transferring the acoustical energy for ultrasonic welding.

Location of the joint relative to where the horn contacts the welding part needs to be as close as possible. Shortening the acoustical path will minimize the loss of energy and produce the higher strength of welding. The minimum path length is less than 6.4 mm (0.25 in.) and is known as the "near-field" weld, and the path length longer than 6.4 mm (0.25 in.) is called the "far-field" weld [1]. It is extremely important for the design of the ultrasonic welding part to use "near-field" weld specifically for unfilled microcellular soft material.

10.1.2.3 Hot Plate Welding. Hot plate welding, or so-called heated tool welding, as the process is well known, can be used as the simplest manual welding method. The plastic will contact the heated tool until a molten bead is formed on each edge. Then, the molten edges of the plastic parts will be quickly brought into intimate contact, using just enough pressure to bring the part into good contact. The two molten edges should be held with just enough pressure to keep the good contact. The proper pressure is critical to weld successfully with hot plate since too much pressure will cause a starved weld bead resulting in lowered strength of joint. On the other hand, the temperature of heated tool needs to be high enough to produce a molten surface of plastic but not so high as to cause degradation of the plastic [1].

10.1.2.4 Spin Welding. Spin welding (also named friction welding) uses frictional heat, which is similar in principle to the vibration and ultrasonic welding methods. The friction heat is generated by rapidly rotating one thermoplastic part against another stationary thermoplastic part. It then causes melting. Once enough melting occurs on the contacting surface of the part, the rotation stops and certain pressure is applied to the weld joint in 1–25 sec. Then, it creates a strong, hermetic seal. It usually requires a joint with a circular shape. However, the new spin welders can be servo-driven, allowing control of final part orientation within one degree. Some models, then, can provide angular spin capability to weld parts that cannot freely rotate the whole circle but can have a reciprocating angular motion. The pressure is supplied by air cylinder with the pressure up to 1.38 MPa (200 psi).

10.1.2.5 Hot Gas Welding. Hot gas welding for plastic parts is similar to metal welding. The difference is that a direct flame will not be used in plastic welding. Compressed air, or nitrogen, heated by passage through a heating element impinges on a plastic welding rod and the work by means of a welder [1]. The weld is made by fanning the gun from side to side while heating the rod and the joint of the part. As the material softens enough, the rod is forced into the joint, applying pressure as the joint cools. It is only used for a thick part and a low-volume production.

10.1.2.6 Laser Welding. Laser welding is the latest welding technology for thermoplastics welding. It is an innovative welding technology based on the "Simultaneous Through-Transmission Infrared (STTIr)" principle. It uses a through-transmission technique where laser light passes through a part that is transparent to the laser wavelength, which is typically near-infrared wavelength [4]. Ideally, this laser wavelength transparent component is on the upper position of the welding parts. The laser light is absorbed by a part as the lower position welding part is filled with carbon black or other colorant. Then, the absorbent part melts quickly and conducts heat to the transmissive part. It will result in welded joints between two parts. Clamp force is necessary for the laser welding and needs to be controlled carefully.

Nearly all thermoplastics may be welded using laser components. Special additives/pigments allow laser welding of dark to dark, as well as transparent to transparent, materials. It is all about the color that is recognized by the wavelength of laser light. Laser welding of thermoplastics depends on many of the same rules of resin compatibility as other processes of thermoplastics. It is more forgiving of resin chemistry or melt temperature differences than most other welding processes.

There are two kinds of basic through-transmission laser welding methods. One is collapse welding, which welds the entire joint width. It is suitable for high-cleanliness applications. This welding mode process tolerates air gaps to about 100 μm or more, so it better compensates for the parts with surface inconsistencies. Another basic welding method is contained welding, which shines laser light on a local area that is smaller than the total wall section of the weld. It eliminates both flash and particulates, and it only needs minimum clamp force. It is capable of producing weld joints that are nearly as strong as the parent material [4].

However, there is a consideration to use this technology for microcellular parts because the cell structure causes refraction of laser light. Theoretically, laser welding should work at least for thick-skin microcellular part if the skin does not have cells. On the other hand, the cells are so small that the colorant in the skin of a microcellular part absorbs the laser light without enough refraction of laser light. Compared to other welding methods, laser welding has the following advantages:

- No particulate is generated
- Vibration-free process
- No wear on the tooling because it is non-contact welding by laser light
- Hermetic seals are achievable
- High-precision welding
- Simple joint geometry
- Optically perfect weld joint area

It will be a new welding method for microcellular foam becuase this technology never works for traditional foam parts. The laser welding may give more flexibility of welding solutions, such as contour welding, simultaneous welding, mask welding, radial welding, and globo welding [5].

10.1.3 Experimental Results for PA 6 and PA 6/6

The studies below were relatively quick screening studies aimed at developing some basic data based on an agreement between Trexel and Branson to have them conduct a limited amount of testing. Based on this, both the solid and MuCell® samples were welded under identical conditions that were deemed typical for the individual resins. Also, no effort was made to change the geometry of the energy director. Unless specified, all data and photos discussed in Section 10.1.3 are taken from reference 3.

Glass-fiber-filled PA 6 and PA 6/6 are two of the more common materials injection-molded with the MuCell® microcellular foaming process. As the technology continues to expand into under-the-hood applications, the need to understand vibration welding and ultrasonic welding becomes more critical. Therefore, we compare the welding performances for PA 6 and PA 6/6 molded as solid and by the MuCell® process. We will also discuss weld line strength and variation across test specimens and will try to understand proper welding performance.

One area of growth is in under-the-hood applications typically in glass-fiber-filled PA 6 or PA 6/6 such as fan shrouds, air intake manifolds, and rocker covers. These applications lend themselves to the MuCell® process because of frequent dimensional issues and overall part mass. Dimensional issues are also a cause for welding problems because they affect the contact pressure between the parts.

Two of the more common welding techniques are ultrasonic for small parts and vibration for larger assemblies. Both methods will be discussed, although vibration welding is the more common for under-the-hood applications.

Two studies were conducted on a 35% glass-fiber-filled PA 6 and a 35% glass-fiber-filled PA 6/6. Test specimens produced in standard solid injection molding and by the MuCell® process were evaluated for weld strength when assembled using both ultrasonic welding and vibration welding. The target density reduction for all samples produced by the MuCell® process was 10%.

The first study was conducted with Branson Ultrasonics using a test configuration known as a coffin, as shown in Figure 10.1a. The coffin mold consists of two parts that, when welded together, form a box. One of the parts has a hole that can be used to pressure test the welded assembly. The materials used for this study were TechnylXCell S218V35 (35% glass-fiber-filled PA 6) and Technyl A218V35 (35% glass-fiber-filled PA 6/6). The samples were ultrasonically welded and the weld strength was evaluated on the assembly first by a pull test and then a pressure test.

The same materials were then tested from Rhodia Engineering Plastics using a bell-shaped part and vibration welding as shown in Figure 10.1b. This part is designed such that two are welded together to give a specimen to be tested for burst strength. In addition, the part is focused on the failure on the weld and not on the material wall.

10.1.3.1 Ultrasonic Welding. The initial step of the ultrasonic welding study was to establish optimum welding conditions. This is defined as having a full weld across the wall thickness but without weld flash. The welding conditions were optimized with adjusting the welding time to either increase or decrease

(a) (b)

Figure 10.1 Test parts for ultrasonic welding. (**a**) Coffin mold [3]. (**b**) Bell mold [3]. (Reproduced with copyright permission of RAFRA.)

TABLE 10.2 Pull Test Result from Ultrasonic Welding [3]

	Solid Parts	MuCell® Parts with 10% Weight Reduction
TechnylXCell S218 V35 XL, 35% glass-filled PA 6	100%	137%
Technyl A218V35, 35% glass-fiber-filled PA 6/6	100%	108%

Source: Reproduced with copyright permission of RAPRA [3].

the amount of molten material created in the welding zone. The suitability of the weld was determined from sectioning assemblies and from looking at the weld joint. A good weld does not have flash outside the walls of the part but does have contact the full width of the walls. The results of the optimization for the PA 6 shows a shorter weld time for the parts produced using the MuCell® process, 0.195 sec as compared to 0.275 sec for solid parts. Since the applied load and frequency were the same, this resulted in lower overall weld energies for the parts produced with the MuCell® process.

However, the short welding time for the microcellular part of PA 6 was not the case with the PA 6/6. The results of the welding optimization showed that the parts PA 6/6 produced with the MuCell® process and those produced with standard solid molding both required the same weld time, 0.200 sec. This resulted in equal weld energy.

The ultrasonically welded samples were evaluated in two separate test configurations: a pull test and a burst pressure test. The pull test is done using a tensile test system. In this case, the upper and lower flanges of the assembly are placed in fixtures, and the specimen is put in tension until the weld fails. The results of this testing in Table 10.2 show that the weld performance of the PA 6 was better than that of the PA 6/6. It also indicated that for PA 6, the parts molded using the MuCell® process had a higher average failure load than

TABLE 10.3 Burst Test Result from Ultrasonic Welding [3]

	Solid Parts	MuCell® Parts with 10% Weight Reduction
TechnylXCell S218 V35 XL, 35% glass-fiber-filled PA 6	100%	100%
Technyl A218V35, 35% glass-fiber-filled PA 6/6	100%	88%

Source: Reproduced with copyright permission of RAPRA [3].

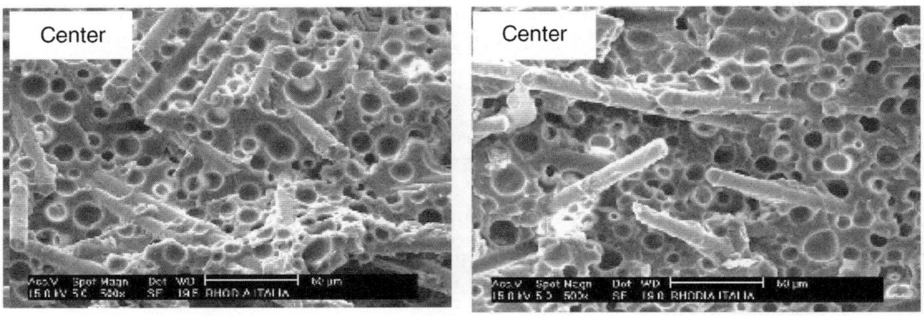

(a) (b)

Figure 10.2 SEM of foamed cores for PA 6 and PA 6/6. (**a**) PA 6 core [3]. (**b**) PA 6/6 core [3]. (Reproduced with copyright permission of RAPRA.)

did the solid parts, which is about 37% higher. Using PA 6/6, it was seen that the average results were the same between the solid parts and the parts from the MuCell® process. The difficulty with the pull test is that the standard deviation is quite large, meaning that there is statistically very little difference between the assemblies made from the solid parts and those from the MuCell® process. The data in Table 10.2 have an average of 8% higher value of the microcellular part compared to the solid part.

The data from the burst testing are of more value due to the repeatability of the data as well as the fact that it more readily simulates the in-use failure condition. The data in Table 10.3 show that the PA 6 samples produced from the MuCell® process had burst performance equal to that the solid samples. There were two observations with the PA 6/6. First, as with the pull testing, the burst pressures were lower than those for the PA 6. Second, there was a 12% decrease in the burst pressure for the MuCell® samples as compared to solid.

Analysis of the ultrasonically welded MuCell® sample shows that the cell structure in the core of the parts molded with both the PA 6 in Figure 10.2a and the PA 6/6 in Figure 10.2b is similar. Both have the cells of 20 μm or less.

In fact, the PA 6 samples have a skin layer that is about three times as thick of the skin layer of the PA 6/6 samples (see Figure 10.3). The skin for the PA

(a) (b)

Figure 10.3 SEM of foamed cores for PA 6 and PA 6/6 [3]. (**a**) PA 6 edge. (**b**) PA 6/6 edge [3]. (Reproduced with copyright permission of RAPRA.)

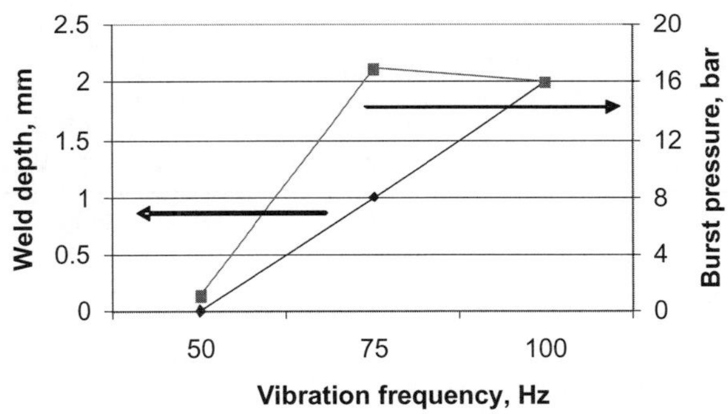

Figure 10.4 Effect of vibration frequency on PA 6/6 [3]. (Reproduced with copyright permission of RAPRA.)

6 samples is almost free of cell structure, as shown in Figure 10.3a. The thicker skin provides more material in the weld area and also, in some instances, can move the cell structure completely out of the weld area. However, the PA 6/6 sample has some cells in and near the skin area.

10.1.3.2 Vibration Welding. The vibration welding study was done using a bell mold. This mold produces a part such that two are put together to form a welded test specimen. The specimens were then tested for burst pressure.

A short study was conducted to evaluate welding conditions and weld depth for parts produced with the MuCell® process at 10% density reduction. It was found that the vibration frequency and weld time both contribute to the formation of a good weld. As vibration frequency was increased from 50 Hz to 75 Hz and to 100 Hz for a given weld time (6 sec), the weld depth increased from no weld formation to 2 mm (see the bottom line in Figure 10.4). The increasing weld depth from high frequency results in a stronger weld, which

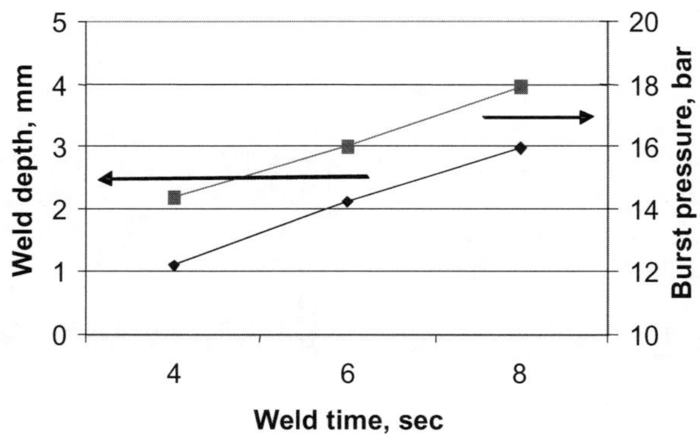

Figure 10.5 Effect of weld time on PA 6/6 [3]. (Reproduced with copyright permission of RAPRA.)

TABLE 10.4 Burst Test Result from Vibration Welding [3]

	Solid Parts	MuCell® Parts with 10% Weight Reduction
TechnylXCell S218 V35 XL, 35% glass-fiber-filled PA 6	100%	88%
Technyl A218V35, 35% glass-fiber-filled PA 6/6	100%	80%

Source: Reproduced with copyright permission of RAPRA [3].

is the top line shown in Figure 10.4. It was also seen that for a given vibration frequency, in this case 100 Hz, weld time effected the weld depth (see the bottom line in Figure 10.5), and therefore, the weld strength increased accordingly (see the top line in Figure 10.5).

The results from the weld study were then used to set welding conditions for the MuCell® samples for the burst testing comparison to the result of solid samples. Using the optimized conditions, it was seen that the PA 6 samples have a little change between the solid parts and the 10% density-reduced MuCell® parts. There is a 12% decrease of the MuCell® part in burst pressure as shown in Table 10.4. However, the difference between the solid parts and the MuCell® parts using the PA 6/6 at a 10% density reduction was 20%, which indicates that the MuCell® parts result in weak welding strength compared to solid parts with the vibration welding method, as shown in Table 10.4.

Looking at the cell structure in the test parts, it was seen that cell structure was into the weld area with both materials. But the cell size distribution was much better with the PA 6 than with the PA 6/6. The cell structure for the PA 6 in Figure 10.6a shows an average size below 20 μm and no cells greater than

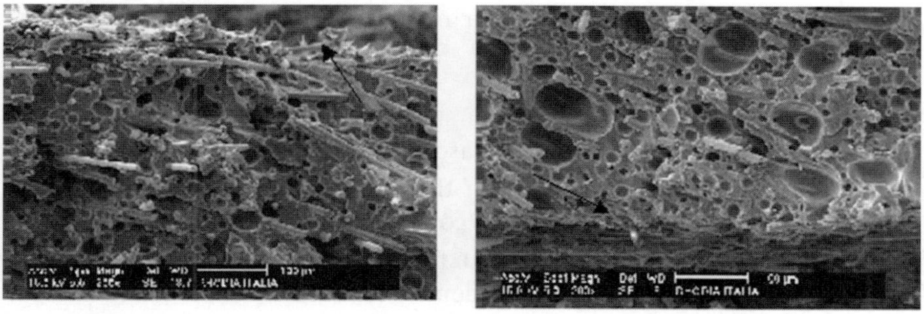

(a) (b)

Figure 10.6 Weld surface comparisons between PA 6 and PA 6/6. (**a**) PA 6 welding surface [3]. (**b**) PA 6/6 welding surface [3]. (Reproduced with copyright permission of RAPRA.)

50 µm. The weld surface is on the top of the sample in Figure 10.6a. The maximum cell size for the PA 6/6 in Figure 10.6b is 80–100 µm. The weld surface is at the bottom of the sample in Figure 10.6b. The maximum cell size in the weld area affects the weld strength, and the big cells in PA 6/6 resulted in weak welding strength as shown in Table 10.4.

In comparing the results of ultrasonic welding and vibration welding, it was seen that the MuCell® samples compared more favorably to the solid samples when assembled by ultrasonic welding. A contributing factor is the energy director. The geometry of the energy director is such that there will be very little cell structure in this area; and when properly designed, the energy should directly represent most of the material that is melted to form the weld in an ultrasonic welding process. Therefore, the material being melted to form the bond is free from cell structure.

In vibration welding, it is difficult to keep the cell structure from the welding area because these areas, by the nature of the process, are flat and relatively wide, about 3–4 mm in width. However, cell structure control can be achieved through control of the SCF level, injection speed, and process temperatures. While different grades of PA 6 and PA 6/6 will react differently to the microcellular foaming process, this typically means nitrogen levels of 0.2–0.3% with higher injection speeds, above 50 mm/sec. The glass-fiber-filled materials also tend to provide better cell structure than the mineral-filled materials.

10.1.3.3 Conclusions. Microcellular foam injection molding is no longer just a development program. Numerous under-the-hood applications in glass-fiber-filled PA 6 and PA 6/6 are commercializing. Some of these applications require either ultrasonic or vibration welding processes to join parts in the final assembly. The work done here looks at the effect of the MuCell® process on the welding process and weld strength. The results show that for a 10%

density reduction in the microcellular molded samples, the following statements hold:

- It is possible to achieve similar burst pressure results between solid parts and microcellular parts using ultrasonic welding with PA 6 materials.
- This is about a 12% decrease in the strength of microcellular PA 6/6 materials with an ultrasonic weld compared the solid materials.
- The drop in burst pressure strength for vibration welded assemblies using solid parts as compared to 10% density reduced MuCell® parts is larger than for ultrasonic welded samples: 12% for PA 6 and 20% for PA 6/6 because the cell structure is different after welding.
- Welding conditions do need to be optimized for microcellular foamed parts. An initial study showed that welding time and vibration frequency had significant effects on vibration weld strength.
- Skin thickness and cell structure in the weld areas affect the weld strength. The greater retention of weld strength with ultrasonically welded MuCell® parts reflects the lack of cell structure in the energy director. SCF level and injection speed are two contributing factors to cell size.

10.1.4 Experimental Results for Other Materials with Ultrasonic Welding

Branson Ultrasonics and Trexel worked together to have more materials using the same test mold that produces specimens for ultrasonic welding studies. These parts were produced in polycarbonate/ABS, 20% talc-filled polypropylene, 20% glass-fiber-reinforced polypropylene, 20% glass-fiber-reinforced PPO/PS, general-purpose polystyrene, and 33% glass-fiber-reinforced PA 6/6.

The mold was a two-cavity family mold that resulted in weight reduction differences between the two cavities. The larger part ran at a 15% weight reduction, and the smaller part ran at an 11% weight reduction. The molded parts were then ultrasonically welded. The parts were tested by putting the weld in tension and measuring the force required to fail the weld. A total of 10 test specimens were run for each material.

10.1.4.1 Experimental Results. Table 10.5 shows the results of the pull testing for the different materials with both solid and microcellular processed parts. The data indicate two general trends: Unfilled, amorphous materials and talc-filled polypropylene show a reduction in the weld strength of the microcellular part. However, the glass-fiber-reinforced materials show an increase in the weld strength of the microcellular part.

The observed improvement in the weld strength for glass-fiber-reinforced materials supports the results of welding studies conducted by other companies on air intake manifolds. In this study, air intake manifolds were produced by the standard injection molding process as well as by the MuCell® process.

TABLE 10.5 Test Result of Ultrasonic Welding Parts [3]

	Solid Parts	MuCell® Parts
Polycarbonate/ABS	490 lb	436 lb
GP Polystyrene	319 lb	214 lb
20% Talc-filled PP	186 lb	106 lb
20% Glass-fiber-filled PP	207 lb	223 lb
20% Glass-fiber-filled PPO/PS	744 lb	816 lb
33% Glass-fiber PA 6/6	568 lb	786 lb

Source: Reproduced with copyright permission of RAPRA [3].

The testing showed equivalent weld strength between parts molded as solid parts and those molded using the MuCell® process.

Further investigation in this area would involve the effects of welding conditions on weld line performance. In addition, welding time may influence the cell structure of the welding if the residual gas still in the cells that is true for N_2 gas unless the sample is completely degassed.

10.2 SURFACE POLISH AND PAINTING

Surface polish is only necessary for some class A surface is required for a microcellular part if the microcellular processing cannot completely remove the swirl marks. This is similar to the structural foam post-mold modification process. The works need to be done in two stages. One is to polish the part surface to remove the asperities and any open bubbles that lie just below the surface. Then, the painting on the polished surface of microcellular part can be carried out.

10.2.1 Surface Polish

An adequate sanding process to polish the part surface is time-consuming and expensive. The general approach is to sand the microcellular part surface with 100- to 150-grit open sandpaper to a depth of about 0.25 mm (0.01 in.). Generally, this is the maximum depth to be removed while polishing the microcellular part since the cell size is smaller and the rough surface is not as deep as the structural foam. The average roughness measured in the PC microcellular part is only 0.023 mm.

Another special technique that is useful to remove the rough surface of foamed part is vapor honing [6]. This technique is only good for some plastics readily to be attacked by certain solvents. For this application, the part is suspended in the pure vapor of the selected solvent. The solvent temperature will be varied depending on the process requirement. The vapor solvates the foam surface, essentially dissolving asperities and rupturing bubble surfaces [6].

10.2.2 Painting

To get the best quality of painting, there are several key factors to be checked before painting.

- The surface preparation not only includes polish to make the surface smooth but also makes sure that the de-gassing process is completed. The gas trapped beneath the paint layer will eventually cause a very small blister. It is particularly important for the microcellular part made with N_2 gas.
- A filler coat of compatible paint is prepared. It usually contains the relatively high filler loading (20–40 wt%). The typical fillers are talc, calcium carbonate, and silicates with particle size in the range of 1–10 μm [6].
- Select the coating that has compatibility with final paint. Also, the coatings may be selected with essentially the same coefficient of expansion as the foamed polymer. It is necessary if the coating is quite thick to avoid the differential interfacial stress that may lead to delamination in the future.
- After the paint coating dries the surface may need to be finish-sanded with 300-grit open sandpaper to a depth of less the 0.025 mm (0.001 in.).

There is a special paint or coating that has been developed for the foam part. It has "breathe" capability. It is porous or has a high permeability to the blowing gases in the plastic [6]. It then can be used for a less strictly de-gassed part. In addition, some polymer is known for difficulty to be coated with impermeable paint, such as ABS and HIPS.

10.3 POST-COOLING

Post-cooling may be necessary for some thick microcellular part. There are different ways to continually cool the microcellular part if it cannot cool completely in the mold, but the skin thickness is thick enough to keep the stiffness of the part to be ejected in the mold and to maintain the shape and dimensions of the part. Then, a cooling conveyor may be used for this post-mold cooling requirement. For less critical parts, or when only minimal cooling is required, an ambient air-cooling unit is good enough. For the faster post-mold cooling, using a chilled air-cooling unit will deliver a regulated flow of chilled air over the part as they pass under the unit.

For extremely fast cooling of a microcellular part, chilled water may need to cool the part outside the mold before the microcellular part completely solidifies without any possible distortion, or post blow from residual gas pressure in the cells.

Sometimes a special fixture is necessary to hold the microcellular part in the positions during post-cooling so that it will not change the final dimensions and shapes after completing the post-cooling process. However, it is an expensive method and is only used for the important part with economic estimation between longer cooling in the mold or post-cooling in the fixture, considering the fact that the machine time is available for more microcellular parts when post-cooling can be carried out.

10.4 DE-GASSING PROCESS

The de-gassing process may take several days, or weeks, to allow the gas to go completely out of the microcellular foamed part. It depends on several factors:

- The thickness of the microcellular part
- The materials of microcellular and the type of gases
- The processing condition that determines the final residual pressure in the cells
- The temperature of the de-gas environment may result in a different de-gassing rate in the microcellular part

It is obvious that the thick part may take a longer time to allow the gas to be completely released. In addition, some plastic materials can retain the blowing gases for extended periods of time. On the other hand, the gases will have a different diffusion rate; for example, CO_2 will be faster diffused out from the molded part than the N_2 gas. Also, the processing conditions may leave different residual gas pressure in the cells after cooling, which is also a factor that influences the de-gassing time.

It is important for the second process, such as putting a label on the microcellular surface. If the de-gassing is not completed, the label will be expanded to separate from the surface. An experiment was carried out at Trexel to test the minimum days to completely de-gass a PBT part made for the printer cartridge that needs to add labels after molding. The results were about more than 2 weeks to de-gas this sample for proper labeling. As the rule of thumb, the N_2 gas in a microcellular part will take weeks to completely be de-gassed. However, CO_2 gas in (a) microcellular part may take days to finish the de-gassing process.

10.5 PROPERTY TEST FOR MICROCELLULAR PARTS

There are no complete standard microcellular material test procedures in industry yet. Since the properties of microcellular materials are between the properties of solid material and the properties of structural foam material, the test procedures can follow the existing standards of both structural foam and solid materials. Here are the several common test procedures already used for microcellular material tests carried out by Trexel Inc. and others [6–8].

10.5.1 Impact Test

This test is performed on discs sectioned from the molded plaques. The sectioning procedure should not have any effect on the results becuase the plaques are impacted at the center. However, the samples made for the impact test need to have a representative cell structure of the microcellular part because the impact strength is strongly related to cell structure. The testing procedure was carried out in compliance with ASTM D3763-95A.

10.5.2 Tensile Test

The tensile test is performed on tensile bars sectioned from the molded microcellular plaques. The good repeatability of the test data and the small standard deviations indicate that exposing the core structure by sectioning the samples from the foamed plaques did not have a significant effect on the results. From the experience of microcellular tensile strength, the skin thickness variations may be the factor that influences the tensile strength the most. The testing procedure was performed in accordance with ASTM D790 guidelines.

10.5.3 Flexural Test

The flexural test is a comparative test (foam versus solid) performed on the molded 6 × 8 in. microcellular plaques. The flexural test uses ASTM D790 as a guide and employs the three-point bend test mode on the MTS Universal Testing machine. It shows that the skin thickness is also important for the better flexural strength of a foamed part because the test that is carried out is the bending test; this creates the maximum stress on the skin, instead of on the foamed core.

10.5.4 Dynamic Mechanical Analysis

This dynamic mechanical analysis test is carried out in order to compare the mechanical properties of solid with foamed parts as a function of temperature (temperature sweep from –50 °C to softening point). The DMA testing is carried out as per ASTM D4065.

10.5.5 Low-Shear Rheology

The low-shear rheology test is conducted on the solid, foamed plaques (with highest weight reduction) as well as on pellets of the virgin resin in order to assess the degree of degradation and evaluate whether the microcellular process affects the rheological properties of the material. The low-shear rheology testing is run on a parallel-plate rheometer manufactured by Rheometric Scientific Inc. in compliance with the ASTM D4440 method.

10.5.6 Thermal Test

The thermal test is used to measure the thermal conductivity of a microcellular foam. The thermal conductivity of the solid and foamed samples is measured using the Holometric Inc. TCA100 machine in accordance with the ASTM E1530 method. It is obvious that the thermal properties of microcellular foam are different from those of the solid.

10.5.7 Shrinkage Test

A comparison between solid and foamed samples is conducted using the CMM measurement system on four holes located at the corners of the molded plaques. The cross-flow shrinkage values are only reported in this work, since they show the highest discrimination and confidence level. For each material, five plaques are tested and the samples were conditioned at 23 °C and 50% R.H. for 48 hr.

10.5.8 Flammability Test

The flammability test for a microcellular part is similar to the solid part test. The 5V device test is performed on the solid and foamed plaques. A 5-in. flame is applied five times at a 20° angle for a period of 5 sec with a 5-sec interval between each application. The tested plaques are placed horizontally, and the flame was applied at the center of the specimen. For each material, three plaques are tested and the samples are conditioned at 23 °C and 50% R.H. for 48 hr. *Passing Criteria*: The flame should self-extinguish within 1 min and the plaque should not drip onto and ignite a surgical cotton placed underneath it.

10.5.9 Acoustical Test

The ASTM C-1050 test method is employed to determine the normal acoustical absorption coefficient of a microcellular foam.

10.5.10 Density Test

The density test of unfoamed and foamed samples are measured with a water displacement technique in accordance with ASTM D792 method. It will use the equation

$$\text{Foam density} = 0.9975\,(W_a/W_w) \tag{10.1}$$

where W_a is the weight of the sample measured in the air and W_w is the weight of the sample measured in the distilled water. Then, a void fraction can be calculated from the result of Equation (10.1). It will be [6]

$$\text{Void fraction} = 1 - (\text{foam density}/\text{solid density}) \tag{10.2}$$

There will be more test methods specified for microcellular material because it has more potential benefits developed for the application and needs to be standardized in the plastic industry. Most of test procedures are adapted from solid material because the microcellular material is used mostly to replace the solid material application.

REFERENCES

1. Schwartz, S. S., and Goodman, S. H. *Plastics Materials and Processes*, Van Nostrand Reinhold Company, New York, 1982.
2. Witzler, S. *Injection Molding Mag.* **April** 29 (2009).
3. Kishbaugh, L., Kolshorn, U., and Bradley, G. Vibration and ultrasonic welding conditions and performance for glass fibre filled PA 6 and PA 6.6 injection moulded using the MuCell® microcellular foaming process, Smithers Rapra Blowing Agents and Foaming Processes Conference, RAPRA, May 2007.
4. Kirkland, T. *Machine Design* **April**, 106–107 (2005).
5. Dukane Web Site, www.dukcorp.com.
6. Throne, J. L. *Thermoplastic Foams*, Sherwood Publishers, Hertford, UK, 1996, pp. 355–357.
7. Trexel Web Site, www.trexel.com.
8. Grellmann, W., and Seidler, S. *Plastics Testing*, Hanser/Gardner Publications, Cincinnati, 2007.

11

MARKETS AND APPLICATIONS OF MICROCELLULAR INJECTION MOLDING

The market analyses for the microcellular injection molding products are presented with both economic and performance benefits. In addition, several typical case studies of the parts successfully made from microcellular injection molding are discussed. The potential applications and research directions are given for the future developments of microcellular injection molding.

11.1 MARKET ANALYSES FOR MICROCELLULAR INJECTION MOLDING PRODUCTS

Microcellular foaming technology was originally conceptualized and invented at the Massachusetts Institute of Technology (MIT) [1, 2] by the market requirements for the material savings of plastic parts. In 1995, Trexel was granted an exclusive worldwide license for the further development and commercialization of the technology. It was named MuCell® as a registered trade name for microcellular foam technology.

The MuCell® microcellular Foam technology is a complete process and equipment technology that enables the production of great high-quality plastic parts. MuCell® technology uses precisely metered quantities of atmospheric gases (nitrogen or carbon dioxide) in any of the three most common thermoplastic conversion processes (injection molding, extrusion, and blow molding) to create millions of nearly invisible microcells in the end product. The fastest and widest usage of MuCell® microcellular foam technology now is injection

Microcellular Injection Molding, Edited by Jingyi Xu
Copyright © 2010 John Wiley & Sons, Inc.

molding. Today there are hundreds of MuCell® parts, most for injection-molded parts in commercial production today around the world in excess of 300 injection molding machines in operation. Examples of MuCell® products include electrical components, electronics connectors, internal business equipments and printer components, a variety of packaging applications, and a broad array of automotive products including HVAC components.

The MuCell® injection molding process is primarily employed in the injection molding process to produce lower-cost precision parts with a consistently high quality and exceptional dimensional stability, where foaming has not historically been deployed [3]. The MuCell® injection molding process has become the leader in microcellular injection molding since 1998. In addition, microcellular foam has been successfully used for glass-fiber-reinforced materials, which brings a unique benefit of fiber disorientation in the foamed core so that an anisotropy problem in glass-fiber-reinforced material has been greatly solved with the microcellular process.

Generally, the microcellular process is significantly different from structural foam based on the cell structure. The small cell size in microcellular foam enables us to use this technology in the thin-wall part applications. Therefore, theoretically, all regular injection-molded parts are capable of being converted for the microcellular process with more or less modification of part and mold designs. The major benefits of the microcellular parts analyzed below are sufficient to drive industry to invest in injection molding for the microcellular process. On the other hand, more potential applications of microcellular parts either for the developing stages or for possible future projects will open more markets for microcellular injection-molding parts.

11.1.1 Low-Cost Products

Generally, microcellular parts are the low-cost products that will be discussed further in Chapter 12. The low-cost product is an important factor in marketing microcellular technology. The uniform internal gas pressure of the microscopic cells provides the cavity pressure needed for final filling, instead of high packing pressure of regular injection molding. This low cavity pressure can be reduced as high as 80% versus conventional molding. On the other hand, the low injection pressure requirement is another obvious benefit of this microcellular process. Then, the low clamp tonnage and injection pressure are the savings of low operation energy and low cost of equipments.

The cycle time of microcellular injection molding can be reduced from 15% up to 50%, and it is the greatest savings in the microcellular process. It is the result of many factors that are important to be considered as the factors of application of the microcellular process:

- The pack-and-hold time is almost zero.
- A reduction in polymer material is about 15%, which reduces the material to be cooled before ejecting.

- Uniform distribution of cells leads to improved dimensional stability and shorter cooling time.
- Fast injection speed results in nucleation and great energy releasing through injection, and it helps to cool material down quickly as well.
- Melt viscosity reductions (typically 20–60%) are well known for gas-laden melt, which results in less shear heating in the mold filling. The low shearing heat leads to shorter cooling time. The low-viscosity material from the gas-laden melt provides the ability to flow into longer and thinner wall sections.

At the beginning of the commercialization for microcellular injection molding, the weight reduction is the major market value. The realistic weight reduction from microcellular part is about 5–15% without significant mechanical property changes of the solid part. This is also an attractive target for the expensive engineering material savings, medical material, and biopolymer. However, given the cost of some cheap materials, such as polyolefin, weight savings is rarely sufficient to drive industry to invest in injection molding for the microcellular process. The applications for the general-purpose materials are those where quality issues can be resolved substantially, cycle time can be reduced, or high-cost material can be converted to the polyolefin microcellular foam.

11.1.2 High-Quality Products

The high-quality product is another important feature that was not recognized at the beginning of commercialization of this technology. The gas swirl surface of the microcellular part is usually the first impression of this product that misleads industry people about the poor surface quality of foamed parts without studying further about the other quality benefits of microcellular parts. The good-quality features of microcellular parts are from several factors, such as low pressure, cell growth to take over shrinkage, inert gas in the molten polymer, and so on.

The inert gas in the molten polymer is good to protect (a) the degrading of plastics during screw recovery and (b) fast injection speed during mold filling. It is well known that the inert gas in the molten polymer greatly lubricates the molten polymer to yield low-viscosity gas-laden material. However, another neglected fact is that the inert gas may protect against further degradation of material because of mechanical shearing. At least for unfilled material the extra high shearing rate in the gate does not cause too much degradation of material because of the combination of effects from both low viscosity and inert gas protection.

The low-pressure molding from the microcellular process is especially beneficial in the insert molding process because the low pressure inside of the mold will not wash the insert away or change its original position.

Similarly, this low cavity pressure will also be good for the in-mold decorating, in-mold labeling, and encapsulation in the prevention deformation and bleed-through.

The pack-and-hold pressure, a major cycle-time limitation and the source of molded-in residual stress in the parts made by conventional injection molding, is eliminated in the microcellular foam process. It is replaced by the cell growth as part of a key process in microcellular injection molding that dramatically solved the quality problems of regular injection molding parts. The quality issues of regular injection molding parts can fall into a number of categories, perhaps the most common being associated with sink marks and dimensional stability. Dimensional stability is also the primary driver behind excessively long cycle times and the use of higher-cost materials [4].

Dimensional compliance is a recognized problem for polyolefin. This is particularly true in long, unsupported walls and flatness. This situation often can be improved with the use the fillers. However, filled material is still difficult to completely eliminate warpage. One way to achieve less warpage part is to extend cycle time beyond what is necessary to cool the part by allowing the mold to serve as a shrink fixture, which is too expensive a way to solve the problem. The microcellular injection molding process, by significantly reducing molded-in stress as well as uneven pack pressures, allows parts to be produced with improved dimension while eliminating the extended cooling time.

Warpage can also be eliminated from using higher-cost material, typically an amorphous resin, either filled or unfilled. However, the microcellular process brings another solution. Where dimensional compliance is the sole selection criterion, considerable cost savings can be gained by simply switching to a polyolefin, such as polypropylene, applying microcellular injection molding.

The application of microcellular processing is also in another quality area: molding behind decorative films and fabrics. The microcellular process allows for a wider range of fabrics and less expensive fabrics because processing temperature can be easily reduced by 20–30 °C, and cavity pressure can be reduced by 75–80%.

The increased toughness of microcellular foam will allow this material to be used in some application that requires impact toughness and energy absorption. For example, the packaging industry will need some tough material to be used for thin-wall parts. Also, the thermal insulation property of microcellular foam will be good for this application as well.

11.1.3 Unique Features of Microcellular Structure

Additional market values of microcellular foam are its sound and thermal insulation properties. The sound insulation may be important for the automotive and entertainment industries. On the other hand, thermal insulation is also of interest to the automotive and other industries.

Microcellular foam used for glass-fiber-reinforced materials contributes a unique benefit of fiber disorientation in the foamed core. If the anisotropy problem in glass-fiber-reinforced material is not tolerated for the finished part, the microcellular process is the possible solution to solve the problem with additional benefits for cycle-time reduction, weight reduction, and operation savings. Hence, the microcellular technology has been successfully applied in the parts made by the short and long glass-fiber-reinforced materials.

11.1.4 Green Products

The inert gas, such as N_2 and CO_2, is truly a green product. As the blowing agent, inert gas will not generate any residual byproduct in the microcellular part that is critical for a green part. Then, the microcellular part made by inert gas is good for recycling programs, and it has no environmental issue of production. As the clean blowing agents used for microcellular parts, these are the solutions of the package productions for food, beverage, and medical application parts.

11.1.5 Others

One more market that has been opened for the microcellular part is injection blow molding. It provides the microcellular foam for the PET bottles with the following benefits:

- It reduces weight by up to 5%.
- The bottles with microcellular parts have significant light barrier characteristics that contribute up to 95% reduction of transmitted light.
- It recycles the clean PET waste with white or silver color from foaming process without adding color concentrates.
- Clean blowing agents, such as CO_2 and N_2 gases, provide the solutions of package production for food, beverage, and medical application parts.
- The foamed surface of containers has a unique surface feel and provides tactile "traction" that minimizes slipping.
- Foamed bottles can withstand hot filling without excessive shrinkage.

There are many special microcellular applications, such as local foam, supermicrocellular foam, overmolding, and low-viscosity material just for easy mold filling and dimension stability without foam. They may have even more potential for the future markets than for the microcellular foam itself.

As a summary, the microcellular process is a proprietary foaming technology that allows the benefits of tradition foaming processes to be applied to the conventional molding markets. The major benefits are the cycle-time reduction, dimensional stability, and weight reduction. For the general-purpose material, weight reduction may not be sufficient economic justification for using the microcellular process, but quality issue and cycle-time savings may be the significant reasons to apply microcellular processes for all materials. Small cells of microcellular parts meet the requirement for regular injection molding market except for the surface quality. Once the surface issue of microcellular part is solved with an economic solution, the microcellular injection molding will be the revolutionary change to cover all markets of the injection molding industry.

11.2 TYPICAL APPLICATIONS: CASE STUDIES

The application of the microcellular process in injection molding industry has been widened with many advances. The typical breakthrough technologies of microcellular injection molding are appearance parts, precise gas dosing control system for small microcellular parts. Also, more and more microcellular processing experts worldwide are the real pushing force for the application of this technologies. The typical application can be discussed in different fields as follows.

11.2.1 Thin-Wall Parts

The thin-wall microcellular process was developed in the initial trial for good microcellular surface and warpage-free part. It is usually has the wall thickness of only 0.5 mm or less. The method filled the mold almost 100% with high injection speed. The result is the same quality of surface as the glass-fiber-reinforced material, along with very nice cell architecture that shows almost 5 μm or less of diameter for all cells with spherical shapes. In addition, all cells are distributed uniformly throughout the part.

The typical application is called Super Light Injection Molding, or SLIM® [5]. SLIM® technology utilizes the MuCell® process (see Figure 11.1), which involves the use of precisely metered quantities of atmospheric gases (nitrogen or carbon dioxide) in the injection molding process to create millions of almost invisible microcells in the finished product. The microcells replace their equivalent volume of plastic, resulting in a reduction of up to 10% in packaging weight without any perceptible difference in the final tub quality.

Veriplast Solutions' objective is the development of innovative packaging solutions, focusing particularly on consumer convenience, food safety, and carbon footprint optimization. Veriplast successfully uses microcellular technology to make its packaging part more competitive. Veriplast combines the microcellular technology with its own Extra Slim Label technology, which is significantly thinner than the market standard and provides an additional environmental benefit by reducing the CO_2 footprint by 30% compared to standard labels.

Similar to the thin-wall container, MuCell® microcellular injection molding technology cuts the finished part weight up to 4–5%, 30% less clamp tonnage,

Figure 11.1 The Rama 500 g of margarine tub made by SLIM technology (Courtesy of Trexel).

and 10% less injection pressure. This microcellular thin-wall technology runs the molding in a small tonnage machine that is impossible for running regular thin-wall molding because of the lack of either the large tonnage requirement or high injection pressure.

A thin-wall study is also discussed in Okamoto's book. In the study the 0.5-mm thin-wall part has 10% weight reduction, with 27% of cycle-time reduction, 36% clamp tonnage reduction, and 20% injection pressure reduction [6]. It is interesting that the cooling time is reduced as well for this type of thin-wall part with microcellular processing [6]. This application was initially run with the issue of mold filling. The existing mold that is produced in a regular process simply cannot fill the corner of the part. The trial was successfully uses microcellular technology not only to fill the mold perfectly with low tonnage but also to reduce the weight percentage with super-microcellular cell structure and nice smooth surface as well. The cell size is about 3–5 μm and is distributed uniformly across the thickness of the thin-wall part [6].

The thin-wall part has a very smooth natural white color surface because the small cell size and uniform cell distribution does not have the traditional swirl of the foamed part. It may be from the contribution of extremely small cell sizes. The thin-wall part was successfully made from PP material. It exhibits very good surfaces, and the swirls tend not to be present or noticeable, particularly with lighter colors. It has been tested for printing quality, and it has a very high surface energy rating (dyne level) and improves printability.

However, the basic requirement for the thin-wall microcellular part is extremely high injection volume rate to fill the mold as soon as possible. This critical injection time is preferred to be within 0.5 sec. If the injection cannot finish in time, the cells will be stretched severely and the cell size will vary from 20–100 μm [7]. Although the cell size in this experiment is still the same as that of a regular microcellular cell structure, the cell size of 100 μm is almost 20% of the total wall thickness of 0.5 mm, so the part properties will be reduced significantly. On the other hand, most failures of thin-wall microcellular injection molding result from the slow acceleration of injection speed, not total injection time. Sometimes the successful thin-wall injection molding is determined by an important parameter of ramping time in addition to the total injection time. It must be ramping as quickly as 0.1 sec or less to avoid the freezing of flow channel because the thin-wall part has a high flow ratio of usually 200 or more. Therefore, even with low-viscosity gas-laden melt for less resistance of mold filling, the biggest challenge for microcellular molding in thin-wall processing is to fill the mold before flow channel freezing. On the other hand, the venting needs to be checked to be able to vent the gas and air in time when the high injection volume rate is used during injection. The interesting morphology of thin-wall molding at extremely high injection volume rate shows not much cell stretching but not the perfect spherical shape either [7].

SLIM® technology provides outstanding benefits in terms of packaging weight reduction and carbon footprint reduction. As a result, SLIM® customers in Europe can save additional money by reducing Eco Tax costs. For example, in Germany for a standard 15-g tub, the 10% weight reduction obtained because of using SLIM® technology is a €200,000 savings in Eco Tax for every 100 million tubs.

The supercell architecture benefits the good insulation properties of microcellular parts for this thin-wall packaging part. It is because of the huge number of micro-close cells in the part with inherently low thermal conductivity.

The Unilever 500-g Rama margarine tub has been awarded the 2008 Deutscher Verpackungs Preis (German Packaging Award) and a WorldStar Packaging Award. The innovative tub design was developed by Veriplast Solutions and uses Veriplast's super-light injection molding technology (SLIM®), which combines the MuCell® microcellular foaming process along with Veriplast's Extra Slim Label, an innovative down-gauged in-mold label.

11.2.2 Automotive Parts

Microcellular injection molding has been used from the world's leading automotive component manufacturers with many successful cases. It is applied as a core plastics manufacturing technology to achieve higher productivity level, vehicle weight reduction for energy efficiency, quality improvements from dimension stability, and significant cost savings [8]. It is proved that the microcellular parts used in automotive components are lighter, flatter, straighter, and more dimensionally stable at extreme operation temperature compared to the regular injection-molded solid parts. The typical application of microcellular parts in the automotive industry is within an extensive range of vehicle systems. For example, the typical interior system parts include instrument panel bezels, retainers, pillars door trims, topper pads, in-mold laminated/decorated articles, overmolded components, mirror brackets, air bag covers, speaker housing, and so on. The typical electrical/electronic systems are power distribution centers, fuse boxes, encapsulated products, housings, wire channels, switches, and connectors. Also, the thermal and energy system and power-train system will include HVAC valves, housings, air cleaner housings, evaporator cases, air flow sensors, air intake manifolds, water inlets, gaskets, fans, shrouds, air distribution housings, canisters, throttle bodies, wheel covers, and radiator end-tanks [8]. With the trends indicating increased resin prices and with growing pressure to reduce overall vehicle weight to meet new higher fuel efficiency standards, the automotive industry will get on board to use more microcellular parts. Some typical parts will be discussed in more detail in the case studies below.

1. Xcell Parts. Rhodia North America is now optimized for use with microcellular (MuCell®) processing technology. It is named Xcell, and these grades are adapted from Rhodia of Technyl Star Pas. These materials will serve in underhood applications such as rocker covers and air intake manifolds. The recent pre-production molding trials show that the Technyl Xcell 6/6 and 6 grades have good mechanical property performance with no compromise in the part's surface finish. It is true low melt viscosity tailored for microcellular processing because it has both (a) low viscosity from the material itself and (b) even low viscosity with gas-laden molten polymer. The higher flow rates create a laminar flow through the mold, coating the cavity walls with a skin that freezes off, preventing any gas from breaking through the surface [9]. It is 10% less dense and can mold thin-walled parts with less injection pressure. It results in less molded-in stress and good rigidity and can withstand high temperatures and impacts. This high-surface-aesthetic part like an engine cover benefits from the weight and warpage reduction provided by microcellular parts, without having an affect on appearance [10].

One of the parts is the radiator end tank made by PA6/6 GF30 from QSC Quality System Control Corporation. The equipment is the Arburg 320-ton injection molding machine, with a 70-mm screw. The conventional molded part often has warpage issues due to the in-molded stress and fiber orientation. Microcellular eliminates inner stress and reduces the fiber orientation. This results in improved dimensional conformity and flatness. A special resin type from Rhodia provides an excellent appearance with microcellular foam in the core.

The economic benefits are as follows:

- Cycle-time reduction about 32%
- Weight reduction about 10% without heat distortion temperature (HDT) change
- Weight reduction about 10% without notched impact change
- Smaller machine requirement savings about 50%

The quality benefits are as follows:

- Warpage reduction about 60%
- Excellent appearance with Rhodia Technyl® XCell™
- Greater design flexibility because of the low viscosity and stable dimension
- Fewer mold interactions and fewer conformance issues due to part dimensions

2. Fan Shroud (GGS Plastic Engineering) [4]. This is made by PAG GF15 MN20 material. The equipment is a Nissei 1000-ton injection molding machine with a 112-mm screw. The part weighs 1160 g, and the hinge has the weight of 16 g. With the microcellular (MuCell®) process, both parts are made in the same mold. An additional U.S. $40,000 savings is from a two-cavity mold and production cost has been saved as well.

The economic benefits are as follows:

- Weight reduction about 8%
- Cycle-time savings about 22%
- Reduction in machine size about 30% saving

The performance benefits are as follows:

- Flatness improved by 50%
- Fatigue-to-failure improved by 400%

3. Thorax Airbag Cover Rear [4]. This is the part made by PP/EPDM material from TRW Automotive Safety System GmbH. The equipment is an Engel 300-ton injection molding machine with a 55-mm screw.

The air bag cover has a complex part geometry with thin-wall sections down to 0.4 mm at the initial breaking line for the air bag release and up to 2.5 mm of nominal wall thickness. The microcellular process allows producing such parts with uniform shrinkage, and the parts are free of sink marks.

The economic benefits are to reduce the machine size by 40% and cycle-time reduction. Quality benefits include elimination of different material shrinkage and sink marks. It has excellent dimensional stability.

4. Jounce Bumper [4]. This is a part made by TPU in a 100-ton vertical press with a 40-mm screw. The molder is by Barre Thomas. It is an anti-vibration rubber and metal component for the automotive industry. To complete rubber solutions, Barre Thomas has developed and patented a jounce bumper in TPU by using microcellular processing.

The economic benefit is cost reduction up to 20% compared to the two-component PUR process.

The quality benefit is full material recycability.

5. Mass Air Flow Meter [4]. This is an intricate component in the car. It is used to control the engine air–fuel mixture and is made with 30% glass-fiber-reinforced PBT. It provides substantial benefits over a conventional injection-molded version. These benefits are listed below.

The economic benefits are as follows:

- Significant cycle-time reduction
- A 10–15% weight reduction within the range of material strength
- A substantial reduction in the peak clamp tonnage requirement because of the low cavity pressure and low mold temperature of microcellular foam
- Estimated direct cost savings about 10%

The quality benefits are as follows:

- Greater dimensional stability at elevated temperature from –4 °C to 93 °C
- Microcellular molded part eliminating the cost of numerous tooling modifications typically needed to profile the mold to achieve proper dimensions
- Significantly improved as-molded dimensions where the conventional-molded solid part was two to five degrees off parallel; however, the microcellular injection-molded part was only less than 0.5 degree off parallel [6]
- Easier assembly of flow tube and carrier due to less warpage/improved consistency

This microcellular part was made by Delphi and was a 2002 SPE Automotive Innovation Award Finalist.

6. Connector of Automotive Traction Control Unit [4]. It is made from the 15% glass-fiber-reinforced PBT. The conventional molded connector wall exhibited up to 1.1-mm warpage. Using microcellular process, the warpage of the connector is reduced to 0.27, which that is well within the specified dimensional tolerance for this part.

This microcellular connector also reduces the bowing by 75% versus conventional molding. It is because the molded-in stress is almost eliminated through the microcellular process.

7. In-Mold Decorated "A" Pillar [4]. This in-mold decorated "A" pillar has a fabric decorating on one side of the part. It has two issues when the regular solid injection molding is used. One is the fabric to be washed away during injection, and another is that two gates must be used to fill the whole length of the mold, which causes the welding line in the middle of the part. Microcellular process solves both problems. First, the low cavity pressure is not strong enough to change the position of fabric during the mold filling.

Also, the low-viscosity gas-laden material is easy to fill the mold, where only one gate is necessary for microcellular processing. Therefore, the microcellular pillar not only solves the problems of solid processing for this part but also brings more benefits. For example, one of the immediate part improvements is to design the pillar 25% thinner to take the advantage of low-viscosity gas-laden material. More benefits from microcellular processing are discussed below.

The economic benefits are as follows:

- Melt temperature reduced from 271 °C to 249 °C
- Weight reduction about 10%
- Injection pressure reduction about 28%
- Sequential valve gate removed because only one gate is used
- No bleed-through needed since low cavity pressure and melt temperature

The quality benefits are as follows:

- Warpage- and sink mark-free
- No weld line in the part

8. A Mirror Bracket. The automotive mirror bracket has been used successfully in microcellular processing technology. It has 28% weight reduction and 50% of cycle time (from 51 sec to 25 sec) without significant mechanical strength drop because it passes the physical property tests. It is made from mineral/glass-fiber-filled PA 6/6.

A return on investment (ROI) on this mirror bracket showed a $0.72 per part savings total. Actually, $0.13 savings per part is due to the reduced weight, and another $0.59 savings per part is from the reduced cycle time. Payback for using this microcellular injection molding technology is approximately 2 months [11].

9. HVAC Parts [12]. The HVAC unit includes some precise parts with tight tolerances. One of them is the evaporator that uses 40% talc-filled polypropylene. The small tonnage machine is used for microcellular processing. With the same mold, the solid part needs an 850-ton injection molding machine to make a good-quality part. However, the microcellular part with the same geometry as the solid part is running only 650 tons of clamp force for better dimensional quality of the parts. It is also demonstrating a 35% cycle-time reduction over conventional molding. The weight of the part is reduced from 145 g to 118 g, which is about 18.6% weight reduction. Also, the tight dimensional tolerances have been reached with the microcellular process, but it was difficult to satisfy the same tolerances with solid processing.

TABLE 11.1 Standard Deviations of Dimensions Related to Different Weight Reductions (30 Piece Samples for Each Data)

	Solid Parts	MuCell® Parts with 5.2% Weight Reduction	MuCell® Parts with 9.2% Weight Reduction
Weight variation (%)	0.019	0.029	0.032
Length dimension variation (%)	0.019	0.012	0.011
Width dimension variation (%)	0.017	0.009	0.009

Source: Courtesy of Trexel, Inc.

Another precise part with tight tolerances is the flat door where the microcellular process made it possible to switch from a 40% glass- and mineral-filled PA 6/6 to a PA compound costing two-thirds as much and based partly on recycled material.

10. Corporate Adaptor. This is the part used for an automatically adjustable outside mirror. The material is Delrin 588P-BK 602 (POM) from DuPont. It has been successfully molded with microcellular processing, having the benefits of the weight and cycle-time reductions. It has up to 16% weight reduction and an 8% cycle-time reduction with a two-cavity mold.

The unique benefit of this part is actually the dimension stability. It is a flat part with width 71.5 mm and length 105.9 mm. The wall thickness is 2 mm with total weight 23.85 g. A dimensional variation study has been carried out to compare the results between microcellular and solid parts. The results are listed in Table 11.1. The dimensional variations in both width and length directions are less than the related dimensional variations of solid parts, and even the weight variation of the microcellular part is higher than the weight variation of the solid part. This is the indication that the overall precision of the microcellular part is much better than the precision of the solid part.

11. Dolphin Skin Technology [13]. A patent pending new technology using microcellular processing combined with overmolding, reversal coining, and using new materials has been developed by Engel, BASF, P-Group, and Trexel. These suppliers team up to bring this new design freedom with less cost to the automotive market, and it was named Dolphin Skin technology. It has the potential to be used for high-quality overmolded components for automobiles, such as instrument panels, center consoles, and glove compartments. It also provides more possible applications for either (a) a soft touch surface with a reinforced back support (b) or a smooth surface with a foamed back layer to remove sink marks. The thermal and sound insulation properties are also important for the potential applications.

The first sample was a passenger car dashboard panel with a soft-touch surface (foam with reversal coining after skin is truly formed with overmolding method), combined with a strong glass-fiber-reinforced PBT/ASA underneath the soft-touch surface. The nonfoamed PBT/ASA reinforced material not only forms a perfect smooth skin surface like that of a solid part but also provides a support underneath foamed layer. The picture of the part morphology is shown in Figure 8.21. The leftmost picture in Figure 8.21 shows the glass-fiber-reinforced solid carrier underneath the foam layer. Then, an interface exists between the foamed layer and the solid carrier, where nice skin forms from the foamed side to contact and bond with solid carrier layer. The mold then cracks open, giving the space for the foaming; thus a cell structure is clearly shown in the foamed core. This is a more controllable foaming process. The rightmost picture in Figure 8.21 shows the interface between the foam and the solid skin, which creates the soft touch surface that is like dolphin skin; this is the part named to emphasize the feature of soft touch surface. It is the requirement for comfortable touch inside of a car, or similar request from other parts. The TPE-E is a special foamable material provided by P-Group with trade name Pibiflex, which is part of a pending patent of this technology. The reversal coining occurs after the forming of solid skin of foamed material, so the surface quality of foamed material is excellent as well. Thus another important feature of this technology is that the foamed soft touch surface is good enough for commercial application. In other words, the foam side can be used as outside surface as well. On the other hand, the other side of this part is smooth surface without sink marks because the cell expansion from the foamed back layer will take care of any dimensional defects. The foamed core made by reversal coining makes foam controllable to be either microcellular or even big cells, thereby making the soft surface even softer beyond the microcellular cell definition. In addition, the foam is uniformly distributed throughout the part because gas-laden material was injected into the mold, 100% filling the mold cavity first just like regular injection molding. Then, the mold cracks open to allow foam to occur without any voids or cell distribution problem. It is also critical for the part with uniform properties that may maintain more average overall strength than the part made by current standard microcellular injection molding.

The first stage of the molding is actually the regular injection molding with a glass-fiber-reinforced PBT/ASA blend (Ultradur S4090 IGX from BASF). Then, the first layer is overmolded in the second stage with the special polyester using microcellular (either MuCell® or chemical blowing agent) process. It gives the flexibility of application with a different foaming technology, either MuCell® or chemical blowing agent. Therefore, the process itself opens the possibilities to have either (a) a perfect surface quality including dimension stability or (b) a foamed surface with acceptable surface quality but much stronger strength because the solid back layer. With the Dolphin process, the possibility opened up for the appearance of a part that is the neck of a bottle for application of microcellular processing. In other words, Dolphin technology offers the potential for use in not only automotive interior applications but also for furniture and sporting goods markets.

The surface skin of foamed co-polyester via the MuCell® microcellular process in Dolphin technology replaces the polyurethane that is normally used in multistage (the dashboard panel used three stages) production operations with different equipments for foam-skinned instrumental panels. It is an efficient way to make the same part but saves cycle time and uses only one machine instead of several different machines to finish the same part.

12. Automotive Intake Manifold Gasket. This is the microcellular part made with 33% glass-fiber-filled PA 6/6 (Zytel 5105-305N, DuPont) with N_2 gas as the blowing agent. The microcellular part was made first, and then thermoset silicone was overmolded on the microcellular part. The microcellular processing eliminates all sink marks and appears to be a free of warpage over a wide range of processing conditions. It also has excellent cell structure, with the average cell size about 7–10 μm.

The economic benefits are as follows:

- Cycle time reduced from 18 sec to 6 sec
- Weight reduction about 17–20%
- Hold stage eliminated
- Clamp tonnage reduced from 150 tons to 40 tons

The quality benefits are as follows:

- Warpage-free, which is critical for the flatness and sealing
- Excellent cell size and uniform distribution with thermal insulation property

13. Automotive Interior Trim [4]. Valeo was a pioneer microcellular process with many successful microcellular parts. The interior trim is one of many automotive parts made by the microcellular process and is made by various materials.

The economic benefits are as follows:

- Significant cycle-time reduction
- A 10% weight reduction within the range of material strength
- A substantial reduction in the peak clamp tonnage requirement from 250 tons to 75 tons

The quality benefits are as follows:

- Improved dimensional stability
- Sink mark-free
- Excellent processability with insert film molding

This microcellular part was a 2002 SPE Automotive Innovation Award Finalist.

14. Automotive Fuse Box. This is a 25%-filled PP copolymer made for an automotive fuse box application. It successfully matches the quality of the part with the reduction of the processing temperature. The results are summarized below.

The economic benefits are as follows:

- Cycle-time reduction
- Weight reduction about 17–20%
- Hold stage eliminated
- Processing temperature reduction up to 56 °C

The quality benefits are as folllows:

- Reduction of wall thickness variation and significant reduction of warpage
- Excellent cell size about 50 μm

15. Other Potential Applications for Automotive Parts. The thermal insulation properties are improved when foaming the parts using the microcellular process [14]. As expected in the conventional foam products, the thermal conductivity decreased as the weight reduction increased from the foam structure. This can be explained by the presence of a larger number of gas close microcells in the part structure. Therefore, microcellular foam provides better thermal insulation properties than does conventional foam because of the unique cell sizes and cell density of microcellular foam.

The function of sound insulation from microcellular structure is another potential application in the automotive industry. It is a new acoustical material for passive noise control in automotive structures. The sound absorption behavior of microcellular foam material is also a unique advantage with porosity structure of microcellular foam. Microcellular material provides better acoustical material than does the conventional foam. The studies carried out by Park and others conclude that a localized behavior is in effect due to the unique microstructure [15]. The modeling work results in the optimal performance of the morphology with higher porosity, lower cell density, and larger cell size [15].

Microcellular foam usually is not used for interior/exterior parts requiring a class "A" surface due to the splaying of gas on the surface. However, the microcellular parts may be polished to have a painted surface, or it may be in-mold decorated using film or fabrics. Several European companies have already used this technology to make automotive parts.

In addition, the microcellular parts used in automotive application have been approved so that the microcellular foam does not change the polymer's resistance to automotive fluids, such as transmission fluid, gear oils, engine oils, coolants, refrigerants, and hydraulic fluids.

11.2.3 Hardware Parts

The hardware parts are widely used microcellular to reduce the cost of material and the advantages of unique microcellular properties.

1. PA 6/6 Cable Ties [11]. This is the application to successfully use PA 6/6 material for a multicavity mold of cable ties molding with microcellular processing. It produces a very good cell structure (about 20 μm) that demonstrates a uniform cell distribution across the whole part. The hold-and-pack stage has been completely eliminated because of microcellular processing. The clamp tonnage is reduced from 3.95 tons/in.^{2} to 2.76 tons/in.^{2} for the thickness of a 1.45-mm microcellular part without any flash. The benefits are summarized below.

The economic benefits are as follows:

- Weight reduction about 8–10% with average cell size 20 μm
- Injection pressure reduced up to 30%
- Smaller machine requirement savings about 30%
- Hold-and-pack stage eliminated to save cycle time

2. Power Tools Base Plate. The microcellular process allows making this part to replace the aluminum in the base plate of the power tool. The material is PA 6/6 with 30 wt% of glass fiber. The microcellular part eliminates the warp and enabling finished parts with all dimensions within required tolerances.

The economic benefits are as follows:

- Cycle-time reduction about 18% (compared with solid molding with dimensions out of tolerance)
- Weight reduction about 8%

The quality benefits are as follows:

- Improvement of dimensional conformity about 70%
- Uses special MuCell® grades of Rhodia (Technyl® Xcell) with significantly improved surface aspect

3. Lock Housing. This is the part made by PP with 20 wt% of glass fiber. It is made by a special TandemMould® that has a stack mold back to back so that it doubles the cavities in a single mold. The microcellular processing reduces the cycle time of this Tandem Mold up to 33%. It again not only makes straight and flat parts but also saves cycle time to increase the productivity.

4. *Soap Dish.* This is made by PP copolymer with 20% tack filler. The major goal of microcellular parts is to save the cost of material. The benefits are summarized below.

The economic benefits are as follows:

- Cycle-time reduction
- Weight reduction about 20% with nice cell size about 50 μm
- Eliminating the hold time and pressure
- Melt temperature reduced up to 56 °C

The quality benefits are as follows:

- Improvement of wall thickness variation down to 0.5 mm
- No warpage

5. *Soles and Sandals.* A closed microcellular structure is produced with the nice cell size from 5 to 30 μm and the overall density of foam is 200 kg/m^3. It is a developing technology for the injection-molded EVA cross-linked microcellular part because it usually is made by compression molding [16]. It is suitable for any application of tough and abrasion-resistant foams. The compression-molded parts have been widely used for flexible packaging, hot melt adhesives, and electrical, medical, and many other applications. The alternative injection molding process provides a more efficient process in lower level of scrap, reduced labor, and cycle time compared to compression molding. Injection molding also provides a higher aesthetic value. The applications of microcellular EVA cross-linked are now mainly for soles, sandals, tires, baby carriages, golf carts, and floating devices. The injection-molded EVA cross-lined microcellular part has more potential to be used for automotive parts, toys, sports protective foams, and so on.

6. *Industry Pallet.* This is the large part made with HDPE with thickness up to 6.4 mm. The previous process for this part is the structural foaming process. The microcellular process improves the mold filling with lower viscosity and much better cell structure than those of the structural foam. The benefit using microcellular processing compared to structural foam process is summarized below.

The economic benefits are as follows:

- Cycle time reduction about 53%
- Weight reduction up to 46% compared to structural foam part with only 13% weight reduction

- Color concentrate dropped from 2% for structural foam to less than 1% for microcellular foam
- Using physical blowing agent N_2 gas instead of chemical blowing agent

11.2.4 Electric Component Parts

1. SEC III Card Edge Connector Housing. This is the typical electric part used for electrical connectors. The material is PBT Valox 420SEO with 30% glass fibers. The melt index is 17 with specific gravity 1.58. It was a successful running microcellular process with fine cell structure of cell size 3–5 μm.

The economic benefits are as follows:

- Cycle-time reduction about 15%
- Weight reduction about 10%
- Smaller machine requirement savings up to 60%

The quality benefits are as follows:

- Warpage-free part
- No sink marks

2. Electrical Enclosure. This is the electrical enclosure made by the PC with 20 wt% of glass fiber reinforcement. It uses the microcellular process to solve the problem of warpage and dimension issues because the box structure causes a warpage problem with solid molding.

3. Electronic Control Module (ECM) Connector [6]. This is also a typical electric part needed for precision of dimensions for connecting pins in the positions during assembly. The material is PBT with glass fibers.

The economic benefits are as follows:

- Cycle-time reduction
- Weight reduction up to 10%

The quality benefits are as follows:

- Warpage-free part
- No sink marks
- Improved flatness
- Pin retention force and standard deviation (SD) improved up to 30%
- Increased assembly yields
- Improved sealing

4. Wire Spool. This is a large part made by ABS material. The benefits are significant, and are listed below.

The economic benefits are as follows:

- Cycle-time reduction about 20%
- Weight reduction about 30%
- Clamp tonnage reduced drastically

The quality benefits are as follows:

- Warpage-free part
- No sink marks
- Improved flatness

5. Printer Chassis [17]. This is a typical complex part with tight tolerance. It is made by 20% of glass-fiber-reinforced polyphenylene oxide (PPO, Noryl® resin). Nitrogen gas was used as the blowing agent to make microcellular foam. This part has a number of bosses, deeply cored sections, and an average wall thickness of 2.5 mm. The flow ratio is approximately 150:1. The conventional molding is limited for the cycle-time reduction because of the dimensional problems. The microcellular part solves the dimensional problems and reduces the cycle time significantly.

The economic benefits are as follows:

- Cycle-time reduction about 42%
- Weight reduction about 10%
- Clamp tonnage reduction up to 50%

The quality benefits are as follows:

- Warpage-free
- No sink marks
- Glass fiber disorientation and less glass fiber breakage increasing the overall part strength

11.2.5 Precision Molding Parts

There are many parts made by injection molding trying the microcellular process for precision control purposes. The microcellular process overcomes all these dimensional obstacles and also reduces the cycle time and weight at the same time. Typical products from the company Empire Precision Plastics include a long list of successful precision parts made with the microcellular process for the stress-free and tight tolerance requirements.

A typical example is a valve with the size same as a penny. It is made by glass-filled polyphenylene sulfide (PPS). The microcellular process maintains the tight tolerance with extra benefits of the microcellular process as listed below.

The economic benefits are as follows:

- Cycle-time reduction about 20%
- Weight reduction about 5%

The quality benefits are as follows:

- Warpage-free
- Sink mark depth reduced up to 80% almost not visible

11.2.6 Medical Parts

The material for medical application is usually expensive. Therefore, the initial microcellular mold trial has the target of saving material. However, in most cases there is only certain weight reduction in the range of 5–15%. The obvious benefit of the microcellular process for medical parts is still from cycle time saving.

On the other hand, the inert gases, such as nitrogen and carbon dioxide, will not react with most chemicals in the plastic melt. Therefore, this actually protects the medial material without any contamination from blowing agents. However, there are no real cases to be discussed here for medical parts, although several medical applications have already been tested in the laboratories.

Several medical parts have been successfully tried with the microcellular process. One of them is the plastic screw that is for fixing the position of human body. Also, one interesting project is to make human bone with microcellular structure because the human bone is naturally hollow. However, more medical instruments and tools made with the microcellular process have been successfully produced.

1. Medical Staple Gun [6]. This is made by Radel PPS. The microcellular molding part has obvious benefits as listed below.

The economic benefits are as follows:

- Cycle-time reduction about 50%
- Weight reduction about 30%
- Clamp tonnage reduction up to 80%
- Mold temperature reduction up to 80% (to cool it quickly without difficulty of mold filling)

The quality benefits are as follows:

- No sink marks
- Warpage-free for tight-fitting tolerance

2. Syringe Plunger Cap. This is made by Kraton G-7722, which is a typical TPE compound used for medical parts. The heavily filled rubber phase created nonuniform cell structure; however, overall distribution is still uniform (see morphology structure in Figure 3.8). The microcellular molding part has obvious benefits as listed below (it is a medical part similar to that in reference 6 but made with Santoprene TPV material).

The economic benefits are as follows:

- Cycle-time reduction up to 30%
- Weight reduction about 20%
- Clamp tonnage reduction up to 40%

The quality benefits are as follows:

- Reduced shore A hardness
- Improved sealing capacity
- Improved compression set
- No oil additives
- No fogging issues

11.2.7 Metal and Ceramic Powder Parts

The power injection molding for the microcellular process is a new technology that is still in the development stage. It has a potential of "significant savings in feedstock consumption" [18]. However, other benefits are (a) the cycle-time savings and (b) stress-free parts from injection before debinding and sintering. The micropores in the metal powder parts will provide insulation properties for both thermal and acoustical. The potential applications will be like jewelry, sporting goods, lightweight structures, and heat-insulating components. Of course, the other general purpose of this application is to save material. The potential application of microporous structures is also a future target for this technology [18].

11.2.8 High-Performance Engineering Materials

This is a unique application for microcellular parts because the microcellular process can foam the high-performance engineering materials that cannot be foamed successfully with conventional foaming technologies. It includes thermoplastic elastomers (TPEs), vulcanizates (TPVs, such as Shell's Kraton rubber, and Advanced Elastomer Systems' Santoprene® TPVs), high-temperature polysulfones, polyetherimides (PEIs), PEEK, and so on. These are expensive materials and have a wide range of application from military to industrial. Microcellular engineering materials may bring another potential to either (a) reduce the cost of engineering material usage or (b) make new application of these engineering materials since unique microcellular structure.

Another good trend of microcellular parts is the high percentage fiber-reinforced material corresponding to high cell density with small cell sizes. It not only saves the costly fiber-reinforced material but also promotes the mechanical properties of the fiber-reinforced part with disorientated fibers around cells [17, 19]. There are several reports from production people that the disoriented fiber distribution makes the mechanical properties of the microcellular part better than those of the solid part. It is because the solid part has a severe fiber orientation problem. This extra benefit of fiber disorientation will increase the applications of reinforced materials with all different fibers.

11.2.9 Special Microcellular Structure

This was another topic regarding the production of open cells of a microcellular part. This needs a very high percentage of gases no matter CO_2 or N_2 gases. It offers special properties with open cell structure compared to closed cell structure in microcellular part.

11.2.10 Microcellular Foamed Bottles

This is a new application in the injection blow molding industry. The first lightweight foamed polyethylene terephthalate (PET) bottle/jar has been made by the injection blow molding process. Although it is the blow molding process, the first stage is injection molding that finishes the microcellular foamed parison. It is marketed under the oPTI℠ (pronounced "opti") brand name. This is an injection blow molding process developed by Plastic Technologies, Inc. (PTI). The oPTI℠ technology is based on MuCell® technology licensed from Trexel, Inc. This application is good for food and beverage brand owners with a broader range of PET container aesthetics and performance capabilities. For example, the process enables white or silvery colored bottles to be made without additives, which may limit the package recycling [4]. Therefore, white color oPTI℠ bottles provide an environmentally friendly option to conventional bottle that uses additions to achieve a similar tint. The white color oPTI℠ can be re-melted and subsequently processed again even with transparent PET without any contaminant.

The flexibility of injection blow molding can even make a nonfoamed preform with 100% mold filling during injection. The nuclei have been formed during injection, and they are compressed because of the 100% filled molding like solid molding. Then, the foaming will occur later during re-heat and blow molding for bottles.

In addition to PET material, the microcellular foamed bottle technology is applied to more materials, such as polylactic acid (PLA) and polyethylene naphthalate (PEN) materials. These materials expand the potential application of oPTI^{SM} technology into a wide range of bottles and jars in food, beverage, personal care, household chemical, and so on, markets. The state-of-the art pilot production and testing capabilities of oPTI^{SM} technology can refer to the web site www.plastictechnologies.com.

11.3 FUTURE RESEARCH TOPICS AND POTENTIAL APPLICATIONS

There is more potential in the future to use microcellular processing technologies for more challenged applications. Some of the future applications were explored by researchers in universities or were accidentally found in the field. Some of them are still in the concept stage, and others may have already become real applications. There are some recent pharmaceutical applications to make pills with microcellular structure to increase the overall surfaces from lots of tiny cells. The increased surfaces of cells in foamed pills will help patients to absorb the medical drug more quickly than solid pills. Many potential applications are on the horizon, and many more new patents are applied every year. However, only real things that have been tested or are already patent pending currently will be discussed in this chapter.

11.3.1 Super-Microcellular Part

This is the ideal microcellular foam that has microcells with the cell size of about 0.05 µm or less. In this range of cell size, the light may go through the cells to keep the foamed transparent material remaining transparent. On the other hand, the mechanical properties will be much better maintained and may even become better for some of mechanical properties, such as toughness. However, currently this cell size can only be made in a batch process with a very careful setup. The processing methodology may need to be reconsidered for making such super-small cells in microcellular parts. It will be the target for the future microcellular foam in industry [2].

11.3.2 Special Functional Part

The special functional part with the microcellular process will be many tailed features targeting the special application requirements. Some typical cases will be discussed in the following.

There are local controlled microcellular foams for special function requirement in the specific location of the part. This local foam can be made with local lights, local heat, or local stress. The whole part may be a microcellular part without foaming and may be cool as a solid part. Then, only local area will be treated with different methods to allow a local microcellular foaming to occur.

The disorientated glass-fiber-reinforced material was successfully made from microcellular foam [4, 19]. The special equipment has been developed at Trexel to process long glass-fiber-reinforced material with microcellular injection molding. Hence, the long glass-fiber-reinforced materials have been the new application case in the list of successful applications of microcellular technology for the automotive industry.

Similarly, the Teflon-filled material can be modified by microcellular foam. The foaming process expels the Teflon filler in the center area, which is the foaming area. Then, most of the Teflon fillers are moved toward the skin; this is a good trend for the printing part, which requires the low-friction filler on the surface only. The trend of Teflon fillers helps to reduce the filler percentage to be added into the same material without gas. It can save cost of expensive filler such as Teflon.

The co-injection process of microcellular foam can be made with selection of two different materials, so the tailor-made properties will result in special application and also open potential new markets. In addition, the surface quality of co-injection microcellular foam is as good as a straight injection molding part that will also expand the microcellular part application into regular injection molding market. Furthermore, the foamed core is sink mark-free and provides good dimension stability to the co-injection microcellular part. It is a very valuable process to be developed for microcellular foam and will be another breakthrough technology in the near future.

11.3.3 Special Processing Using SCF

This was a benefit accidentally found in the microcellular foaming process. In a critical gas dosage at certain processing conditions, a nonfoamed part can be made with controlled gas percentage in the molten polymer, with almost 100% of mold filling. The holding stage is still not required for this special processing for easy mold filling not for foaming. However, it is still possible to make 2% or less weight reduction in a short flow ratio. There will be almost no foam, but a gas-laden molten polymer provides easy mold filling and also warpage and is sink mark-free with good surface finish. This special processing using SCF in the molten polymer is summarized as follows:

- Class A surface finish has been made because there is almost zero foam on the skin and core.
- Mold-filling capability increases significantly.

- There are less warpage and no shrinkage with stable dimension because residual gas pressure remains in the part until the gas is completely de-gassed.
- Clamp tonnage is decreased because of the low viscosity of gas-laden molten polymer and mold filling percentage less than 100%.
- Injection pressure is less than that of a solid because the mold filling is less than 100% and viscosity of gas-laden melt is low.
- High injection speed is necessary because the short injection time can protect any cell growth in the flow front of mold filling.
- Filled material is the better material to be benefited the most using this technology because of the surface finish of the molded part. The gas-laden melt with near 100% mold filling can produce the same quality of surface finish as the regular molding part with the filled material without gas.
- There is more of a demand for a sufficient gas vending opening for this technology.
- The gas-laden material allows the colder mold temperature and the cooling may be faster than regular solid material molding.

Up to now, the microcellular injection molding technology has been successful not only with regard to the original goal of saving materials and better foaming parts but also with regard to a lot more. All of the benefits of microcellular processing dramatically widened the markets for microcellular parts. More and more applications of microcellular processing are in regular injection molding to solve molding and dimensional issues that are not related to foaming at all. Therefore, the future of microcellular injection molding will expand into the regular injection molding more and more and will eventually become an important technology in regular injection molding industry and market.

REFERENCES

1. Martini-Vvedensky, J. E., Suh, N. P., and Waldman, F. A. U.S. Patent No. 4,473,665 (1984).
2. Suh, N. P. *Innovation in Polymer Processing*, edited by James F. Stevenson, Hanser/Gardner Publications, Cincinnati, Chapter 3, 1996.
3. Xu, J., and Pierick, D. *J. Injection Molding Technol.* **5**, 152–159 (2001).
4. Trexel Web Site, www.trexel.com.
5. Dvorak, P. *Machine Design* **January**, 115 (2007).
6. Okamoto, T. K. *Microcellular Processing*, Hanser/Gardner Publications, Cincinnati, 2003, pp. 118–120.
7. Wang, J., Lee, W. S., Yoon, J. D., Park, C. B. *SPE ANTEC Tech. Papers*, 2168–2172 (2008).

8. Lee, J. H., and Kishbaugh, L. How Microcellular Foam Molding Changes the Cost Structure of Injection Molded Automotive Components: A Review of the Process and Automotive Applications, *Technical Conference, SAE*, 2002.
9. Material Thoughts, *Modern Plastics Worldwide* **October**, 86 (2006).
10. Hoffman, J. M. *Machine Design* **September**, 118 (2006).
11. Pierick, D., and Jacobsen, K. *Plastics Eng.* **57**, No (5), 46–51 (2001).
12. Knights, M. *Injection Molding Mag.* **March**, 41–42 (1999).
13. Maniscalco, M. *Injection Molding Mag.* **April**, 30–33 (2007).
14. Sabic Innovative Plastics Web Site, http://www.sabic-ip.com/.
15. Serry Ahmed, M. Y., Atalla, N., and Park, C. B. *SPE ANTEC, Tech. Papers*, 1109–1112 (2008).
16. Lee, J. *SPE ANTEC Tech. Papers*, 2060–2064 (1997).
17. Kishbaugh, L. A., Levesque, K. J., Guillemette, A. H., Chen, L., Xu, J., and Okamoto, K. T. U.S. Patent No. 7,364,788 B2 (2008).
18. Chitwood, A. *Injection Molding Mag.* **April**, 92 (2001).
19. Xu, J. *SPE ANTEC, Tech. Papers*, 2158–2162 (2008).

12

COST SAVINGS FOR MICROCELLULAR INJECTION MOLDING

This detailed cost analysis of microcellular injection molding presents the overall economic advantages of this technology. Typical cost savings include the cost reductions from energy, materials, mold, equipment, and the processing. The creation of microcellular structures brings a wide array of benefits including reduced weight of the part, reduced material and additives usage, reduced or removed warpage and sink mark, and reduced production cycle time.

Generally, benefits of the microcellular process are from reductions of the melt viscosity due to the presence of the supercritical fluid (SCF) in the melt. The initial usage of SCF is the purely a foaming agent. However, the SCF also acts as a plasticizer reducing the polymer melt viscosity significantly, which can enable a reduction in cycle time, injection pressure, melt temperature, and clamp tonnage. The microcellular process with SCF is reportedly capable of producing parts with good dimensional stability as well as eliminating sink marks and warpage. It is also reported that these benefits could be achieved without dramatically compromising the material properties, as is the case sometimes with structural foam. On the other hand, SCF under pressure will be the initial driving force for the injection. The fast energy release is one of the key factors resulting in cycle-time reduction. All these factors are the additional cost savings for microcellular injection molding.

However, there are some disadvantages of microcellular processing and other costs to be considered during cost savings calculation. Similar to

Microcellular Injection Molding, Edited by Jingyi Xu
Copyright © 2010 John Wiley & Sons, Inc.

structural foam, the microcellular process produces gas swirls on the surface of the molded parts, although the gas swirls are removable through some technologies as part of special microcellular technologies (see Chapter 8). The process requires some equipment modification, which may add an additional 10–15% to the cost of the machine of the regular injection molding. The equipment modification includes the plasticizing assembly that consists of a modified barrel, a special designed screw (24–28 *L*/*D*, as usual, but possible 22 *L*/*D* with limit of short recovery stroke), a shut-off nozzle for high pressure (high pressure sealing to prevent pressure loss and material drool), and a specially designed check valve in the front of the screw. On the other hand, the control system needs to be changed with software modification for gas dosing unless a chemical blowing agent is used for foaming. Tooling modifications (runner, gate designing, venting considerations, mold draft angle, and undercut in the core for the part to stay on the core during ejection, etc.) may be required for some applications. In addition, this process package also consists of (a) a supercritical fluid (SCF) system and (b) gas injectors if a physical blowing agent is selected, instead of a chemical blowing agent [1].

One more factor of cost analysis that needs to be considered is the license fee for using this technology with gas as the blowing agent. Anyone using microcellular injection molding usually needs to get a license and secrecy agreement from Trexel first. Trexel is the exclusive developer of the MuCell® microcellular foam technology and has an extensive portfolio of patents in the United States, Canada, Europe, Japan, Korea, and Asia. Trexel's primary business is the supply of MuCell® Systems for the production of foamed injection-molded and extruded articles. It also provides engineering support, training, and other services, as well as the equipment integral to the MuCell® process [2].

12.1 COST SAVINGS OF MATERIAL

Basically, there are two different materials to be discussed. One is the base material, such as plastics, compounds, metal powders, and so on. Another is additive, such as colorant, nucleation agent, UV protection agent, blowing agent, and so on. However, a blowing agent is specific agent only for foaming process so that it will be discussed separately from other additives. Therefore, there are three different material savings to be discussed in the following.

12.1.1 Base Materials

It is the initial goal of microcellular process to save on base material (usually plastics or some compounds) cost [3, 4], which is still one of the advantages of the microcellular process. Generally, about 70% of the cost of a plastic part is the base material cost. Usually, up to 15% weight reduction is reasonable for the applications with microcellular technology. However, in most cases this base material saving for the microcellular part can be as much as 5–10%.

The amount of material saved with microcellular processing will be significant either with the great quantity of the parts or with the high cost of material. The high-cost materials are expensive engineering materials, nanocomposites, biopolymer, and medical materials.

For example, the well-known thin-wall microcellular technology called SLIM® will have material saving of about 6–7% weight reduction for each small container. The amount of material saved from these small containers seems small for an individual part but can be significant when container numbers reach 200,000,000 per year. In addition, in Europe, material saving can also bring an extra saving on taxes. For example, SLIM® technology as the leader of thin-wall microcellular technology provides outstanding benefits in terms of packaging weight reduction and carbon footprint reduction. As a result, SLIM customers in Europe can save additional money by reducing Eco Tax costs. Specifically, the cost calculation shows that in Germany for a standard 15-g tub, the 10% weight reduction of the part is a €200,000 saving in Eco Tax for every 100 million tubs [5].

The costs of engineering material and alloy are much higher than those for general-purpose material. For example, the ABS material cost is only half or 30% of the ABS/Nylon alloy. It is obvious that the high-performance engineering material used to reduce the weight in microcellular processing is so expensive that it is worth the effort to reduce the weight through the microcellular process. As an example, the cost of unfilled polysulfone material is more than 10 times as much as the cost of general-purpose HDPE. Glass-fiber-reinforced material is also expensive compared to the same material without reinforced fibers. On the other hand, the exciting trend for saving the material for microcellular foam with reinforced material is much higher than that for the unfilled material. This is because the reinforced or filled material has a better microcell structure in comparison to the unfilled same material. In addition, the reinforced material has property improvements due to the fiber disorientation from cell expansion during the foaming process. The material improvement with reinforced filler or fiber will further allow more weight reduction than will the unfilled same material. Saving money from material is crucial for survival in the very competitive market, because the prices of plastics will continue to rise with incessant increase in oil prices until new energy sources are readily used.

The newly developed materials, such as nanocomposites, medical materials, and biopolymers, are all expensive materials that are worth using in the microcellular process for reducing the material usage as much as possible. Even for a small volume of production, 5–10% weight saving will be a significant cost reduction for these special materials.

There is also metal powder injection molding in the development stage for micropores of the metal powder parts. It was reported to save "significant

TABLE 12.1 Resin Price Range for the Recommendation of the Major Microcellular Benefits

Less than $1/lb: Focus on Cycle-Time Reduction	$1/lb to $5/lb: Focus on Cycle Time Reduction, and Weight Reduction	$5/lb to $10/lb: Focus on Weight Reduction	More than $10/lb; Focus on Weight Reduction
PP, HDPE, LDPE, LLDPE, HMW-HDPE, EVA, PVC, SAN, olefin elastomer, PS, EPS,	ABS, acetal, Acrylic, PC CAB, CAP, EVOH, nylon 6, nylon 6/6, glass-filled nylon, PPE/PPO, PPS, TPU, styrene-acrylic, polyamide, and polyester	PTFE, FEP, Glass- or Mineral-filled LCP, glass-filled polyetherimide, glass-filled polysulfone, silicones	PFA, FEP, ETFE, CTFE, PVDF, carbon-fiber-filled LCP, PEEK, PEK, polyamidemide

feedstock consumption" [6] because gas generates many microcells in the final parts.

In Table 12.1, the recommendations are given per price range of base materials. It is obvious that for the expensive materials, such as unit cost of PEEK up to $33 to $44/lb, it is worth making more efforts to save material to use in the microcellular process. In addition, more weight reduction usually means cycle-time reduction as well because there is more energy released during nucleation and cell growth. The other materials ranging from cheap ones (less than $1/lb) to expensive ones (more than $10/lb) are also recommended for microcellular process use, with the focus on both weight reduction and cycle-time reduction.

In addition, the base material saving has a long-term effect on the environment because most polymer materials are made from the products of petroleum. Therefore, the benefit of protecting the environment is additional market value for microcellular processing.

12.1.2 Blowing Agents

The inert gases of CO_2 and N_2 are from the atmosphere, so they are the environmentally friendly and cheap resources of physical blowing agents. This kind of inert gas does not have any pollution for the environment, and it has been widely used in the microcellular foaming industry. There are several cases showing that the inert gases in the barrel will help to decrease the degrading of the material in the shearing. That will bring the benefit for protecting shearing damage of materials during screw recovery, so the machine may run faster without degrading the materials in the barrel. It is important for some shear-sensitive materials, such as biopolymer and medical plastics, to be able to run with the microcellular process without extra material degrading in the recovery process.

Material saving was analyzed by switching from a chemical blowing agent (CBA) to a physical blowing agent in reference 10. Nitrogen gas costs about U.S. \$0.11/kg and carbon dioxide is approximately U.S. \$0.37/kg. However, the chemical blowing agent costs about U.S. \$4.4/kg to \$12/kg. Assuming that 0.5% nitrogen gas and 0.2% CBA make the same microcellular part, the cost of unit plastic of the microcellular part with nitrogen gas is at most 21% of the cost of unit plastic of the microcellular part with CBA. In addition, CBA may have some chemical reaction with the additives in plastics and leave some chemical residuals in the plastic part that may not be allowed for FDA-regulated parts for medical and food industries. On the other hand, the CBA may not be good for high-processing-temperature materials, while inert gases have no limit for the high-processing-temperature materials.

12.1.3 Additives

The increased toughness of microcellular foam allows the microcellular material to replace or reduce the usage of impact modifiers in polymers or composites. For example, the impact modifier may not be necessary for the microcellular RPVC–wood-flour composite part [7]. The cross-linked impact modifier increases the diffusion of CO_2 gas in the RPVC. The un-cross-linked modifier in the polymer also promotes the rate of gas diffusion because of the greater softening effect of the un-cross-linked modifier. The softened RPVC increases the acceleration of gas loss in the wood-flour–RPVC compound. Therefore, only a small portion of the gas in the composite of softened RPVC has been utilized for the nucleated cell growth because of an accelerated gas loss.

Consequently, impact modifiers are not a necessary ingredient in the RPVC–wood-flour composites because microcellular foam will increase the toughness of the RPVC material, creating an effect that is similar to that of an impact modifier.

In addition, the transparent material has a white-pearl-like color after foaming. The fragmentation of bubbles on the surface of the microcellular part leads to many tiny microscopic gas bubbles that result in small dents in the foamed surface. Through different light reflection properties of these dents, they appear brighter and affect the silver shining streaks on the surface. If the mold temperature is kept higher, the smooth surface with this white color in the microcellular part is made to satisfy the necessary surface quality and inherent white color part. In general, the microcellular foam is usually opaque without the need to introduce the pigments such as titanium dioxide [4]. Therefore, microcellular parts can save the cost of coloring products, such as colorant of a white color or a light color. On the other hand, the color concentrates are not only very expensive but can also cause a degrading problem from the low molecular weight of the carrier material of the color concentrator.

On the other hand, the surface roughness of a microcellular part is about more than 20 μm. This is about 20 times higher than the roughness of the smooth surface of a solid part. A suited and effective mold technology for improving the visibility of the surface roughness of the microcellular part is to use a structuring mold cavity surface, also known as a textured mold surface. It is helpful because of two possible improvements. One is that the texture with continual grooves will vent the escaping gas from the surface of the microcellular part better. Another is that the texture surface can avoid the sliding action between the skin of the microcellular part and the mold surface and, then, can reduce the bubble shearing on the surface, which can consequently reduce the stretching of the surface bubbles. This will result in less broken bubbles on the surface and will reduce the big silver streaks on the surface. In addition, different textures can conceal more or less the surface defects. A carefully designed texture will superpose the rough surfaces of microcellular parts so that the appearance of silver streaks can be concealed by a diffuse backscatter [8]. Then, the cost of color additives can be saved without extra polish or paint on the microcellular part.

12.2 MOLD

Microcellular processing is possible either to use aluminum material or to use the same steel material of mold with a longer lifetime than regular molding. It is because the strength requirement of mold is reduced significantly by low injection pressure and low cavity pressure of microcellular molding.

Using aluminum material for the microcellular mold has well-known benefits of a faster cooling rate from the high heat transfer coefficient of aluminum. In addition, the low cost of aluminum material and cheap manufacturing are also attractive benefits of using the aluminum tool. The aluminum alloy tool is usually made 40% thicker than the steel tool due to the low strength of aluminum alloy. The modulus of elasticity in aluminum alloy is only 30% of that of steel. However, the total weight of aluminum molds is about 50% less than steel mold even if aluminum mold is designed with thicker sections. The aluminum material permits a 5–10 times higher cutting speed in comparison to steel material. Even with a high cutting speed allotted for the aluminum tool, the residual stresses in the aluminum tool is very little due to the high heat transfer of aluminum. Hence, aluminum tools sustain the high precision of molds. Furthermore, the aluminum tool can use high eroding speeds about 6–8 times that of steel in an EDM process as well. However, the aluminum mold may only run 250,000 parts as a predicted tool life for regular injection molding because the hardness of aluminum tool is only about 5 Rockwell C. Overall the aluminum tool for microcellular processing is a very good choice. There are some heavy-duty aluminum alloys for low-pressure microcellular processing that will have a longer lifetime as an aluminum tool if final injection pressure is controlled carefully. The modern injection molding machine guarantees more uniform load distribution that will allow more and more of these aluminum alloys to be employed for microcellular injection molding. The typical heavy-duty aluminum alloys are 7075-T6 and 7029-T6. This aluminum alloy improves thermal conductivity almost four times better than tool steel. Overall, the cost analysis for the aluminum tool must be related to the lifetime, easy cutting operation, and cycle-time reduction to estimate the reasonable benefit in using the aluminum tool.

Zinc alloy is another choice if the microcellular molding requires running with a low volume of the production. Zinc alloy is characterized by a high thermal conductivity of about 105 W/(m·K). Therefore, zinc alloys are the substitute materials of aluminum to be used for inserts fitted into the steel base. The most common zinc alloys for the mold manufacturing are Zamak™ and Kirsite™.

Sintered metal can be used for microcellular processing mold as well. It was reported that sintered metal mold has excellent air and gas vending benefits from the micropores in sintered material. However, it may only be used for the production of 100,000 parts with unfilled or lightly filled material. The maximum part size is limited to 4–10 in^2, with a hardness of less than 50 Rockwell C.

Different processing methods will have different mold costs as well. The major three different molding methods are gas counterpressure molding, co-injection molding, and low-pressure molding, respectively. A gas counterpressure mold always has the highest cost among all three methods of microcellular processes. With molds for the gas counterpressure technique, the costs incurred for sealing the mold are strongly dependent on the part geometry, either slightly or significantly higher than the other two molds. Then the multi-components mold, such as co-injection, has the cost between the gas counterpressure mold and the low-pressure mold. This is because the high-quality mold surface is required for the multicomponents mold. Therefore, low-pressure molding uses the lowest cost of molds, such as the aluminum material discussed above. The multicomponents mold usually requires a 3–5% higher cost than the low-pressure mold. However, the cost of a gas counterpressure mold might be up to 23% higher than the low-pressure mold.

The microcellular part has extremely stable dimensions after molding. It basically overcomes the disadvantage of large shrinkage in the unfilled plastics. Then, it is possible to lower the mold precision grade to reduce the cost of mold manufacturing.

12.3 EQUIPMENT

Currently, most microcellular capable equipments will be directly or indirectly from Trexel Inc. because all OEMs supplying this MuCell® microcellular injection molding machine, or any retrofit injection molding machine with a

MuCell®-capable package, must have a license from Trexel Inc. More detailed information can be found from Trexel Inc. at www.trexel.com. The details discussed in this chapter are the cost saving of microcellular processing equipment without considering the license fee and gas pump unit cost.

It is obvious that the clamp tonnage saving is one of the benefits for thin-wall microcellular parts. The SLIM® technology using a smaller machine has been successfully applied for thin-wall microcellular processing without using a larger injection molding machine. Therefore, the immediate saving from SLIM® technology is the cost saving to avoid using the big clamp tonnage machines. In addition, the long-term cost saving from SLIM® technology is a low-energy operation and a long lifetime of any stressed component in clamp unit and mold because of low pressure used in microcellular injection molding for thin-wall parts [5].

This cost saving from the microcellular process is also truly from energy saving. Usually, more than 90% of the energy costs in regular injection molding is due to electricity usage, but only 5–10% of the energy is truly used in polymer processing [9]. In other words, running the microcellular-capable molding machine can have large cost and energy savings without affecting the quality of plastic processing. This is because the microcellular process lowers both the clamp force and injection pressure without changing the qualities of polymer melting, mixing, and gas dosing.

Also, using large machine for small products is inherently wasteful [9]. The microcellular process provides not only a cost saving of the low-energy running conditions but also a cost saving to run big mold in small machine because of the low injection pressure and low cavity pressure. In addition, the microcellular part will run up to 15% less materials than the solid part, so the screw recovery energy is also less than the solid part recovery energy. A solid part made with a 500-ton machine with struggle to fit the corners in the mold can now successfully run in a 200-ton machine for the microcellular part. When selecting the clamp system, one important consideration is how much space of machine is available to fit, for example, a big mold in a small machine, not how larger tonnage is required for the low-injection-pressure microcellular injection molding.

In addition, the operation cost to run the low-tonnage clamp unit is also energy saving because the friction force for a moving platen is proportional to the weight of the mold and platen. On the other hand, the inertial force from mold and platen movement is low for lightweight platen and mold in the microcellular process as well. From the part-wearing point of view, the low tonnage will increase the lifetime of the critical parts, such as tie-bars, bushings, and sliders under the moving platen in a clamp unit. However, saving energy from the clamp unit itself is not enough because the hydraulic system for the injection molding machine is designed with the most requirements from Injection unit. Usually, the highest injection pressure and the highest injection linear speed are the highest peak amount of energy usage in the whole cycle. Focusing on reducing the highest energy requirement of all actions performed by a machine will help to reduce the energy of operation effectively.

Cost saving from the benefits of low tonnage for microcellular injection molding was an interesting topic regarding of the use a thin platen and the lower weight of components in the machine. It definitely reduces the cost from the thin platen because the material of casting iron is charged by the weight of the platen. On the other hand, it may use the existing machine for the microcellular process. Then, a large mold may be used in small tonnage clamp units if the platen is wide enough. In most cases, a tie-bar-less machine provides the best solution because there is no tie-bar restriction to fit an extra wide mold.

Actually, low injection pressure has a benefit of large saving in the total power specified based on the peak injection load even in a very short time. It can be calculated with the following simple relationship:

$$\text{Peak energy} = \text{Injection pressure} \times \text{Injection velocity} \tag{12.1}$$

The machine total power input has been specified based on the peak energy requirement calculated from Equation (12.1) with high percentage overload safety factor. For microcellular processing the peak energy will be reduced significantly, and it may change the rule of machine design to determine the total power input of the injection molding machine.

A similar study for the energy savings of microcellular processing was carried out by W.H. Fuller Company in Glastonbury, CT, USA. They concluded that the microcellular process would generate substantial savings from energy. For example, by using a 400-ton injection molding machine there will be 298,698 kWh of power saved each year with microcellular processing. Depending on the location in the United States and the different cost of electricity, the savings can translate into US$10,000 to US$30,000 per year for one machine [10].

The reasonable comparison of costs between solid and microcellular parts should use a cost per part. MuCell® microcellular processing has tested the same parts on the same machine with a 42-ton molding machine between solid and microcellular processes, respectively. The results of the tests were typical examples of a small-tonnage machine with a normal-size screw. It is different from the results in Table 12.2 for a large-tonnage machine with an oversize screw. The conclusion for the MuCell® process in a small molding machine was that the running amps only reduced 2% for the microcellular process. However, the cost saving percentage per piece is up to 36% from microcellular processing [10]. On the other hand, this comparison was made with the same tonnage machine and the same injection unit. To run the machine at low request for microcellular processing is not economic, and then the results from this test may not reflect the real potential cost saving.

In addition, the different machines may have different energy saving margins. For example, the toggle clamp system definitely contributes more

TABLE 12.2 Energy Consumption Comparison Between Solid (Data in Parentheses) and Foam (1500 Metric Tons, 160 mm, 24:1 *L/D* Screw, Material 20 lb of HDPE)

Microcellular Machine Energy Saving	Actions of Machine	Time Analysis (sec)	Energy (Wh)	Distribution Percentage in Total Energy Consumption (%)
	Cores in	0.7 (0.7)	3.0 (3.0)	0.48 (0.36)
	Closing	1.70 (1.70)	69.97 (69.97)	11.20 (8.28)
	Locking	0.50 (0.50)	11.38 (11.38)	1.82 (1.35)
60% reduction: 10.48 Wh	Clamping force buildup	0.24 (0.60)	10.48 (26.19)	1.68 (3.10)
(15% weight reduction) 50% injection pressure reduction: 44.47 Wh	Injection unit	2.98 (3.50)	44.47 (114.05)	7.12 (13.50)
100% reduction: 58.09 Wh	Post-pressure unit	0 (5.00)	0 (58.09)	0 (6.88)
12% energy saving with microcellular recovery: 49.43 Wh	Feeding unit with hydraulic motor	17.60 (18.00)	364.41 (413.84)	58.31 (49.00)
50% reduction: 10 Wh	Cooling time	10 (20.00)	10 (20.00)	1.60 (2.37)
Only 40% clamp force needs to be reduced back to zero: 5.83 Wh energy saving	Clamping force reduction	0.56 (1.40)	3.89 (9.72)	0.62 (1.15)
No change	Unlocking	0.80 (0.80)	5.56 (5.56)	0.89 (0.66)
No change	Opening	1.60 (1.60)	59.10 (59.10)	9.46 (7.00)
No change	Cores out	1.00 (1.00)	4.34 (4.34)	0.69 (0.51)
No change	Ejector forward	1.00 (1.00)	2.89 (2.89)	0.46 (0.34)
Possible lightweight part removing energy saving about 6 Wh	De-molding	4.00 (4.00)	34.00 (40.00)	5.44 (4.74)
No change	Ejector back	0.50 (0.50)	1.45 (1.45)	0. 23 (0.17)
Total 26% energy saving with microcellular processing; 29% cycle-time reduction	Total	43.18 (60.8)	624.94 (844.58)	100 (100)

saving to the clamp system. In most toggle clamp units of the injection molding machine, the injection unit takes most of the power in a very short time. The total power of the machine is determined by this power accordingly with allowance of 50%, or up to 60% of overload in the electric motor in short time, although the electric motor manufacturer only allows having 20% overload in the motor for offering a normal warranty. There will be more energy saving from this kind of machine with microcellular processing. Therefore, the low injection pressure may take an even higher percentage of saving compared to low clamp tonnage saving because the toggle itself is an energy saving system. In addition, the toggle is a self-locking mechanism, so there is no difference between microcellular and regular molding for toggle locking tonnage. However, only the peak load during injection on the tie-bar of toggle system will influence the clamp system load every cycle. It may be good for a microcellular process with a low injection load to (a) reduce the peak of pulse of fatigue load in the whole clamp system and (b) help for longer lifetime of tie-bars or other heavily stressed parts in the clamp unit. The best energy-saving injection unit is the one with the hydraulic accumulators for the high injection speed but low injection pressure.

The conventional measure of process energy is still useful to calculate the specific energy consumption (SEC) for the microcellular injection molding process. SEC is simply the number of kWh per unit weight of material to be processed, and the unit is kWh/kg. As long as the same method of SEC is used in solid processing, it is a good tool to estimate the difference of SEC between microcellular and solid processes. The detailed benchmarking of injection molding energy efficiency can be found in reference 11.

12.4 PROCESS

The great benefit of microcellular processing is actually from the cycle time saving. Every action in a whole cycle time is intended to save as much cycle time as possible because the reduction of cycle time increases the yield. It may be divided into several key factors below. The most popular plastic materials, such as PE, PP, PS, PVC, and so on, are focused on the most in cycle-time reduction because the material price is low (see Table 12.1). However, if the weight reduction is allowed for strength maintaining and processing allowance, then we should take advantage of this as well.

12.4.1 Cooling Process

This is a traditional method in that the foam is treated as a thermal insulator. Therefore, a foam process usually increases the cooling time and consequently increases the cycle time. However, the microcellular processing has quick cooling inherently because of the fast energy release during the nucleation. The initial fast injection at a certain thickness of the part is usually performed at a point that is thinner than the thickness of structural foam. A single-phase solution in the barrel is under high pressure before injection begins. Then, the single-phase solution is injected into an empty mold with much faster injection volume rate than with structural foam. The fast cooling from the microcellular process may be from quick pressure release that is also an energy release during the mold filling. In addition, the gas pressure in the cells pushes the part in contact with the cold mold surface tightly to cool it more quickly than regular solid molding because it shrinks during the cooling period, and the surface of the solid part does not always touch the cold mold surface tightly.

In most cases, cooling time saving is the significant cost reduction. It is because cooling time is about half of the total cycle time for regular injection molding, and it also uses a water source intensively to waste the natural resource of water for industrial use. It is a particularly significant saving of water cost if the cooling water in the mold is the treated water.

The fast cooling is a profitable parameter that is worth reviewing carefully to bring the cost down. Several factors resulting in fast cooling are summarized in the following:

- Cold water temperature in the mold cooling channels is absolutely necessary. It is usually 10 degrees lower than that of regular molding if it is allowed for the balance between crystallinity in skin and surface finish.
- High Reno's number over 5000 in the mold cooling channels helps cooling efficiently. This high Reno's number is calculated in the mold only. A molding machine usually requires big water pipes before water pipes connect the mold. However, it may not be necessary for the water channel inside of the mold to be large because the most efficient water cooling in the mold needs turbulent flow with the Reno's number recommended above.
- Use high heat transfer material in the mold, such as aluminum alloy, BeCu alloy, and copper alloy. It is specifically important to have this special metal in the hot spot where it is difficult to arrange good cooling channels inside, and heat transfer is the only method for cooling in this area.
- Fast injection is also necessary for fast cooling of the microcellular process because of instant energy release, and, consequently, high enough nucleation to create many cells.

When the microcellular part becomes very thick, the microcellular process may lose a quick cooling benefit. It is because a slow heat transfer in thick foam becomes a truly thermal insulator whose insulation effect overcomes the benefit from quick energy release. Then, the microcellular processing cannot cool fast enough, and the cycle time becomes extra long—even longer than the same thickness of solid part. It is difficult to calculate exactly how thick this limit should be at certain conditions of processing. The rule of thumb is that the upper limit of maximum thickness of microcellular foam without a significant insulation effect is about 4 mm.

On the other hand, foamed material becomes softer because of the foamed surface. Unless the ejecting area is increased to have less contact stress on the foamed surface during the ejecting process, the microcellular part needs more time to cool the surface to get enough stiffness and hardness of foamed surface before ejecting. This may influence the cooling time saving if the ejecting system does not change in response to the reducing contact force of the ejectors.

12.4.2 Packing Stage

The cycle time is reduced without a packing stage because it is replaced by a cell growth function. In addition, cell growth is good for overcoming the shrinkage of the part during cooling. Although theoretically the packing stage is not needed at all for microcellular processing, sometimes the machine hydraulic system and control system may need packing to act in a very short time to switch from injection pressure smoothly to a certain level of hydraulic pressure of back pressure that will immediately be needed for screw recovery action after injection. However, the electric machine does not have this requirement that can set up the packing time absolutely zero.

12.4.3 Injection Time

It is well known that the cycle time is reduced with fast injection. The fast injection is essential for microcellular injection molding. In other words, the machine may run most time at the upper limit of injection speed. However, injection time for the microcellular process is usually short, so the injection time may only have a smaller percentage of the whole cycle time. Therefore, saving from injection time may not be significant compared to other parameters of processing. The consequence of fast injection time may not be the saving of cycle time but instead the possible saving of cooling time and better cell structure because the fast energy release rate and better nucleation result from fast injection.

12.4.4 Screw Recovery Time

The screw recovery time always needs to be shorter than the cooling time. Ideally, in a hydraulic screw driving system the screw recovery time must end just before the mold is ready to be open or before the cooling stage ends. The microcellular process benefits from the weight reduction because it cuts the shot size of the microcellular part smaller than it does for the solid part. It is more critical for the screw recovery time to end before mold opening because of the short cooling time for thin-wall molding. When the screw recovery time finishes before cooling ends, the cycle-time reduction in thin-wall molding is positively reachable.

12.4.5 Demolding Time

Historically, demolding of a microcellular part was a neck of a bottle to improve cycle time of microcellular processing. Demolding fully depends on (a) the drafting angle of molds and (b) the efficiency of the ejection system. The reason is that the microcellular part will expand more from cell growth than from plastic shrinkage during cooling. Thus, it may cause more friction force between the mold and the part in the cavity from cell expansion. In other words, the cell growth is so strong that the part stays in the cavity instead of staying in the core of the mold. In most cases the core of the mold is in the half of the moving platen where the ejecting system is equipped with the moving half of the mold. Therefore, the microcellular parts must stay in the core of the mold, not in the cavity, to have the benefit of cycle-time reduction.

On the other hand, once the mold design is determined with a special drafting angle in the mold, and some measures (such as undercuts in the core of mold) are taken to keep the microcellular part on the core of the mold during mold opening, the demolding becomes easy for microcellular foam because it does not have any packing stage to add more residual stress that may cause demolding difficulty as well as solid molding difficulty.

The foamed surface is softer than a solid surface, so it may be a difficult for the quick demolding because the ejecting speed is required to be slow at the beginning without impact action from ejectors on the part. This factor needs to be considered during the cycle-time saving analysis.

Another difficult location of part demolding is the cold sprue of microcellular processing. Unless it can be redesigned to be short and strong enough to hold the demolding force, the cold sprue is not recommended for microcellular processing. Either hot runner or valve gate is the best choice of microcellular injection molding. The cost of valve gate can be paid back quickly from cycle-time saving.

12.4.6 Special Microcellular Processing—Dolphin Skin Technology

There are several special microcellular processes that will save a significant amount of costs from the comprehensive aspects. A typical special microcellular process is Dolphin skin technology [12]. It makes a dashboard panel in a single operation in one machine. The original process to make the same dashboard panel is extremely time-consuming and needs three separate stages in three different machines. The original process starts with the separate production of the body and the foamed soft-touch outer skin. Subsequently, a lamination of the body needs to be bonded with the outer skin. The Dolphin skin process greatly promotes the production with less cycle time in one machine and makes the part more attractive in terms of both appearance and properties. The dimension stability of Dolphin technology brings the perfect fitting tolerance control and saves the extra cost of assembly.

12.4.7 Tiny Cells for Thin-Wall Molding

A thin-wall foamed part cannot be made by using structural foam because the bubble is so big in structural foam that a thin-wall part is no longer useful once the big voids are present in the thin-wall part. However, a microcellular part with small cells can be made in a thin-wall part without significant change of properties. In addition, a thin-wall microcellular part can save more cooling time through fast injection even if solid thin-wall processing itself has a very short cooling time.

12.4.8 Low-Viscosity Melt

Low viscosity of gas-rich material provides another processing benefit for microcellular molding. Many parts are either with a thin wall or with a long flow ratio. They are failed to run in the regular injection molding machine because of the reasons of either thin wall or long flow ratio. However, the microcellular process solves this problem with low-viscosity gas-laden material, along with self-expansion from cell growth to fill the details of mold in the final mold-filling flow. Therefore, the thin-wall microcellular part can save the cost from quick cooling and can use less energy from both low injection pressure and low clamp tonnage.

12.4.9 Others

On the other hand, management is really another factor that controls the optimization of the machine settings to maximize the cycle-time reduction. It is inexpensive and reduces all costs related to the errors of administration and organization [9]. Many errors are simply from the misunderstanding of this new technology of the microcellular process.

12.4.10 Summary of the Rules of Thumb for Cycle-Time Saving

For overall sequences of a machine, it is important to check some rules of thumb for maximizing the cycle-time saving:

- Injection time is short enough to be in the range of freezing time of the part in the thickness direction at the coldest mold temperature setup.
- Screw recovery and gas dosing process must be finished in the range of cooling time with hydraulic screw drive. However, with electric screw drive, the screw can rotate with an independent electric motor and can

be run in the whole cooling stage, mold opening, and closing period except for the injection stage for microcellular injection molding.

- The rule of thumb of shot volume of screw in the microcellular process is to have shot volume no larger than 20% of the total shot size if the machine cycle time is less than 6 sec. It will match the requirement of minimum material residence time in the barrel in each cycle.
- Keep fast injection speed as the priority of mold trial to maximize the cycle-time reduction. The short injection time not only saves the cycle time, but also promotes the nucleation to have more cells; then, more heat inhibits nucleation. All of them will more or less contribute to the results of short cycle time.

12.5 DIMENSIONAL STABILITY

The part made by the microcellular process also provides a consistently high quality and exceptional dimensional stability, where the foamed part has not historically been deployed. It means that there will be less postprocessing requirements for the microcellular part to do the final assembly, dimension tolerance control, and correction by tempering or machining the molded microcellular parts. The details of dimensional stability saving will be discussed in the following.

12.5.1 Fixture for Holding Dimension Tolerance

Sometimes a fixture is necessary to hold the part in the post-cooling stage to finish the final cooling with the forced restriction to get the correct final dimensions. This is not only costly but also leaves some residual stresses in the molded part because the molded part is forced to hold certain dimensions, and the residual stress in the part during constrained cooling period is frozen in the part. These residual stresses may exist in the final cold part after the release from the post-cooling fixture. All of these costly measurements are not necessary once microcellular injection molding process has been used. This is good savings not only for cooling time but also for postprocessing cost, expensive fixture, and setup time to hold the part correctly in the fixture.

12.5.2 Post Heat Treatment

If the part is held at high pressure to maintain the precise dimensions, it may leave some residual stresses in the part. One way to release the residual stresses is to use heat treatment to slowly get rid of the residual stresses inside of the part. The microcellular part does not need any hold stage or any pack pressure at all. Therefore there is no residual stress to be removed after molding. This stress-free microcellular part brings another cost saving by eliminating the post heat treatment.

12.5.3 Post Machining

The high-precision plastic part may need post machining, or polishing, to match tight tolerances of assembly. This is not only costly but also needs extra caution to eliminate the machining stress in the part as much as possible. Most plastics cannot cut fast, and they need some fixture to hold them correctly for machining. The microcellular part keeps the dimension precisely without any post-machining requirement, and it reduces this possible extra cost.

12.5.4 Sorting by Dimensional Group for Assembly

If the part has some dimension deviations that must be sorted into several dimension groups, then it is still usable in a certain dimension range of sorting groups. Microcellular dimension will be very stable, and the deviation will not be as large as that of the solid part because it copies the dimension of the mold perfectly by the cell expansion during cooling. Hence, the sorting that works for the matched parts assembly may never be needed for microcellular parts.

12.6 PROPERTY IMPROVEMENT FOR MICROCELLULAR FOAM

This will make full use of the advantages of the microcellular part because property improvement will promote the market of microcellular application and will increase the microcellular part value as well. The potential value increases that result from property improvement of microcellular parts are listed below.

12.6.1 Thermal Insulation

Thermal insulation of the microcellular part is an obvious advantage because microcellular foam is almost exclusively a close cell structure. The important difference between microcellular foam and conventional foam is the uniformity of that microcellular part that provides the uniform thermal insulation properties across the whole microcellular part.

12.6.2 Impact Absorption

There are several reports for the impact absorption of the injection-molded part improved by microcellular processing. The small cell size is the key factor to get this benefit [13]. There will be no significant damage if the surface of the microcellular part is hit by a heavy object as an impact load. It only damages the shallow layers near the surface but does not damage the deeper inside of cell structure in the microcellular part.

12.6.3 Acorsmatic Insulation

The similar principle of thermal insulation from microcellular structure, along with the sound absorption behavior of microcellular foam material, is also a unique advantage of the porosity structure of microcellular foam. It provides better acoustical material than does the conventional foam. The modeling work results in the optimal performance of the morphology with higher porosity, lower cell density, and larger cell size [14]. It is a potential good acorsmatic insulator material for the automotive industry.

12.6.4 Light Weight

A microcellular structure the microcellular part makes a structural part with light weight. A unique characteristic of the microcellular part is to save material while maintaining a much higher strength than that of conventional foam. The microcellular structure can support much more load than conventional foam because the cell size is so small.

12.6.5 Stress-Free Part

Lack of stress is another unique feature of the microcellular part because it results in better dimension stability of a molded part. This is the benefit for both operation cost saving and performance promotion of the microcellular part. It will make the mold design of the part, much easier and flexible for the optimizing design to maximize the cost saving of the microcellular part by eliminating the packing stage. The dimension stability of the microcellular part saves mold trial time because there are no requirements for optimizing holding time and pressure. Therefore, there are no extra pressure holding energy and holding time for the microcellular part.

A stress-free part is also important for some materials that are sensitive for the residual stress after molding. Microcellular molding fills the mold finally by the cell expansion so that there is no need to pack the part. It results in no pressure gradient from the gate to the end of mold filling, as well as no residual stress because the mold does not need to be finally filled by the injection pressure. On the other hand, the quick injection speed helps to fill the mold in a short time, so the overall heating history and cooling history for material in the entire part are uniform. This stress-free part will eliminate the postprocess to relax the residual stress or to simply make the part lifetime longer.

There is another potential stress in the transition zone between skin and foamed core in the microcellular part. However, the most of the stress will form in the final packing stage that is at high pressure, and the viscosity of material in the transition becomes higher because cooling occurs near the skin. For regular molding, it most of the residual stress will be created in the part during packing. The mold temperature needs to be kept high, which increases the cycle time and is not the economic solution. However, the stress is not the issue for the microcellular part because the packing stage is eliminated in the microcellular process.

12.6.6 Tailor-Made Properties by Combining Different Materials

The co-injection process of microcellular foam can be made with the selection of two different materials, so the tailor-made properties will result in special application. In addition, the surface quality of co-injection microcellular foam is as good as that of a straight injection molding part. Furthermore, the foamed core provides good dimension stability to the co-injection microcellular part and enables it to be without sink marks.

The application also expands to the recycling business, which will use the recycled material as the foamed core between virgin materials and skin. The foamed core can be not only the scrap of the same material of skin but also other material or mixture of different materials as long as they have acceptable adhesion to the virgin skin materials. As the heterogeneous nucleation requirement, the mixed recycled material may even help for good cell structure of the core in the co-injection process. On the other hand, some of combination is simply for using low-cost material in the center core with expensive material or functional material as skin. In this way the plastic industry can save material and go to the next generation of production to make more green parts, which is good for environmental conservation worldwide.

12.6.7 Recyclable Waste Material

The special injection blow-molding parts are marketed under the oPTI^{SM} (pronounced "opti") brand name. This is an injection blow-molding process developed by Plastic Technologies, Inc. (PTI). The oPTI^{SM} technology enables white or silvery-colored bottles to be made without additives that may limit the package recycling [2]. Therefore, white color oPTI^{SM} bottles provide an environmentally friendly option without using additives to achieve a similar tint. The white color oPTI^{SM} can be re-melted and subsequently processed again even with transparent polyethylene terephthalate PET without any contaminant. Similar recycling may be applied for white color (the color is from the foaming process and not from additives) injection molding parts. It is not only saving but is also the future way to obey the rules to protect against environment contamination by some plastics.

12.7 ANALYSIS OF RETURN ON INVESTMENT (ROI)

The analysis of return on investment (ROI) will be available from a lead microcellular technology company, Trexel Inc. [2]. It calculates the savings per year and determines the payback time for the initial investment for the specific case of switching from regular injection molding to MuCell®

microcellular injection molding. The calculation is based on the cycle time or material savings and initial investment of a new microcellular machine [10].

There is another cost-saving analysis based on retrofitting of the existing machine to switch from regular molding to microcellular molding [15]. The retrofitted microcellular molding machine is considerably less expensive than the newly manufactured microcellular injection molding machine. However, performance of the retrofitted microcellular molding machine will be largely based on the existing machine itself.

12.8 COMPARISON OF THE COSTS WITH DIFFERENT PROCESS METHODS

There are several special processing methods introduced in Chapter 8. The cost comparison among different processing methods is interesting, and the data are based on the structural foam that is a close reference for microcellular foam. The conclusions for the cost in the order of most expensive to least expensive are: counterpressure structural foam, co-injection structural foam, and regular foam. The details are discussed as follows [16]:

- The machine price for co-injection is the most expensive price. Then, the gas counterpressure equipment is the second highest price. The ratio among co-injection, gas counterpressure, and regular structural foam is 1.5:1.04:1.
- The machine hourly cost for all three processing equipments are in an order similar to that of the cost ratio above, and it shows 1.28:1.05:1.
- The part weights of all three different processing methods indicate that the heaviest part is from gas counterpressure because it compresses the part in the mold by gas pressure in the cavity and creates lest number of cells. The ratio of the weight among gas counterpressure, co-injection, and structural foam is about 1.07:1:1.
- Cycle-time ratio among three processing methods is 2:1.43:1 among gas counterpressure structural foam, regular structural foam, and co-injection structural foam. The gas counterpressure process has high resistance of gas pressure in the cavity and creates the heaviest part with extra materials and creates fewer cells in the part that may need longer injection time and longer time to cool more mass of material. It is very interesting to see that the co-injection structural foam process has the shortest cycle time, which is even shorter than the cycle time of regular structural foam. It is because of the thickness of structural foam since the foam at certain thickness will create a thermal insulator to extremely delay the cooling time. It is also from the quality of the surface and better heat conductivity of virgin material skin of co-injection. On the other hand, the co-injection process uses the lowest amount of chemical blowing agent in the foam

core, which may contribute to the short cycle time as well. This ratio may not be relevant for microcellular foam because the thickness is much different from that of microcellular foam and structural foam. Overall, the cycle-time differences are significant among three processing methods.

- The mold cost ratio is 1.28:1.03:1 among gas counterpressure structural foam, co-injection structural foam, and regular structural foam, respectively.
- The total cost ratio for all three processing methods is 1.58:1.26:1.00 for regular structural foam, gas counterpressure, and co-injection, respectively. The highest total cost is for the regular structural foam. The gas counterpressure has the second highest cost processing. The co-injection process is the least expensive, which doesn't even include possible regrind material in the foamed core between virgin material skins that will save even more total cost for each part.

12.9 CASE STUDY OF ENERGY SAVING

The following energy analysis is the case study on the real machine. The process itself in this case almost maximizes the productivity for a solid processing with the possible shortest cycle time. Based on these real solid processing data, a microcellular processing is predicted by the reference data from another microcellular molding database. In other words, it shows a conservative cost analysis case that is fair to be an example of estimation for how much energy can be saved from microcellular processing compared to the maximized solid molding production. In addition, this analysis is simply for processing using a chemical blowing agent only, so there is no energy consumption involved from the gas pump unit.

Table 12.2 shows a typical modern full hydraulic clamp unit of an injection molding machine running under very aggressive processing conditions. It uses an oversized screw in a small injection unit, so the injection volume rate also increases by a large-diameter screw with the same linear injection speed in a small injection unit (the same oil volume flow rate in a hydraulic system will provide a fast injection speed with a low injection pressure in a small injection cylinder but large diameter of screw). It is a common approach to use a big screw with low injection pressure requirement to save cost. It is also the same approach as that used in a microcellular injection molding machine because a combination of low injection pressure and a high injection volume rate is absolutely necessary for microcellular processing. Possible energy saving from using microcellular processing is analyzed based on the data in Table 12.2.

The first three actions in Table 12.2 are cores in, mold closing, and clamp half-nuts (typical two-platen clamp system) locking, which are the common actions for both solid and microcellular moldings. Therefore, no energy is saved from these three actions. The first action of energy saving related to microcellular processing in Table 12.2 is the clamp force buildup, or the clamp tonnage saving. An assumption is made with 60% tonnage saving in a 15% weight reduction microcellular process. Then, there are two parameters to be calculated. One is that only 40% of tonnage needs to be built up, so only 40% of the regular clamp tonnage time is needed ($0.6 \times 40\% = 0.24$ sec). Another one is that only 40% of the oil pressure is needed to build this low tonnage ($26.19 \times 40\% = 10.48$ Wh with low tonnage built up with the same time period as normal tonnage). This is the result when neglecting any heat loss and efficiency change without running the full load of the machine. Also, the energy in a whole period of tonnage maintenance will be calculated as cooling time because it almost simultaneously occurs when cooling time begins after a very short injection time.

The next possible energy saving is the carriage unit moving forward because microcellular processing does not have a sprue break. There may be a 5-Wh energy saving. However, this should not count as microcellular processing energy saving because the solid molding can run the machine without a sprue break, too. Therefore, this action is eliminated in Table 12.2 (refer to Table 7.10 for the solid part molding in a whole cycle).

Then, it is the energy saving of injection unit that is really the big saving to be focused on not only for processing but also for machine power system design. During processing, the same calculation will be carried out on the clamp unit. However, there are also two parameters to be considered. One is the 15% weight reduction from the shot size, so only 85% of the injection stroke of solid molding is needed ($114.05 \times 85\% = 96.94$ Wh). The other is that only 50% of the oil pressure is needed to build low injection pressure ($85.54 \times 50\% = 44.47$ Wh with low injection pressure at short injection time). Then, the final energy consumption in the injection period for the microcellular process is 44.47 Wh.

It is important to know that the total machine power (electric motor for the hydraulic pumps) is determined from the product of injection speed and pressure. The total power requirements of an injection unit are usually much higher than that of any other unit in the injection molding machine. If the whole machine is designed based on the injection pressure required by the microcellular process, the total machine power should be smaller than a regular injection molding machine. In most cases the microcellular process will be carried out in an existing machine modified with a microcellular-capable package. Then, the best energy-saving approach in the retrofit project for the microcellular process is to use an oversized screw in a small injection unit.

It is obvious that zero energy is needed in the packing stage of the microcellular process. The packing stage in a post-pressure unit is eliminated for the microcellular process, so it immediately saves 58.09 Wh of energy.

Feeding is a very complicated stage for the energy analysis. There are several factors to be considered when calculating the fare energy savings for both solid and microcellular processes, including the following:

- Back pressure during screw recovery is a direct parameter controlling the energy consumption in the recovery stage. The back pressure required for regular molding is usually 3.45 MPa (500 psi). However, the back pressure for the microcellular process is about 6.9 MPa (1000 psi) as the minimum back pressure to maintain gas dosing pressure in the solution.
- Output rate for microcellular screw recovery is usually only 80–90% of the output rate of solid screw recovery. The average percentage of the microcellular recovery output rate is 85% of the regular recovery output rate. This will be used for theoretical calculation that may satisfy most cases from small screw to big screw.
- Microcellular plasticizing may have a 10% torque reduction, and even the high back pressure is used during screw recovery because of the low viscosity of gas in the melt.
- The microcellular process will melt the plastics mostly from the mechanical energy because such high back pressure is needed for maintaining gas dosing pressure; therefore, less heating energy is needed from the heat band on the barrel. Using mechanical energy in the screw to melt the plastics is actually the most efficient way for plasticizing.
- The viscosity ratio of a microcellular single-phase solution (gas-laden molten polymer) to the solid melt (molten polymer without gas) is about 0.8. Usually a 10–30% viscosity drop with gas in the molten polymer is common, so an average percentage of viscosity drop for gas-laden melt will be about 20%.

Based on the facts above, the proposed formulae to calculate the energy consumption for microcellular will be

$$E_m = (E_s) \times (1 - R_{\text{core}}) \times (1.15) \times (0.9) \tag{12.2}$$

where E_m is the energy consumption for microcellular processing, E_s is the energy consumption for solid processing, and R_{core} is the weight reduction percentage. Then, the ratio of the recovery rate (output rate) of a solid to the microcellular part is about 1.15, which means that the microcellular output rate will be 15% smaller than the solid output rate. The last term in Equation (12.2) is the torque saving percentage from the microcellular process, which that is usually 0.9. In this way the energy used for the microcellular process during screw recovery is about 364.41 Wh, which is still lower than the energy consumption of solid screw recovery in this case. There is a prediction of about 49.43 Wh, or 12% energy decrease with microcellular recovery. The recovery time of the microcellular process is almost the same as that of the solid process, which is the original solid recovery time with 15% less shot volume produced; however, there is a 15% slower recovery rate of microcellular because the high back pressure for gas dosing. On the other hand, after reviewing the total power used in each action of injection molding in Table 12.2, the highest energy consumption is actually from the feeding stage, not from the injection stage because the oversize screw is used in this small injection unit. In addition, the feeding uses the highest percentage of energy because it lasts for a long period of time to produce a very large shot volume (large shot size is prepared in the fast cycle of the microcellular process). This long screw recovery time is the best approach for microcellular processing. It is also energy saving and allows the screw recovery period to overlap with the whole cooling period because little energy is used during the mold cooling period. In the machine used in this energy calculation, three pumps are feeding the screw recovery, and only one small pump maintains the pressure on the clamp for the cooling period.

The next energy calculation is from the cooling stage. It is simple that the oil is returned to the oil tank and there is only very little oil flow to compensate for the oil leaking in clamp tonnage maintained during cooling stage. For microcellular processing, data showing a 50% cooling time saving is reasonably good. Therefore the final calculation energy for cooling is about 10 Wh.

More energy is saved from a clamp force reduction. Energy savings come from a short time force reduction because the low tonnage is used, and there is a low clamp force to be reduced compared to full tonnage of clamp force to be reduced. The final low clamp tonnage reduction is about 5.83 Wh.

Finally, the demolding will have 15% energy saving if the low weight is considered for the moving load reduction for the robot lifting the parts vertically because the production of lifting load and the distance moved vertically equals work. However, there is no time saving for robot demolding.

The predicted energy saving from the microcellular injection molding machine is significant in this example. Both the total cycle time and total energy in Table 12.2 show a significant saving. Therefore, a large microcellular injection molding machine with very aggressive processing conditions can save up to 29% of energy for each microcellular part compared to the solid part. However, the impressive saving is also in the cycle-time saving, which is up to 26% compared to the solid molding process. The total energy saving must connect the cycle-time saving because the screw recovery maximizes the energy used to match the short cycle time for both solid and microcellular processes.

This is the proper way to calculate how much energy is used in each action of the microcellular molding process. The longest cycle time and the highest energy consumption among all actions of molding machine can be found through the analysis above. Then, the efforts may be focused on the longest cycle time and the highest energy consumption action to reduce it.

One more thing that needs to be emphasized in this energy analysis is the high percentage of injection energy versus the low percentage of clamp energy. The ratio of the energies of injection period to the clamp period is over 4. The ratio in toggle clamp machine may be higher than 4. This means that the least amount of energy required for injection (even it is short) is at least four times as high as the energy required for clamp tonnage building up. The feeding energy in this case is special in that it will be almost half of the total energy. The normal screw size is only about 20–30% of the overall energy consumption. This is also the case in thin-wall and packaging injection molding machines with very high feeding energy to match the cycle time requirement. However, it will be a mandatory requirement of oversize screws in microcellular injection molding machine. Usually, the cooling time is the longest period of injection molding because it is about 50% of the whole cycle time. The microcellular process reduces the cooling time significantly, usually by 10–40%. It is a challenge for the feeding screw of injection molding machine to provide a higher output rate to satisfy the rule that the screw recovery must finish before the end of the cooling period of the molding unless an electric screw driver is used. The suggestions for further energy saving for the machine in Table 12.2 are to use (a) an electric screw driver for a high-output required screw recovery and (b) hydraulic accumulator bottles in a modified hydraulic system to save more energy.

The cost calculation for the total saving is simple. It saves the cost in several aspects if the data in Table 12.2 are used for one million of the parts. The cost-saving estimation is discussed below:

- Labor cost is about $15/hr. The 26% cycle-time saving for one million parts will be $73,517 (total 4901 hr saving per millions parts with microcellular at labor cost $15/hr).
- Electric bill is reduced by $17,571 (219.64 Wh saving for each part, and one million parts to be total 219,640 kWh with the electric cost $0.08/kWh).
- Material of HDPE cost saving is up to $2,490,000 (15% material saving for 20 lb HDPE part, and total 3,000,000 lb of HDPE material saving for one million parts with HDPE cost $0.83/lb).
- Machine maintenance cost saving is about $4901 (total cycle-time saving 4901 hr total per millions parts with cost of maintenance $1/hr).
- Other cost savings including water, supervising, and management overhead, mold, machine load saving, and the lifetime increase of the parts because of low load, and so on, are not listed here. On the other hand, the machine hours of 4901 can be sold to another contract to make production more flexible, and the products may be delivered 4901 hours earlier than the contract due day. It may make the customer happy and may build a better trust relationship with the customer.

In addition, based on the data shown in Table 12.2, the total parts that can be produced in 1 hr in this machine are 83 for a microcellular part, but 59 for a solid part. It is another significant saving for the cost because the production per unit time is increased.

REFERENCES

1. Xu, J., and Pierick, D. *J. Injection Molding Technology* **5**, 152–159 (2001).
2. Trexel Web Site, www.trexel.com.
3. Martini-Vvedensky, J. E., Suh, N. P., and Waldman, F. A. U.S. Patent No. 4,473,665 (1984).
4. Suh, N. P. *Innovation in Polymer Processing*, Chapter 3, edited by James F. Stevenson, Hanser/Gardner Publications, Cincinnati, 1996.
5. Dvorak, P. *Machine Design* **January**, 115 (2007).
6. Chitwood, A. *Injection Molding Mag.* **April**, 92 (2001).
7. Matuana, M. L., and Mengeloglu, F. *J. Vinyl Additive Technol.* **7**(2), 67–75 (2001).
8. Michaeli, W., and Cramer, A. *SPE ANTEC, Tech. Papers*, 1210–1214 (2006).
9. Robin, K. *Mod. Plastics Worldwide* **July**, 36–37 (2006).
10. Okamoto, T. K. *Microcellular Processing*, Hanser/Gardner Publications, Cincinnati, pp. 137–140, 2003.
11. Robin, K. *Plastics Technol.* **June**, 25–26 (2009).
12. Maniscalco, M. *Injection Molding Mag.* **April**, 30–33 (2007).
13. Shimbo, M., Higashitani, I., and Miyano, Y. *J. Cell. Plastics* **43**, 157–167 (2003).
14. Serry Ahmed, M. Y., Atalla, N., and Park, C. B. *SPE ANTEC, Tech. Papers*, 1109–1112 (2008).
15. Xu, J., Cardona, J. C., and Kishbaugh, L. A. U.S. Patent No. 7,615,170 B2 (2009).
16. Eckardt, H. *J. Cell. Plastics* **23**, 555–592 (1987).

NOMENCLATURE

LOWERCASE ROMAN CHARACTERS

a_C	gas concentration as horizontal shift factor on the viscosity change
a_{cube}	length of the cube to be measured for calculation of surface area of foamed part
a_{fs}	coefficient in linear equation of flexural modulus
a_p	pressure shift factor for viscosity change under the pressure
a_T	horizontal shift factor for temperature coefficient of the viscosity, it is also determined experimentally
w	width of microcellular model
c	overall thickness of microcellular model
b	unloaded length of beam at each end
b_C	gas concentration as vertical shift factor on the viscosity change
b_{fs}	constant in linear equation of flexural modulus
b_T	vertical shift factor for temperature coefficient of the viscosity; it is also determined experimentally
d	diameter of nozzle tip, or valve gate, orifice
$d1$	diameter of piston rod of injection cylinder
$d2$	diameter of injection cylinder bore
d_{cell}	the average diameter of the cells
d_g	diameter of valve gate opening
d_{g1}	diameter of stem in valve gate
d_{g2}	diameter of the valve chamber in valve gate

Microcellular Injection Molding, Edited by Jingyi Xu
Copyright © 2010 John Wiley & Sons, Inc.

d_i	cell diameter in the defined same size group
d_j	diameter of orifice in gas injector
d_{l1}	bubble diameter before stretching in the shearing field
d_{l2}	bubble diameter after stretching in the shearing field
$\bar{d}_n$	number average cell diameter
$\bar{d}_w$	weight average cell diameter
d_v	clearance between ID of annular orifice and OD of poppet stem in gas injector
d_{scr}	displacement of screw during PVT test
d_s	diameter of poppet stem in orifice area of gas injector
d_{vol}	absolute volume reducing during PVT test
dp/dt	pressure drop rate
e_f	axial width of flight
f	frequency of atomic or molecule lattice vibration
f_0	frequency factor for homogeneous nucleation
f_1	frequency factor for heterogeneous nucleation
f_{pt}	pressure drop rate function
k	Boltzmann's constant
k_c	constant to be determined by experiment with gas concentrate C
k_g	adiabatic exponent
κ_{h}	Henry's law constant (solubility)
l	thickness of plastic that is cut by the flights in wiping section, or the thickness of plastic for gas diffusion
m_g	mass of the gas molecule
m_{p0}	power-law coefficient
m_{ref}	value of m_T at reference temperature T_{ref}, it is determined experimentally
m_T	consistency index, otherwise known as the viscosity at unit shear rate
m_v	mass of the shot size under the pressure
n	power-law index
n_b	number of gas molecules in a bubble nucleus
n_g	number of valve gates or nozzles
n_i	number of cells with equivalent diameter
n_j	the number of cells seen in a SEM picture
r	radius of bubble in nucleation
r_c	the ratio of cell width over height
s_n	$1/n$
t	skin thickness of microcellular part
$\tilde{t}$	reduced time
t_b	time since the first heterogeneous nucleation has occurred
t_c	the core thickness on the microcellular foam part
t_d	SCF diffusion time
t_{gv}	opening time of gas injector during dosing
t_j	average morphological cell wall thickness

t_r average residence time of SCF doing material in mixing and wiping sections of screw
u_d absolute velocity of melt in axial direction of screw
v_1 volume of the gas
y^* reduced lagrangian coordinates

UPPERCASE ROMAN CHARACTERS

A the area of the micrograph of SEM to be used to count cell density
$A1$ area of piston rod of injection cylinder
$A2$ area of piston at blind side of injection cylinder
A_{ab} surface area of the additive particle-bubble interface
A_{ap} surface area of the additive particle–polymer interface
A_{bp} surface area of the bubble
A_s area of outside diameter of screw
B liquid bulk modulus
B_T constant for Arrhenius temperature relation
C gas concentrate
C_0 concentration of gas molecules in solution
C_0' concentration of gas molecules in solution of mixed model
C_1 concentration of heterogeneous nucleation sites
C_{melt} specific heat of plastic melt
D outside diameter of screw; outside diameter of plunger
D_2 nominal diameter of ID in ring; OD of step on seat of screw tip
D_a gas (SCF) diffusivity coefficient
D_b outside diameter of blister ring of screw
D_c average diameter of cells
D_f critical filler size
D_i root diameter of a screw located in the port of barrel for liquid agent injection
E_a activation energy for the flow
E_g the activation energy of gas g for either diffusion or permission
E_m energy consumption of microcellular processing
E_p activation energy through polymer p for either diffusion or permission
E_s energy consumption of solid processing
H Henry's law constant
H_0 preexponential constant for Henry's law constant
H_3 depth of metering zone in the screw
H_d depth of gas dosing zone
H_e depth of mixing channel in dosing zone
J_n rate of bubble nucleation
L length between supports of the platen
L_1 length between supports of the standard platen

L_2	length between supports of the microcellular platen
L_3	length of metering zone in the screw
L_a	overlap length between ID of ring and OD of pilot step of seat of screw tip
L_{flow}	longest flow length from gat to the end of mold flow
L_j	size of droplet (assume the droplet with the orifice diameter)
L_{ori}	length of orifice
L_v	length of the orifice in gas injector
M	magnification factor of SEM picture
M_b	number of gas molecules per unit volume
M_w	molecular weight of the gas
$\bar{M}_w$	weight average molecular weight
$\bar{M}_n$	number average molecular weight
M_i	molecular weight
N	number of cells
N_0	number of available sites for nucleation
N_f	number of flights in wiping section
N_{hom}	homogeneous nucleation rate
N'_{hom}	homogeneous nucleation rate in the presence of heterogeneous nucleation
N_{het}	heterogeneous nucleation rate
N_{mix}	number of the channels in one mixing section
N_i	mole fraction of the that has a molecular weight M_i
N_j	number of cells per volume of cm^3
N_{sf}	reference cell density for standard foam, and it will be calculated with average cell size 100 μm
N_s	screw rotation speed
P_0	pressure at the entrance of nozzle orifice
P_1	pressure at inlet of orifice
P_2	pressure at outlet of orifice
P_a	atmospheric pressure
P_b	internal gas pressure
P_{cr}	critical gas pressure
PDI	polydispersity index of the cell sizes
P_g	gas pressure
P_g^*	reduced bubble pressure
P_{ld}	gas charged pressure
P_m	melt pressure
$P_{\max}$	maximum injection pressure
P_{oil}	pressure of hydraulic oil
P_{ref}	reference pressure at certain temperature and 1 atm pressure
P_{reg}	injection pressure at regenerative operation setup
P_s	pressure in molten plastic near the gas injector
P_t	pressure in the single-phase solution accumulated in front of screw tip
Q_a	volume flow rate for leaking in the screw tip

Q_g	flow rate of gas in gas injector
Q_p	flow rate of hydraulic oil for injection cylinder
Q_{part}	volume of the part
Q_v	screw output volume rate
R	bubble radius in the unit cell concept
R_0	initial bubble radius
R^*	dimensionless instantaneous bubble radius
$\dot{R}^*$	time derivative of dimensionless instantaneous bubble radius R^*
R_{core}	ratio of weight reduction of foamed core
$R_{\text{core-f}}$	ratio of weight reduction of foamed core for the flexural strength model
R_c	average radius of annular channel of nozzle rheometer
R_f	local instant dosing (%)
R_{fs}	strength ratio of foam to solid
R_g	universal gas constant
R_{gf}	real weight reduction ratio of the core material without filled material
R_i	intensify ratio of gas local dosing
R_{iz}	Izod impact strength ratio of foamed part to solid part
R_{no}	outside radius of annular channel in the nozzle rheometer
R_{ni}	inside radius of annular channel in the nozzle rheometer
R_{scf}	overall ratio of weight of gas to the weight of the part
R_{tu}	ratio of tensile strength of unfilled foam to unfilled solid
R_u	ratio of an axial velocity difference between the plastic melt forward movement related to the fixed position in the barrel and the screw backup movement during screw recovery
R_w	weight reduction ratio of the whole microcellular part
S	outer radius of the cell in the unit cell concept
S_0	radius of the shell within which one cell grows
S_f	number of multichannel sections
S_n	number of striations divided by distributive mixing element with multi-channel
T_{cr}	critical gas temperature
T_0	temperature at entrance of nozzle orifice
T_f	polymer melt freezing temperature (melt point for crystal polymer; soft point for amorphous polymer)
T_g	polymer glass transition point
T_m	polymer melting point
T_{poly}	melt absolute temperature
T_{ref}	reference melt temperature at the shear zero condition
T_w	period of wiping
U	energy stored in gas
U_m	absolute velocity of melt in axial direction of screw
U_t	absolute axial velocity of the screw moving back during recovery
V	linear speed of injection

V_b	volume of the bubble nucleus
V_g	percent volume fraction of the cells
V_{gas}	volume of released gas from the flow front
V_{free}	free air shooting velocity driving by the gas stored energy U
V_{ori}	velocity of the injection in the nozzle orifice at Δr position
$\dot{V}$	volumetric flow rate
$\dot{V}_g$	gas volume flow rate in gas injector
V_m	volume of mixing section of screw
V_{mold}	volume of molding
V_p	injection speed at operation set with maximum pressure
V_r	volume of runner system
$\dot{V}_{\text{vent}}$	total venting flow rate in the mold
W	total load on the platen
W_1	load on the standard platen
W_2	load on the microcellular platen
W_a	weight of the sample measured in the air
W_w	weight of the sample measured in the distilled water
W_f	width of flights in wiping section
W_{gf}	weight percentage of filled material before foaming
W_i	mole fraction of the that has a molecular weight W_i
W_t	frequency of wiping
W_x	weight percentage of gas in molten polymer
X_c	crystallinity in semicrystalline material
X_a	solubility of the gas in amorphous portion of the semicrystalline material

LOWERCASE GREEK CHARACTERS

α	gas (SCF) diffusivity
β	pressure coefficient for the shift factor of pressure on the viscosity change
$\dot{\gamma}$	apparent shear rate
γ_{ab}	interfacial tensions of the solid particle–bubble
γ_{ap}	interfacial tensions of the solid particle–polymer
γ_{bp}	interfacial tensions of the bubble–polymer
$\dot{\gamma}_c$	average shear rate in the clearance between top of flight and barrel
$\dot{\gamma}_s$	average shear rate in the screw channel
γ_w	wall shear rate
η_0	the zero-shear viscosity which is a function of temperature and pressure
η_c	modified viscosity of molten plastic in the clearance between top of flight and barrel
η_s	modified viscosity of molten plastic in screw channel

η_T	non-Newtonian gas-laden polymer viscosity
θ	the mold temperature
θ_{ld}	angle of deformed bubble with the horizontal line (original angle is 90°)
θ_w	wetting angle
μ	Newtonian viscosity
μ_g	gas viscosity in the gas injector
ρ_c	cell density
ρ_{foam}	density of foam
ρ_g	density of gas
ρ_{matrix}	gas density in the polymer–gas matrix
ρ_{melt}	melt density
ρ_{poly}	density of unfoamed polymer
ρ_{sf}	density of filled solid material
δ_b	clearance between inside diameter of barrel and outside diameter of blister ring
δ_r	clearance in radius between ID of ring and OD of step in screw tip
δ_s	clearance between outside diameter of screw and inside diameter of barrel
ϕ	helix angle of flight in the screw
ϕ^*	reduced form of the gas concentration potential function
σ_b	bubble surface tension
τ	time of melt touching the mold—that is, approximately the hold time plus cooling time
τ_r	shear stress on any surface at position of $R_c + \Delta r$ and $R_c - \Delta r$
τ_{rr}^*	dimensionless normal stress in radial direction
$\tau_{\theta\theta}^*$	dimensionless normal stress in circumferential direction
τ_w	wall shear stress in the nozzle viscometer
τ^*	model parameter in Gross-WL Fequation
λ	relaxation time of polymer

UPPERCASE GREEK CHARACTERS

ΔG	activation energy
ΔG^*	maximum free activation energy
ΔG_{hom}	change in Gibbs free energy for homogeneous nucleation
ΔG_{het}	change in Gibbs free energy for heterogeneous nucleation
ΔL_b	axial length of blister ring
ΔP_{ori}	pressure drop across the nozzle orifice
ΔT_{ori}	melt temperature increase in the nozzle orifice
ΔP	the pressure of the gas in the bubble
ΔP_r	pressure drop across the ring of screw tip
ΔP_{gas}	gas pressure drop across the length of orifice of gas injector
ΔV	absolute volume change (%) at different pressure and temperature of melt

APPENDIX A (CHAPTER 7)

PRESSURE DROP RATE *dp/dt* FORMULA

The pressure drop rate dp/dt can be easily derived with the geometric parameters of nozzle (or gate) orifice diameter d_g, screw diameter D, and the linear injection speed of screw. Assume that the injection speed V is linear and constant; then, a constant injection volume rate Q may be written as

$$Q = \frac{\pi V D^2}{4} \qquad (\mathrm{m^3/sec}) \tag{A.1}$$

Assuming that there are n nozzles (or gates) opening simultaneously, then the average flow velocity V_n inside of nozzles (or gates) can be expressed in the form

$$V_n = \frac{4Q}{\pi n d_g^2} \qquad (\mathrm{m/sec}) \tag{A.2}$$

Then, the average time dt to pass the nozzle tip orifice (or gate) with L_n length may be defined as

$$dt = \frac{L_n}{V_n} \qquad (\mathrm{sec}) \tag{A.3}$$

Microcellular Injection Molding, Edited by Jingyi Xu
Copyright © 2010 John Wiley & Sons, Inc.

Equation (A.2) is used for replacing V_n in Equation (A.3), and Equation (A.3) becomes

$$dt = \frac{\pi n L_n d_g^2}{4Q} \qquad \text{(sec)} \tag{A.4}$$

Substituting for Q in Equation (A.4) and rearranging, we obtain

$$dt = \frac{n L_n d_g^2}{V D^2} \qquad \text{(sec)} \tag{A.5}$$

Assuming that the material melt viscosity μ is Newtonian behavior and with isothermal condition, then the dp can be derived as

$$dp = \frac{128 Q \mu L_n}{n \pi d_g^4} \qquad \text{(Pa)} \tag{A.6}$$

Substituting for Q in Equation (A.6) gives

$$dp = \frac{32 D^2 \mu L_n V}{n d_g^4} \qquad \text{(Pa)} \tag{A.7}$$

Equation (A.7) is divided by Equation (A.5), and we find

$$\frac{dp}{dt} = \frac{32 \mu V_j^2 D^4}{n_g^2 d_g^6} \qquad \text{(Pa/sec)} \tag{A.8}$$

Equation (A.8) is equal to Equation (7.2). When n_g is equal to 1, it becomes Equation (7.1).

APPENDIX B (CHAPTER 7)

CLAMP LOAD *W* VERSUS SUPPORT DISTANCE *L* FOR THE PLATENS AT THE SAME DEFLECTION

The beam deflection with the load model shown in Figure 7.20 can be calculated by

$$f = \frac{WL^3(5 - 24a^2 + 16a^4)}{384EI(1 - 2a)} \tag{B.1}$$

where f is deflection (mm), W is the total load (N), L is the total length between supports (mm), E is the modulus of elasticity (N/mm^2), I is the moment of inertia of the beam section (mm^4), and a is the fraction of length of the beam at each end, which is not loaded.

$$a = \frac{b}{L} \tag{B.2}$$

If the ratio of loaded length of mold to the total length of supports is 0.65, for both platen #1 of a standard machine and platen #2 of a microcellular machine, a_1 will be the same as a_2, and it is $a_1 = a_2 = 0.175$. Also, if the material is the same for both platens, then $E_1 = E_2 = E$. If the width and thickness for both platens are the same, then the moment of inertia I is kept the same for both

Microcellular Injection Molding, Edited by Jingyi Xu
Copyright © 2010 John Wiley & Sons, Inc.

platens, $I_1 = I_2 = I$. Finally, if the deflection f is the same for both platens, then we have

$$f_1 = f_2 \tag{B.3}$$

Then,

$$\frac{W_1 L_1^3 (5 - 24a^2 + 16a^4)}{384EI(1-2a)} = \frac{W_2 L_2^3 (5 - 24a^2 + 16a^4)}{384EI(1-2a)} \tag{B.4}$$

Therefore,

$$W_1 L_1^3 = W_2 L_2^3 \tag{B.5}$$

It is equal to

$$\frac{W_1}{W_2} = \left(\frac{L_2}{L_1}\right)^3 \tag{B.6}$$

Equation (B.6) is the same as Equation (7.20).

APPENDIX C (CHAPTER 5)

TENSILE STRENGTH RATIO OF FOAM TO SOLID

With the same physical model in Figure 5.2 and the tensile load theory, we know

$$F_f = \varepsilon_1 (E_s A_s + E_f A_f) \tag{C.1}$$

where F_f is the tensile load on the whole foamed part (solid skin plus foamed core), E_f is the elastic modulus of the foam, E_s is the elastic modulus of the solid, A_f is the cross-section area of the foam, A_s is the cross-section area of the solid, and ε_1 is the elongation per unit length of the whole foamed part at tensile load F_f.

Based on the physical model in Figure 5.2, we can get

$$A_f = (h - 2t)(w - 2t) \tag{C.2}$$

$$A_s = wh - (h - 2t)(w - 2t) \tag{C.3}$$

where w is the width of section of part, h is the thickness of section of part, and t is the thickness of the solid skin.

For the solid part the tensile load becomes

$$F_s = \varepsilon_2 E_s wh \tag{C.4}$$

Microcellular Injection Molding, Edited by Jingyi Xu
Copyright © 2010 John Wiley & Sons, Inc.

where F_s is the tensile load on the solid part and ε_2 is the elongation per unit length of the solid part at tensile load F_s.

The elastic modulus ratio of foam to solid is

$$\frac{E_f}{E_s} = \left(\frac{\rho_f}{\rho_s}\right)^2 \tag{C.5}$$

where ρ_f is the foam density and ρ_s is the solid density without fillers. Also, define a weight reduction ratio R_{core} as a real weight reduction of core, and it is also a density reduction of foamed core to solid:

$$R_{\text{core}} = \frac{\rho_s - \rho_f}{\rho_s} = 1 - \frac{\rho_f}{\rho_s} \tag{C.6}$$

Assuming that the elongations per unit length for both the solid part and the foamed part are the same ($\varepsilon_1 = \varepsilon_2$), the tensile strength ratio R_{tu} of foamed part to solid part can be derived from Equations (C.1), (C.4) and (C.6). After substituting for R_{core}, A_f, and A_s in Equation (C.1), we obtain

$$R_{tu} = \frac{wh - (h-2t)(w-2t) + (1-R_{\text{core}})^2(h-2t)(w-2t)}{wh} \tag{C.7}$$

Equation (C.7) is equal to Equation (5.5).

Similarly, Equation (C.5) for filled material can be given with the R_{core} replaced by R_{gf} [see Appendix D, Equation (D.3)], along with the power index 2 for fillers and 1.2 for glass fiber.

APPENDIX D (CHAPTER 5)

REAL WEIGHT REDUCTION CALCULATION

For unfilled material the real weight reduction R_{core} can be calculated based on the whole part weight reduction R_w that is defined as a ratio of foamed part (solid skin plus foamed core) weight to the solid part weight. It can be written as

$$R_w = \frac{(h-2t)(w-2t)\rho_f + [(wh-(h-2t)(w-2t)]\rho_s}{wh\rho_s} \tag{D.1}$$

Substituting the relationship among R_{core}, ρ_s, and ρ_f [from Equation (C.6)] in Equation (D.1) and rearranging, we obtain

$$R_{\text{core}} = \frac{wh(1-R_w)}{(h-2t)(w-2t)} \tag{D.2}$$

R_{gf} is defined as a real weight reduction ratio of core material without filled material. If we know the weight percentage of filled material W_{gf} prior to foaming, and $W_{gf} \times R_{\text{core}}$ is the weight percentage of core material after foaming, then the real weight reduction ratio of core material without filled material can be written as

Microcellular Injection Molding, Edited by Jingyi Xu
Copyright © 2010 John Wiley & Sons, Inc.

$$R_{gf} = \frac{\rho_s - (1 - R_{\text{core}})\rho_{sf}(1 - W_{gf} R_{\text{core}})}{\rho_s} \tag{D.3}$$

where ρ_{sf} is the density of filled solid material. Equation (D.3) is equal to Equation (5.7).

APPENDIX E (CHAPTER 5)

FLEXURAL STRENGTH RATIO OF FOAM TO SOLID

Flexural strength model can be derived by the bending stiffness formula. For solid unfilled material the bending stiffness D_s is given by

$$D_s = E_s \frac{wh^3}{12} \tag{E.1}$$

For foamed unfilled material the bending stiffness D_f will be

$$D_f = E_s \frac{wt^3}{6} + E_s \frac{wt(h-t)^2}{2} + E_f \frac{h(h-2t)^3}{12} \tag{E.2}$$

The relationship between E_s/E_f and R_{core} can be established from Equations (C.5), and (C.6). Then, a bending stiffness ratio of the foamed part to the solid part for unfilled material can be written as

$$\frac{D_f}{D_s} = R_{fu} = \frac{2t^3}{h^3} + \frac{6t(h-t)^2}{h^3} + (1 - R_{\text{core}})^2 \left(\frac{h-2t}{h}\right)^3 \tag{E.3}$$

where $D_f/D_s = R_{fu}$ is the bending stiffness ratio of the foamed part to the solid part.

Microcellular Injection Molding, Edited by Jingyi Xu
Copyright © 2010 John Wiley & Sons, Inc.

Equation (E.3) is the same as Equation (5.9). D_f/D_s is replaced by R_{fu} in Equation (E.3). Usually the bending stiffness is named as flexural strength. Therefore, flexural strength is used in this chapter.

The filled material will have a linear relationship between E_s/E_f and R_{core}, and the bending stiffness ratio of foamed part to solid part R_{ff} is

$$R_{ff} = \frac{2t^3}{h^3} + \frac{6t(h-t)^2}{h^3} + (1 - R_{gf})\left(\frac{h-2t}{h}\right)^3 \tag{E.4}$$

Equation (E.4) is the same as Equation (5.11).

APPENDIX F (CHAPTER 6)

RELATIONSHIP BETWEEN U_M AND U_T

In addition of all previous assumptions for Equation (F.1), assume an axial velocity U_s that is the velocity related to the screw that has diameter D and root diameter D_i and is constant during the whole recovery stroke of the reciprocating screw. Based on the continuum principle, we know that the volume rate of accumulating plastics in front of the screw is equal to the volume rate that passed through the screw itself:

$$\frac{\pi}{4}D^2U_t = \frac{\pi}{4}(D^2 - D_i^2)U_s \tag{F.1}$$

It can be simplified as

$$U_s = \frac{D^2}{D^2 - D_i^2}U_t \tag{F.2}$$

Then, the absolute plastic melt velocity U_m related to the barrel in the axial direction is given by

$$U_m = U_s - U_t \tag{F.3}$$

Microcellular Injection Molding, Edited by Jingyi Xu
Copyright © 2010 John Wiley & Sons, Inc.

After substituting for U_s in Equation (F.3), we obtain

$$U_m = \left[\left(\frac{D^2}{D^2 - D_i^2}\right) - 1\right] U_t \qquad \text{(F.4)}$$

Equation (F.4) is the same as Equation (6.5).

With the same procedures above, the volume of the flights are subtracted from Equation (F.4), and then Equation (F.3) can be easily derived.

APPENDIX G (CHAPTER 9)

VISCOSITY MODEL FOR ANNULAR CHANNEL OF NOZZLE RHEOMETER

The annular channel of the rheometer has a more complex model than does the cylindrical channel, and it is not available in the literature. The viscosity model of the rheometer with an annular channel can be derived with procedures similar to those used for the cylindrical channel of the rheometer.

The flow in the annular channel is assumed symmetric about the center of the annular channel. Then, the shear stress is equal on the inside diameter R_{ni} and the outside diameter R_{no} in the annular channel. The average radius is defined in the center between the inside radius and the outside radius:

$$R_c = \frac{R_{no} - R_{ni}}{2} \tag{G.1}$$

where R_c is the average radius of the annular channel of nozzle rheometer, R_{no} is the outside radius of the annular channel in the nozzle rheometer, and R_{ni} is the inside radius of the annular channel in the nozzle rheometer.

Assume an small increment radius equally away from the average radius. A force balance of this small annular element can be analyzed with the shearing force on the annular surfaces and the force of pressure on the cross-section area:

Microcellular Injection Molding, Edited by Jingyi Xu
Copyright © 2010 John Wiley & Sons, Inc.

$$\tau_r(2L_{\text{ori}}\pi)((R_c+\Delta r)+(R_c-\Delta r))=\Delta P_{\text{ori}}\pi\left((R_c+\Delta r)^2-(R_c-\Delta r)^2\right) \tag{G.2}$$

where L_{ori} is the length of the orifice, τ_r is the shear stress on any surface at position of $R_c + \Delta r$ and $R_c - \Delta r$, and ΔP_{ori} is the pressure difference in the length of $L_{\text{ori.}}$

$$\Delta P_{\text{ori}} = P_1 - P_2 \tag{G.3}$$

where P_1 is the pressure at the inlet of the orifice (at the upstream of L_{ori}) and P_2 is the pressure at the outlet of the orifice (at the downstream of L_{ori}).

Therefore, the wall shear stress τ_w is

$$\tau_w(2L_{\text{ori}}\pi)(R_{no}+R_{ni})=\Delta P_{\text{ori}}\pi(R_{no}^2-R_{ni}^2) \tag{G.4}$$

The wall shear stress can be solved from Equation (G.4), and it is

$$\tau_w = \frac{(R_{no}-R_{ni})(P_1-P_2)}{2L_{\text{ori}}} \tag{G.5}$$

Similarly, shear stress can be derived from Equation (G.2):

$$\tau_r = \frac{\Delta P \Delta r}{L_{\text{ori}}} \tag{G.6}$$

On the other hand, τ_r can be written as

$$\tau_r = \eta_T \dot{\gamma}^n = \eta_T \left[\frac{\partial V_{\text{ori}}}{\partial(\Delta r)}\right]^n \tag{G.7}$$

where V_{ori} is the velocity of the injection in the nozzle orifice at Δr position, η_T is the non-Newtonian gas-laden polymer viscosity, and $\dot{\gamma}$ is the apparent shear rate.

From Equations (G.6) and (G.7), more relationships can be derived:

$$\dot{\gamma} = \left(\frac{\Delta P \Delta r}{L_{\text{ori}}\eta_T}\right)^{s_n} = \frac{\partial V_{\text{ori}}}{\partial(\Delta r)} \tag{G.8}$$

where $s_n = 1/n$.

Assume that the viscosity η_T is constant at the same position of L_{ori}. Solve Equation (G.8) for velocity in the nozzle:

$$V_{\text{ori}} = \left(\frac{\Delta P}{\eta_T L_{\text{ori}}}\right)^{s_n} \frac{1}{s_n+1}\left[\left(\frac{R_{no}-R_{ni}}{2}\right)^{s_n+1} - \Delta r^{s_n+1}\right] \tag{G.9}$$

The volume flow rate $\dot{V}$ can be derived from Equation (G.9):

$$\dot{V} = 2\pi R_c \int_0^{(R_{no}-R_{ni})/2} V_{\text{ori}} d(2\Delta r) = 4\pi R_c \int_0^{(R_{no}-R_{ni})/2} V_{\text{ori}} d(\Delta r) \tag{G.10}$$

After substituting for V_{ori} in Equation (G.9), it becomes

$$\dot{V} = \frac{4\pi R_c}{s_n + 2}\left(\frac{\Delta p}{\eta_T L_{\text{ori}}}\right)^{s_n}\left(\frac{R_{no} - R_{ni}}{2}\right)^{s_n+2} \tag{G.11}$$

Finally, a wall shear rate with Robinowitch correction for non-Newtonian fluid is derived:

$$\gamma_w = \frac{(2n+1)\dot{V}}{\pi R_c n (R_{no} - R_{ni})^2} \tag{G.12}$$

Equation (G.12) is the same as Equation (9.5).

APPENDIX H (GLOSSARY)

ABS: Acrylonitrile–butadiene–styrene polymer.

Adaptive pressure control: Adapting to the pressure change and keeping the pressure difference the same during processing.

Alloy: A polymer alloy for the physical admixture of two or more polymers.

Amorphous material: Material showing no sharp melting point and only a range of soft temperature.

Back pressure: The preset pressure in front of the screw tip during screw recovery.

Ball valve: A check valve either in the front of a screw or in the middle of a screw that uses a ball as seal element during the closed position.

Batch process: The process carried out in a sealed high-pressure chamber with temperature control to reach thermodynamic instability, thereby producing foam.

Blister ring: A restriction element in the middle of a screw to create a pressure difference across the blister ring locally to prevent gas leak back.

Blowing agent: Physical or chemical additives (either gas or chemicals) to generate gas for forming the cells in the foam.

Blowout: Surface deformation by residual gas pressure.

Cell: The hollow bubbles, or small holes, in the part as a foam structure.

Microcellular Injection Molding, Edited by Jingyi Xu
Copyright © 2010 John Wiley & Sons, Inc.

CBA (chemical blowing agent): Additive that will decompose upon heating to generate gas that is usually either carbon dioxide or nitrogen.

Class A surface: Part surface with good smoothness and glossy.

Coalescence: Bubble collapse, thereby forming a big void.

Co-injection (or sandwich) molding: A process to completely encapsulate an inner core with an outer skin in the same molding cycle.

Copolymer: Polymer with more than one repeating structure.

CPET: Crystallized polyethylene terephthalate.

Critical pressure: Pressure above which a gas will become supercritical fluid (SCF).

Critical temperature: Temperature above which a gas will become supercritical fluid (SCF).

CSM (clean surface molding): Mold surface coated with polyimide for class A surface of a molded part.

Dimensional stability: Amount of dimensional change measured over time through thermal cycling and loading for the finished part.

Dimples: Small depressions in part.

Dispertive mixing: The stress-related mixing with the reduction of the substance size.

Dissolution: Gas dispersion within a solution.

Distributive mixing: Moving the mixing substance from place to place, with no size change of the substance.

Dolphin skin technology: A passenger car dashboard panel with a soft-touch surface (foam with reversal coining after skin is truly formed with overmolding method); a strong glass-fiber-reinforced PBT/ASA underneath the soft-touch foam surface.

Ejector: Component in the mold to push a molded part out of the mold.

Elephant surface: Rough surface like elephant skin caused by cold material flow on the cold mold surface.

Endothermic: Chemical reaction involving bond-breaking where energy is absorbed from surroundings.

EPDM: Ethylene–propylene–diene monomer rubber.

Ergocell: Registered trademark of Demag (now Sumitomo-Demag), Inc., which adds gas in a dynamic mixer.

EVA: Ethylene vinyl acetate.

Exothermic: Chemical reaction involving bond-making where energy is released from its surroundings.

Feed section: The first section of a plasticizing screw that coveys the solid material and builds the pressure of the solid in the screw.

Filler: Any material, usually an inert substance, added into polymers.

FIM: Foam injection molding.

Foam: Material containing small holes with reduced density compared to the same solid material.

Free-flow-front roughness: The roughness formed in the free-flow front during mold filling.

Gas-assist molding: A part with hollow channel generated by pressurized inert gas, usually N_2 gas, inside of the part.

Gas counterpressure molding: Material to be injected into the sealed mold under the consistent resistance from gas counterpressure in the mold.

Gas diffusivity: The measure of the rate at which gas moves through substances.

Gas solubility: The measure of a gas-dissolving potential in a liquid.

GPPS: General-purpose polystyrene.

HDPE: High-density polyethylene.

Heterogeneous nucleation: Cell nucleation in a not pure material.

HIPS: High-impact polystyrene.

Hold phase: Molding stage after the packing stage, in which pressure is maintained in the mold by a screw or a plunger until the gate in the mold is frozen.

Homogeneous nucleation: Cell nucleation in a pure material.

Homopolymer: Polymer with a single repeating structure.

Hot runner: Heated runner system.

HVAC: Components used in the automotive industry for heating, ventilating, and cooling.

Hydraulic: Engineering science pertaining to liquid pressure and flow.

Idle phase: Screw stops rotation after finishing the recovery until the beginning of injection.

Injection stage: Portion of molding cycle in which the plastic is injected under pressure at a certain speed, or speed profile, into the mold.

Injector: Usually means an instrument that can be used to inject gas into the barrel.

Intellectual property: All protected information and technologies that including patents, patent applications, trademarks, know-how, and trade secrets.

Interface: The area between solid skin and foamed core in a foam structure.

Interface roughness: The sheared bubble to be broken on the surface and to form a rough surface.

Internal blister: Surface defect that forms over a void and has a smooth inner surface.

LCP: Liquid crystal polymer.

License: Legal contract permission for using company's intellectual property by another with the payment for a royalty.

MeltFlipper: A balanced runner system with equal flow length for each runner.

Metering section: The third section of a plasticizing screw that pumps the molten material at a constant output rate at a certain back pressure.

Microcellular material: Foam with microcells, usually where the cell size is 100 μm or smaller.

Middle check valve: A restricted element in the middle of a screw to separate the metering section from the wiping section and prevent gas flow back to the hopper.

Mixing section: A location of mixing element in the screw.

Molecular weight distribution: The degree of polymerization, which is the number of structurally recurring units in the chain.

Monomer: The small reactant molecules.

Morphology: Microstructure and shape of the material.

MuCell®: Registered trademark of Trexel Inc.; this uses the reciprocating screw in an injection molding machine for microcellular processing.

MWD (molecular weight distribution): An average value of the weights of the individual molecules.

Notched Izod: Impact strength test using a notched sample to perform the test.

Nucleating agent: Additive used for enhancing cell nucleation in foam.

Nucleation: A few molecules of blowing agent aggregate and come out of the molten polymer to form nuclei where the cells may grow.

OEM (original equipment manufacturer): Final equipment manufacturer.

Optifoam: Registered trademark of Sulzer Chemtech; this adds the gas into the nozzle.

oPTI: Brand name for a microcellular injection blow molding process developed by Plastic Technologies Inc.

Overlapping (also called Overmolding): Similar to the co-injection molding, except the sandwich structure is replaced by two layers that overlap each other.

PA: Polyamide, also known as nylon.

Pack phase: Portion of the molding cycle directly after plastic injection into the mold to fill the mold finally with 100% volume.

Parting line: Locations in the mold with two mold halves contacting together.

PBT: Polybutylene terephthalate.

PC: Polycarbonate.

PCL: Polycaprolactone.

PDI: Polydispersity index.

PE: Polyethylene.

PEEK: Polyetheretherketone.

PEI: Polyetherimide.

PET: Polyethylene terephthalate.

PGA: Polyglycolic acid.

PHBV: Polyhydroxybutyrate-valerate.

Physical blowing agent: Gas or liquid used in foaming.

PLA: Polylactic acid.

PLC: Programmable logic controller.

PMMA: Polymethyl methacrylate.

Pneumatic: Engineering science pertaining to air pressure and flow.

Polyolefin: Aliphatic hydrocarbon polymer.

POM: Polyoxymethylene.

Post blow: Similar to blowout.

PP: Polypropylene.

PPO: Polyphenylene oxide.

PPS: Polyphenylene sulfide.

Pre-closing screw tip: The non-return element in the screw tip closed automatically before injection.

Pressure transducer: Measurement instrument for reading the pressure.

ProFoam®: Registered trademark of IKV; this feeds the gas through hopper.

PS: Polystyrene.

PSU: Polysulfone.

Pulse SCF dosing: Quick opening and closing of an SCF injector.

PVA: Polyvinyl acetate.

PVC: Polyvinyl chloride.

PVT: The acronym for pressure, volume, and temperature.

Reciprocating screw: Screw that rotates and moves in an axial direction simultaneously in the barrel.

Reversal channel: A reversal flight or channel used for smaller extruder screws as a pressure-restrictive element in the middle of the screw, similar to the middle check valve.

Reversal coining (also known as Expandable mold or Breathing mold): Mold is filled 100% full first, and then the mold to will crack open with a controlled distance when enough skin is formed and the core is still hot and soft enough to make foam.

RHCM (rapid heat cycle molding): A heating and cooling mold surface processing technology from Ono Sangyo.

Rheology: The study of flow characteristics of fluids.

Rheometer: An instrument to measure the rheology data.

RPVC: Rigid polyvinyl chloride.

Rupture disc: Metal thin disc that breaks at set up pressure for safety.

Satellite system: SCF delivery units that share the same center pump.

Saturation: Maximum amount of SCF dissolved into the molten polymer.

SCF (supercritical fluid): The state of matter above a critical temperature and critical pressure.

SEM: Scanning electron microscope.

Screw recovery: Portion of the molding cycle where the plasticizing screw is melting and conveying the melt to the set-up shot size.

Screw tip: A nonreturn valve is located at the downstream end of the screw.

Semicrystalline material: Material has crystalline structure that melts with sudden viscosity reduction and crystallizes upon cooling.

Short shot: Injection molding part without 100% filling the mold.

Shut-off nozzle: AN open and closed channel where to control the channel between the barrel and the mold.

Single-phase solution: A mixture of gas and molten polymer where the gas is dissolved completely in the molten polymer.

Sink mark: A surface defect of a molded part with significant depression or concavity.

SLIM: Super Light Injection Molding, a trade name of thin-wall molding registered by Veriplast Solutions.

Solid-fed: The pumping capability of screw determined by the feeding section.

Solid material: Unfoamed material.

Structural foam: Foam made by low-pressure injection molding process.

Super-microcellular: Just filled the mold almost 100% with high injection speed and to get ultra-small cell size without weight reduction.

Surface blister: Surface defect that forms over a void.

Surface defect: Excessive swirls, surface wrinkling, and cold flow.

Swirl: The features of the microcellular foam that is the bubble broken on the surface and leave the roughness there.

TEM: Transmission electron microscopy.

Thermodynamic: The study of the effects of work, heat, and energy on a system with a large scale of observation.

Thermoplastics: Plastic material that undergoes no permanent change on heating and that can be melted and cooled repeatedly.

TPE: Thermoplastic elastomer.

TPO: A modified PP-based material.

Transition section (or compression section): The second section in a plasticizing screw that compresses the solid material and melts most of the plastics in this section.

Two-closing stage screw tip: The screw tip to be closed with two stages.

Valve gate: Valve between the hot runner system and the mold cavity.

Vent: Thin passageways cut from mold cavity to release the gas or air trapped in the mold during mold filling.

Viscosity: Resistance to the flow of liquid.

Warpage: Difference in dimensions for the molded part from the mold.

Weld line: Location of the join for the two flow fronts during mold filling.

Wiping section: Extruder screw has the flights to wipe the gas injector face for gas dosage.

Xcell: Special PA 6/6 and 6 grades developed by Rhodia Technyl have good mechanical property performance with no compromise in the part's surface finish. It is true low melt viscosity tailored for microcellular processing.

INDEX

Microcellular Injection Molding, Edited by Jingyi Xu
Copyright © 2010 John Wiley & Sons, Inc.

N

O

P

TP 156 .M65 X83 2010
Xu, Jingyi.
Microcellular injection
molding
WITHDRAWN
CAL POLY SLO
KENNEDY LIBRARY
CALIFORNIA POLYTECHNIC STATE UNIVERSITY LIBRARY